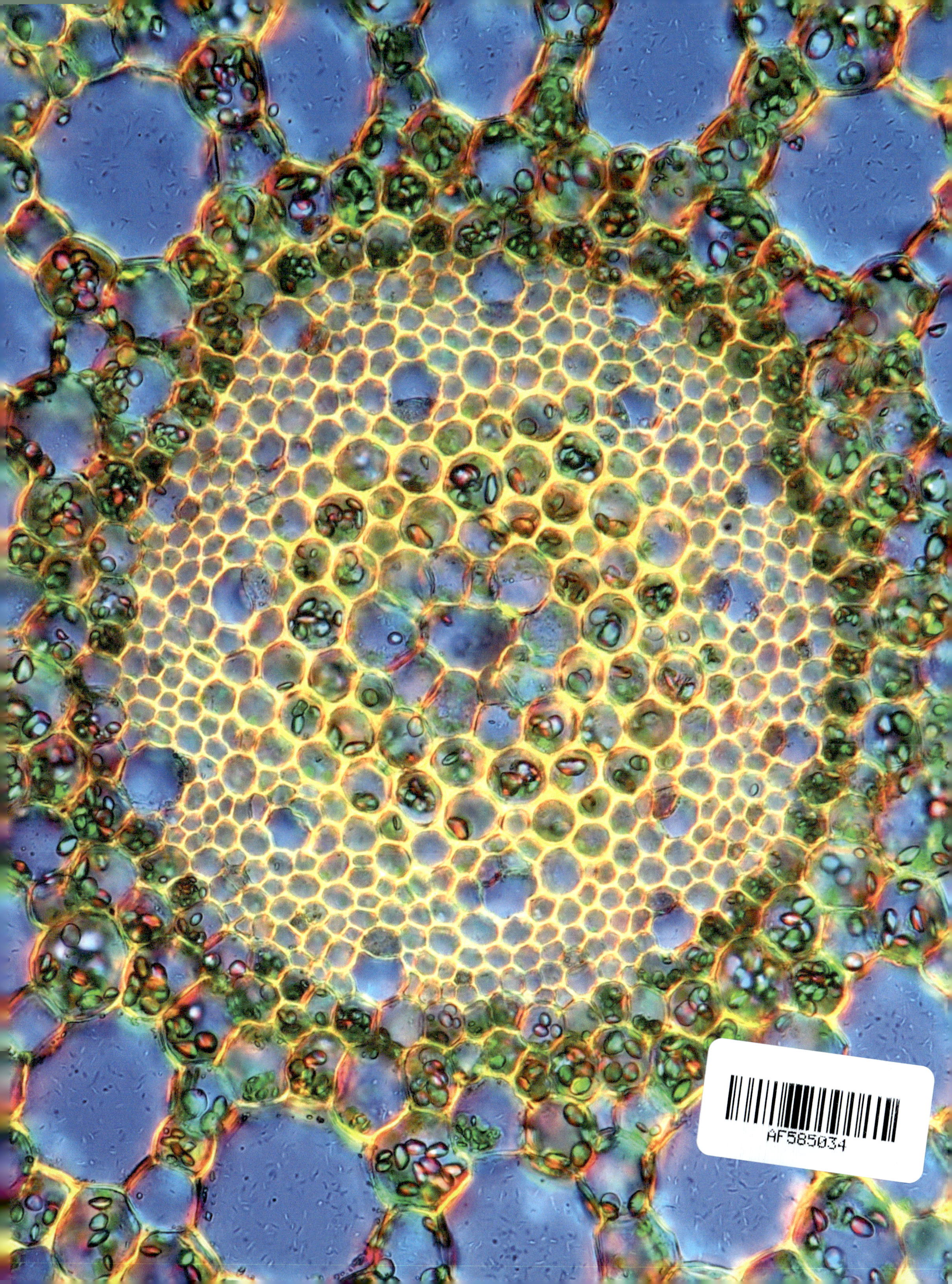

Pearson Australia
(a division of Pearson Australia Group Pty Ltd)
459–471 Church Street, Level 1, Building B,
Richmond, Victoria 3121
PO Box 23360, Melbourne, Victoria 8012
www.pearson.com.au

First published 2016 by Pearson Australia
2027 2026 2025 2024
1 2 3 4 5 27 26 25 24

Publisher: Alicia Brown
Development Editors: Zoe Hamilton, Antonietta Anello
Project Manager: Shelly Wang
Production Manager: Elizabeth Gosman
Editors: Marcia Bascombe, Aptara
Proof Reader: Marcia Bascombe
Index: Brett Lockwood
Designer: Anne Donald
Copyright & Pictures Editor: Sian Human
Desktop Operators: Lauren Statham, Ben Galpin, Aptara
Illustrator: DiacriTech
Printed in Australia by Pegasus Media and Logistics

National Library of Australia Cataloguing-in-Publication entry (paperback)
Creator: Rickard, Greg, author.
Title: Pearson science SB9 / Greg Rickard [and fourteen others]
Edition: 2nd Edition.
ISBN: 9781488656903 (paperback)
Series: Pearson science SB; 9.
Notes: Includes index.
Target Audience: For secondary school age.
Subjects: Science—Study and teaching (Secondary)
Science—Textbooks.
Dewey Number: 500

Pearson Australia Group Pty Ltd ABN 40 004 245 943

Disclaimer/s
The selection of internet addresses (URLs) provided for this book/resource was valid at the time of publication and was chosen as being appropriate for use as a secondary education research tool. However, due to the dynamic nature of the internet, some addresses may have changed, may have ceased to exist since publication, or may inadvertently link to sites with content that could be considered offensive or inappropriate. While the authors and publisher regret any inconvenience this may cause readers, no responsibility for any such changes or unforeseeable errors can be accepted by either the authors or the publisher.

Some of the images used in *Pearson Science* might have associations with deceased Indigenous Australians. Please be aware that these images might cause sadness or distress in Aboriginal or Torres Strait Islander communities.

Practical activities:
All practical activities, including the illustrations, are provided as a guide only and the accuracy of such information cannot be guaranteed. Teachers must assess the appropriateness of an activity and take into account the experience of their students and facilities available. Additionally, all practical activities should be trialled before they are attempted with students and a risk assessment must be completed. All care should be taken and appropriate protective clothing and equipment should be worn when carrying out any practical activity. Although all practical activities have been written with safety in mind, Pearson Australia and the authors do not accept any responsibility for the information contained in or relating to the practical activities, and are not liable for any loss and/or injury arising from or sustained as a result of conducting any of the practical activities described in this book.

PEARSON Australian Curriculum
Writing and Development Team

Anna Bennett
Differentiation Consultant, Victoria

Ian Bentley
Former Head of Science, VCE exam and trial exam writer
STEM investigation developer, Victoria

Christina Bliss
Teacher, VCE assessor Biology
Author and question writer, Victoria

Donna Chapman
Science Laboratory Technician
Safety consultant, Victoria

Dr Warrick Clarke
Curriculum Writer, Science Communicator and Australian Post-Doctoral Research Fellow at UNSW
Author, NSW

Jacinta Devlin
Science and senior Physics Teacher
Coordinating Author, Victoria

Julia Ferguson
Author and reviewer
Education Officer, Earth Science Western Australia, Western Australia

Bob Hoogendoorn
Exam-style question writer
VCAA Exam Assessor, Chemistry
Former senior Chemistry teacher, Victoria

Penny Lee
Science Laboratory Technician
Safety consultant, Victoria

Louise Lennard
Head of Science. Former Industrial scientist, Author and STEM investigation developer, Victoria

Greg Linstead
Former Head of Science, Education Department of WA Curriculum Writer, Author, Western Australia

Bryony Lowe
Director of Numeracy Improvement.
Former Head of Science and Region Teaching & Learning Coach
Author and reviewer, Victoria

David Madden
Science Learning Area Manager at QCAA, Former Head of Science, Author and reviewer, Queensland

Fran Maher
Bioscience educator, Science teacher
Formerly a bioscience researcher
Author, Victoria

Rochelle Manners
Science and mathematics teacher.
Co-ordinating author Teacher Companion, Queensland

Shirley Melissas
Teacher Librarian and Author, Development Editor, Victoria

Tamsin Moore
Science and senior Psychology teacher, Author
Western Australia

Natalie Nejad
Head of Science, Science and mathematics teacher
STEM investigation developer, Victoria

Malcolm Parsons
Education consultant, former teacher
Author, Victoria

Greg Rickard
Teacher, Former Head of Science
Coordinating Author, Victoria

Lana Salfinger
Teacher, Head of Science, IB Workshop leader in MYP Sciences
Author, Western Australia

Maggie Spenceley
Former Teacher, Curriculum Writer
Queensland Studies Authority
Author, Queensland

Jim Sturgiss
Teacher, Former Coordinating analyst (NSW Department of Education) reporting NAPLAN, Senior test designer for Essential Secondary Science Assessment (ESSA) Author Thinking Scientifically questions, New South Wales

Craig Tilley
STAV Trial Exam Coordinator, VCAA Exam Assessor
Author and exam-style question writer, Victoria

Jo Watkins
Chief Executive Officer, Earth Science Western Australia, Author and reviewer, Western Australia

Dr Trish Weekes
Science Literacy Consultant, New South Wales

Rebecca Wood
Science educator and tutor, Author, Victoria

Table of Contents

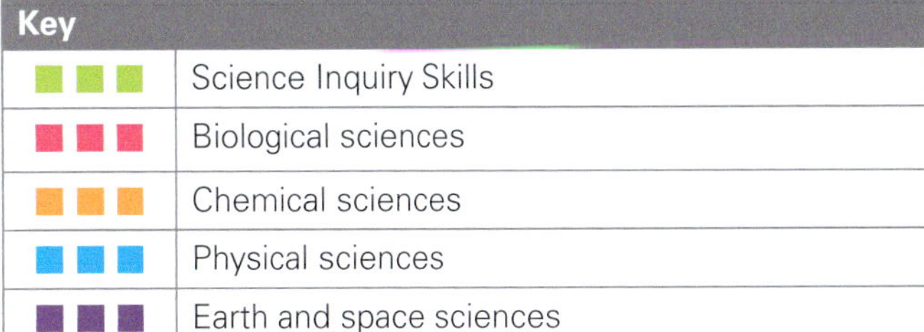

Key
Science Inquiry Skills
Biological sciences
Chemical sciences
Physical sciences
Earth and space sciences

Go to your **eBook** to access this **STEP UP chapter** as well as:

- Activity Book worksheets
- Answers
- Teacher support notes

Key	
■■■	Science Inquiry Skills
■■■	Biological sciences
■■■	Chemical sciences
■■■	Physical sciences
■■■	Earth and space sciences

How to use this book • STUDENT BOOK

Pearson Science 2nd edition has been updated to fully address all strands of the new **Australian Curriculum: Science** which has been adopted throughout the nation. Since some states have tailored the Australian Curriculum slightly for their own particular students, the coverage of the new **Victorian Curriculum: Science** is also captured in this new edition. We address inclusion by clearly indicating the additional content which enables flexibility to determine the approach, as well as the added bonus of an option to engage with **extension** and **revision** opportunities.

All aspects of the student books have been thoroughly reviewed by our **Literacy Consultant Dr Trish Weekes** and the result is **more accessible** content, **enhanced scaffolding** and **strengthened question and instructions sets**. The design is updated to improve the readability and navigation of the text.

In this edition, we retain a flexible approach to teaching and learning. A careful mix of **inquiry**, **STEM** and a range of **practical investigations**, along with **fully updated** content reflect the dynamic and ever-changing nature of scientific knowledge and developments. Combined with the improved and enhanced sets of questions, this series provides a rich assortment of choice, supporting a **differentiated approach**.

An integrated and research-based approach to science education, which ensures every student has engaging, supportive and challenging opportunities.

Be set

The **chapter opening page** sets a context for the chapter, engaging students through questions that get them thinking about the content and concepts to come. The chapter learning outcomes are provided in student friendly language and give transparency and direction for the chapter. Each chapter is divided into self-contained modules. The **module opening page** includes an introduction that places the material to come in a meaningful context.

CHAPTER 5 Electromagnetic radiation

Have you ever wondered ...
- why objects appear different colours?
- how heat gets from the Sun through empty space?
- how night-vision goggles work?
- how your radio works?

After completing this chapter you should be able to:

178

MODULE 5.2 The visible spectrum

The visible spectrum is the rainbow of colours that combine to form white light. Visible light is just a small band of the frequencies that make up the electromagnetic spectrum. This is the band of electromagnetic radiation that our eyes can detect.

science 4 fun

Polarisation

Colour

189

Be interested

Stunning and relevant **photos and illustrations** are purposefully selected to build understanding of the text. Students know when and how they should engage with artwork as each image is clearly referenced from within the text to develop understanding. Captions for every artwork, along with labels for more difficult images, build further meaning and understanding.

Seeing in colour

Primary colours

190

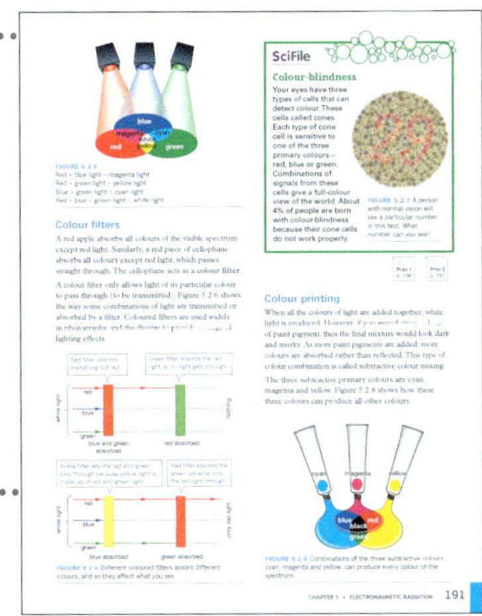
SciFile

Colour-blindness

Colour filters

Colour printing

191

Be inventive

The **STEM4fun** activities are simple STEM-based applications. Students are given an open-ended problem and asked to create, design or improve something. These problems require students to draw on their acquired knowledge and skills, but are more about the process than the actual solution.

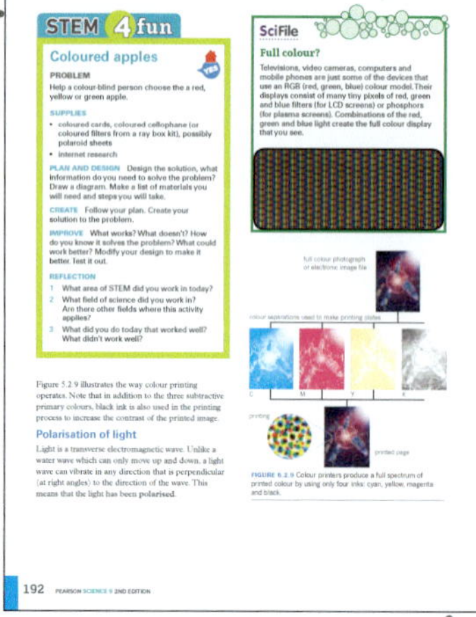

Be inquiring

Science4fun are inquiry based activities. They pre-empt the theory and get students to engage with the concepts through a simple activity that sets students up to 'discover' the science before they learn about it. Broadly speaking, they encourage students to think about what happens in the world and how science explains this.

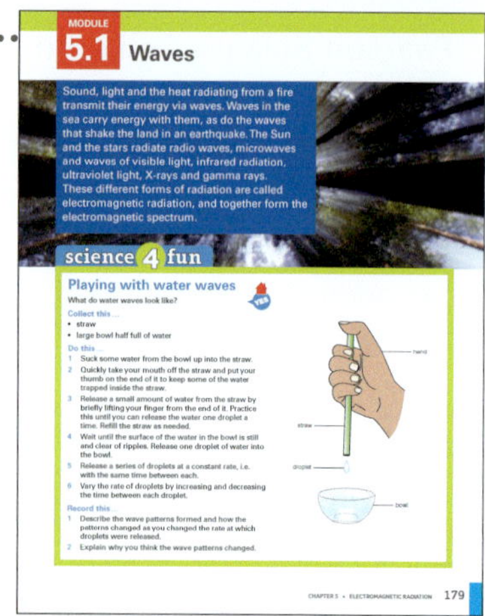

Be inspired

Working with science career profiles cast a spotlight on the diversity of career opportunities available through science with a focus on future science directions, STEM and women in science. Profiles include questions that to relate to the topic.

Be amazed

The **Science as a Human Endeavour** strand is addressed throughout the modules as well as in spreads. Many of the spreads have a special focus on Australian Scientists and highlight exciting developments, innovations and discoveries across all science fields. This feature also includes questions to help students build connections with the content they are learning and the relevance of these contributions.

Be skilled

Skill Builders outline a method or technique and are instructive and self-contained. They step students through the skill to support science application.

Be guided

Worked examples scaffold problems and techniques with a new thinking and working approach to guide students through solving problems and applying techniques to master and practice key skills.

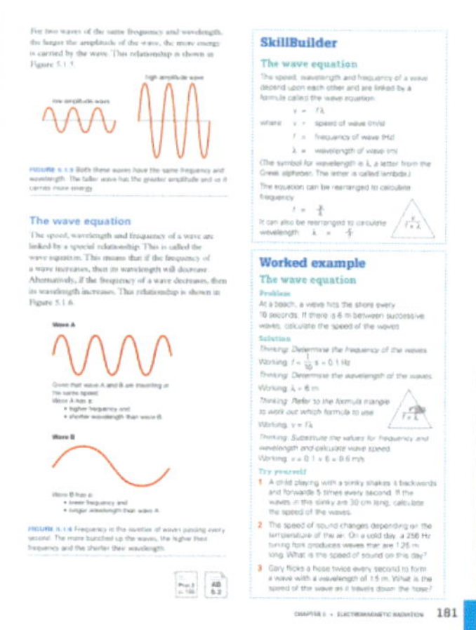

Be confident

Each module concludes with a comprehensive **module review** set that checks for understanding of key concepts and ideas developed through a carefully prepared range of Blooms categorised questions. Students enjoy the benefit of checkpoint opportunities to engage with module review questions at key points throughout the module.

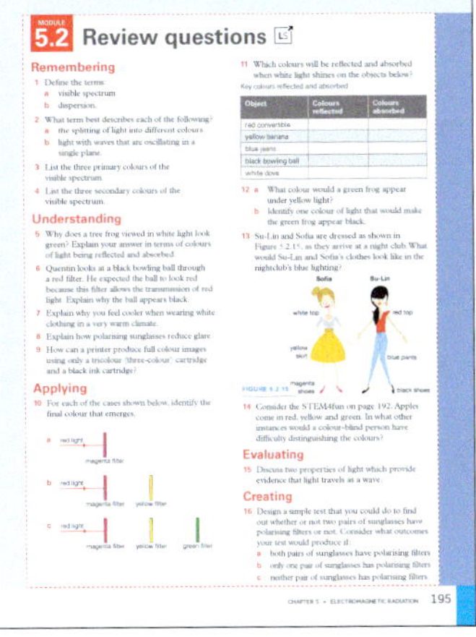

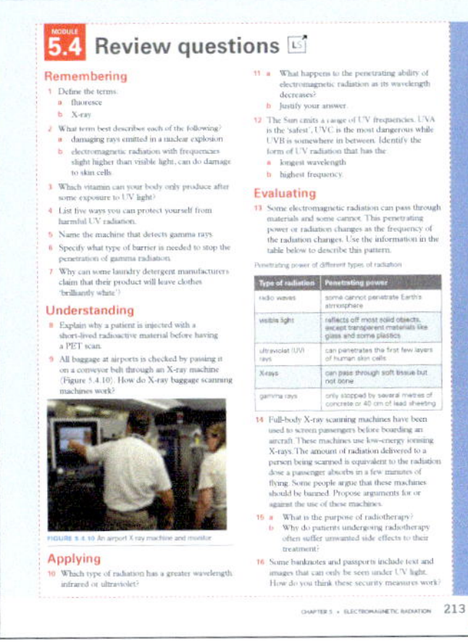

Be investigative

Practical investigations are placed at the end of each module. New Student Design Investigations and STEM inquiry tasks provide students with opportunities to plan investigations, design and trial their plans to seek answers and solve problems. A timing suggestion assists with planning, whilst safety boxes highlight significant hazards. Full risk assessments, safety notes and technician's checklist and recipes provided via ProductLink and eBooks.

Practical investigation icons appear throughout the modules to indicate suggested times for practical work. An icon will also appear to indicate where a SPARKlab alternative is available.

Prac 1
p. 175

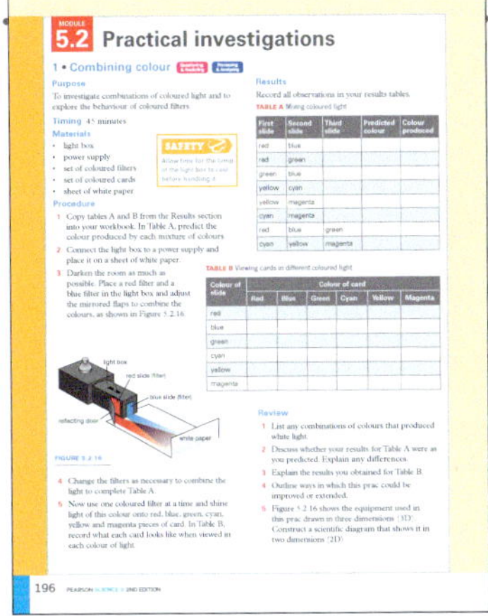

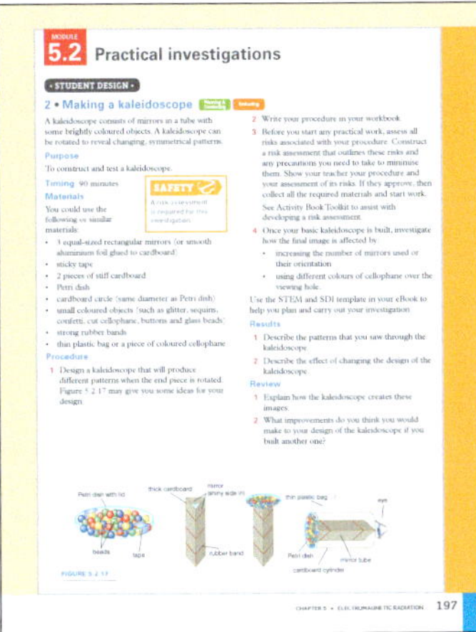

Be extended

Each chapter concludes with an improved and richer assortment of questions organised within the Blooms structure, that bring together the learning of concepts from across a chapter. Apply knowledge and skills to answer questions, engage in fresh new opportunities for **inquiry** and extend into **research** to take your learning to a new level with the enhanced **Chapter review**.

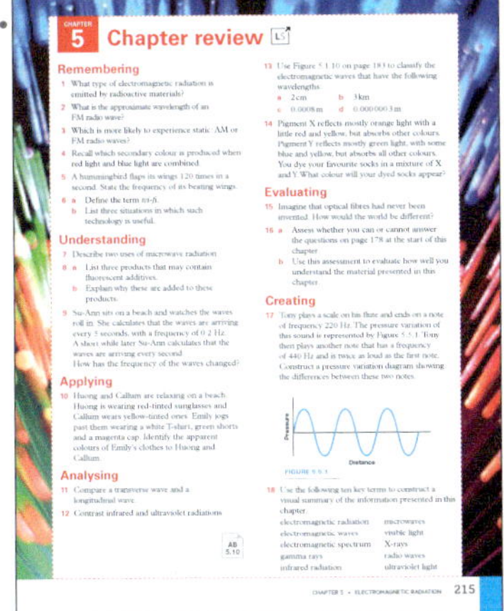

Be a thinker

Following the chapter review are **thinking questions** relevant to the chapter. These test students' science and interpretive skills.

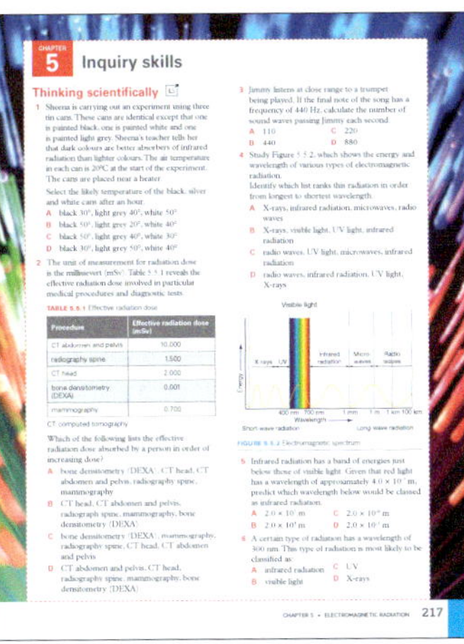

Be supported

Every chapter concludes with an illustrated **glossary** that is an easy reference for additional support in comprehension of key terms. All key terms are bolded throughout the chapter.

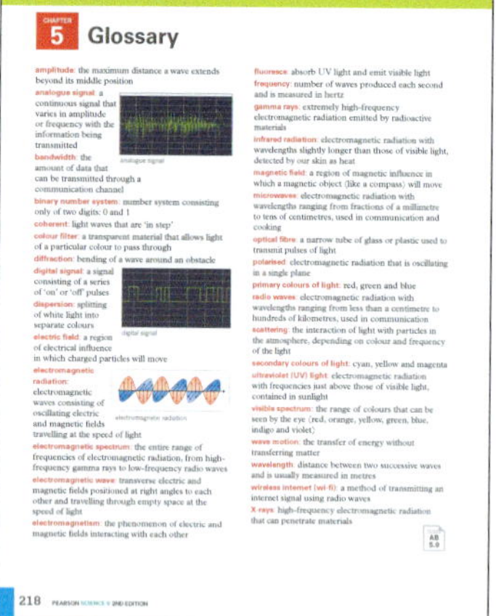

Be reinforced

The **Activity Book** provides a set of worksheets for every student book chapter, giving lots of opportunities for practice, application and extension. Reference **Activity Book icons** indicate when the best time is to engage with a particular worksheet.

AB 4.2

Be progressed

Lightbook Starter contains **complementary sets of questions** for the module and chapter review sets from within the **student book**. This serves as alternate or additional assessment opportunities for students who enjoy the benefit of **instant feedback**, **hints** and **auto-correction** when engaging with this cutting-edge digital **formative** and **summative assessment** platform. Questions are all **tracked** against curriculum learning outcomes, making **progress** monitoring simple. A handy icon indicates the best time to engage with Lightbook Starter.

Be prepared

Focussed on supporting the greater **diversity of learners and pathways**, a 'step-up' program has been developed to launch students into senior sciences, in addition to the 'core' science program. A series of **step up chapters**, written by experienced senior science teachers, have been developed with the view to providing all students with best chance of success.

The **Year 9 Student Book** features a step up chapter on **Psychology**. The **Year 10 Student Book** includes step up chapters for **Biology**, **Chemistry** and **Physics**. These chapters are referenced from the print text and are provided in full via the **eBook**. The eBook also contains **worksheets** specific to supporting the application and development of skills and knowledge from within the text.

All Year 10 Student book chapters include a new series of **Exam Style questions** to provide students practice and exposure in preparation for examinations.

Student pathways

Pearson Science SB9 is designed to cater for a range of student abilities and prepare students for future pathways they may plan to pursue.

Pearson Science SB9		Focus of chapter	Pathway	
Chapter 1	Scientific inquiry skills	Curriculum science skills: a reference tool for use all year	■■■	Year 9 and all senior sciences
Chapter 2	Materials	Curriculum strands and elaborations	■■■	Chemistry
Chapter 3	Reaction types		■■■	Chemistry
Chapter 4	Heat, sound and light		■■■	Physics
Chapter 5	Electromagnetic radiation		■■■	Physics
Chapter 6	Electricity		■■■	Physics
Chapter 7	Body coordination		■■■	Biology
Chapter 8	Disease		■■■	Biology
Chapter 9	Ecosystems		■■■	Biology
Chapter 10	Plate tectonics		■■■	Earth sciences
Chapter 11	Psychology step-up	Extension	■■■	Senior psychology

Pearson Science 2nd edition Teacher Companion

The Teacher Companion makes lesson preparation easy by combining full-colour student book pages with teaching strategies, ideas for class activities and fully worked solutions. All of the Activity Book pages are also included and are complete with model answers.

Be prepared

The **Chapter preview** provides an overview for planning purposes, including things to be aware of and organise ahead of commencing. The **pre-prep** also has an indicator of the time allocation to complete the chapter.

Be an expert

A further improved Teacher Companion places the support of **experts** alongside every Pearson Science 2e teachers, featuring wrap-around teaching and learning strategies and support from:

- **Literacy Consultant: Dr Trish Weekes**
- **Differentiation Consultant: Anna Bennett**
- **School laboratory technicians: Penny Lee and Donna Chapman**

Be confident

All practical activities have been trialled, reviewed, amended and replaced as necessary to ensure teachers and students can undertake practical activities that are tested, work and will yield effective results. Suggested replacement materials and equipment provided to make science more accessible.

Full risk assessments, safety notes and technician's checklist and recipes provided. Pracs and risk assessments have been updated to reflect new regulations around safety and materials in school science classrooms.

Be informed

Full **answers** including suggested findings and possible answers to practical activities, fully worked solutions and support for open-ended research, inquiry and STEM activities.

Pearson Science Lightbook Starter

Lightbook Starter offers a **digital formative and summative assessment tool** with **hints**, **instant feedback** and **auto-correction** of responses. Students and teachers also enjoy the visibility of learning through a **progress tracker** which shows student achievement against curriculum learning outcomes. Lightbook Starter provides questions with the most sophisticated auto-correction of answers.

Be ready

Commence each chapter with questions to establish a baseline for each student around prior knowledge. The **'before you begin'** section includes useful preparatory material with **interactive** resources to **activate prior knowledge** and **reteach key concepts**.

Be in control

Lightbook starter is written to enable teachers and students to use this digital assessment tool as an **alternative** (or additional practice) **to student book questions**. The Lightbook Starter structure mirrors the student book question set, thereby providing a complimentary alternative to the student book questions. This supports a fully integrated approach to digital assessment and feedback.

Be assisted

Module review questions (with **hints** and **solutions**), help students **check for understanding** of learning, revise and provide useful **formative assessment** to help teachers identify areas of weakness, great for lesson planning. These serve as a touchpoint throughout the chapter and students benefit from auto-corrected responses which provide **instant feedback** and support.

Be assessed

The **Chapter Review** in the student book has a complimentary **assessment** set in Lightbook Starter. Use this as an alternative to a class test at the end of a topic.

Be reflective

An integrated **reflection** set supports students in considering their progress and future areas for focus.

Be tracked

Enjoy seeing progress through the learning outcomes updated instantly in the **progress tracker**.

LightbookStarter 

Pearson Science eBook

Pearson eBook enables viewing and interaction with the student book online or offline on any device: PC or Mac, Android tablet or iPad and interactive whiteboard. This eBook retains the integrity of the printed page whilst offering easy to access resources, support and linked activities that will engage your students at school and at home.

The eBooks provide a fully integrated, digital learning platform. Enjoy the benefits of having the following digital assets and interactive resources at your fingertips:

- New interactive activities and lessons
- New Untamed Science videos
- Web destinations
- Student investigation templates and teacher support
- New STEP UP Student Book and Activity Book chapters with answers at Years 9 and 10
- Full answers to all Student Book and Activity Book questions
- SPARKlabs
- Risk assessments
- Full teaching programs and curriculum mapping audits
- Chapter tests with answers

Pearson Science ProductLink

Additional student and teacher resources are available free when you purchase **Pearson Science 2nd Edition**. To access, visit **www.pearsonplaces.com.au** and log in. Click on 'Toolkit' then select 'ProductLink' and browse your title.

Professional Learning, Training and Development

Did you know that Pearson also offers teachers a diverse range of training and development product-linked learning programs? We are dedicated to supporting your implementation of Pearson Science, but it doesn't stop here.

Our courses align closely with Pearson Science 2nd Edition and offer an in-depth learning experience, combining both practical and theoretical elements, enabling you to implement the resource effectively in your classroom.

Find out more about our product-linked learning, workshops, courses and conferences at **Pearson Academy www.pearsonacademy.com.au**

Acknowledgements

We thank the following for their contributions to our text book:
The following abbreviations are used in this list: t = top, b = bottom, l = left, r = right, c = centre.

ACARA

All material identified by (AC) is material subject to copyright under the Copyright Act 1968 and is owned by the Australian Curriculum, Assessment and Reporting Authority 2017.

ACARA neither endorses nor verifies the accuracy of the information provided and accepts no responsibility for incomplete or inaccurate information. In particular, ACARA does not endorse or verify that:

- The content descriptions are solely for a particular year and subject;
- All the content descriptions for that year and subject have been used; and
- The author's material aligns with the Australian Curriculum content descriptions for the relevant year and subject.

You can find the unaltered and most up to date version of this material at http://www.australiancurriculum.edu.au/. This material is reproduced with the permission of ACARA.

VCAA

The Victorian Curriculum F–10 content elements are © VCAA, reproduced by permission. Victorian Curriculum F–10 elements accurate at time of publication. The VCAA does not endorse or make any warranties regarding this resource. The Victorian Curriculum F–10 and related content can be accessed directly at the VCAA website.

Cover: Science Photo Library/Marek Mis

123RF: 123rf.com, pp. 8, 350br; algre, p. 269t; Wang Aizhong, p. 124; Elnur Amikishiyev, p. 18r; Charles Brutlag, p. 15; cookelma, p. 87t; Sandra Cunningham, p. 2t; dolgachov, p. 25t; goodluz, p. 17; elvan mehmed halil, p. 97b; Dmitriy Krasko, p. 307tr; Volodymyr Krasyuk, p. 200l; Marlon Ornek, p. 351r; James Steidl, p. 141b; subbotina, p. 151; vlue, p. 130c; Wavebreak Media Ltd, p. 311b; woodoo007, p. 274.

AAP: AP Photo/The Family Dental Center via KUSA-TV via The Denver Post/AP via AAP, p. 211tl; APTN, pp. 422t, 430; Joe Castro, p. 287t; CSIRO, p. 205; Annaliese Frank, p. 367t; Dean Lewins, p. 213.

ABC: pp. 204r, 232b; Australian Broadcasting Corporation Library Sales/Clint Jasper, p. 368.

Age Fotostock: Sylvain Grandadam, p. 80l.

Alamy Stock Photo: Advert. Park, p. 58br; AF archive/Alamy Stock Photo, p. 211bl; Ian Allenden, p. 1, 34–37; Archivio World 1, p. 405r; Ashley Cooper pics, p. 330t; Kitch Bain, p. 209tr; Blend Images, p. 272b; blickwinkel/Hecker, p. 380l; Bon Appetit, p. 167tr; BrazilPhotos.com, p. 115t; BSIP, pp. 111t; BSIP SA, pp. 208, 310c; Scott Camazine, p. 307bl, p.346l; John Cancalosi, p. 364b; Christine Osborne Pictures, p. 3br; CN Boon, p. 272c; Cosmo Condina North America, p. 198t; David Tipling Photo Library, p. 355; Dinodia Photos, pp. 378bc, 390cr; FALKENSTEINFOTO, p. 324l; David Foster, p. 365b; GL Archive, p. 115br; Sigrid Gombert, p. 114t; Spencer Grant, p. 68; Graphic Science, p. 376br; Daniel Grill, p, 329t; GUILLAUME P./BSIP, p. 320l; Martin Harvey, pp. 378tr, 380r; Tracy Hebden, p. 157; imageBROKER, p. 352b; INTERFOTO, p. 295; Sebastian Kaulitzki, p. 155; Jake Lyell, p. 259t; Nick Lylak, p. 242bl; Mary Evans Picture Library, p. 323; MaximImages, p. 167bl; mediacolor's, p. 137t; Mediscan, pp. 211tr, 322bl; Purestock, p. 275; Robert Read, p. 322tl; RGB Ventures/SuperStock, p. 410r; Matt Smith, p. 377l; Kumar Sriskandan, p. 55; Stocktrek Images, Inc., p. 317b; Top-Pics TBK, p. 354r; travel images, p. 255l; Nick Turner, p. 391, 427–429; Andrew Twort, p. 111b; David Wall, p. 141t; Carol & Mike Werner, p. 153tr; A. T. Willett, p. 182b.

Auscape International Photo Library: David Parer & Elizabeth Parer-Cook, p. 378bl; Wayne Lawler, p. 378tl.

Australian National University: Stuart Hay, p. 417b.

Black Diamond Images: Courtesy of Black Diamond Images, p. 376tr.

CSIRO: p. 50l.

Canan Dagdeviren: pp. 212l, 212r.

Department of Health and Ageing: National Health and Medical Research Council, p. 329.

DK Images: Frank Greenaway, p. 321b; Clive Streeter, p. 74tr.

Dreamstime: John Anderson, p. 353r; Timothy Boomer/Sergey Galushko, p. 63; William Casey/Dreamstime.com, p. 214; Jacek Chabraszewski, p. 334bl; Christian Delbert, p. 231br; Lane Erickson, p. 231tr; Tim Hester, p. 403; Jim Hughes, p. 80br; Jamin1121, p. 379b; Jwk1, p. 303tr; Lianem, p. 131r; p. 150b; Andrei Malov, p. 385; Yaroslava Polosina, p. 379tl; Russplaysguitar, p. 179; Seaphotoart, p. 365t; Sinelyov, p. 130b; Stocksnapper, p. 269b; Ulga, p. 128l.

Earthbyte Group: Dietmar Müller, School of Geosciences, The University of Sydney, pp. 399br, 399l, 399tr.

Flickr: Craig Howell, licenced under Creative Commons Attribution 2.0 (CC BY 2.0), p. 423l.

Fotolia: agnormark, p. 367b; arsdigital, 147l; Awe Inspiring Images, p. 244 (LDR); Christophe Baudot, p. 378br; bmf-foto.de, p. 193; eag1e, p. 230; Firma V, p. 24; Fotos 593, p. 414; Fotoschlick, p. 244 (LED); karandaev, p. 5; Witold Krasowski, p. 250; lemonmeringue, pp. 178, 215–217; leodoc63, p. 350bl; Michelle Meiklejohn, p. 296r; Federico Rostagno, p. 40t.

Geoscience Australia: Map from Blewett, R.S. (ed.) 2012. Shaping a nation: a geology of Australia. Geoscience Australia and ANU E Press, Canberra. http://epress.anu.edu.au/titles/shaping-a-nation, © Commonwealth of Australia (Geoscience Australia) and the Australian National University 2016. Licenced under a Creative Commons Attribution 4.0 International Licence (CC BY 4.0)., p. 421tr.

Getty Images: Anatomical Travelogue, p. 282tl; Juergen Berger, pp. 308bc, 346tr; Neil M Borden, p. 113l; John Crux Photography, p. 76; Dr P Marazzi, p. 308b; Anthony Mercieca, p. 101b; Hank Morgan, p. 211bl; Nancy Nehring, p. 307br; Omikron, p. 166; Photo Researchers, p. 89l; Mark A Schneider, p. 209bl; Victor De Schwanberg, pp. 57bl, 57br; Science Photo Library, pp. 65t, 78l, 281; John Smith, pp. 75, 119–121; Paul Souders, p. 405bl; Penny Tweedie, pp. 381t, 382; Jim Varney, p. 335b; Charles D. Winters, pp. 57tr, 88, 123bc.

GravityLight: gravitylight.org, pp. 259b, 260.

ImageFolk (was Amana/Corbis): Auscape/UIG, p. 376tl; Craig Ingram, p. 376bl; Ragnar Th. Sigurdsson, p. 405tl.

iStockphoto: p. 307tl; Jeffrey Daly/istockphoto.com, p. 254; Fedels, p. 127b; Anders Sellin, p. 127t.

Ixom: Ixom Operations Pty Ltd, p. 11.

Greg Linstead: pp. 242br, 242tr, 413.

Lochman Transparencies: Marie Lochman, p. 349r.

Mainstream Data Ltd (Newscom): Chris Barry, p. 64b.

Anna Meadows: p. 286.

NASA Images: p. 409t; SA/courtesy of nasaimages.org, p. 416b; Jeff Schmaltz, MODIS Rapid Response Team, NASA/GSFC, p. 398bl.

National Computational Infrastructure (NCI) Vizlab: Drew Whitehouse, p. 421b.

National Library of Australia (NLA): Lycett, Joseph, Aborigines using fire to hunt kangaroos, 1817, National Library of Australia / nla.pican2962715s20, p. 381b.

National Oceanic & Atmospheric Administration (NOAA): Der Spiegel/NOAA, p. 422b; Computerized digital images and associated databases are available from the National Geophysical Data Center, National Oceanic and Atmospheric Administration, U.S. Department of Commerce, http://www.ngdc.noaa.gov/, p. 392.

Paul Nylander: p. 252r.

Pearson Education: Awe Inspiring Images, p. 252l; Peter Bernik, p. 331t; Martyn F Chillmaid. © Pearson Education Ltd. 2014, p. 222; Trevor Clifford/Pearson Education Ltd, pp. 237b, 237t; Goodluz, p. 12r; HL Studios, p. 171; Jupiter Images. Brand X, p. 209br; Nikita G. Sidorov, p. 31; Sozaijiten, pp. 6, 16.

Pearson Australia: Alice McBroom, pp. 118, 229tr.

PEARSON CD: pp. 149t, 227t, 267b.

Greg Rickard: p. 156t.

Science Photo Library SPL: p. 308tc; Andrew Lambert, p. 66b; Andrew Lambert Photography, pp. 64t, 74br, 79b, 190b, 190t, 231bl, 244 (Diode); Paul Avis, p. 110t; John Bavosi, pp. 66t, 90l, 90r; Juergen Berger, pp. 312b, 325; Ian Boddy, p. 306; Dr John Brackenbury, p. 318t; Massimo Brega, The Lighthouse, p. 324r; Wladimir Bulgar, p. 244 (transistor); Dr Jeremy Burgess, p. 98t, 123tc; Martyn F. Chillmaid, pp. 33, 87b, 89r, 244 (thermistor); Dr Ray Clark & Mervyn Goff, p. 201l; A.B. Dowsett, p. 311c; Wim van Egmond, p. 352tl; Eye of Science, p. 271tr; Simon Fraser, p. 91t; Dr David Furness/ Keele University, p. 169r; Mark Garlick, p. 220b; Steve Gschmeissner, pp. 169l, 271tc, 308tl; Gusto Images, pp. 200r, 219, 264–266; Roger Harris, p. 294, 303br; Health Protection Agency, p. 210l; IBM Research, p. 40b; Richard Kail, p. 116r; Frans Lanting/Mint Images, p. 130t; Dr Najeeb Layyous, p. 141c; Dr P. Marazzi, pp. 320c, 330b; Dr David M. Martin, p. 312t; Chris Martin-Bahr, p. 128cr; MDA Information Services, p. 407l; MEDIMAGE, p. 273t; Dr P Menzel, p. 321t; Cordelia Molloy, p. 92l; © NASA/Science Photo Library, pp. 39, 72, 73; NOAA PMEL Vents Program, p. 106; RIA NOVOSTI/S, p. 287b; NREL/US DEPARTMENT OF ENERGY, p. 57 tl; Claude Nuridsany & Marie Perennou, p. 357b; Otis Historical Archives, National Museum of Health and Medicine, pp. 110b, 123b; David Parker, p. 408t; Alfred Pasieka, p. 243b; Pixologisticstudio, p. 101t; POWER AND SYRED, p. 232t; Phillipe Psaila, pp. 99, 374; Paul Rapson, p. 125t; David Scharf, p. 307tc; Science Photo Library, pp. 13b, 318bl; Sovereign ISM, pp. 282br, 283t, 303l; Sputnik, p. 112r; p. 114b; St Mary's Hospital Medical School, p. 330c; Volker Steger, p. 58tr; TEK IMAGE, p. 243t; Trevor Clifford Photography, pp. 227b, 228l; UC Regents, National Information Service for Earthquake Engineering, p. 408b; UK Crown copyright courtesy of FERA, p. 18l; University Corporation for Atmospheric Research, p. 423r; US Army, p. 112l; US Navy, p. 406t.

Science Source: Charles D. Winters, pp. 57tr, 88, 89l, 123bc.

Narelle Sheean: pp. 56l, 56r.

Shutterstock: pp. 10b; p. 139tc; p. 334r; 2xSamara.com, p. 299; aleks.k, p. 348bl; Zakharov Aleksey, p. 366l; Potapov Alexander, p. 184b; Artwork studio BKK, p. 29; asife, p. 163t; auremar, p. 168; Nina B (geothermal), p. 257; Darren Baker, p. 12l; Anton Balazh, p. 54 (polished); Marcin Balcerzak, p. 66bc; Jim Barber, p. 113r; Joe Belanger, p. 58tl; Gerald Bernard, p. 95; bluehand, p. 65b; Bochkarev Photography, p. 334t; Bokica, p. 273b; Stephen Bonk, p. 354tl; Steve Bower, p. 363; John Carnemolla, p. 377r; Jeff Cleveland, p. 354bl; Alessandro Colle, p. 415t; Raphael Daniaud, p. 139br; Ethan Daniels, p. 357t; Dark Moon Pictures, p. 221; Christian Delbert, p. 54 (lightbulb); denira, p. 58bl; Dikiiy, p. 210tr; dotshock, p. 20b; Dustie, p. 242tl; Elena Elisseeva, p. 54 (grill); EpicStockMedia, pp. 268, 300–302; ever, p. 214l; Everett Historical, p. 36; fckncg, p. 56t; Anton Foltin, p. 348t; Domenic Gareri, p. 139bc; gnatuk, p. 320r; Lukas Gojda, p. 46c; Happy Stock Photo, p. 150tr; Jiri Hera, p. 2b; IDN, p. 335t; Bogdan Ionescu, p. 249; irabel8 (Wave), p. 257; Jes2u.photo, p. 319; Jodie Johnson, p. 189t; Alexander Kalina, p. 361; Panos Karas, p. 107; Sebastian Kaulitzki, pp. 318bl, 346b; Matej Kastelic (Wind), p. 257; Judy Kennamer, p. 139t; Natalia Klenova, p. 328t; kochanowski, p. 310b; koi88, p. 116l; Perati Komson, p. 220t; Irina Kozorog, p. 353bl; Piotr Krzeslak, p. 7r; S. Kuelcue, p. 182tl; Yuriy Kulik (Solar)., p. 257; Lagui, p. 59r; Lakeview Images, pp. 78r, 123t; Landscape Nature Photo (Hydro), p. 257; Lana Langlois, p. 322r; Scott Latham, p. 108; Lighthunter, p. 54 (wire); litn, p. 409b; Ralph Loesche, p. 360; Robyn Mackenzie, p. 54 (coins); MaksiMages, p. 80tr; Petr Malyshev, p. 66tc; Oleksiy Mark/ Shutterstock.com, p. 244 (capacitor); Doug Meek, p. 238t; MilanB, p. 252c; MillaF, p. 188; Monkey Business Images, p. 142; Mopic, p. 189b; nexus 7, pp. 4, 41tl; Nuk2013, p. 7l; Sabino Parente (Nuclear); Michael Pettigrew, pp. 352tr, 390br; Jure Porenta, p. 52; Pablo Prat, p. 184t; Ingrid Prats, p. 348br; Przemyslaw, pp. 347, 387–389; qcontrol, pp. 305, 342–345; RamonaS, p. 156b; Jaggat Rashidi, p. 210br; Alexander Raths, p. 333; shyshak roman, p. 332; Roomanald, p. 244 (resistor); Roger Rosentreter, p. 350tr; Jason Patrick Ross (Tidal), p. 257; Nicram Sabod, p. 194; Roman Samokhin, p. 136; science photo, p. 182tr; Serp, p. 46t; Hank Shiffman, p. 251; kowit sitthi (biomass), p. 257; Dmitrij Skorobogatov, p. 311t; StevenRussellSmith, p. 13t; Stubblefield Photography, p. 54 (ship); suravid, p. 40t; Svetislav1944, p. 41br; tobkatrina, p. 54 (cans); Mike Truchon, pp. 353tl, 390l; Martin Turzak, p. 328b; VanderWolf Images (Fossil fuel), p. 257; Yakobchuk Vasyl, pp. 21t, 279; Nils Versemann, p. 14t; Krivosheev Vitaly, p. 41bl; wavebreakmedia, pp. 9, 317t; Brooke Whatnall, p. 350tl; Peter J. Wilson, p. 375; Zdravinjo, p. 30; Zerbor, p. 331l.

The Department of the Prime Minister and Cabinet: Permission to reproduce the Commonwealth Coat of Arms granted by the Department of the Prime Minister and Cabinet., p. 366r.

Thinkstock: benmm, p. 96t; DoroO, p. 92r; Eraxion, p. 304r; FourOaks, p. 231tl; John Foxx, pp. 122r, 180; GlobalP, pp. 91b, 128br; Jeffrey Hamilton, p. 74tl; John Howard, p. 79t; Julijah, p. 128tr; Jupiterimages, p. 293; Jupiterimages/ Thinkstock, p. 255r; korionov, p. 67; Oleksiy Mark/Thinkstock, p. 267t; TABoomer/ Thinkstock, p. 198b; Dina Trifonova, p. 122l; James Woodson/Thinkstock, p. 253; Cathy Yeulet, p. 77.

UNAVCO/Plate Boundary Observatory (PBO): p. 410l.

United Nations: From Statistical Yearbook for Asia and the Pacific 2011, by ESCAP, © 2011 United Nations. Reprinted with the permission of the United Nations., p. 343.

Vestergaard: pp. 336, 337l, 337r.

Western Australian Museum: p. 3bl.

CHAPTER 1

Scientific inquiry skills

Have you ever wondered …

- why scientific inquiry is important?
- how to develop a research idea?
- how to judge when data is reliable?
- how to present scientific research?

After completing this chapter you should be able to:

- develop research questions and hypotheses that can be investigated using a range of inquiry skills
- identify independent, dependent and controlled variables
- consider safety requirements, assess risk and address ethical issues
- use repeat trials to improve accuracy, precision and reliability
- construct and use a range of graphs, keys, models and formulas, to record and summarise data from investigations
- collect and process qualitative and quantitative data and distinguish between discrete and continuous data
- analyse patterns and trends in data, including describing relationships between variables
- identify inconsistencies in data and sources of uncertainty then suggest improvements to procedure
- draw conclusions that are consistent with evidence and relevant to the research question.

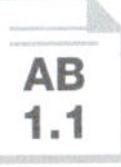

MODULE

1.1 Planning investigations

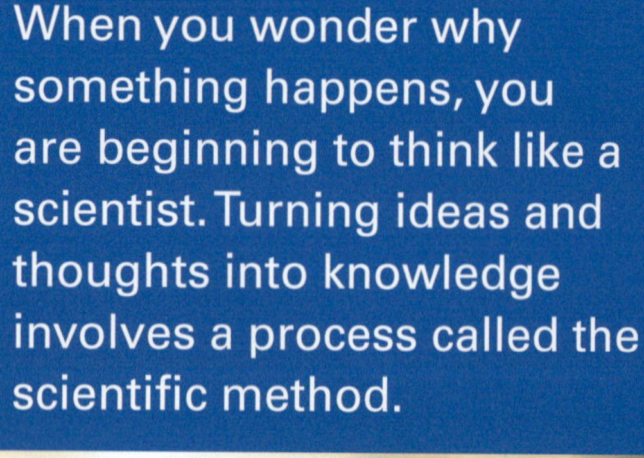

science  fun

Making little things bigger

YES

What happens to gummy bears when left in water?

Collect this ...

- glass or jar about 250 mL in size
- water for the jar
- Haribo® gummy bear

Do this ...

1 Add water to the jar until it is approximately ¾ full.

2 Place one Haribo® gummy bear in the water.

3 Leave overnight.

Record this ...

1 Describe what happened.

2 Explain why you think this happened.

Scientific investigation can take different forms. Scientists may conduct experiments or fieldwork where they test a hypothesis, collect and analyse first-hand data and reach a conclusion. Scientists also carry out investigations through research of scientific information in books, journals and other sources. An investigation may also involve a combination of both experiments and research.

In this chapter, you will focus on doing scientific investigation through experiments. When you wonder why something happens, you are beginning to think like a scientist. Turning ideas and thoughts into knowledge involves a process called the scientific method.

The scientific method

The scientific method, shown in Figure 1.1.1, is a practical way of asking and answering scientific questions. It is a process that uses observation and experimentation to make new discoveries, find out how the world works and to build up scientific knowledge. The scientist in Figure 1.1.2 is using a variety of approaches in her investigation.

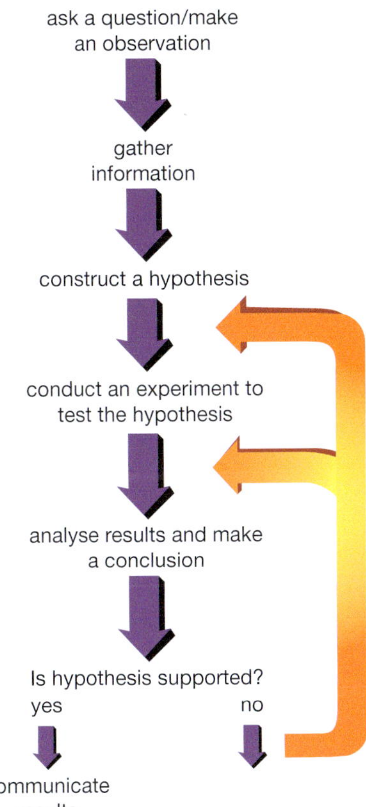

FIGURE 1.1.1 The scientific method includes a number of steps that help scientists follow a process to research and answer questions.

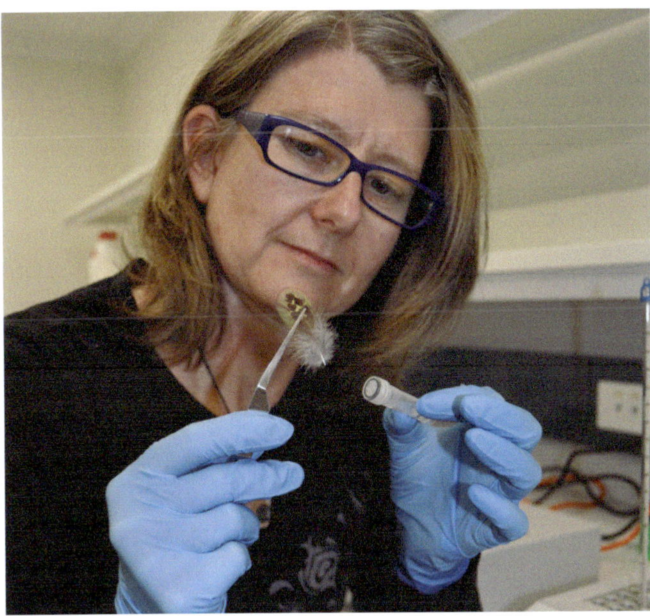

FIGURE 1.1.2 Careful observation in the field and analysis of DNA extracted from feathers helped scientists at the Museum of Western Australia confirm that the night parrot (*Pezoporus occidentalis*) is not extinct, as previously thought.

Observations and questions

All scientific inquiry is based on an idea or a question that needs to be answered. This question is often based on an **observation**, such as noticing that ice melts quicker in a hot drink than in a cold drink.

Observation is an important part of the scientific process. It includes using your senses and a wide variety of instrument and laboratory techniques that allow closer and more accurate observation. Observation allows scientists to collect and record data which provides the basis to construct a hypothesis to be tested.

Developing a question

One way of answering a question is through conducting an **experiment**. When choosing a topic of investigation for an experiment:

- choose a question you find interesting
- start with a topic which you know something about, or where you have some clues about how to perform the experiment
- make sure your school laboratory has the resources for your experiment
- choose a topic that can provide data that is clear and able to be measured.

The question should be one sentence that clearly describes the purpose of the experiment. An experimental hypothesis must be able to be tested.

The question, aim/purpose and hypothesis are interlinked. It is important to note that each of these can be refined as the planning of the investigation continues. For example, if you are investigating the drying time for clothes on a line like the one in Figure 1.1.3, then this may be phrased as a question or as an aim as shown in Table 1.1.1 on page 4.

FIGURE 1.1.3 Clothes drying on a line

TABLE 1.1.1 Relationship between question, aim and hypothesis

Question / Aim / Hypothesis	Examples
Question: this is a sentence that needs an answer and must end with a question mark ('?').	What is the effect of wind speed on the time taken to dry clothes in the shade when the temperature is 20°C?
Aim/Purpose: a sentence summarising what will be investigated. It states the purpose of the experiment starting with a 'to' verb. Be very specific about its purpose (aim).	To determine the effect of wind speed on the clothes drying time when the temperature is 20°C.
Hypothesis: a possible outcome of the experiment. It is an educated guess based on previous knowledge on which you can make a prediction for the results of the experiment. The form of a hypothesis is 'if X happens, then Y will happen'. Formulate a question first and it will lead you to a hypothesis when you: • are able to reduce the question to measurable variables. • can suggest a possible outcome of the experiment.	If the wind speed increases (X), the time taken for the clothes to dry will decrease (Y).

Experimental variables

A **variable** is a factor that can change (or 'vary') and may affect the result of the experiment. A good experiment has carefully **controlled variables** and tests only one variable at a time. This is called a **fair test**. Table 1.1.2 describes independent, dependent and controlled variables in an experiment to melt ice cubes (Figure 1.1.4).

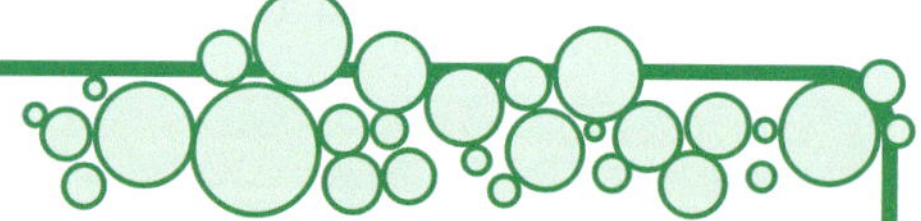

A hypothesis or a theory?

A hypothesis is really just an educated guess of the outcome of an experiment. If, after many different experiments, one hypothesis is supported by all the results obtained so far, then this explanation can be given the status of a **theory**. A theory is a model that fits the available evidence and predicts the outcome of an experiment. One theory you may have heard of is the theory of evolution.

TABLE 1.1.2 Variables in an experiment to test how quickly an ice cube melts in hot water

Type of variable	Description	Example
independent variable	• is changed in a systematic way • what you change (the cause of the change) • observe effect on dependent variable as this variable is changed.	Temperature of the water could be the variable being changed. The experiment would be repeated with the water heated to different temperatures: 20°C, 30°C, 40°C, 50°C and so on.
dependent variable	• the variable being observed and measured • what you observe (the effect of the change) • the variable that changes when the independent variable is changed • provides the data for experiment.	The time taken for the ice to melt is the dependent variable.
controlled variables	• all other variables in the experiment besides the dependent and independent variables • what you keep the same (the things that do not change) • kept the same in all tests that are part of the experiment.	Other possible variables in this experiment include the amount of water used, the size and shape of the container and the size of the ice cube.

FIGURE 1.1.4 Ice cubes will melt when placed in water.

Constructing a hypothesis

A **hypothesis** is an educated guess based on evidence and prior knowledge to answer the question. The hypothesis must be worded so that you can test it. To do this, you will need to identify the **dependent variables** and **independent variables**.

Testing a hypothesis

A good hypothesis should be a testable. This means that the independent variable can be changed and the resulting change in the dependent variable can be measured. It may be written like this:

If X happens, then Y will happen.

A testable hypothesis may look like the three examples below:

If the water temperature increases, then a block of ice placed in the water will melt faster.

If the amount water decreases, then the ice cubes will melt more slowly.

If the size of ice cubes placed in water decreases, then the ice will melt more quickly.

Not all hypotheses are written in exactly the same way, but they should all clearly identify the change in the independent variable and the expected change in the dependent variable.

A correctly written hypothesis will clearly state exactly what you expect will happen in the experiment.

Hypothesis checklist

If you have written a good hypothesis, then you should be able to answer *yes* to the following questions:

- Is the hypothesis based on information contained in the question?
- Does the hypothesis include the independent and dependent variables?
- Can the independent variable be changed?
- Can changes in the dependent variable be measured?

To be a fair test, an experiment should only test one hypothesis at a time.

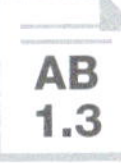

Writing a hypothesis from an inference

Another way to write a hypothesis is to make an **inference** and then convert that inference into a hypothesis. An inference is a logical idea that comes from an **observation**. For example, many parts of Australia have a dry season during which grass changes from being green to being brown or yellow. One observation is that the grass does not turn brown as quickly near the edges of a road, but remains green for much longer (Figure 1.1.5).

FIGURE 1.1.5 Grass growing on the side of the road remains green.

An inference can be made based on this observation and using what information is known about how grass grows. Some inferences that may explain why grass growing near the edge of the road remains green in summer are:

- The road insulates the grass roots from the heat and cold.
- The grass near the road receives runoff when it rains.
- People do not walk on grass near the road.
- The soil under the road remains moist while the other soil dries out.
- More earthworms live under the road than under the open grass.

For the first inference in the list above, the hypothesis might be:

If the temperature of the grass roots were measured, then the temperature of the grass roots under the road would be cooler than the grass roots beside the path.

All of the hypotheses from these inferences can be tested, either by more observations or by taking measurements in multiple experiments, to find out which hypothesis is supported.

Prac 1
p. 8

MODULE 1.1

Review questions

Remembering

1 Define the terms:
   a observation
   b dependent variable
   c independent variable
   d research question.

2 What term best describes each of the following?
   a an inference that can be tested by an experiment
   b test to determine whether or not a hypothesis is supported
   c the variables that are kept consistent during an experiment.

Understanding

3 Write a hypothesis for each of these to test whether:
   a carrot seeds or tomato seeds germinate quicker
   b sourdough, multigrain or white bread goes mouldy the quickest
   c dogs like dry food or fresh food better.

4 Read the hypothesis then answer the following questions.

   If the water temperature increases, then an ice block placed in the water will melt quicker.
   a What is the independent variable for the experiment?
   b What is the dependent variable for the experiment?
   c List three other variables that would need to be controlled.

Applying

5 An experiment to determine the effect of sunlight on plant height was carried out. The plant and experimental variables are shown in Figure 1.1.6. Copy Table 1.1.3 and identify each variable by placing a tick in the correct column. The first has been done for you.

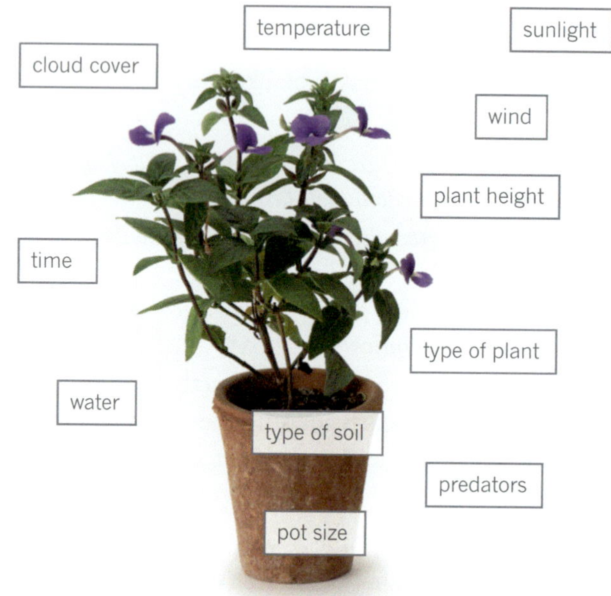

FIGURE 1.1.6

TABLE 1.1.3

	Independent variable	Dependent variable	Controlled variable
type of plant			✓
plant height			
type of soil			
pot size			
temperature			
sunlight			
wind			
cloud cover			
time			
water			
predators			

MODULE 1.1

Review questions

6 Identify the independent and dependent variables for each of the following hypotheses.

a If a cup of hot chocolate has a lid on it, then it will stay hot longer.

b Aquatic plants produce more oxygen in warmer water than in cold water.

c Increasing the intensity of exercise will increase a person's breathing rate.

7 Isla wants to test how many drops of water can sit on top of a coin before the surface tension breaks and the water spills. Identify as many variables as you can for the:

a independent variable

b dependent variable

c controlled variables.

8 A scientist observes that the human eye responds to sudden increases in light by decreasing the diameter of the pupil (Figure 1.1.7). He wonders if this response would change for light of different colours. He ran an experiment to investigate this.

a What would the independent variable be and how could it be changed?

b What would the dependent variable be? How could this data be collected?

c Write a hypothesis for the experiment.

d List three variables that need to be controlled and identify how these could be controlled. Describe the effect on the experiment of not controlling one of these variables.

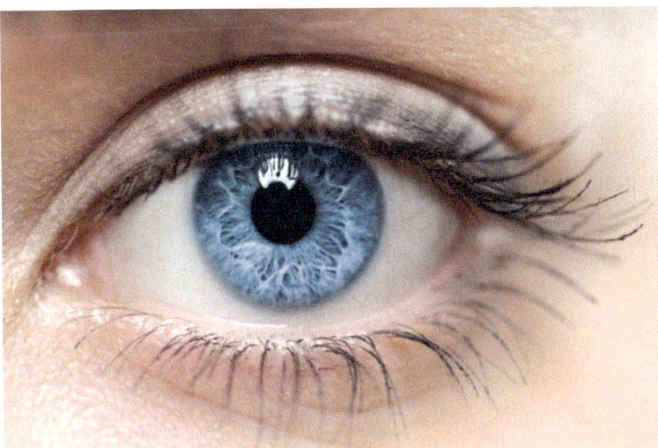

FIGURE 1.1.7 The human eye responds to an increase in light intensity by decreasing the diameter of the pupil. Is the response different for light of different colours?

Analysing

9 Explain the difference between an inference and a hypothesis.

10 Read the following paragraph then write an appropriate question and hypothesis for the experiment.

When watching the tennis I have noticed that the tennis balls get changed regularly. I have also heard that Wilson supplies the Australian Open with 48 000 tennis balls each year! When players start a match in the Australian Open, they are supplied with six new balls, which are used for the five-minute warm-up and the first seven games of the match. I wonder if this is because the balls heat up over the duration of a game and this affects their bounce.

Evaluating

11 Three statements are given below. One is an observation, one is an inference and one is a hypothesis. Identify which is which and give a reason for your choice.

a If there is more water in the soil, then the grass will be greener in these areas.

b The grass growing next to the path is green.

c The grass next to the path gets more water than the grass that is further from the path.

MODULE

1.1 Practical investigation

• STUDENT DESIGN •

1 • Melting ice

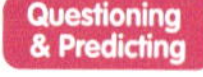

Purpose

To test whether adding substances to water will change how quickly ice melts.

Hypothesis

Once you have decided which investigation to perform, write a hypothesis in your workbook.

Timing 45 minutes

Materials

- ice
- water
- choose from: sugar, salt, sand, flour, pebbles, other substances
- beaker
- stopwatch

A risk assessment is required for this investigation.

Procedure

1 Design an experiment that will answer one of the following questions.
- Will dissolving a substance in water change the melting rate of ice more than adding an insoluble substance?
- Will adding the same amount of different soluble substances have the same effect on the melting rate of ice?
- Will adding different amounts of the same substance have different effects on the time it takes an ice block to melt?

2 Write your procedure in your workbook. Include a diagram of your design for your experiment.

3 Before you start any practical work, assess your procedure. List any risks that your procedure might involve and what you might do to minimise those risks. Show your teacher your procedure and risk assessment. If they approve, then collect all the required materials and start work.

See Activity Book Toolkit to assist with developing a risk assessment.

Use the STEM and SDI template in your eBook to help you plan and carry out your investigation.

Results

Record your results and observations in your workbook.

Review

1 Construct a conclusion for your investigation.

2 Assess whether your hypothesis was supported or not.

MODULE 1.2

Risks, working safely and ethics

There are many opportunities for practical work in science, in the laboratory and outside in the field. Planning these activities is important but so is your safety and the safety of others. For this reason, you must consider the potential risks.

Identifying hazards and assessing risks

A **hazard** is anything that could cause harm. Laboratories are full of hazards such as dangerous chemicals, electrical equipment or sharp instruments. Every time you interact with a hazard, you take a **risk** that you could get hurt.

Up until now, your teachers and laboratory technician have assessed the risks for every experiment that you have done in the laboratory. As a young scientist, you must now begin to write your own **risk assessment**. The purpose of the risk assessment is to:

- identify hazards
- assess the risk by working out how likely each hazard is to cause harm
- control the hazard by identifying ways to reduce the chance of something going wrong or by making sure that the potential harm is minimised.

To identify risks, think about the experiment or activity that you will be doing:

- where you will be working—in the laboratory, a classroom or outside in the field
- how you will be using equipment
- materials that you will be using such as chemicals, organisms or parts of organisms
- how these materials and organisms will be disposed of
- the clothing that you are wearing.

Figure 1.2.1 shows a flow chart of how to consider and assess the risks involved in a research investigation.

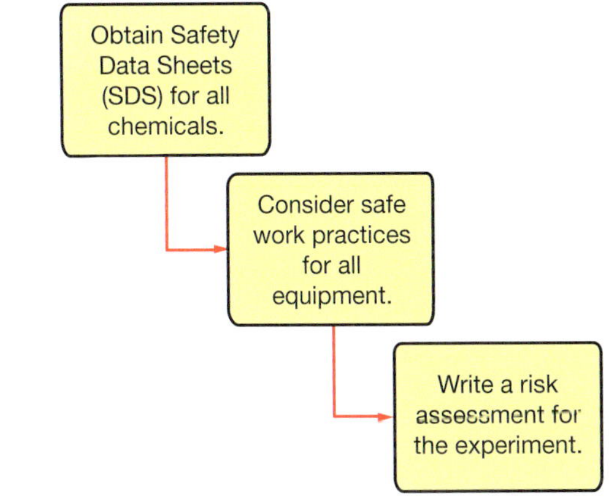

FIGURE 1.2.1 Steps involved in identifying risks

Chemical safety

When you are working with chemicals in the laboratory or at home, it is important that you are aware of the risks involved.

Laboratory chemicals can enter the body in three ways:

- **ingestion**—chemicals that have been swallowed enter the stomach, and may be absorbed into the bloodstream. They can interfere with body processes.
- **inhalation**—chemicals that are breathed in can cross the thin cell layer of the alveoli in the lungs and enter the bloodstream.
- **absorption**—some chemicals are able to pass directly through the skin.

When working with any type of chemical you should:

- identify the chemical codes and be aware of the dangers they are warning about
- become familiar with the relevant Safety Data Sheet (SDS), formerly known as the Material Safety Data Sheets (MSDS)
- use safety glasses and protective clothing such as chemical-resistant aprons or laboratory coats
- wipe up any spills
- wash your hands thoroughly after handling the chemical
- put the chemical away properly after use.

Chemical codes

Hazardous chemicals should always have a warning symbol or **chemical code** on the label or container. These chemical codes are known as HAZCHEM codes. Some common HAZCHEM codes and their meanings are shown on the diamond warning signs in Figure 1.2.2.

Each HAZCHEM sign has a distinctive colour, symbol and code number to make it easy for people to quickly identify the type of hazards involved.

It is important to understand the HAZCHEM codes found on the chemicals so that you can ensure that you use them safely and dispose of them in a way that will not hurt anybody else. As Figure 1.2.3 shows, trucks that carry chemicals display HAZCHEM signs.

FIGURE 1.2.3 Trucks transporting hazardous substances, such as flammable liquids, must have hazard symbols attached.

SciFile

Flea and tick protection

Cat and dog owners often protect their pets from ticks and fleas by placing a few drops of a pesticide solution on the back of the pet's neck. It is then absorbed and enters the bloodstream to kill existing parasites and to make the animal resistant to further infestation.

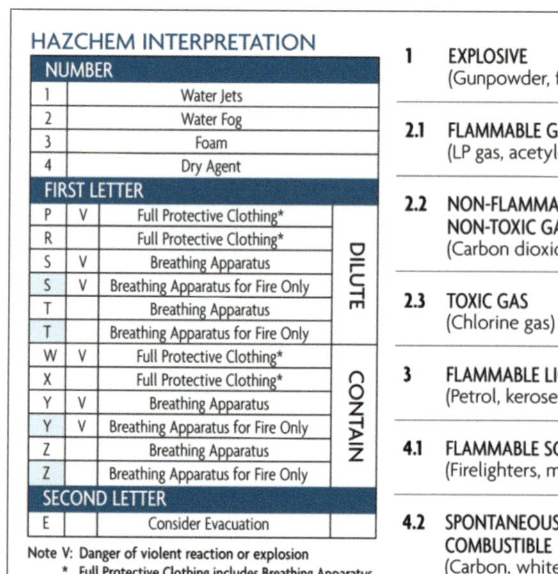

HAZCHEM INTERPRETATION

NUMBER			
1		Water Jets	
2		Water Fog	
3		Foam	
4		Dry Agent	
FIRST LETTER			
P	V	Full Protective Clothing*	DILUTE
R		Full Protective Clothing*	DILUTE
S	V	Breathing Apparatus	DILUTE
S	V	Breathing Apparatus for Fire Only	DILUTE
T		Breathing Apparatus	DILUTE
T		Breathing Apparatus for Fire Only	DILUTE
W	V	Full Protective Clothing*	CONTAIN
X		Full Protective Clothing*	CONTAIN
Y	V	Breathing Apparatus	CONTAIN
Y	V	Breathing Apparatus for Fire Only	CONTAIN
Z		Breathing Apparatus	CONTAIN
Z		Breathing Apparatus for Fire Only	CONTAIN
SECOND LETTER			
E		Consider Evacuation	

Note V: Danger of violent reaction or explosion
* Full Protective Clothing includes Breathing Apparatus

- 1 EXPLOSIVE (Gunpowder, flares)
- 2.1 FLAMMABLE GAS (LP gas, acetylene)
- 2.2 NON-FLAMMABLE NON-TOXIC GAS (Carbon dioxide)
- 2.3 TOXIC GAS (Chlorine gas)
- 3 FLAMMABLE LIQUID (Petrol, kerosene)
- 4.1 FLAMMABLE SOLID (Firelighters, matches)
- 4.2 SPONTANEOUSLY COMBUSTIBLE (Carbon, white phosphorus)
- 4.3 DANGEROUS WHEN WET (Calcium carbide)
- 5.1 OXIDISING AGENT (Calcium hypochlorite)
- 5.2 ORGANIC PEROXIDE
- 6 TOXIC (Arsenic)
- 7 RADIOACTIVE MATERIAL (Uranium)
- 8 CORROSIVE (Hydrochloric acid)
- 9 MISCELLANEOUS DANGEROUS GOODS (Dry ice, asbestos)
- MIXED CLASS LABEL (For road transport)

FIGURE 1.2.2 Common HAZCHEM codes

Safety Data Sheets

Each chemical substance has a **Safety Data Sheet**, or **SDS** (Figure 1.2.4). An SDS contains important safety and first aid information about each chemical you commonly use in the laboratory. For example, if the products of a reaction are toxic to the environment, then you must pour your waste into a special container and not down the sink. An SDS must state:

- the name of the chemical
- the chemical and generic names of its ingredients
- the chemical and physical properties of the substance
- health hazard information
- precautions for safe use and handling
- the manufacturer's or importer's name, Australian address and telephone number.

The SDS provides employers, workers and health and safety representatives with information needed to safely manage the risks associated with using hazardous substances.

For example, the following information is advice from the SDS regarding the first aid needed if you are exposed to hydrochloric acid:

> **HYDROCHLORIC ACID FIRST-AID MEASURES**
>
> **Inhalation:** remove from exposure, rest and keep warm.
>
> **Ingestion:** rinse mouth thoroughly with water immediately. If rapid recovery does not occur, obtain medical attention.
>
> **Skin:** wash affected areas with copious quantities of water immediately.
>
> **Eye contact:** immediately irrigate with copious quantity of water for at least 15 minutes. Eyelids to be held open. If rapid recovery does not occur, obtain medical attention.
>
> **First-aid facilities:** maintain eye wash fountain and drench facilities in work area.
>
> **Other information:** for advice, contact a Poisons Information Centre on 131126 or a doctor.

1. IDENTIFICATION OF THE MATERIAL AND SUPPLIER

Product Name:	**HYDROCHLORIC ACID - 20% OR GREATER**
Recommended use of the chemical and restrictions on use:	Precursor for generation of chlorine dioxide gas used in water treatment.
Supplier:	Ixom Operations Pty Ltd
ABN:	51 600 546 512
Street Address:	Level 8, 1 Nicholson Street Melbourne 3000 Australia
Telephone Number:	+61 3 9665 7111
Facsimile:	+61 3 9665 7937
Emergency Telephone:	**1 800 033 111 (ALL HOURS)**

Please ensure you refer to the limitations of this Safety Data Sheet as set out in the "Other Information" section at the end of this Data Sheet.

2. HAZARDS IDENTIFICATION

Classified as Dangerous Goods by the criteria of the Australian Dangerous Goods Code (ADG Code) for Transport by Road and Rail; DANGEROUS GOODS.

This material is hazardous according to Safe Work Australia; HAZARDOUS SUBSTANCE.

Classification of the substance or mixture:
Corrosive to Metals - Category 1
Skin Corrosion - Sub-category 1B
Eye Damage - Category 1
Specific target organ toxicity (single exposure) - Category 3

SIGNAL WORD: DANGER

Hazard Statement(s):
H290 May be corrosive to metals.
H314 Causes severe skin burns and eye damage.
H335 May cause respiratory irritation.

Precautionary Statement(s):

Prevention:
P234 Keep only in original container.
P260 Do not breathe mist / vapours / spray.
P264 Wash hands thoroughly after handling.
P271 Use only outdoors or in a well-ventilated area.
P280 Wear protective gloves / protective clothing / eye protection / face protection.

FIGURE 1.2.4 Part of a Safety Data Sheet (SDS) for concentrated hydrochloric acid. The SDS alerts the user to any potential hazards when using the chemical.

Personal protective equipment

When working in the laboratory, you should wear safety glasses, a lab coat or heat- and chemical-resistant protective apron (Figure 1.2.5). If you are handling organisms or toxic or corrosive chemicals, then you should also wear disposable gloves. These clothing items help to keep you safe and are called **personal protective equipment (PPE)**.

FIGURE 1.2.5 A lab coat or protective apron, gloves and safety glasses are essential items of personal protective equipment in the laboratory.

Safety outdoors

Sometimes investigations and experiments need to be done outdoors (Figure 1.2.6). There are additional risks associated with working outdoors and it is important to consider ways of eliminating or reducing these risks.

For example, fieldwork in a national park may include risks of:

- sunburn
- exposure due to hot or cold weather
- dehydration
- insect and animal bites
- allergic reactions
- getting lost.

FIGURE 1.2.6 Researchers need to consider the additional risks associated with fieldwork.

Some possible measures for reducing risks include:

- wearing sunscreen, a hat and sunglasses
- wearing suitable clothing to protect against heat or cold
- taking drinking water
- using insect repellent
- bringing a first-aid kit
- carrying a GPS unit
- carrying a mobile phone.

Ethics in the laboratory

Ethics are a set of principles by which your actions can be judged morally acceptable or unacceptable. Every society or group of people, including scientists, has its own principles or rules of conduct. For example, experimenting on humans with new drugs is considered unethical. Before any drug to be used by humans is released, thorough testing needs to take place. In the past, there have been incidences where drugs have been used without sufficient testing. Occasionally, there have been serious consequences. For example, in the 1950s thalidomide was taken during pregnancy but caused severe birth deformities in the baby born soon after.

Using animals

Working with living organisms can be very interesting and informative, but scientists must follow very strict guidelines when doing so. There are many important factors that you need to consider when collecting data using living things (Figure 1.2.7).

FIGURE 1.2.7 Keeping animals in the laboratory is a great way to observe developments such as the lifecycle of a butterfly.

Before an experiment can be carried out on animals, it must be assessed by an ethics committee to ensure that the animals will be properly cared for. Your school will have a system to ensure that it follows local and state legislation regarding animal welfare and the prevention of cruelty to animals.

When using animals for laboratory work, you should:

- try to reduce the number of animals involved or even replace them by a simulation.
- treat animals with respect and care (Figure 1.2.8). The welfare of the animal must be the most important factor to consider when determining the use of animals in experiments.
- refine the experiment to eliminate any potential for harm to the animal. If at any time the animal being used in your experiment becomes distressed, then the experiment must stop.

Prac 1
p. 16

FIGURE 1.2.8 Using animals in science is a great way to learn but they must be treated with respect and care.

Collecting data using people

When considering an experiment that includes the use of people, the following guidelines must be followed:

- Participant details must be kept confidential. You cannot reveal people's experiment results to anyone without the participant's permission.
- Participants should be told the purpose of the experiment and what they will be expected to do, so that they can give consent. That is, they must give permission with full understanding.
- Participants must agree to participate and should be able to withdraw from the experiment at any time.

Seeking an alternative

Before you start an experiment, an investigation or field work, it is important to consider the risks involved. Sometimes it may not be possible to reduce or eliminate the risks. Sometimes, the risks remain too high even when using PPE or making a change in the procedure to incorporate safety measures. In these cases, your best option may be to seek an alternative chemical, equipment or procedure to allow a similar experiment to be conducted with less risk.

Working with Science

ANIMAL WELFARE INSPECTOR

FIGURE 1.2.9 The RSPCA employs animal welfare inspectors.

Animal welfare inspectors are responsible for investigating reports of animal cruelty, neglect or medical emergency. They investigate cases and make decisions for the welfare of the animals involved. They conduct research about people who have been reported as mistreating animals, searching for animals on council registers, contacting vets to obtain veterinary records and looking at relevant legislation. From this information, animal welfare inspectors must assess the best course of action and determine if there are likely to be any safety or ethical concerns when working on the case.

Animal welfare inspectors carry out rescues, assist in emergency situations and work with police to prosecute offenders who have mistreated animals. Many situations involve providing information and support to people who need help taking care of their animals. Communication skills are an important part of the job. Animal welfare inspectors often give talks at schools or community events and advise government personnel and police about local animal welfare issues.

An animal welfare inspector's job is both challenging and rewarding. They are confronted with many difficult cases but also have the satisfaction of improving the lives of countless animals. Most animal welfare inspectors have completed a Bachelor of Science specialising in Animal Science or Agricultural Science but this qualification isn't necessary to start out in this career. Certificates and Diplomas in animal handling, care and welfare will help you gain skills that are important in this field.

FIGURE 1.2.10 Pam Ahern is the founder of Edgar's Mission, an animal sanctuary in rural Victoria. She can be seen here with rescue animals, Ruby the sheep dog, and pigs, Thumbelina and Leon Trotsky.

These qualifications can lead to employment with organisations such as animal rescue services, veterinary hospitals, animal shelters, wildlife care, quarantine services or the RSPCA (Figure 1.2.9). They can even lead to setting up your own animal sanctuary like the one set up by Pat Ahern, shown in Figure 1.2.10.

Review

1. Why do you think inquiry skills are important in an animal welfare inspector's job?
2. There are many ethical concerns around animal welfare cases. List ethical issues that an animal welfare inspector would need to consider when carrying out investigations.

MODULE 1.2

Review questions

Remembering

1 Define the terms:
   a inhalation
   b absorption.

2 What term best describes each of the following?
   a the process by which chemicals enter the body through the stomach
   b this is performed to identify, assess and control hazards.

3 Name three places where you would expect to find HAZCHEM warning signs.

Understanding

4 Outline three ways a laboratory chemical could enter the body and how you might prevent each of these occurring.

5 Louis is working in the laboratory and has spilled a chemical substance with the label in Figure 1.2.11 on himself. What would be the appropriate thing to do?

FIGURE 1.2.11

6 a What does SDS stand for?
   b Explain the reasons for having an SDS for each chemical used in the laboratory.

7 Outline the guidelines that must be followed when considering an experiment that collects data from people.

Applying

8 Refer to the Safety Data Sheet extract on page 11. If you were using HCl in an experiment, identify the:
   a personal protective equipment that you should wear
   b safety precautions that you should take to ensure your safety and the safety of those around you.

Analysing

9 Compare two of the warning signs used for chemicals in your school laboratory.
   a How are the hazards they indicate different?
   b How are they similar?

Evaluating

10 Why do you think it is important to have guidelines when using people in experiments?

11 The code of practice for the use of animals in school experiments can be summarised using 'three Rs'. What do you think these 'three Rs' could be?

12 Carefully examine your school science laboratory. Assess the laboratory by identifying:
   a the safety measures in place to eliminate risks and hazards
   b possible measures that would improve the safety of students and teachers in the laboratory.

Use a T-chart to note the safety measures and the improvements to safety.

Safety measure	Safety improvement measure

MODULE
1.2 Practical investigation

• STUDENT DESIGN •

1 • Replacing animals in dissections

Planning & Conducting | Communicating

Every year across the globe, millions of dead animal specimens are dissected, being cut open and examined in school science classes. For example, frogs were commonly used for dissection (Figure 1.2.12). However, there are now many virtual dissections available as an alternative. Can these online dissections provide an equivalent opportunity to experience animal anatomy?

Purpose

To find a suitable online dissection or computer model that provides basic understanding of anatomy and physiology.

Timing
45–60 minutes

Materials

- a computer, tablet or laptop

Procedure

1 Use web search techniques to find an online dissection tool that provides a basic understanding of animal anatomy.

2 Share the tool you have found with your peers.

Hint

Try using the search term *digital dissection*.

Review

1 Draw a PMI chart to describe the plus, minus and interesting aspects of doing a digital dissection.

Digital dissections		
Plus	Minus	Interesting

2 Discuss whether this is a suitable alternative to using dead animals for dissection in the classroom.

FIGURE 1.2.12 In the past, frogs have been used for scientific experiments and dissections.

MODULE
1.3 Conducting investigations

The experiment is the testing phase of the scientific method. Every experiment should be designed carefully so that when other scientists repeat the experiment they will get similar results.

How effective is your sunscreen?

YES

Have you ever wondered if sunscreen actually works and what it does?

Collect this ...

- piece of dark cardboard or construction paper (black, blue or purple)
- sunscreen (one that goes clear—not a zinc oxide sunscreen)
- pen or texta that can write on dark cardboard

Do this ...

1. Draw a line in the middle of the card and write 'Sunscreen' at the top of the left side and 'No sunscreen' at the top of the right side.
2. Cover the palms and fingers of your hands with a thin layer of sunscreen and press your sticky hands down on the 'Sunscreen' side of the card.
3. Allow the sunscreen to dry on the cardboard.
4. Place the cardboard in full sun for a day.

Record this ...

1. Describe what happened.
2. Explain why you think this happened.

Designing experiments

Before the details of an experiment can be decided, a number of important steps need to be completed.

- Background information on the topic needs to be researched.
- Whether this experiment has been conducted by other researchers needs to be determined.
- The question and hypothesis needs to be determined.
- The independent, dependent and controlled variables need to be identified.
- Potential safety hazards need to be recognised.

Next, write a clear and detailed description of how the experiment will be conducted. To do this, you will need to answer the following questions:

- How will the independent variable be changed?
- How will the dependent variable be measured?
- How will the other variables be controlled?

Fair tests

A fair test is an experiment in which only one variable is tested at a time. This ensures that the results of the experiment can be used to either support or refute (not support) the hypothesis.

For example, if a scientist were testing whether or not fertiliser affects plant growth, then two groups of seedlings would be required—an **experimental group** and a **control group** (Figure 1.3.1). Seedlings in the experimental group would be given the fertiliser, while seedlings in the control group would be given no fertiliser. All other variables such as the soil the seedlings were potted in, the volume of water given and the amount of sunlight available would be the same for both groups (these are controlled variables). If at the end of the experiment the experimental plants were larger than those that were not given fertiliser, then the difference between the two groups of plants was most likely due to the presence of fertiliser.

FIGURE 1.3.1 Conducting a fair test experiment with an experimental group and a control group

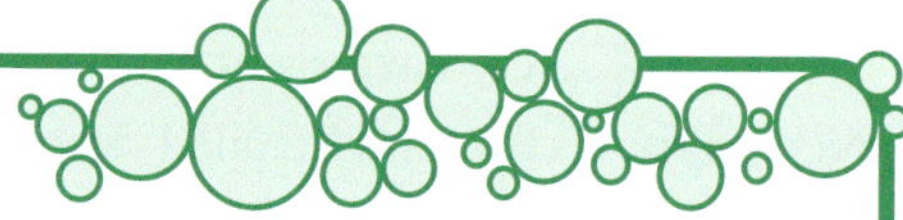

Scifile

The placebo effect

Have you ever taken a tablet for a headache and felt better straight away? This is a phenomenon called the placebo effect. Your condition can be improved just because you have the expectation that taking pain medication will be helpful. When new drugs are tested the control group is given a placebo, which may just be a sugar tablet to determine if the new drug is indeed making a difference. If patients on the new drug improve significantly more than those in the control group who were given the placebo then it can be concluded that the medication is effective.

This conclusion can only be made because the only difference between the two groups of plants was the independent variable. In this case, this was the fertiliser.

Whenever possible, results of an experiment should involve both qualitative observation and measurement. A scientist would record the appearance of the plants involved the fertiliser experiment, but would also measure such things as the height of the plant, length of stems, number of new leaves, or calculate the mass of the plants. This is important because measurements are easier to compare than descriptions.

Prac 1
p. 23

Reliability

An experiment is considered reliable if it gives the same results every time it is performed. Reliability of results can be increased through **repeat trials** and **replication** of the experiment (Figure 1.3.2).

Conducting repeat trials means doing the same experiment many times to ensure that the results can be reproduced. If there are small variations between the results for the repeat trials, this may show that there are some measurement errors. If the results are significantly different, then it suggests that there may be some significant variables that have not been properly controlled.

FIGURE 1.3.2 Reliability of results is increased through replicating experiments.

STEM 4 fun

Minute timer

PROBLEM

Can you time 1 minute without a timer?

SUPPLIES

- any or all of: paper, cardboard, string, paperclips, drawing pins, glue, small coins, paper cup, stopwatch (for trialling only)

PLAN AND DESIGN Your challenge is to build a timing device that will allow you to measure as close as possible to 1 minute without a timer of any sort. Your group will be allowed time to build your device and trial it before being tested.

Design the solution. What information do you need to solve the problem? Draw a diagram. Make a list of materials you will need and steps you will take.

CREATE Follow your plan. Build and test your timing device.

IMPROVE What works? What doesn't? How do you know it solves the problem? What could work better? Modify your design to make it better. Test it out.

REFLECTION

1 What area of STEM did you work in today?
2 How did you use mathematics in this task?
3 What did you do today that worked well? What didn't work well?

Replication is the creation of duplicate experimental set-ups, so that the experiment can be run more than once *at the same time*. The duplicate experiments are called **replicates**. Once again, if there are significant discrepancies in results for replicates, then it can give information about the extent to which the variables were properly controlled.

In the fertiliser experiment, having only one experimental plant and one control plant would not be sufficient to demonstrate the reliability of the results as there may be random differences between individual plants. For example, one of the plants may naturally grow more slowly or more quickly. This is natural variability between individual plants and cannot be controlled. If the scientists used only one individual in their experiment, then this could lead to misleading results. If 10 plants were used in the experimental group and 10 plants used in the control group, then the average change in size over the period of the experiment could be calculated. Any individual differences in plants would then be unlikely to affect the result.

AB 1.4

Writing a procedure

To conduct a scientific experiment correctly, the procedure used must be reliable and in sufficient detail to allow other scientists to repeat the experiment. It is also helpful to include a diagram so that other scientists can see how the experiment was set up. An example of an experimental procedure is shown in Figure 1.3.3 on page 20.

The procedure of an experiment is usually described as a list of numbered steps that describes the process that was followed as exactly as possible.

science 4 fun

Making sandwiches

Can you write a procedure?

Collect this ...

- two slices of bread
- butter or non-dairy spread
- knife
- slice of cheese
- plate

Do this ...

1 Write a step-by-step procedure for making a cheese sandwich.
2 Get a brother or sister, a friend or someone in your house to follow your instructions EXACTLY!

Record this ...

1 Describe what happened.
2 Explain why you think this happened.

Purpose

To investigate the effect of pH on seedling growth.

Hypothesis

If the soil pH is increased, then seedling growth will increase.

Procedure

1 Germinate twenty pea seeds on damp cotton wool and choose twelve with a height of about 12 mm.
2 Plant a seedling in each of twelve pots of the same size. For each pot, use 80 g of quality potting mix, and water with 10 mL of tap water. **Safety note**: ensure that gloves and a mask are worn when handling potting mix, as it may contain harmful microbes.
3 Label each pot with the pH treatment the soil will receive: four pots at pH 5, four pots at pH 7 and four pots at pH 9.
4 Weigh each pot to the nearest 0.1 g. Draw up a data table and record the results for each pot in the column for day 0.
5 Reweigh the seedlings in their pots 2 days later. Record the results for each pot in the column for day 2.
6 Immediately after weighing, give each plant 10 mL of water at the appropriate pH according to the label on the pot.
7 Repeat steps 5 and 6 every 2 days for the next 10 days.
8 Keep plants in the same position where light is available to maintain lighting conditions.
9 Repeat steps 1–8 twice to reduce the chance of variability between trials.

FIGURE 1.3.3 An experimental procedure

Prac 2
p. 24

Collecting data

The values collected for the dependent variable during an investigation are called **data**. When planning an investigation, it is important to consider the type of data that will be collected and how best to record the data.

When taking a measurement, you need to:

- check the measuring instrument. Does it read zero correctly? Does it produce accurate readings?
- look straight at the measuring scale. Your eyes should be at right angles to the scale as in Figure 1.3.4.
- check your measurement. Did you read the scale correctly? Did you record the measurement correctly into your results?
- check that the numbers of your data are written in a consistent format. For example, round up or down to the nearest whole number or to one, two or three decimal points.

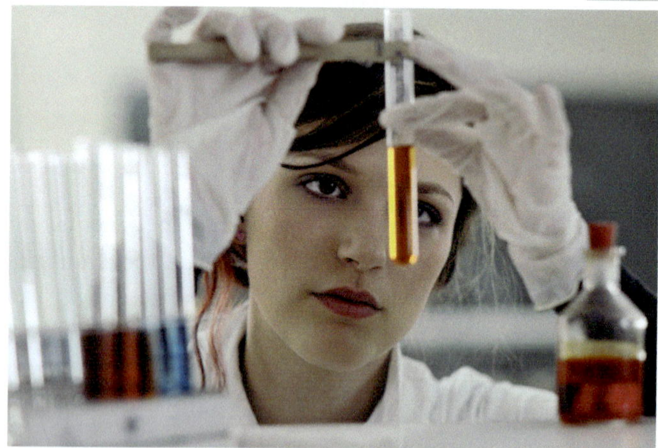

FIGURE 1.3.4 It is important to take accurate measurements to ensure that you have reliable data.

When collecting data, it is also important to avoid personal bias that might affect the results. A good scientist works hard to be **objective** (free of personal bias) rather than **subjective** (influenced by personal views). The results of an experiment must be clearly stated and must be separate from any discussion of the conclusions that are drawn from the results.

Types of data

It is important to recognise that there are different types of data that can be collected in a scientific investigation. Data may be quantitative or qualitative (Figure 1.3.5). Quantitative data may also be described as continuous or discontinuous. Figure 1.3.6 describes the categories of data.

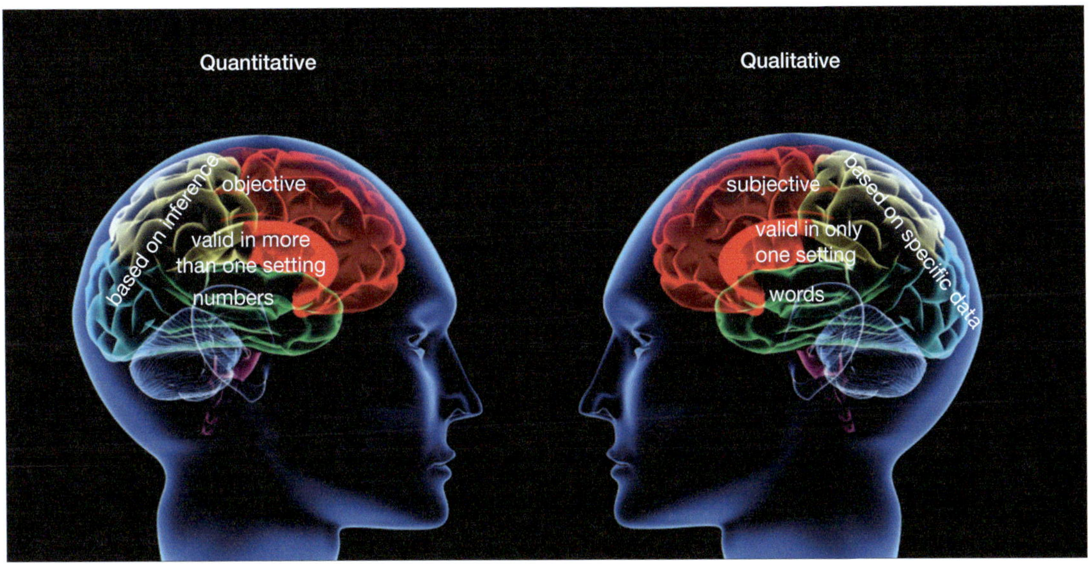

FIGURE 1.3.5 Differences between qualitative and quantitative data

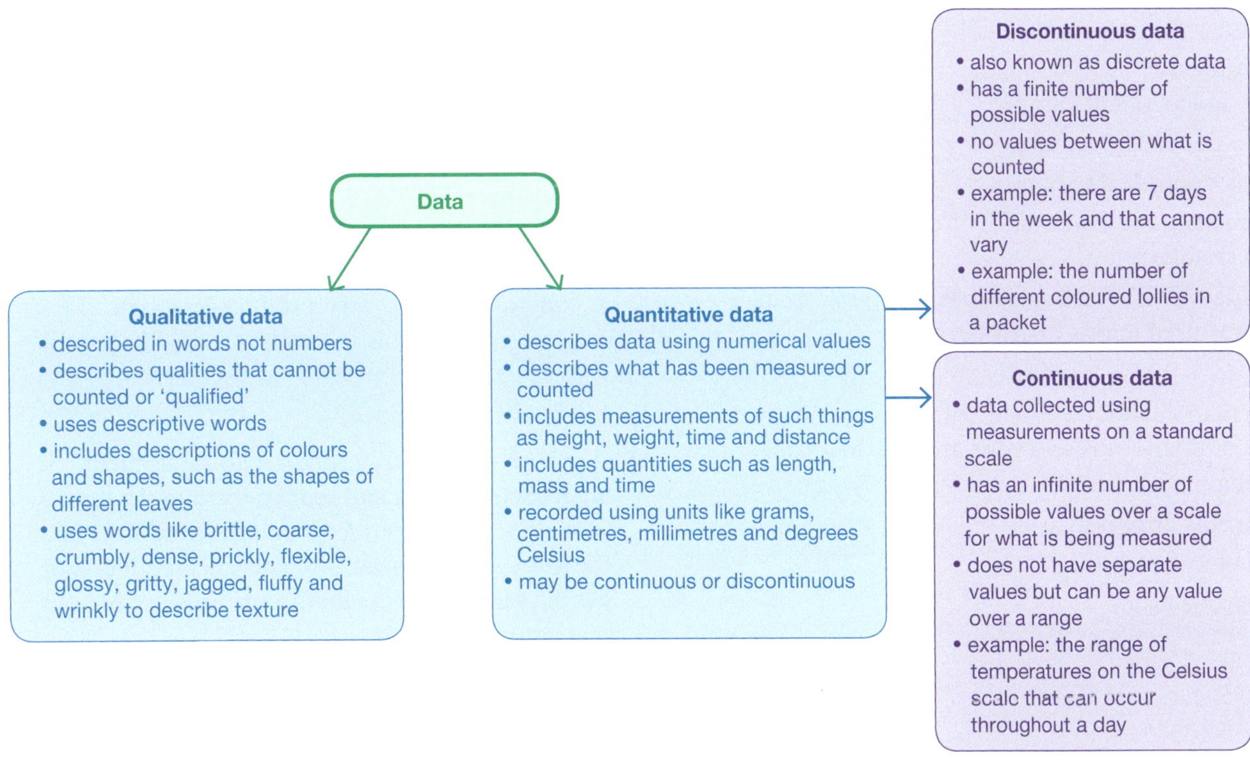

FIGURE 1.3.6 Categories and examples of each type of data

MODULE

1.3 Review questions

Remembering

1 Define the terms:
  a fair test
  b replicate.

2 What term best describes each of the following?
  a collecting multiple data sets by performing an experiment again after the initial test
  b an experiment in which only the independent variable is changed and all other variables are controlled or measured.

3 What four things do you need to check when taking measurements?

4 List the steps that should be undertaken before the experimental method can be decided.

Understanding

5 Why is replication of an experiment necessary?

6 Explain why an experimental procedure needs to be written clearly and in sufficient detail.

Applying

7 Consider the seedling growth experiment in Figure 1.3.1 on page 18.
  a State the:
    i independent variable
    ii dependent variable
    iii controlled variables.
  b Explain the importance of controlling these variables.
  c Identify whether repeat trials or replication was used in this experiment.

8 Identify whether the following pieces of information about a cup of coffee are qualitative or quantitative. Present your answers in a table.
  a temperature 82°C
  b frothy appearance
  c volume 180 mL
  d strong taste.

Analysing

9 What is the difference between qualitative data and quantitative data?

10 Identify why the following experiment would not provide reliable data.

Purpose

To determine if adding glycerol to detergent will make bigger bubbles.

Hypothesis

If glycerol is added to detergent, then the bubbles produced will be larger.

Procedure

1 Put some detergent in a beaker and then add water.
2 Use a bubble blowing device to blow bubbles.
3 Put some detergent in another beaker and add 10 mL of glycerol then add water.
4 Use a bubble blowing to device blow bubbles.
5 Record which mixture produced the largest bubbles.

Evaluating

11 Propose a reason why a hypothesis can only support or refute the data but not prove that it is correct.

12 When conducting experiments on living organisms, replicates are often used. Propose reasons for using replicate experiments when working with living organisms.

13 A company does its own product research.
  a Determine if this research is more likely to be objective or subjective.
  b Justify your choice.

14 A group of students carries out an experiment, analyses the results and reports them to the rest of the class. Other students in different classes in the same school repeat the experiment, but do not get the same results as this group. Propose reasons that could explain this.

MODULE 1.3

Practical investigations

1 • Sugar, carbon dioxide and yeast

Questioning & Predicting | Evaluating

Yeasts are microscopic organisms that feed on sugars and starches. They turn this food into usable energy and release carbon dioxide as a result. This process is known as fermentation. The carbon dioxide gas made during fermentation is what makes bread so soft.

Purpose

To determine if yeast will produce more carbon dioxide bubbles when sugar is added to the mixture.

Hypothesis

Which yeast mixture will produce more carbon dioxide—the one with or the one without sugar? Before you go any further with this investigation, write a hypothesis in your workbook.

Timing

40 minutes

Materials

- 2 teaspoons of dry yeast
- 1 teaspoon of sugar
- warm water
- 2 × 150 mL conical flasks
- 2 rubber stoppers with hole in the middle and a side arm attached
- 2 lengths of rubber tubing to fit onto the side arm
- 2 × 250 mL beakers
- 2 stirring rods

SAFETY

After handling yeast, wash hands before leaving the laboratory.

Procedure

1. Label conical flasks 'yeast and sugar' and yeast only/no sugar.
2. Add 50 mL of warm water to each conical flask.
3. Add 1 teaspoon of yeast to each conical flask.
4. Add 1 teaspoon of sugar to the conical flask labelled 'yeast and sugar'.
5. Gently swirl the yeast and water mixture in both conical flasks with a stirring rod to combine the contents.
6. Attach the arm and rubber tubing to the rubber stopper and set up the equipment as shown in Figure 1.3.7.
7. Carefully place the rubber stopper into the top of the conical flask. Ensure it is firmly sealed.
8. Add 200 mL of water to each beaker.
9. Put the rubber tubing into the beaker and ensure it is under the water.
10. Wait 10–20 minutes for the yeast to activate.
11. Count the number of bubbles generated per minute over a period of 5 minutes from when the first bubbles appear.

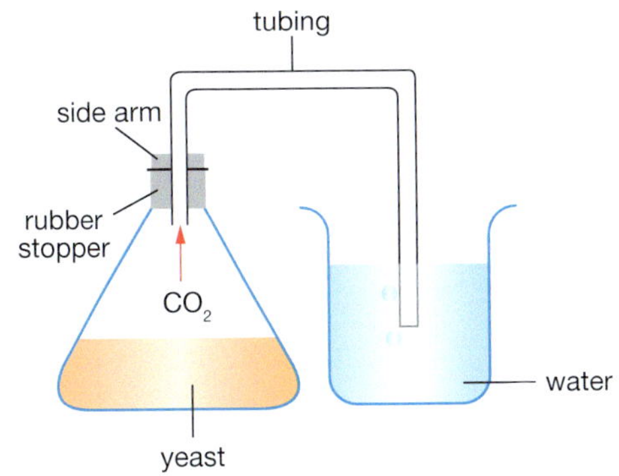

FIGURE 1.3.7

Results

1. Record the number of bubbles produced each minute for a period of 5 minutes.
2. Design a suitable table to record this data.
3. Calculate the rate of gas produced in each flask.

Review

1. Compare the rate at which the two flasks produced gas.
2. Assess whether your hypothesis was supported or not.
3. Propose a reason why one of the flasks produced more gas than the other.

Extension

Design your own method by making one of the following factors the independent variable:

- water temperature
- mass of sugar added
- type of sugar added
- pH of solution added to the yeast.

MODULE
1.3 Practical investigations

• STUDENT DESIGN •

2 • Bigger bubbles

Purpose

To design a super bubble mix.

Timing 40 minutes

Materials

- glycerol
- detergent
- water

- 250 mL beaker
- 2 × 10 mL measuring cylinder
- pipette
- device for blowing bubbles—can be made using pipe cleaners or wire

SAFETY

Avoid touching eyes with bubble mix, wash thoroughly with water if contact occurs.

Procedure

1 Design a procedure to determine which combination of glycerol, detergent and water will produce the largest bubbles.

2 Write your procedure in your workbook.

3 Before you start any practical work, assess your procedure. List any risks it might involve and how you can minimise those risks. Show your procedure and risk assessment to your teacher. If they approve, then collect all the required materials and start work.

See the Activity Book Toolkit to assist with developing a risk assessment.

Use the STEM and SDI template in your eBook to help you plan and carry out your investigation.

Review

1 a Was it difficult to collect data for this experiment?

  b What type of data did you collect?

2 a Which mixture produced the largest bubbles?

  b Why do you think this happened?

MODULE

1.4 Presenting and evaluating data

After an experiment, you will usually write a report about what you did and what you found out. In your report you will describe and analyse the collected data, evaluate the procedure, discuss any improvements that should be made and give a conclusion about the relationship between the independent and independent variables.

Presenting data

After you have completed your experiment, the data collected will need to be organised and presented appropriately. There are a number of ways to **process** and present data, including tables, graphs, flow charts, pictures or diagrams. The best way to visualise data depends on the type of data that has been collected.

Presenting data in tables

Tables provide an accurate record of the numerical values and allow you to organise your data. Tables usually present data in rows and columns, and can vary in complexity according to the nature of the data.

The simplest form of a table is a two-column chart. The first column should contain the independent variable (the one that is changed in a systematic way) and the second column should contain the dependent variable (the one that may change in response to the changes in the independent variable).

As you can see in Figure 1.4.1, tables should have the following features:

- descriptive title including the independent and dependent variable
- column headings (including the units)
- the independent variable placed in the left column
- the dependent variable placed in the right column/s.

Tables can be much simpler than the one shown in Table 1.4.1 on page 26, depending on the data being collected.

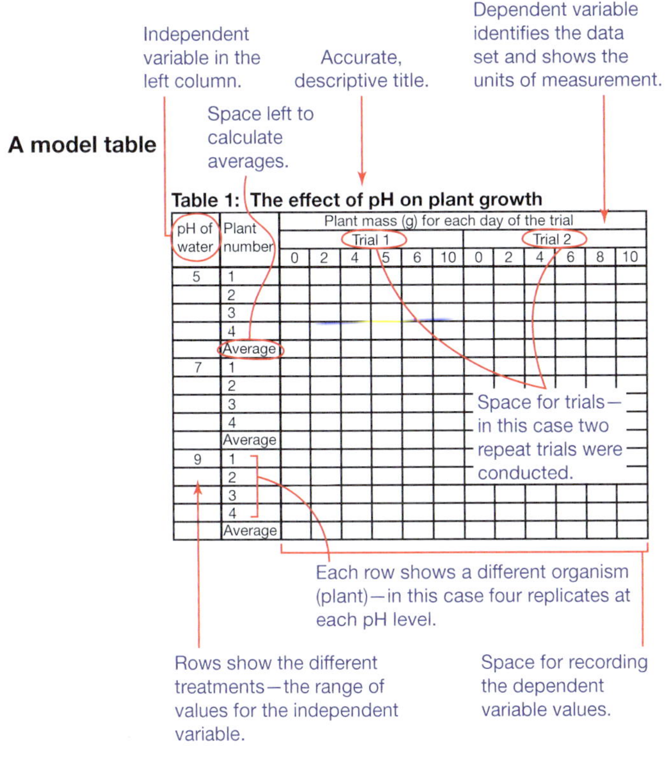

FIGURE 1.4.1 Features of a good table

The same rules always apply though. The independent variable (water temperature) is always shown in the first column and the dependent variable (time taken for sugar to dissolve) is shown in the second column. Units of measurement are included in the column headings only (Water temperature and Time taken for sugar to dissolve) and not with the data entered into the table.

TABLE 1.4.1 The effect of water temperature on the time taken for sugar to dissolve

Water temperature (°C)	Time taken for sugar to dissolve (seconds)			
	Trial 1	Trial 2	Trial 3	Average
20				
40				
60				

Presenting data as a graph

Graphs are a very useful way of presenting data visually to display any patterns or trends that may not be visible from a table. It is usually appropriate to include both a table and a graph in your report.

There are several types of graphs, including line graphs, bar or column graphs and pie charts. The best one to use will depend on the nature of the data. The general rules for drawing graphs are shown in Figure 1.4.4.

A **column graph** is the most appropriate type of graph for discrete (discontinuous) data. Consider the following example: a scientist records whether or not students eat breakfast before coming to school. This variable has two discrete categories and should therefore be graphed using a column graph. When drawing a column graph it should have bars of equal width, with a space between each bar as shown in Figure 1.4.2.

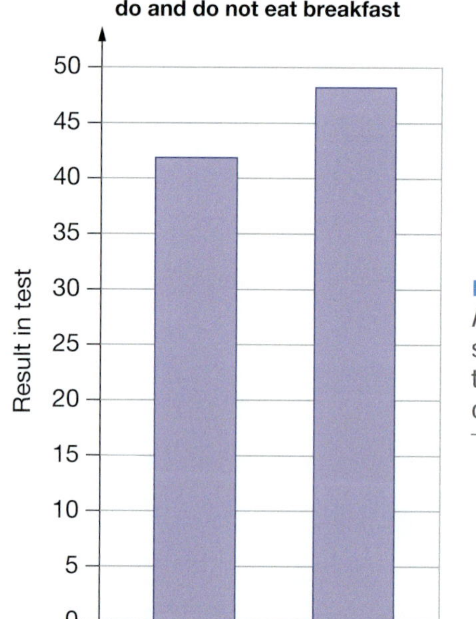

FIGURE 1.4.2 A column graph should be used to represent discontinuous data.

A **line graph** is a good way of representing continuous quantitative data. In a line graph, the values are plotted as a series of points on the graph. A line can then be drawn from each point to the next, as shown in Figure 1.4.3. In an experiment in which a scientist records the number of eagles observed each month for a year, then time is a continuous variable so it is appropriate to use a line graph.

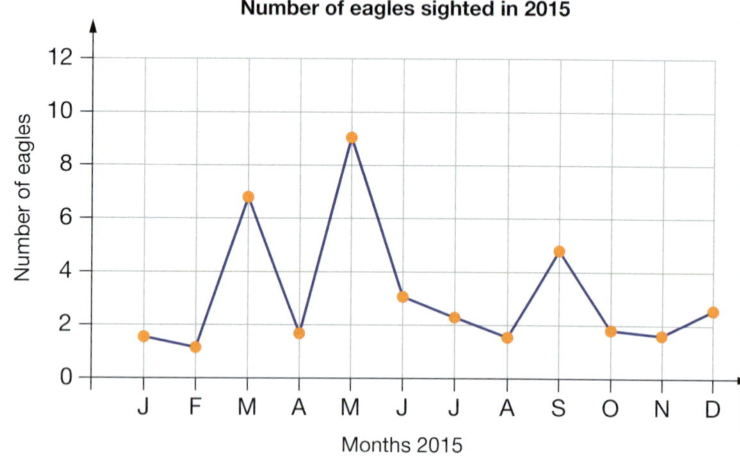

FIGURE 1.4.3 A line graph should be used to represent continuous data.

Drawing a graph correctly requires careful attention. The following are general rules to follow when drawing a graph. These points are summarised in Figure 1.4.4.

General rules for correctly drawing a graph

Scale:

- Evenly distribute scale on both axes—a scale is a number line so you must always write a sequence of numbers with equal intervals between them on each axis.
- Only use a scale break if it is necessary because it is not possible to draw a long enough *y*-axis on the page.

Axis:

- Represent the independent variable on the horizontal *x*-axis.
- Represent the dependent variable on the vertical *y*-axis.
- Match the length of the axes to the data.
- Clearly label each axis with both the variable it represents and the unit in which it is measured.

Legend:

- Include a key or legend to show what the colours and symbols on the graph represent.

Title:

- Use a descriptive title that includes both the independent and dependent variable.

Size:

- Use two-thirds to three-quarters of the space on graph paper.

Data:

- Only draw a line to zero if zero is actually part of the data collected.
- Use small symbols such as circles or squares for data points.

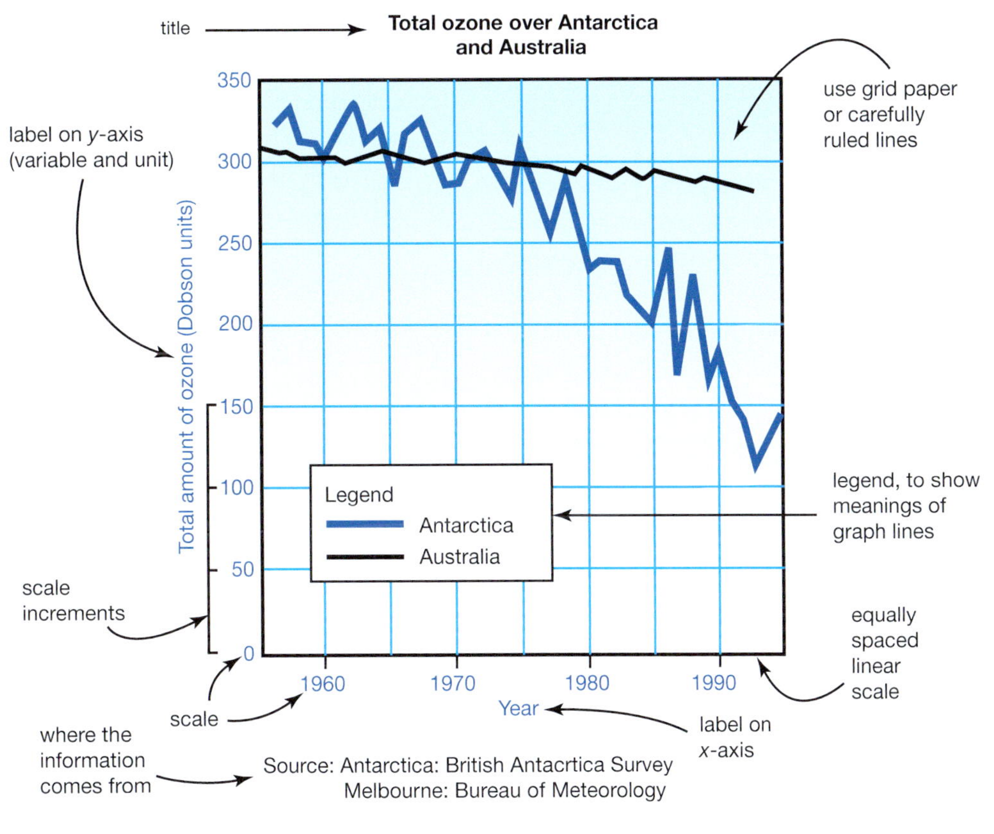

FIGURE 1.4.4 General rules to follow when drawing a graph

Interpreting data

After you have collected your data and represented it with a diagram, you will need to interpret your results. Clearly state whether a pattern, trend or relationship was observed between the independent and dependent variables. This is where your research on the topic becomes important—you should have an idea of what sort of relationship to expect. For example, you might expect that the experimental group of plants given fertiliser (independent variable) will grow larger than control group of plants given none (dependent variable). So your data would show a relationship between plant growth and fertilisation, and a relationship between plant growth without fertilisation. Relationships should be stated clearly and concisely.

- Refer to the measurements as **evidence**, and draw conclusions from the data. Be sure that conclusions are supported by the data from the experiment and not just based on what you expected to happen.
- Refer to the evidence you have collected using phrases such as 'the data shows that …', 'this is supported by …' and 'it can be inferred from the data that …'.

AB 1.5

Descriptive statistics

Data can be organised using **descriptive statistics**. Descriptive statistics are used to summarise, organise and describe data obtained from research. This allows data to be more easily interpreted. Descriptive statistics can be used to analyse both quantitative and qualitative data. Descriptive statistics include percentages, graphs and measures of central tendency. It is good practice to use a measure of central tendency to provide a clearer understanding of the data.

Measures of central tendency

Measures of **central tendency** (sometimes also called measures of central location) are single values that allow you to describe the central position in a set of data (Table 1.4.2). The mean, median and mode are all measures of central tendency.

TABLE 1.4.2 Measure of central tendency: mean, median and mode

Mean	Median	Mode
The **mean** (or average) is the sum of the values divided by the number of values.	The **median** is the 'middle' value in an ordered list of values.	The **mode** is the value that occurs most often in a list of values. This measure is particularly useful for describing qualitative or discrete data.
For example, the mean of 3, 7, 9, 10 and 11 is (3 + 7 + 9 + 10 + 11) ÷ 5, which is 8.	For example, the median of the seven values 5, 5, 8, 8, 9, 10, 20 is the fourth value, which is 8.	For example, the mode of the values 0.01, 0.01, 0.02, 0.02, 0.02, 0.03, 0.04 is 0.02.

Percentage change

Percentage change applies to increases and decreases relative to the control or the starting point of the measurement.

For example, data was collected in an experiment that investigated the osmotic strength of different solutions. Four sets of dialysis tubing (a semipermeable membrane), each containing a different solution, were suspended in a beaker of saline (sodium chloride) solution. The procedure for preparing the dialysis tubing is seen in Figure 1.4.5. The mass of the dialysis tubing was measured at the start and after 24 hours. The results are shown in Table 1.4.3.

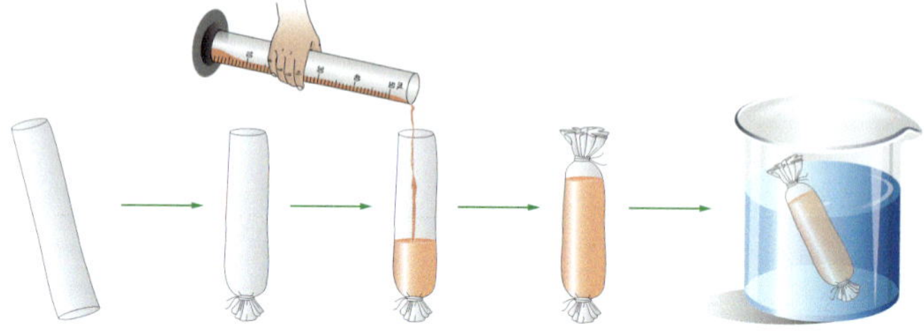

FIGURE 1.4.5 Procedure for preparing the dialysis tubing

TABLE 1.4.3 Table showing the % change in mass of dialysis tubing in 24 hours

Sample number	Original mass (g)	Mass after 24 h (g)	% mass change
1	20.55	20.89	1.65
2	20.01	21.94	9.65
3	21.25	22.09	3.95
4	20.55	20.32	−1.12

The percentage change in mass is calculated with the equation:

$$\text{\% mass change} = \frac{\text{final mass} - \text{original mass}}{\text{original mass}} \times 100$$

Calculating percentage change accounts for natural variation and/or errors in the replicates within your experiment, or for the same experiment repeated by others. In Table 1.4.3, the starting mass is not identical in each sample, perhaps due to errors in measuring the volume put into the tubing as seen in step 3 of Figure 1.4.5. Although the final mass for sample 3 is the greatest, the percentage change is less than sample 2 because the original mass was higher.

Calculating percentage mass change shows that sample 2 has the greatest osmotic effect.

Percentage difference

The **percentage difference** (also often expressed as a fraction) is a measure of the precision of two measurements. It is calculated by working out the difference between the two measurements and dividing by the average of the two measurements:

$$\text{percentage difference} = \frac{\text{measurement 1} - \text{measurement 2}}{\text{average of measurements}}$$

For example, if your two measurements were 25 cm and 24 cm, you would calculate percentage difference as follows:

$$\begin{aligned}\text{percentage difference} &= \frac{(25 - 24)}{(25 + 24)/2} \\ &= \frac{1}{24.5} \\ &= 0.041 \times 100 \\ &= 4.1\%\end{aligned}$$

Range

The **range** is simply the difference between the highest and lowest values in a data set. Figure 1.4.6 shows the range of hearing frequencies for different animals. From the diagram you can clearly see the range for human hearing and how this compares to different animals. This type of visual representation of the data clearly shows the differences in the sample set.

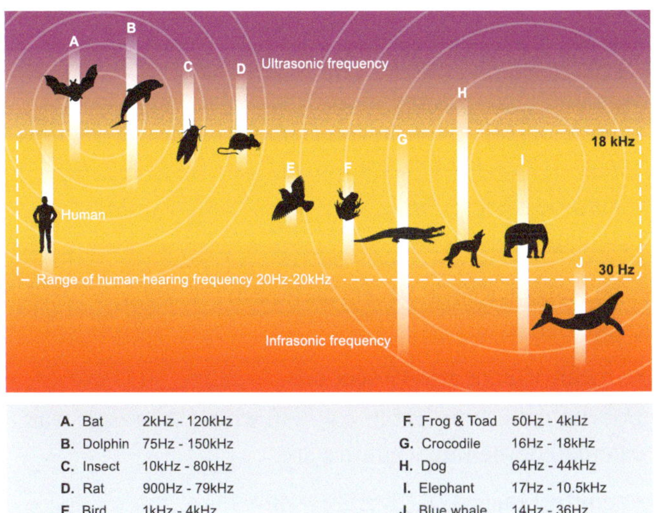

FIGURE 1.4.6 Animal hearing frequency range. Hearing range describes the range of frequencies that can be heard by humans and other animals.

Table 1.4.4 shows the measurements taken for five different plants after treatment with a plant hormone. To determine the range for the values in Table 1.4.4, you would subtract the smallest value from the largest value. Notice how an abnormally large or small value in the data set makes the variability appear high. If one value appears way out of range, such as plant 1 in the hormone-treated group, it is considered an **outlier** and can be deleted from the calculations. The range for the hormone-treated plants would then be 378 – 320 = 58. This illustrates the importance of having a sample size that is large enough to limit the effect of anomalies in the data set.

TABLE 1.4.4 The range of measurements taken for five different plants after treatment with a plant hormone

Plant	1	2	3	4	5	Mean	Range
Hormone-treated plants (mm)	158	378	320	377	363	319.2	378 – 158 = 220
Untreated-control plants (mm)	140	135	170	171	193	161.8	193 – 135 = 58

Uncertainty in measurement

When averaging repeat measurements, the **uncertainty** should be reported alongside your average. Uncertainty results from errors and represents a realistic range within which the true value is likely to be. A simple way to calculate the uncertainty is the range divided by 2.

For example, if an experiment were conducted to measure the length of time it takes to convert a substrate to a product in an enzymatic reaction, and three replications of the experiment produced the times 2.50, 3.47 and 2.81 seconds, the average time taken would be 2.93 seconds. The uncertainty would be calculated as follows.

The result showing the mean and uncertainty is expressed as:

- mean = 2.93 ± 0.49 seconds.

For the data set in Table 1.4.5, in which the range was calculated, the uncertainties are:

- control plants 161.8 ± 29.0 mm
- hormone-treated plants 359.5 ± 29.0 mm (with the outlier removed).

Presenting processed data in tables

Table 1.4.5 shows the relationship between temperature and mean transpiration rate. It displays transpiration data in a processed format, because several values have been averaged to calculate the mean.

TABLE 1.4.5 Relationship between temperature and mean transpiration rate

Temperature (°C)	Mean transpiration rate (mL/g/h)
15	0.038
25	0.043
35	0.059
45	0.074

Table 1.4.6 is an improved version of the data in Table 1.4.5, because it includes the uncertainty in the processed data.

TABLE 1.4.6 Relationship between temperature and mean transpiration rate (with uncertainty)

Temperature (°C)	Mean transpiration rate (mL/g/h)
15	0.038 ± 0.002
25	0.043 ± 0.001
35	0.059 ± 0.001
45	0.074 ± 0.0015

Evaluating the method

In your report you should acknowledge any possible sources of error that could not be eliminated. Even the most accomplished scientists are unable to eliminate error completely.

There are a number of different types of errors that can occur when you make measurements. Being aware of these errors is the first step to eliminating them and ensuring the validity of your results.

Systematic errors occur because of the way that an experiment has been designed. They will make the results consistently high or low. A systematic error can occur if the measuring instrument is not calibrated correctly or if you make the same mistake every time you take the measurement. This will mean that the measurements will be incorrect in the same way throughout the experiment.

A common form of systematic error is called 'zero error'. For example, if your bathroom scales read 5 kg when you are not standing on them, then it is likely that any measurement made with these scales will be 5 kg heavier than it should be (Figure 1.4.7). Systematic errors are easiest to spot when you have an idea of what the correct measurement should be.

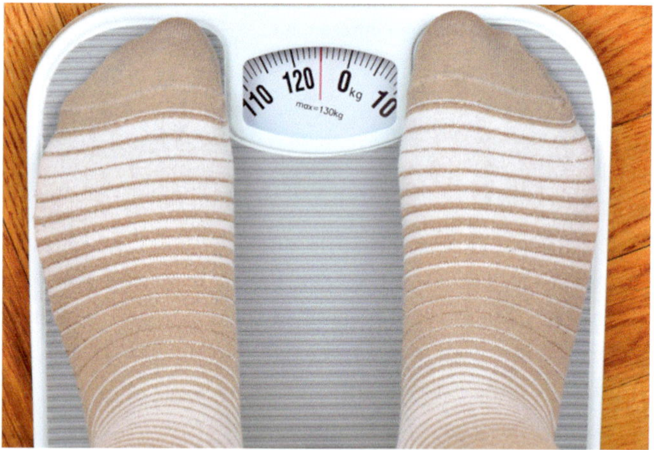

FIGURE 1.4.7 Before stepping onto bathroom scales they should read zero. A reading above or below zero will result in an incorrect reading called a systematic error.

Random errors are unpredictable errors that can occur in all experiments. They occur because no measurement can be absolutely exact. Random errors are due to unpredictable fluctuations in the equipment or inconsistencies in the way you have interpreted the readings.

Random errors can be detected when, for example, two readings for the same measurement appear as different numbers in the data. The effect of random errors can be reduced by repeating each measurement several times and taking an average of these results.

Systematic and random errors often occur if you have overlooked or were unable to control a variable that should have been controlled.

Mistakes are not errors. Whereas errors are unavoidable, mistakes can be avoided with care (Figure 1.4.8). Examples of a mistake are forgetting to press a button on a stopwatch, spilling some liquid when measuring volume or pressing the wrong calculator buttons.

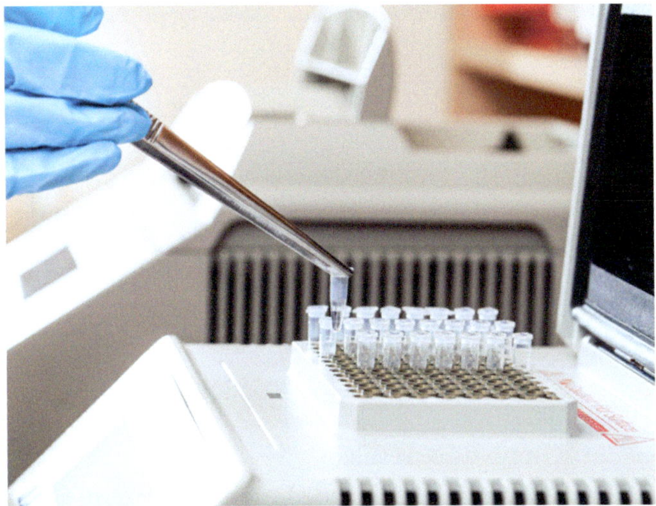

FIGURE 1.4.8 It is important to work carefully to avoid making errors.

After you have identified any problems with the data you have collected, discuss how things could be done differently in the future to improve the method. Often this can be done by repeating measurements, collecting more data or controlling other variables.

Discussion of results

The discussion section of the report includes two important features. First, it explains the results of the investigation. Second, it explains the significance of the experiment and whether the data supports the hypothesis. In this section include:

- an explanation of what the results mean—the patterns, relationships that results show
- reference back to the question and check if it has been answered
- whether the results support a theory
- whether the results were what was expected
- any new questions that arise out of the results
- any qualifications or defects in the experiment design—possible sources of error, how the experiment could be improved.

Writing a conclusion

Your **conclusion** should be one or two paragraphs that link your evidence to your hypothesis. It should provide a carefully considered response to your research question based on your results and discussion. You should clearly state whether your hypothesis was supported or not. The conclusion briefly restates the main results and explains the significance of the findings.

Do not provide irrelevant information or introduce new information in your conclusion.

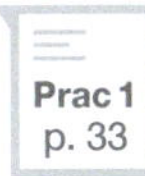

Prac 1
p. 33

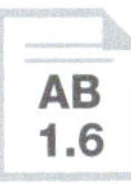

AB
1.6

MODULE
1.4 Review questions

Remembering

1 Define the terms:
  a random error b column graph.

2 What term best describes each of the following?
  a a summarising statement that links the evidence gathered to the hypothesis
  b raw data presented in rows and columns.

Understanding

3 A student measured the temperature outside every hour from 6 am to 6 pm. What would be the best graph for this type of data?

Applying

4 An experiment was conducted to find out whether salt dissolves more quickly in hot water than cold water.
  a Identify the independent variable.
  b Identify the dependent variable.
  c In a table representing the data from this experiment, which variable would go in the first column?

5 Two groups of students measured the time it took for ice cubes to melt on a sunny day. Their data is recorded in the Table 1.4.7.

TABLE 1.4.7 Raw data

Group	Time taken for ice cube to dissolve (min)					
	Trial 1	Trial 2	Trial 3	Trial 4	Trial 5	Trial 6
A	11.4	10.9	11.8	10.6	1.5	11.1
B	25	27	22	26	28	23

  a Both sets of data below contain errors. Identify which set is more likely to contain systematic error and which is more likely to contain random error. Explain your answers.
  b Use the data set to draw an appropriate graph.

6 Identify whether the following are mistakes, systematic errors or random errors.
  a A student spills some solution during a **titration**.
  b The reported measurements are above and below the true value.
  c A weighing balance has not been calibrated.

Analysing

7 A student conducted an experiment to determine the effects of temperature on seed germination. She placed 50 seeds of wheat in a gauze cloth and heated them in water at various temperatures for two minutes. She then placed the seeds on moist cotton wool and kept them in the dark for seven days. She recorded the germination rate of the seeds in Table 1.4.8.

TABLE 1.4.8 Effect of temperature on seed germination

Temperature °C	Number of seeds germinating
10	0
20	30
25	42
35	40
40	27
50	10
60	2

  a What graph would be best to display this data?
  b Graph these results.
  c What can be concluded from these results?

Evaluating

8 A scientist conducted an experiment by asking 50 people to eat a piece of chocolate every day for two weeks. At the end of the experiment, all 50 people gained weight. The scientist concluded that eating chocolate causes weight gain.
  a Is this conclusion valid?
  b Justify your answer.

Creating

9 Create a checklist in your own words that will help you to remember all of the important points you need to include in a table and a graph. You can use words, diagrams or a flow chart. Once complete, get your checklist checked by your teacher then stick it into the front of your workbook.

MODULE 1.4 Practical investigation

• STUDENT DESIGN •

1 • Gas collection

Questioning & Predicting | Evaluating

When acid is added to calcium carbonate the gas carbon dioxide is produced (Figure 1.4.9).

Purpose

To design an investigation that will change the rate at which carbon dioxide gas is produced.

Hypothesis

Once you have decided which investigation to perform, write a hypothesis in your workbook.

Timing 45 minutes

Materials

- Choose from:
- calcium carbonate powder and chips
- hydrochloric acid 0.1 M, 0.5 M, 1 M

- test-tubes
- conical flasks
- rubber stopper with a side arm attached
- rubber tubing that will fit onto the side arm
- 250 mL beakers
- water bath

SAFETY

A risk assessment is required for this investigation.

Procedure

1. Design an experiment that will answer one of the following questions.
   - Will changing the size of the particles of the calcium carbonate change the rate at which carbon dioxide is produced?
   - Will changing the concentration of the acid change the rate at which carbon dioxide is produced?
   - Will adding different amounts of the same substance have different effects on the rate at which carbon dioxide is produced?
   - Do small temperature changes have an effect on the rate that carbon dioxide is produced?
2. Write your procedure in your workbook. Include a diagram of your design for your experiment.

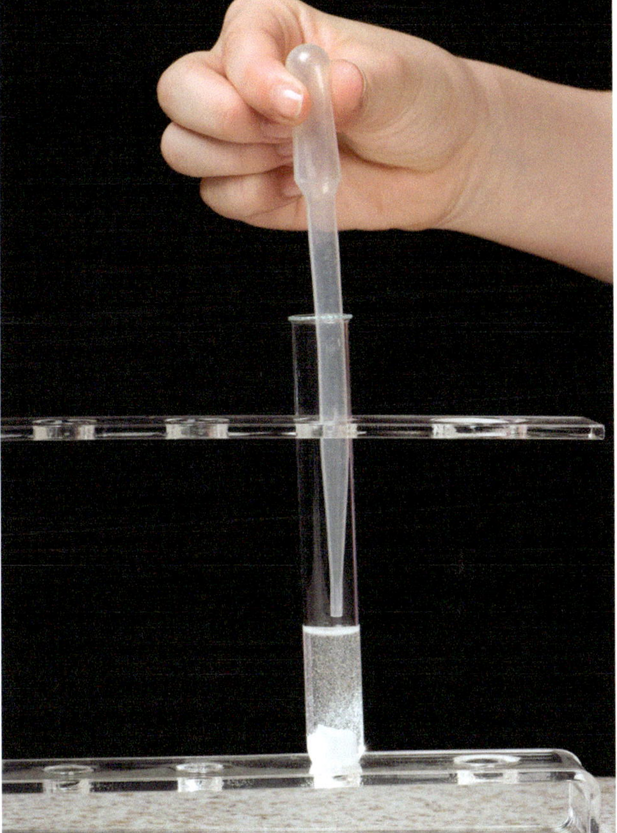

FIGURE 1.4.9 Carbon dioxide gas is produced during an acid–carbonate reaction.

3. Before you start any practical work, assess your procedure. List any risks that your procedure might involve and what you might do to minimise those risks. Show your teacher your procedure and risk assessment. If they approve, then collect all the required materials and start work.

See the Activity Book Toolkit to assist with developing a risk assessment.

Use the STEM and SDI template in your eBook to help you plan and carry out your investigation.

Results

Record your results and observations in your workbook.

Review

1. Construct a conclusion for your investigation.
2. Assess whether your hypothesis was supported or not.

CHAPTER 1

Chapter review

Remembering

1 Define the terms:
   a variable
   b quantitative data
   c reliability.

2 What term best describes each of the following?
   a a statement outlining what is being investigated
   b a prediction of the outcome of the experiment based on prior knowledge or research
   c all of the variables that must be kept constant during the investigation.

3 Copy and complete the following sentences.
   a A ________ __________ is used when comparing data in an investigation to represent discrete data.
   b A _______ __________ is used with continuous __________ data.

4 Appropriate protective equipment should be used when conducting laboratory experiments. List the protective equipment that is available in your school laboratory.

5 What factors are likely to cause errors in an experiment?

Understanding

6 What is the most suitable type of graph for the following data?
   a body temperature measured every two hours
   b the populations of different types of lizard found in an area of bushland.

7 Explain what is meant by the term *controlled variable*.

Applying

8 Use an example to distinguish between the terms *independent variable* and *dependent variable*.

9 For each of the following hypotheses, select the dependent variable.
   a The concentration of lead in water will be higher in storm water close to an industrial site than in drinking water.
   b The pH of commercially available sparkling mineral water will be lower than commercially available non-sparkling mineral water.

10 Identify the independent variable, the dependent variables and three variables that would be needed to be controlled to investigate each of the following hypotheses.
   a If an elastic band is wet, then it will not stretch as far as a dry elastic band.
   b If a cup of hot chocolate has a lid on it, then its temperature will decrease more slowly than when it is uncovered.

Analysing

11 If you spilled on yourself a chemical substance with the label in Figure 1.5.1 on it, what would be the appropriate thing to do?

FIGURE 1.5.1

Evaluating

12 Select the best hypothesis, and explain why the other options are not good hypotheses.
   A If light and temperature increase, the rate of photosynthesis increases.
   B The transpiration rate of a plant is affected by temperature.
   C If rock salt is broken into smaller pieces, it will dissolve more quickly.

CHAPTER 1

Chapter review

13 Everyone uses paper towel at some stage to clean up mess. But have you ever considered when using paper towel is it better to use it folded or flat? Your task is to design an investigation to find the answer.

- a Identify the following variables.
  - i independent variable
  - ii dependent variable
  - iii controlled variables.
- b Write a hypothesis for your experiment.
- c Draw a simple diagram for your experiment.
- d Use numbered steps to describe your procedure.

14 Have you ever noticed that you need to walk faster to keep up with some people whereas you have to decrease your pace to walk with others? This may be due to the leg length of the person you are walking with. A pedometer is an instrument that is often used by joggers or walkers to tell them the distance they have gone. On some pedometers you need to enter your height to get an accurate reading. Design an investigation to test just how much faster or slower different people walk, and see if you can use the relationship between a person's walking pace and their height to estimate your own height.

15 What conclusion do you think could be drawn from the graph in Figure 1.5.2?

16 Scientist Dr Julie Jones noticed that on the hills where there were plants growing there was less erosion than on the hills with no plants. Dr Jones suggested that growing plants on an incline would help to slow soil erosion. Your task is to design an investigation to find the answer.

- a Identify the following variables.
  - i independent variable
  - ii dependent variable
  - iii controlled variables.
- b Write a hypothesis for your experiment.
- c Draw a simple diagram for your experiment.
- d Use numbered steps to describe your procedure.

Creating

17 Design an investigation to determine how temperature will affect the elasticity of a rubber band. As part of your design, complete the following.

- a At what temperature do you think elastic bands will have the greatest stretch? Write a hypothesis for your experiment.
- b Draw a diagram to show how you intend to carry out your experiment.
- c Describe the procedure you intend to use.
- d Identify the independent, dependent and controlled variables.
- e How will you make the test fair?

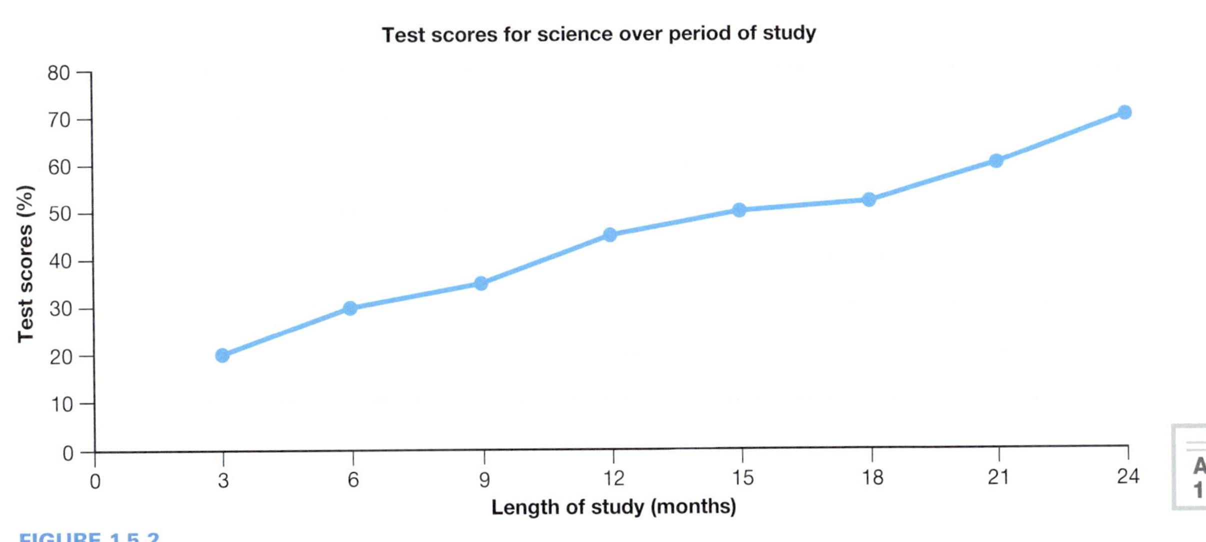

FIGURE 1.5.2

AB 1.8

CHAPTER 1

Inquiry skills

Research

1 Planning & Conducting | Processing & Analysing

Benjamin Franklin was one of the Founding Fathers of the United States of America. He was also a writer, oceanographer, inventor and scientist. One very famous experiment that it is said he performed is 'the kite experiment'. It might have looked something like that shown in Figure 1.5.3.

- a i Investigate the kite experiment. Describe what it is and how it is proposed that it was performed.
  - ii What was the aim of the kite experiment?
  - iii What hypothesis was being tested?
  - iv What conclusion could be drawn from the results?
- b i It is said that Franklin only did the experiment once. How does that affect the validity of any conclusions to be drawn?
  - ii Consider the descriptions of the experiment and the conditions under which it was purportedly performed. Is it likely that the experiment was performed exactly as described? Explain your reasoning.

Present your findings in digital form.

2 Planning & Conducting | Communicating

Find out about a current topic of scientific research at CSIRO.

- a Where is the research being performed?
- b Who are the scientists performing the research?
- c What is the aim of the research?
- d What practical benefits could be derived from this research?

Present the information as a poster encouraging people to support the work of the CSIRO.

3 Planning & Conducting | Communicating

Research the Australian code for the care and use of animals for scientific purposes.

- a Identify the purpose of the code for the care and use of animals for scientific purposes.
- b What types of animals does this code include?
- c Who has to abide by this code?

Present your findings as a pamphlet.

4 Planning & Conducting | Communicating

Research the placebo effect.

- a Define the term *placebo*.
- b Identify how placebos are used in medical research.
- c How does the placebo effect work?

Present your findings as a pamphlet.

FIGURE 1.5.3 A painting of Benjamin Franklin conducting his kite experiment

CHAPTER 1

Inquiry skills

Thinking scientifically

1 Katy was investigating how changing the concentration of hydrochloric affects the rate at which marble chips dissolve. Identify the potential hazards in her experiment. More than one answer may be correct.

A hydrochloric acid, as it is an irritant

B hydrochloric acid, as it is corrosive

C marble chips, as they are harmful

D marble chips, as they are corrosive.

2 As part of her experiment, Katy used the following equipment: a measuring cylinder, a glass beaker, 100 g of small marble chips, three different concentrations of hydrochloric acid, a weighing balance, a glass stirring rod and a stopwatch. Which pieces of equipment will help Katy collect and record data accurately? More than one answer may be correct.

A stopwatch

B concentrations of acid

C measuring cylinder

D weighing balance.

Katy's results are shown in Table 1.5.1 and Figure 1.5.4.

TABLE 1.5.1

Concentration of hydrochloric acid (M)	Mass of marble chips (g)	Time taken for marble chips to dissolve (s)
0.5	10	98
1	10	57
1.5	10	22

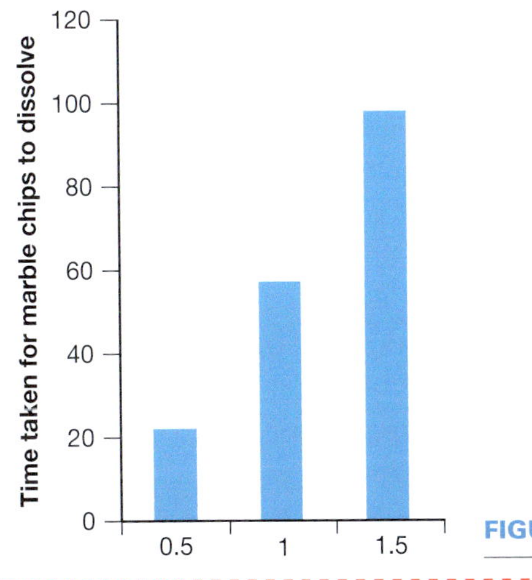

FIGURE 1.5.4

3 There are errors that may have occurred in this experiment—for example, the weighing balance giving a false reading. What type of error is this known as?

A systematic error

B zero error

C human error

D mistake.

4 What is missing from Katy's graph?

A title on y-axis and units on x-axis and a graph title

B title on x-axis, units on x-axis and y-axis, and a graph title

C labels on x-axis and a graph title

D labels on y-axis and a graph title.

5 Which of the following describes the trend in Katy's results bar chart?

A As the concentration of acid decreases, the rate at which marble chips dissolve decreases.

B As the concentration of acid decreases, the rate at which marble chips dissolve increases.

C As the concentration of acid increases, the rate at which marble chips dissolve decreases.

D As the concentration of acid increases, the rate at which marble chips dissolve increases.

6 How could Katy improve this investigation if she were to do it again?

A She could use a more precise weighing balance.

B She could repeat the experiment to get more reliable results.

C She could repeat the experiment to get more valid, accurate results.

7 Which of the following correctly describes the term *reliability*?

A how close a measurement is to the true value

B the closeness of two or more measurements to each other

C the ability to consistently reproduce results.

CHAPTER 1

Glossary

absorption: substances that pass through the skin

aim: a sentence summarising what will be investigated

central tendency: single values that allow you to describe the central position in a set of data

chemical code: warning symbol or HAZCHEM code on the label or container

column graph: a graph that shows the value of the dependent variable by the height of the column

continuous data: data measured within a range

control group: the experimental conditions of the control group are identical to the experimental group, except that the independent variable is also kept constant

controlled variable: the variable kept constant throughout an experiment

data: experimental results, often in the form of numbers or written observations

dependent variable: the variable you are measuring; it changes as the independent variable changes

descriptive statistics: used to summarise, organise and describe data obtained from research

discontinuous (discrete) data: data that can be counted

ethics: a set of principles by which your actions can be judged morally acceptable or unacceptable

evidence: results that can be used in support of statements being made

experiment: testing out a hypothesis under controlled conditions to examine its validity

experimental group: the experimental conditions of the experimental group are identical to the control group, except that the independent variable is changed

fair test: an experiment where one variable is changed; one variable is measured and all other variables are controlled

hypothesis: a statement about the relationship between two variables which can often be tested experimentally; an 'educated guess'

independent variable: a variable that is changed in a systematic way in an experiment

inference: a conclusion reached on the basis of evidence, an educated guess

ingestion: swallowed

inhalation: breathed in

line graph: a type of graph that is good for representing continuous quantitative data

objective: free of personal bias

observation: closely monitoring something or someone

outlier: abnormally big or small value in the data set

percentage change: applies to increases and decreases relative to the control or the starting point of the measurement

percentage difference: a measure of the precision of two measurements

personal protective equipment (PPE): clothing items that help to keep you safe when doing experiments

processed data: data that has been manipulated in some way, often mathematically

qualitative data: data recorded as words or descriptions

quantitative data: data recorded as numbers

random error: an error that affects experimental results in an unpredictable way

range: the difference between the highest and lowest values in a data set

reliability: the ability to consistently reproduce results

repeat trials: collecting multiple data sets by performing an experiment again after the initial test

replicates: duplicate experiments

replication: when duplicate sets of an experiment are run at the same time

research question: a statement describing in detail what will be investigated

risk: the chance of injury or loss

risk assessment: a systematic way of identifying potential risks

safety data sheet: an information sheet that contains important information about the hazards in using a substance and how it should be handled and stored

subjective: influenced by personal views

systematic error: error that affects all experimental results in the same way; occurs because of the way that the experiment has been designed

theory: a scientific explanation supported by all the experiment results obtained so far

titration: the process of adding one solution to another in a controlled environment to observe a reaction

uncertainty: a realistic range within which the true value is likely to be; calculated as the range/2

validity: how well an experiment and its results meet the requirements of a fair test

variable: a factor or condition that can change the result of an experiment

AB 1.7

CHAPTER 2

Materials

Have you ever wondered ...

- how scientists can study the structure of an atom without being able to look inside?
- why jewellery body piercings are usually made of stainless steel?
- what the difference is between an acid and a base?

After completing this chapter you should be able to:

- describe and model the structure of the atom in terms of the nucleus, protons, neutrons and electrons
- compare the mass and charge of protons, neutrons and electrons
- investigate the historical development of models of the structure of the atom
- investigate the work of scientists such as Ernest Rutherford, Pierre Curie and Marie Curie.

MODULE

2.1 Atoms

The universe is made up of millions of different substances. All of these substances are made up of building blocks known as atoms. Different types of atoms can combine with each other to form new substances. Understanding atoms helps scientists create materials for use in LCD screens, lasers and solar cells.

Atomic building blocks

Look around you and you will see thousands of different materials—paper, plastic, wood, glass, skin and many more. All these different materials are made up of tiny building blocks known as **atoms**.

Atoms are so small that they cannot be seen with even the most powerful optical microscope. To see atoms, scientists must use a special type of microscope known as a scanning tunnelling microscope or STM. Figure 2.1.1 shows an image of silicon atoms taken with an STM.

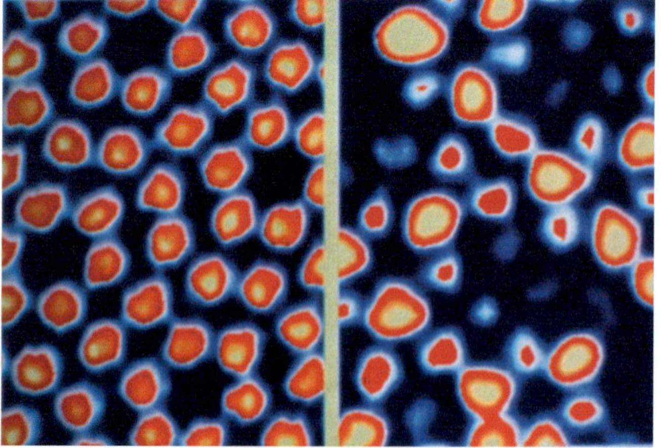

FIGURE 2.1.1 Billions of silicon atoms like these stick together to form a silicon crystal.

There are 118 known types of atoms and only 98 of these are found naturally on Earth. The remaining 20 types of atoms must be made in a laboratory.

Each type of atom is given its own chemical symbol that is usually made up of one or two letters. Often, the chemical symbol is related to the name of the atom. For example, the symbol for hydrogen is H and the symbol for carbon is C, the symbol for magnesium is Mg, while chlorine is Cl. However, sometimes the symbol does not appear to be related to the name of the atom. This is because the symbol has come from the atom's name in another language, usually Latin. For example, the chemical symbol for sodium is Na, which comes from its Latin name *natrium*. Similarly, the chemical symbol for potassium is K, which comes from its Latin name *kalium*.

Atoms in elements and compounds

Atoms can either combine with other atoms to form clusters of atoms known as **molecules** or form large grid-like structures known as **crystal lattices**.

For example, the water shown in Figure 2.1.2 is made up of molecules. Every water molecule is identical and contains two hydrogen atoms (H) and one oxygen atom (O).

In contrast, a grain of beach sand is a crystal lattice with the chemical formula SiO_2. This lattice is made of silicon (Si) and oxygen (O) atoms. The number of atoms in the lattice depends on the size of the grain of sand.

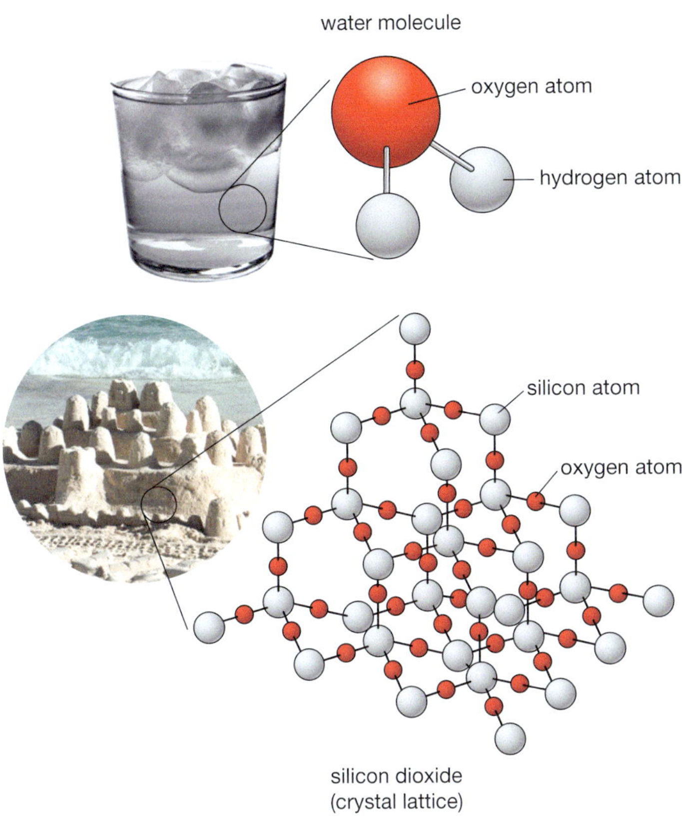

FIGURE 2.1.2 Atoms can form molecules like the water molecule, or large crystal lattices like that of silicon dioxide in beach sand.

Elements

If a substance is made up of just one type of atom, then it is referred to as an **element**.

Molecular elements are made up of small molecules like the oxygen, phosphorus and sulfur molecules shown in Figure 2.1.3. Carbon is a unique element because carbon atoms can form extremely large molecules. A buckyball is made up of 60 carbon atoms (C_{60}) in the shape of a soccer ball, while a carbon nanotube can have thousands of carbon atoms forming a long cylinder.

Carbon is also the only non-metallic element that can form crystal lattices. The diamonds found in jewellery and the graphite in pencil 'leads' are two forms of carbon crystal lattices. Metallic elements such as copper and gold always form crystal lattices. Figure 2.1.4 compares these two types of lattices.

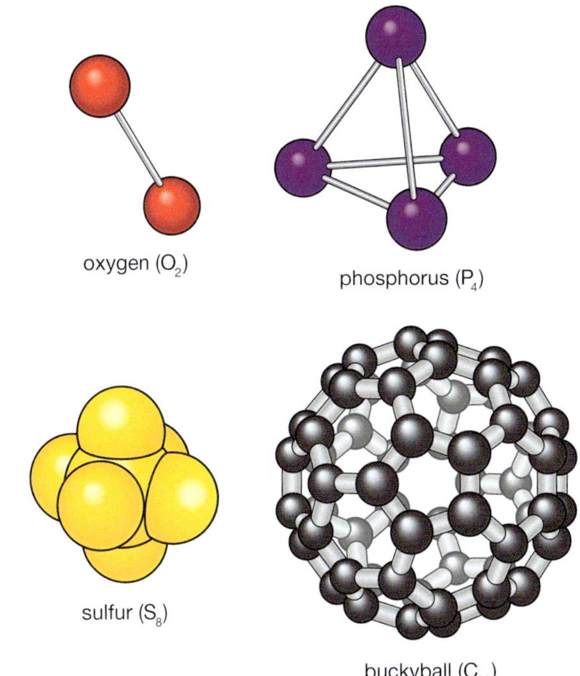

FIGURE 2.1.3 In molecular elements, each molecule is made up of just one type of atom.

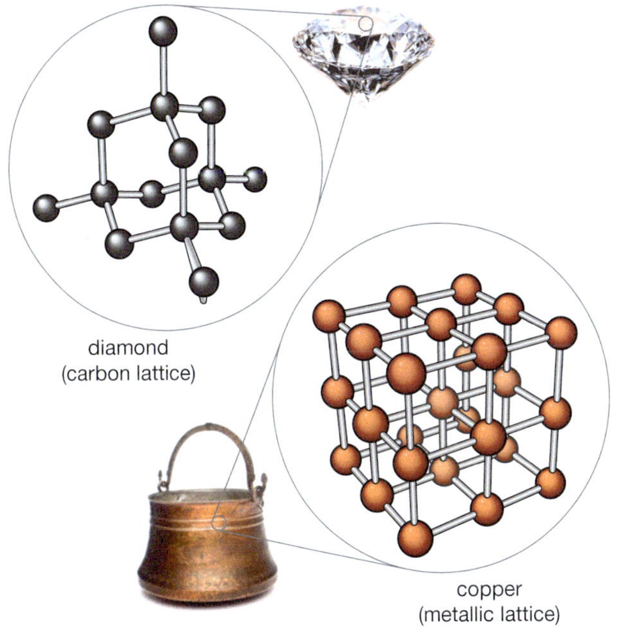

FIGURE 2.1.4 Elements that exist as crystal lattices contain many atoms of the same type, arranged in a grid-like structure.

Compounds

If a substance is made up of different types of atoms, then it is known as a **compound**. The molecules that make up compounds range from small to very large. Glucose is a simple sugar that is the main energy source for animals and plants. Glucose is a molecular compound made of molecules, each molecule being made up of just 24 atoms. You can see it in Figure 2.1.5. In contrast, a single molecule of DNA inside one of your cells is made up of billions of atoms and can be stretched to over a metre in length.

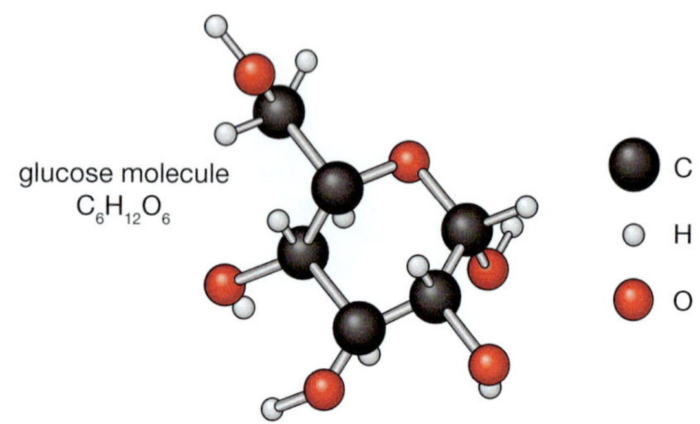

FIGURE 2.1.5 Glucose is a molecular compound because it is made up of carbon, hydrogen and oxygen atoms.

Many compounds form crystal lattices. Table salt is a crystal lattice of sodium (Na) and chlorine (Cl) arranged into a three-dimensional grid. Its structure is shown in Figure 2.1.6.

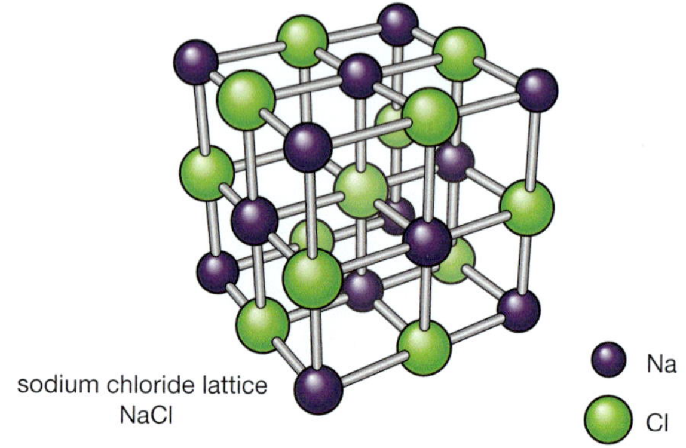

FIGURE 2.1.6 Sodium chloride forms a crystal lattice—it is a compound because it contains more than just one type of atom.

Inside atoms

The atoms that make up elements and compounds were once thought to be hard and unbreakable. Today, scientists know that atoms are made up of even smaller particles known as **subatomic particles**. Each atom is made up of three types of subatomic particles: **protons**, **neutrons** and **electrons**.

As Figure 2.1.7 shows, the protons and neutrons form a cluster that sits at the centre of the atom. This cluster is known as the **nucleus**. The electrons are much smaller and lighter and move very fast to form an **electron cloud** that surrounds the nucleus.

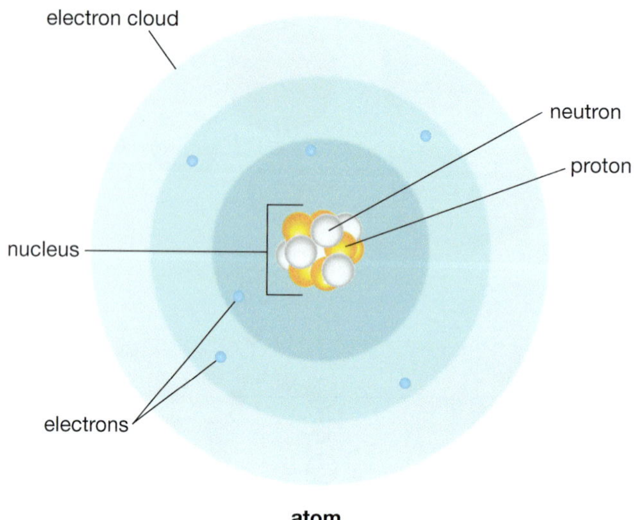

FIGURE 2.1.7 Atoms are made up of subatomic particles known as protons, neutrons and electrons.

Table 2.1.1 summarises some of the important properties of protons, neutrons and electrons. Protons and neutrons are similar in mass. However, protons have a positive electric charge (+1) while neutrons have no electric charge. Electrons are approximately 1800 times lighter than protons and neutrons, and have a negative electric charge (–1). In an atom, the number of electrons equals the number of protons. Therefore the negative charge of the electrons balances the positive charge of the protons, making the atom's charge neutral.

TABLE 2.1.1 Properties of subatomic particles

Subatomic particle	Location	Mass compared with the mass of an electron	Electric charge
proton	nucleus	× 1800	+1
neutron	nucleus	× 1800	0
electron	electron cloud around the nucleus	× 1	–1

The negative charge of electrons causes them to be attracted to the positively charged protons in the nucleus. This is because opposite electric charges attract each other, a bit like the way opposite poles of a magnet attract each other. This attractive force is known as electrostatic attraction. This attraction stops the electrons from straying too far from the nucleus but is not enough to trap and pull the electrons completely in.

While opposite charges (+/–) attract each other, like charges (+/+ or –/–) repel each other. This force is called electrostatic repulsion. All protons have a positive charge and so all the protons in the nucleus should repel each other. This would cause the nucleus to split apart. However, there is an even stronger force holding the protons together—this force is called the nuclear force.

STEM 4 fun

Scale of an atom

PROBLEM

How big is an atom and how would you represent its size?

SUPPLIES

- science reference source
- metre ruler or tape measure
- calculator

PLAN AND DESIGN Research the size of a typical atom compared to the size of the nucleus. If you and three of your friends were to act as the protons and neutrons in a nucleus, how far away would another friend need to be to form the electron shell?

CREATE Follow your plan. Work out your solution to the problem.

IMPROVE What works? What doesn't? How do you know it solves the problem? What could work better? Modify your design to make it better. Test it out.

REFLECTION

1 What area of STEM did you work in today?
2 What field of science did you work in? Are there other fields where this activity applies?
3 How did you use mathematics in this task?

SciFile

The atomic universe

Approximately 98% of the atoms in the universe are either hydrogen (H) or helium (He) atoms. These atoms make up the Sun and the stars. The other types of atoms make up only 2% of all the atoms in the universe.

Atomic number and mass number

The number of protons in the nucleus determines the type of atom it is and what element it belongs to. For example, all gold atoms contain 79 protons while all oxygen atoms contain eight protons. The number of protons in an atom is called the **atomic number**.

The number of neutrons does not affect which element the atom belongs to, but it does affect the atom's mass. The number of protons plus neutrons in an atom is called the **mass number**. These numbers are often written alongside the chemical symbol. For example, an atom of sodium (Na) can be shown as:

From this one symbol, you can calculate the number of protons, neutrons and electrons in the sodium atom:

- The number of protons is the atomic number, 11. So there are 11 protons in the nucleus.
- The number of electrons is equal to the number of protons. So there are 11 electrons spinning in a cloud around the nucleus.
- The number of neutrons is the mass number minus the atomic number: 23 – 11 = 12. So there are 12 neutrons in the nucleus.

Electrons and the nucleus

The number of electrons surrounding the nucleus of an atom is exactly equal to the number of protons in the nucleus. As a result, atoms are charge **neutral**. This means that atoms have no electric charge—the positive charge of the protons is exactly balanced by the negative charge of the electrons.

Although each electron is 1800 times lighter than a proton, the electron clouds can be 100 or even 1000 times larger than the nucleus. This means that if the nucleus were the size of a golf ball, the electrons would form clouds the size of a football stadium and the electrons would be the size of a single grain of sand. It also means that most of an atom is empty space.

SkillBuilder

Writing atomic symbols

To show the mass number and atomic number of an atom, scientists write an **atomic symbol**. The atomic symbol for helium is:

$$\text{mass number} \rightarrow {}^{4}_{2}\text{He} \leftarrow \text{chemical symbol}$$
$$\text{atomic number} \rightarrow 2$$

This shows the chemical symbol for helium (He), with the mass number above and the atomic number below. From this symbol it is possible to work out the number of neutrons in the nucleus by subtracting the atomic number from the mass number.

Number of neutrons = 4 – 2 = 2

It is also possible to work out the number of electrons, which is equal to the atomic number:

Number of electrons = atomic number = 2

In this way, the atomic symbol can be used to obtain a complete description of the subatomic particles of the helium atom, shown in Figure 2.1.8.

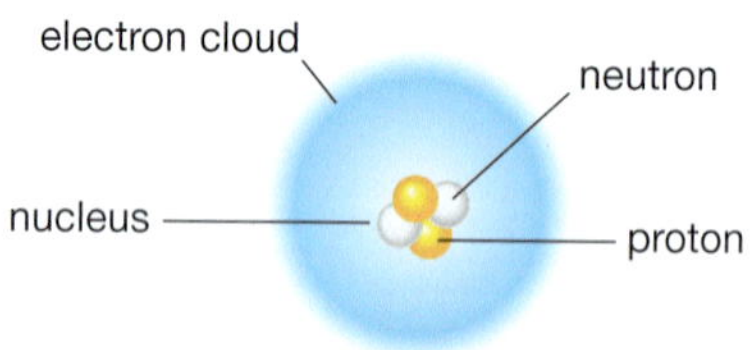

FIGURE 2.1.8 This helium atom has two protons and two neutrons. So its atomic number is 2 and its mass number is 4. Helium also has two electrons forming its electron cloud.

The Rutherford experiments

In 1909, the New Zealand scientist Ernest Rutherford (1871–1937) discovered that the nucleus only takes up a small fraction of the space inside an atom. In a famous experiment, Rutherford fired a beam of helium nuclei (alpha particles) at a thin sheet of gold foil. This is shown in Figure 2.1.9.

Rutherford found that most of the alpha particles passed straight though the foil. However, surprisingly, some alpha particles bounced straight back. Rutherford concluded that an atom was mostly made up of empty space with a small, positively charged nucleus in the centre. It was these nuclei that caused the alpha particles to occasionally bounce back in Rutherford's experiment.

Worked example

Writing atomic symbols

Problem

Determine the number of protons, electrons and neutrons in:

$${}^{39}_{19}\text{K}$$

Solution

Thinking: Determine the number of protons by looking at the atomic number.

Working: number of protons = atomic number = 19

Thinking: Determine the number of electrons by looking at the atomic number.

Working: number of electrons = number of protons = atomic number = 19

Thinking: Calculate the number of neutrons by subtracting the atomic number from the mass number.

Working: number of neutrons = mass number – atomic number = 39 – 19 = 20

Try yourself

1 How many protons are there in the iron atom shown below?

2 How many electrons are there?

3 Calculate the number of neutrons.

$${}^{56}_{26}\text{Fe}$$

AB 2.2

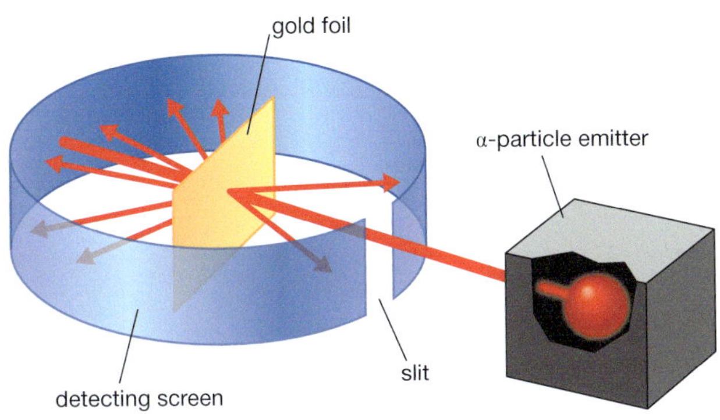

FIGURE 2.1.9 In Rutherford's famous experiment, a beam of helium nuclei (alpha particles) was fired at gold foil. Most of the alpha particles went straight through the foil but only a small number were deflected. Rutherford concluded that atoms are mostly empty space with a small, positively charged nucleus and a large negatively charged electron cloud.

Ions

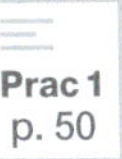
Prac 1 p. 50

Atoms can lose or gain electrons to become electrically charged particles called **ions**. If an atom loses electrons, then it has more protons than electrons. This gives the atom a positive charge. The ion formed is known as a **cation**. If an atom gains electrons, then it has more electrons than protons. It is now negatively charged and is known as an **anion**.

The symbol for an ion is the same as the chemical symbol for the atom but with the charge of the ion added to it. For example, when a sodium atom (Na) loses one electron, the ion forms a cation. This gives it a charge of +1 and so the symbol of the sodium ion is Na^+. Similarly, the symbol Mg^{2+} indicates that a magnesium ion is formed when a magnesium atom (Mg) loses two electrons.

A similar system is used to show negative charges for anions. When a chlorine atom gains one electron, it forms an ion with a charge of –1. Its symbol is Cl^-. If an oxygen atom gains two electrons, then it becomes an anion with a charge of –2 and its symbol is written as O^{2-}. Unlike with cations, the name of the anion changes slightly by adding *-ide* to the end of the atom name. So the chlorine atom becomes the chloride ion and the oxygen atom becomes the oxide ion.

Ions can be formed in many situations. They are commonly formed when some substances are dissolved in water. Not all substances will form ions when dissolved and you can determine which substances form ions by passing an electric current through the solution. Substances that form ions when dissolved will conduct electricity because the charged ions are free to move through the liquid, carrying the electric current with them. For example, salt (sodium chloride, NaCl) dissolves because water molecules break up the lattice. This releases sodium ions (Na^+) and chloride ions (Cl^-) into the solution. The presence of these ions in the solution means that salt water is a conductor of electricity.

In contrast, sugar does not release ions when it dissolves in water—there is nothing in the solution that can conduct electricity.

Prac 2 p. 51

science 4 fun

Electrostatic attraction

Can you use electrostatic force to stick a balloon to the wall?

Collect this ...

- balloon
- head of clean, dry hair

Do this ...

1. Inflate the balloon and tie a knot in it.
2. Rub the balloon vigorously on the hair.
3. Gently place the balloon in contact with a wall and see if it will stay.

Record this ...

1. Describe what you saw.
2. Explain why you think this happened.

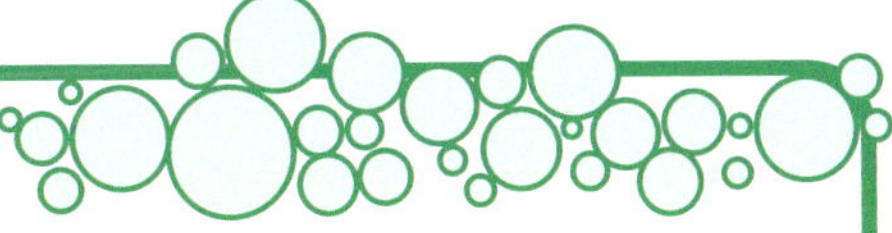

SciFile

Ernest Rutherford

Rutherford was lousy at mathematics and wasn't even very good at experiments! However, he was a hard worker. He started working on the structure of the atom only after others convinced him that his early work on getting radio to work was pointless because they thought that radio had no practical applications. He thought that physics was the only real science and compared chemistry to stamp collecting! Despite this, he won the Nobel Prize for Chemistry in 1908 for his work on radioactivity.

SCIENCE AS A HUMAN **ENDEAVOUR**

Use and influence of science

History of the atomic model

The internal structure of an atom cannot be seen with any microscope. Therefore, scientists must rely on indirect observations to build a model of what is inside an atom (Figure 2.1.10). As technology has advanced, scientists' understanding of atoms has increased and the atomic model has evolved.

FIGURE 2.1.10 Model of atoms in a lattice

Table 2.1.2 shows the historical development of atomic models. Over time, the models of an atom have become more accurate as scientists have developed new techniques for observing subatomic particles.

TABLE 2.1.2 Historical development of models of the atom

Year	Observation and theory	Model
600–500 BCE	The ancient Greeks believed that all matter was made up of only four fundamental elements: earth, fire, air and water. This was the basis of the continuum model, which predicted that regardless of the number of times you halve a piece of matter, it can always be broken down into even smaller pieces.	Continuum model
460–370 BCE	Greek philosopher Democritus suggested that matter was not continuous but was made up of tiny, solid and unbreakable particles. He was the first to use the term *atomos* meaning 'indivisible', from which the word *atom* comes.	Solid-ball model
1803	English chemist John Dalton proposed experiments based on the atomic theory of matter, with elements and compounds. In Dalton's theory, all matter is made up of hard, indivisible spheres.	
1904	British scientist Joseph John Thomson (J.J. Thomson) discovered the electron and its negative charge in 1897. However, Thomson knew that there must also be a source of positive charge in the atom to balance the negative charge. Therefore, in 1904 he proposed the plum pudding model. In this model, an atom is thought of as a round ball of positive charge with negatively charged electrons embedded in it (like plums or sultanas in a plum pudding).	Plum pudding model

SCIENCE AS A HUMAN ENDEAVOUR

Year	Observation and theory	Model
1904	Hungarian scientist Philipp Lenard described atoms as mostly empty spaces filled with fast-moving 'dynamides'. These were neutrally charged particles made up of a heavy positive particle stuck to a light negative particle.	Dynamide model
1909–1911	In 1909, New Zealand scientist Ernest Rutherford performed an experiment where he fired a beam of positively charged alpha particles at gold foil. He found that while most of the alpha particles went through the foil, a small number bounced back. This led to the development of a nuclear model of the atom in 1911. In this model most of the mass is contained in a small positive nucleus surrounded by a large space occupied by negative electrons.	Nuclear model
1913	Danish scientist Niels Bohr modified Rutherford's model and proposed that electrons can only travel along certain pathways around the nucleus, called orbits. As a result, this model is sometimes called the planetary model.	Planetary model
1932	English scientist James Chadwick discovered the neutron, showing that the nucleus was not just a mass of positive charge but a cluster of positively charged protons and charge-neutral neutrons.	Planetary model with neutrons
1932–today	Today, scientists have concluded that the position of an electron in an atom can never be known exactly. This means that it is impossible for electrons to revolve around the nucleus in specific orbits as suggested by Niels Bohr. Instead, the electrons form clouds around the nucleus. Scientists can predict the shape of these clouds but never the exact location of electrons within them.	Electron cloud model

REVIEW

1 Name the scientist who discovered the existence of the atomic nucleus.

2 Compare the model proposed by Niels Bohr with the motion of the planets around the Sun.

3 Explain where the term *atom* came from.

AB 2.3

4 Calculate how long it took to discover the neutron after the discovery of the electron.

5 Propose a reason why the neutron was the last of the subatomic particles to be discovered.

MODULE **2.1** # Review questions

Remembering

1 Define the terms:
  a element
  b compound
  c neutron
  d ion
  e atomic number.

2 What term best describes each of the following?
  a the total number of the protons and neutrons in a nucleus
  b a positively charged ion
  c the cluster of protons and neutrons at the centre of an atom
  d the area in an atom occupied by electrons
  e large grid-like structure.

3 For each of the subatomic particles electrons, protons and neutrons, state:
  a its charge
  b its relative mass
  c where it is located in the atom.

4 Which of the following are elements and which are compounds?
  a carbon (C)
  b water (H_2O)
  c silicon dioxide (SiO_2)
  d sulfur (S_8)
  e sodium chloride (NaCl).

5 Which of the following elements are made of molecules and which are crystal lattices?
  a oxygen (O_2)
  b copper (Cu)
  c diamond (C)
  d phosphorus (P_4).

6 Name the force that:
  a attracts electrons to the nucleus
  b makes protons repel each other
  c holds protons together in the nucleus.

Understanding

7 Explain why an atom is charge neutral.

8 a Outline Rutherford's experiment with gold foil and alpha particles.
  b Explain how he deduced that the atom was largely empty space.

9 Outline how a:
  a magnesium atom Mg becomes a magnesium ion Mg^{2+}
  b chlorine atom Cl becomes a chloride ion Cl^-.

10 Identify if the following substances (a–h) are elements or compounds.

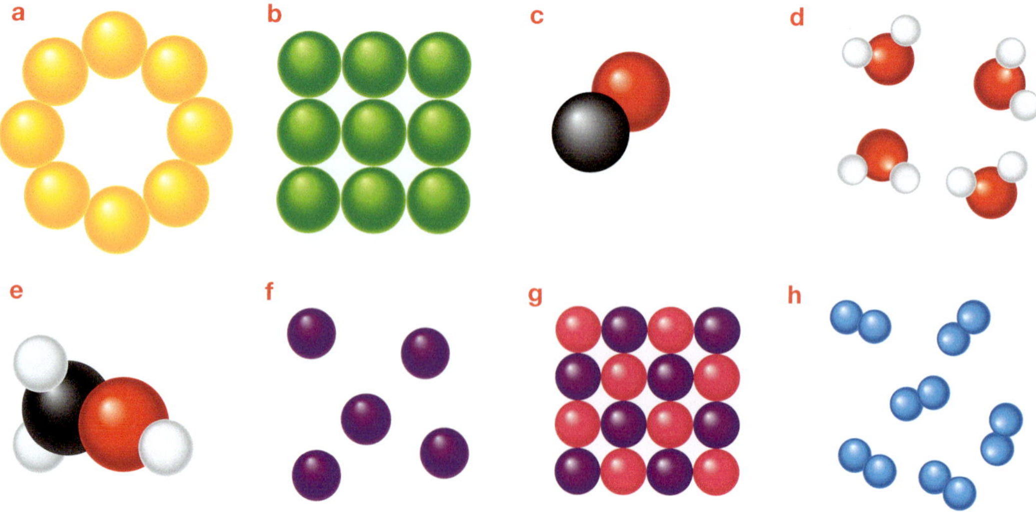

MODULE 2.1

Review questions

Applying

11 Use the chemical formulas to identify whether the following are elements or compounds.

- **a** $C_6H_{12}O_6$
- **b** C_{60}
- **c** Fe
- **d** $MgCl_2$
- **e** H_2SO_4

12 What is the atomic number and mass number of each of the following atoms?

- **a** An oxygen atom with 8 protons, 8 neutrons and 8 electrons
- **b** A calcium atom with 20 protons, 20 neutrons and 20 electrons
- **c** A gold atom with 79 protons, 114 neutrons and 79 electrons
- **d** A uranium atom with 92 protons, 146 neutrons and 92 electrons.

13 Use atomic numbers and mass numbers to write atomic symbols in the form $^{56}_{26}Fe$ for each of the atoms in question 12.

14 For each of the situations i–iv described below:

- **a** name the ion formed
- **b** identify its charge
- **c** identify the symbol for the ion formed.
  - **i** Lithium (Li) loses an electron.
  - **ii** Aluminium (Al) loses three electrons.
  - **iii** Fluorine (F) gains an electron.
  - **iv** Sulfur (S) gains two electrons.

Analysing

15 Compare the five lightest atoms by copying and completing the following table.

Comparing the five lightest atoms

Atom	Atomic number	Mass number	Number of protons	Number of neutrons	Number of electrons	Atomic symbol
hydrogen	1			0	1	$^{1}_{1}H$
helium		4		2		
lithium	3			4		
beryllium	4	9				
boron						$^{11}_{5}B$

16 Calculate the number of protons, neutrons and electrons in the following atoms.

- **a** $^{4}_{2}He$
- **b** $^{16}_{8}O$
- **c** $^{28}_{14}Si$
- **d** $^{207}_{82}Pb$
- **e** $^{238}_{92}U$

17 Compare a sodium atom Na with its ion Na^+.

Evaluating

18 Why do you think scientists use atomic symbols instead of their names when communicating with other scientists?

19 A balloon becomes charged when you rub it against dry hair. What do you think happens to cause this?

MODULE 2.1 Practical investigations

1 • Experimenting like Rutherford

Questioning & Predicting | Evaluating

Purpose

To use indirect observation to estimate the size of an unseen object.

Timing
30 minutes

Materials

- large cereal box with the top and bottom open
- objects of various shapes and sizes that can fit inside the box
- 5 marbles

Procedure

1. This activity requires you to work in pairs.
2. Place the open cereal box on the desk as shown in Figure 2.1.11.
3. One person places an object in the box without the other person seeing the object.
4. The other person then rolls the five marbles through the box and tries to estimate the size of the object.
5. Record your estimates in a table like the one in the Results section. Compare them to the real size of the object.
6. Repeat this process three more times, so that each member of the pair has two turns at determining the size of the hidden object.

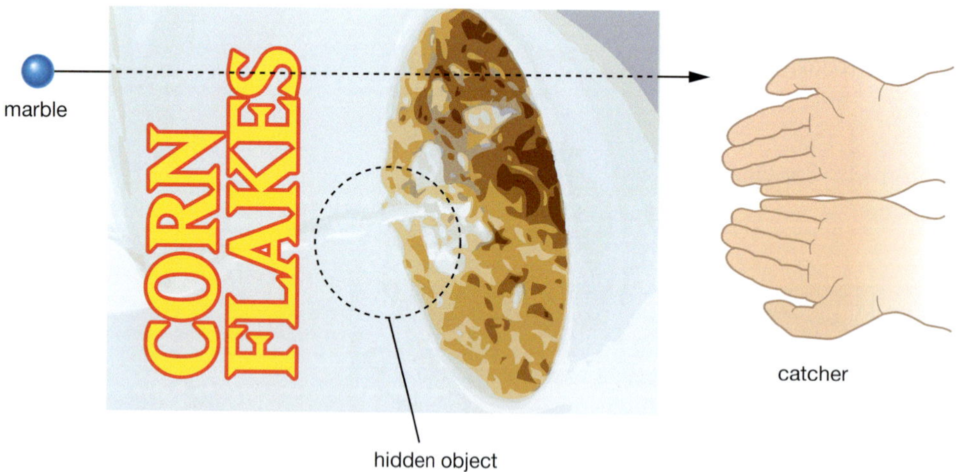

FIGURE 2.1.11 Overhead view of activity set-up

Results

Record your measurements in a table like the one below.

Estimated and real sizes

	Estimated size	Real size
Object 1		
Object 2		

Review

1. How is this experiment similar to Rutherford's experiment?
2. What factors might have influenced the accuracy of your estimates?
3. What other properties of the object may be determined by indirect observation using this technique?

MODULE 2.1 Practical investigations

2 • Detecting ions by indirect observation

Evaluating

To save time and equipment, the seven beakers in this experiment could be split amongst two or more groups, who would then share their results.

Purpose

To determine whether common household compounds form ions.

Hypothesis

Which solutions do you think will contain ions—distilled water, salt water, a solution of sugar, coffee or tea, vinegar or vegetable oil? Before you go any further with this investigation, write a hypothesis in your workbook.

Timing 45 minutes

Materials

- distilled water
- salt water solution
- sugar (sucrose)
- tea bag
- coffee
- vinegar
- vegetable oil

SAFETY

A risk assessment is required for this investigation. Refer to the SDS of all chemicals when constructing your risk assessment.

- 7 × 250 mL beakers
- ammeter
- wires with alligator clips
- carbon electrodes

Procedure

1 Copy the table from the Results section into your workbook.

2 Use the wires to connect the voltage source, ammeter and electrodes in a circuit (Figure 2.1.12).

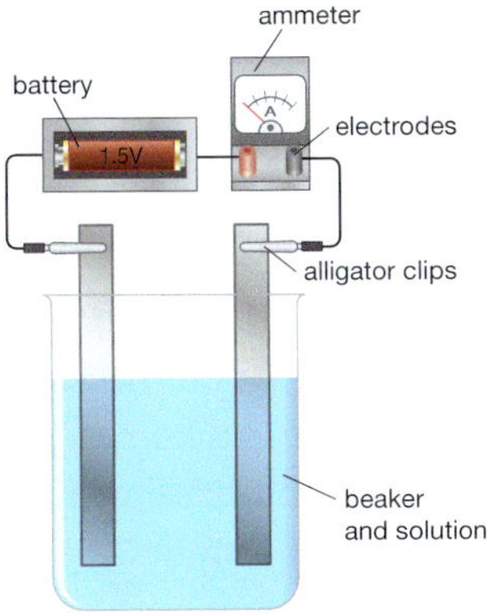

3 Three-quarters fill the seven beakers with different liquids and solutions:
beaker 1: distilled water
beaker 2: salt water solution
beaker 3: sugar solution
beaker 4: coffee solution
beaker 5: tea solution
beaker 6: vinegar solution
beaker 7: vegetable oil and distilled water

FIGURE 2.1.12

4 Dip the electrodes in the distilled water in beaker 1. Record any current that flows through the ammeter.

5 Rinse the electrodes with distilled water.

6 Dip the electrodes into the salt water solution in beaker 2. Once again, record any current that flows.

7 Repeat for all the other beakers, rinsing the electrodes in distilled water between each test.

See the Activity Book Toolkit to assist with developing a risk assessment.

Extension

8 Repeat step 4 but try a globe in place of an ammeter. Will the globe light up?

Results

Record all your measurements in a table like this one.

Ions in household compounds

Solution	Current detected? (Yes/No)	Ions present? (Yes/No)
distilled water		
salt water solution		
sugar solution		
coffee solution		
tea solution		
vinegar		
vegetable oil		

Review

1 List all the solutions in which ions were present and all the solutions in which ions were not present.

2 Explain why a current flowing indicates the presence of ions.

3 a In the cases where no current flowed, do you think the compounds form atoms, molecules or lattices in solution?
b Justify your answer.

4 a Construct a conclusion for your investigation.
b Assess whether your hypothesis was supported or not.

MODULE

2.2 Metals, non-metals and metalloids

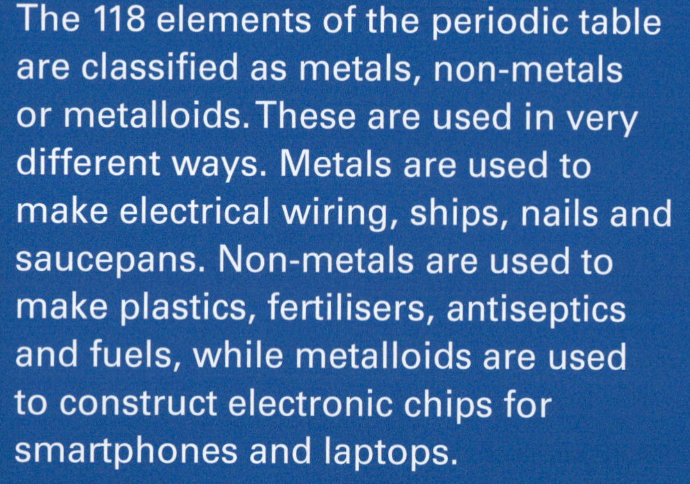

The 118 elements of the periodic table are classified as metals, non-metals or metalloids. These are used in very different ways. Metals are used to make electrical wiring, ships, nails and saucepans. Non-metals are used to make plastics, fertilisers, antiseptics and fuels, while metalloids are used to construct electronic chips for smartphones and laptops.

science 4 fun

Rust away!

Can you get steel to rust in one day?

Collect this ...

- steel wool (plain, with no soap)
- vinegar
- liquid bleach
- screw-top glass jar

Do this ...

1. Put a lump of steel wool in the bottom of the screw-top jar.
2. Pour in enough water to cover the steel wool.
3. Add a little vinegar and a little bleach.
4. Screw on the top of the jar and check what happens to the steel wool over the next day.

Record this ...

1. Describe what happened.
2. Explain why you think this happened.

Elements

Elements are substances that are made up of only one type of atom, each atom having exactly the same number of protons in its nucleus as the next atom. This gives each element its own distinctive atomic number. For example, carbon (symbol C) is an element because all of its atoms are carbon atoms. Each carbon atom has 6 protons in its nucleus, giving carbon an atomic number of 6. Likewise, the element gold (Au) has an atomic number of 79 and so every gold atom contains 79 protons.

The periodic table

There are 118 different elements and therefore 118 different types of basic atoms. The **periodic table** is a list of all 118 known elements, arranged in order of their atomic number. As the periodic table in Figure 2.2.1 shows, elements are classified according to their properties as metal, non-metal or metalloid.

There are roughly four times as many metals as there are non-metals and metalloids in the periodic table. However, in the universe the number of non-metallic atoms is far, far greater than the number of metallic atoms. This is because stars are made mainly of the non-metals hydrogen and helium.

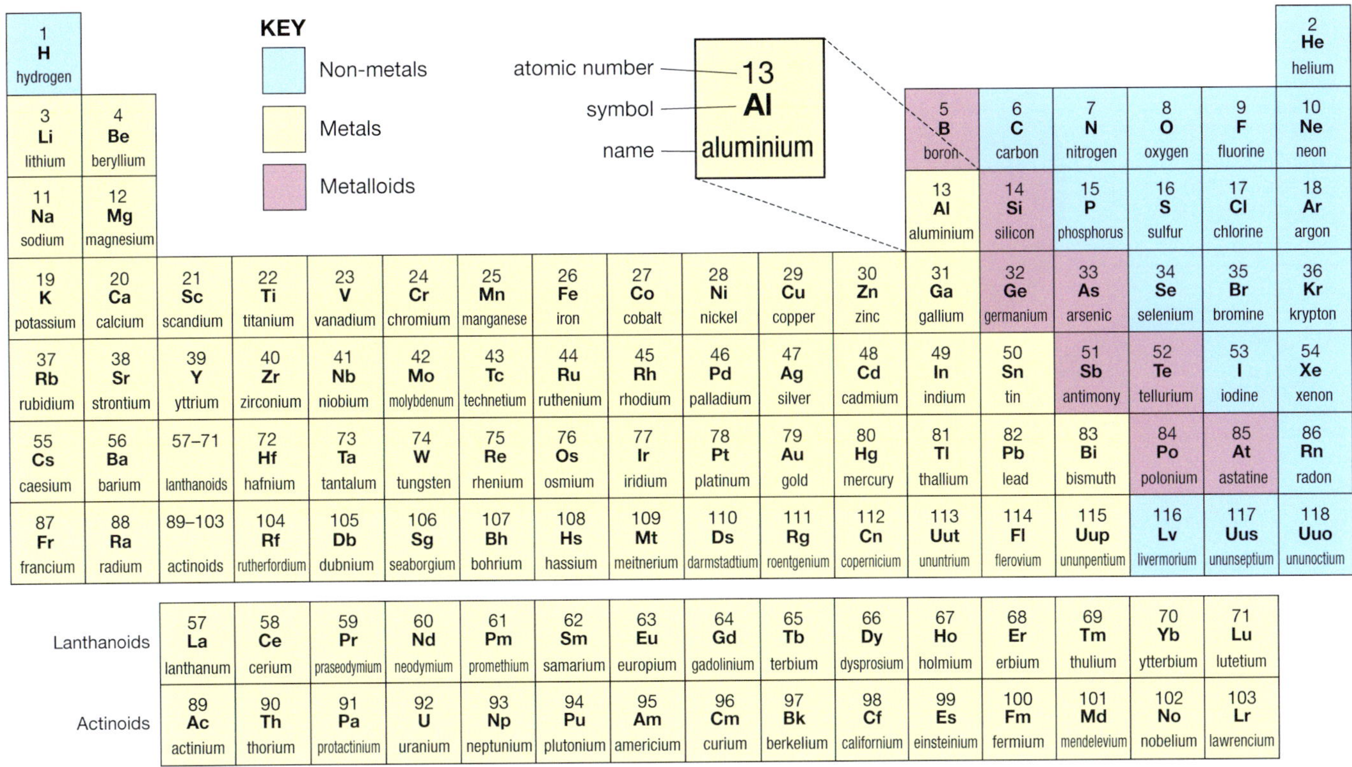

FIGURE 2.2.1 The periodic table displays all 118 known elements.

Metals

Metals are elements that are:

- **lustrous**—they shine when polished
- **malleable**—they can be bent into new shapes without breaking
- **ductile**—they can be stretched into wires.

These are just three of the physical properties that have made metals very valuable to humans throughout history. They form the basis of much of our technology and art, from horseshoes, swords, electrical wiring and the frames of skyscrapers to jewellery, statues and the gold leaf on paintings.

Figure 2.2.2 on page 54 outlines the physical properties shared by the metallic elements.

Pure metals

Table 2.2.1 on page 54 shows metals that are often used as pure elements. However, most metals cannot be used in their pure form. This is because they have properties that make them impractical. For example, most pure metals are too soft to be made into anything useful.

Alloys

Most of the metals around you are not pure elements but are alloys. An **alloy** is a metal (known as the **base metal**) combined with small amounts of other elements. The properties of the new alloy are usually an improvement over those of the base metal. For example, **steel** is much stronger and harder than its iron base metal, allowing it to be used in everything from paperclips, staples, nails and screws to cars, ship hulls and the frames of bridges and skyscrapers. Steel is an alloy of iron with small amounts of carbon added to it. Different amounts of carbon produce different steel alloys:

- Wrought iron contains almost no carbon and is the closest alloy to pure iron.
- Mild steel has only 0.5% carbon.
- Hard steel or tool steel has about 1% carbon.
- Cast iron has between 2.4% and 4.5% carbon. Cast iron is strong but **brittle**, shattering easily if hit or dropped.

Metals are dense. Almost all metals are denser than water and so will sink when dropped into it. The only exceptions are lithium (Li), sodium (Na) and potassium (K). These float on water.

Metals are thermal conductors. They pass heat easily along and through them.

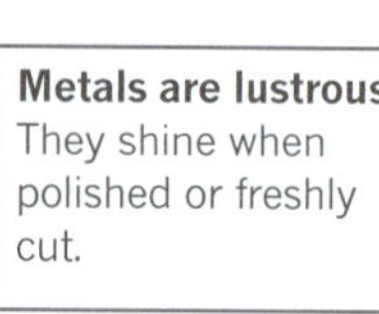

Metals are electrical conductors. They pass electricity along and through them.

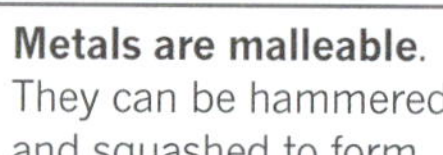

Metals are lustrous. They shine when polished or freshly cut.

Metals are malleable. They can be hammered and squashed to form new shapes.

Metals are ductile. They can be stretched and drawn into long thin wires.

Most metals are solid at room temperature. (Mercury (Hg) is an exception because it is a liquid.)

FIGURE 2.2.2 The physical properties of metals

TABLE 2.2.1 Pure metals and their uses

Pure metal	Uses	Properties that make the metal particularly suited to its use
aluminium Al	overhead electricity cables, saucepans and cans, aluminium foil	excellent conductor of heat and electricity, extremely light, non-toxic
copper Cu	electrical wiring, water pipes	excellent electrical conductor, easily stretched into wires
lead Pb	flashing around windows and roofs to stop water entry	very soft and easily bent, resists corrosion
mercury Hg	clinical thermometers, barometers, mercury switches	liquid at room temperature, expands rapidly when heated, leaving tubes clean once it retreats, leaving no traces
sodium Na	nuclear reactor coolant, street lamps (as a vapour)	good conductor of heat, melts at 98°C, allowing molten sodium to flow along pipes in the reactor
tin Sn	coating for steel cans used for storing food	stops steel from rusting, doesn't react with food, non-toxic
zinc Zn	coating for iron and steel (galvanised iron)	is more reactive than iron and so protects it from rusting

Steel can be further improved by adding chromium and nickel to it. This addition produces rust-resistant **stainless steel**. Stainless steel is used in hot, wet and salty environments that would cause rapid rusting of other types of steel. This is why stainless steel is used in kitchens, on ships, for surgical instruments and for jewellery for body piercings like those in Figure 2.2.3.

FIGURE 2.2.3 High-grade stainless steel doesn't rust and so is ideal for body piercings.

SciFile

Gold isn't always gold!

Australian 'gold' \$1 and \$2 coins contain 92% copper, 6% aluminium, 2% nickel and no gold. The 'silver' coins are 25% nickel, 75% copper and no silver. In contrast, the first circular 50-cent coins of 1966 were 80% silver. Eventually, this made them far more valuable as metal than as a coin!

Pure gold is so soft and fragile that any jewellery made from it would soon break. For this reason, silver or copper are added to it to create a stronger alloy. The **carat** scale measures the amount of pure gold in jewellery, with pure gold rated as 24 carat. Jewellery is often 18 carat, meaning that it is $\frac{18}{24}$ (three-quarters or 75%) gold.

Other alloys are shown in Table 2.2.2.

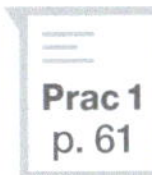
Prac 1 p. 61
AB 2.4

AB 2.5

TABLE 2.2.2 Alloys and their uses

Alloy	Composition	Uses	Advantages
brass	70% Cu, 30% Zn	hinges, door handles, fittings on boats and ships, musical instruments, e.g. trumpets and trombones	• good appearance • doesn't corrode much • stronger than its base metal (copper)
bronze	95% Cu, 5% Sn	statues, ornaments, bells	• good appearance • doesn't corrode easily • sonorous (makes a pleasant ringing sound when struck) • harder than brass • stronger than its base metal (copper)
duralumin	96% Al, 4% Cu, traces of Mg and Mn	aircraft frames	• very light • stronger than its base metal (aluminium)
solder	60–70% Sn, 30–40% Pb	joining metals together, electrical connections, low-friction bearings	• easy to melt • easy to use
cupronickel	75% Cu, 25% Ni	'silver' coins (5, 10, 20 and 50 cents)	• hard wearing • looks like silver
EPNS (electroplated nickel silver)	46–63% Cu, 18–36% Zn, 6–30% Ni	plated onto cutlery, plates and bowls	• looks like silver • cheaper than silver • resists corrosion
dental amalgam	43–54% Hg, 20–35% Ag, 10% Cu, 2% Zn, traces of Sn	tooth fillings	• hardens slowly after being mixed

SciFile

Mag wheels

Mag wheels (alloy wheels) are made from an alloy of magnesium and aluminium. This alloy is much lighter than the steel normally used for car wheels, making the car handle better. The alloy also conducts heat away from the brakes better than steel, keeping the brakes cooler and improving their performance.

Working with Science

ARTIST AND DESIGNER OF MEDICAL TECHNOLOGIES

Leah Heiss

Leah Heiss is a Melbourne-based artist and designer (Figure 2.2.4). She brings her artistic vision and knowledge of design to the world of science and medical technology. Leah creates medical devices that are both functional and beautiful. Medical devices, such as hearing aids and insulin delivery systems for diabetics, are often unattractive, bulky and impersonal, yet necessary for many people's wellbeing. Leah hopes that her designs give users more discreet and attractive options for health management.

FIGURE 2.2.4 Leah Heiss is an artist and designer who creates medical devices that are both functional and beautiful.

For her designs, Leah combines a variety of materials, such as magnetic liquid, electricity-conducting textiles, optic fibres, resin and silver, with manufacturing technology such as 3D printing. She has used these materials and many others to design a variety of artistic and functional products. Examples include a necklace that removes arsenic from drinking water, emergency jewellery that communicates the wearer's medical needs, a device that enables hearing aid users to adjust their hearing program to suit their environment and diabetes jewellery that delivers insulin pain-free through micro needles (Figure 2.2.5). Leah's innovative approach to design opens up many exciting possibilities for users and developers of medical technology.

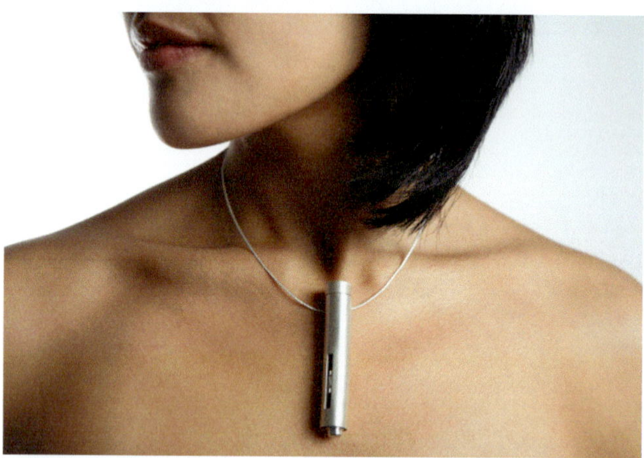

FIGURE 2.2.5 Diabetes jewellery designed by Leah Heiss. The Diabetes Neckpiece is an applicator device that applies small, circular discs covered in micro needles that deliver insulin through the skin.

Review

1 How does Leah bring artistic design and science in her work? Why is this important?
2 List the medical devices that Leah has created. For each, indicate why the device is needed and, if possible, how it functions.

Non-metals

Most non-metals are found naturally as gases in the air. A few are solids found in the Earth's crust, such as the sulfur that occurs around volcanoes. The physical properties of non-metals are very different from those of metals. You can see these properties in Figure 2.2.6.

Non-metals are poor conductors of heat and electricity.
They are thermal and electrical insulators.

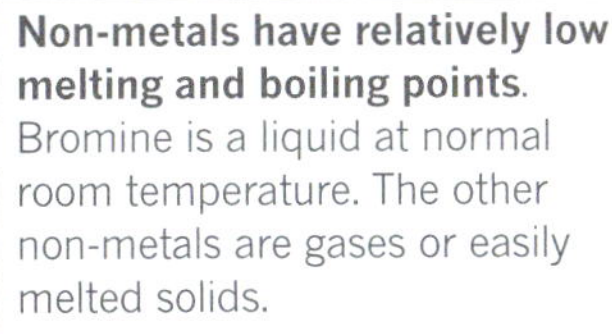

Non-metals are brittle.
Solid non-metals tend to crumble into powders.

Non-metals are dull.
They have little or no shine.

FIGURE 2.2.6 The physical properties of non-metals

Carbon

Carbon is an unusual element because its atoms combine with other carbon atoms and with atoms of other elements (usually hydrogen and oxygen) to form lattices, long chains and rings. Over 90% of all known compounds contain carbon. Some of these compounds are essential to life on Earth. Carbon exists in molecules in every living thing and anything that was once part of a living thing.

Pure carbon exists in several different forms, called **allotropes**. Three common allotropes are:

- amorphous carbon
- diamond
- graphite.

These are shown in Figure 2.2.7 on page 58.

SciFile

Carbon wheels!

In 2010, Deakin University in Geelong (Victoria) and research firm CFusion released the world's first car wheel constructed from a *single* carbon fibre. Because it is incredibly light yet strong, the wheel promises to dramatically enhance car performance.

Graphite
A soft, slippery solid that conducts electricity. It is an excellent lubricant and forms the electrodes in many batteries and the connection brushes in electric motors.

The grey 'lead' in pencils is a graphite/clay mix.

FIGURE 2.2.7 Some of the forms in which carbon exists

Prac 2 p. 62 Prac 3 p. 63

Metalloids

Metalloids act like non-metals in most ways. However, they also have some properties that are more like those of metals. Most importantly, metalloids are semiconductors, meaning that they can conduct electricity under certain conditions.

This ability has made silicon and germanium ideal materials from which to build electronic components like the one shown in Figure 2.2.8. These components are used in devices such as laptops, LED TVs and smartphones.

SciFile

Diamond destruction!

The English scientist Sir Humphry Davy (1778–1829) demonstrated that diamond was a form of carbon by burning a diamond that belonged to his wealthy wife! All that was left was carbon dioxide. Temperatures of about 800°C are required to convert diamond to graphite. Unfortunately it's much, much harder to turn graphite into diamond.

FIGURE 2.2.8 This electronic microprocessor chip is constructed from the metalloid silicon.

SCIENCE AS A HUMAN ENDEAVOUR

Use and influence of science

Cactus power for electric cars

Materials scientists from CSIRO and Hanyang University in Korea have developed a membrane inspired by the specialised water-retaining qualities of cactus skin (Figure 2.2.9).

FIGURE 2.2.9 Materials scientist from the CSIRO, Dr Cara Doherty, examines a cactus-inspired membrane.

FIGURE 2.2.10 A synthetic membrane, inspired by the water-retaining qualities of cactus skin, shows promising results for efficient water management in fuel cells for electric cars.

Cacti are adapted to survive in hot, dry conditions, holding as much water as they can. Their thick skin has specialised pores and is critical to their survival (Figure 2.2.10). During the day, cacti close the pores on the surface of their skin to prevent water loss. At night, when it is cooler, cacti open the pores to absorb moisture from the air. The scientists investigated the characteristics of the cactus skin at a molecular level to understand how they could mimic its specialised features in synthetic membranes. Synthetic membranes are membranes that are produced artificially rather than naturally.

One of the applications of synthetic membranes is in the fuel cells that power electric cars. Like cacti, these fuel cells often have to work in hot, dry conditions. One of the biggest challenges for electric car technology is the efficiency of their fuel cells in these conditions. In order to work, fuel cells need to be constantly hydrated. The current technology requires a water reservoir and humidifier to be placed next to the fuel cell. The humidifier is a device that keeps the atmosphere around the fuel cell moist. However, this takes up a lot of energy and space in the car. The cactus-inspired membrane offers an effective alternative.

The membrane works in a similar way to cactus skin, with pores that close when it is hot and dry, and that open when it is cool and humid. This allows the fuel cells to stay moist without an external water reservoir and humidifier. The membrane allows fuel cells to function up to four times more efficiently, allowing them to power the car four times longer than at present. This material also has potential uses for other technologies that require hydrated membranes. Examples are water treatment and gas separation devices.

REVIEW

1 Why do you think research and development in technologies like electric cars are important?

2 What are the qualities of cactus skin that inspired scientists in their design of fuel cells?

3 Imitating the systems or structures of living organisms to solve human problems is called biomimicry. Using the cactus skin as inspiration helped scientists develop the synthetic membrane for fuel cells. List any other inventions you think may have been inspired by nature.

MODULE 2.2

Review questions

Remembering

1 Define the terms:
  a lustrous
  b malleable
  c brittle.

2 What term best describes each of the following?
  a the main metal in an alloy
  b the measure of the amount of pure gold
  c able to be stretched into wires.

3 How many different elements are there?

4 List the names and symbols of three metals, three non-metals and three metalloids.

5 What is the only metal that is liquid at normal room temperature?

6 Arrange the different types of steel, from the lowest carbon content to the highest.

7 For stainless steel, name the:
  a base metal
  b added metals that give it rust resistance.

Understanding

8 Explain why most metals sink in water.

9 Why is gold rarely used in its pure form?

10 Explain why the slipperiness of graphite makes it ideal for use in grey-lead pencils.

Applying

11 Identify two physical properties that make metals the ideal material from which to construct electrical wires.

12 Identify the metal common to both the alloys brass and bronze.

13 Calculate the fraction and percentage of pure gold in a:
  a 12-carat gold ring
  b 9-carat gold nose stud
  c 22-carat gold chain.

14 Wood, paper and food scraps all burn, leaving charcoal and ash behind. This suggests that they all have the same basic element in them. Identify what that element is.

15 Iron and steel rust in the presence of water and oxygen. Use this information to predict how much rusting would occur to steel in the:
  a body of a car left in the desert
  b hull of a sunken ship buried in mud so dense that there is no oxygen in it.

16 Below is a list of different atoms. Their element symbols have been replaced with the letters A–E.

$^{12}_{5}A$ $^{25}_{12}B$ $^{14}_{7}C$ $^{13}_{5}D$ $^{25}_{15}E$

  a State how many different elements are represented in the list.
  b Use the periodic table on page 53 to identify the different elements represented in the list.

Analysing

17 Compare the number of elements that are metallic, non-metallic and metalloids.

18 Classify the following as normally properties of metals or non-metals:
  a ductile
  b normally gas or liquid
  c dense
  d malleable
  e brittle
  f lustrous
  g dull
  h most are solid
  i thermal and electrical insulators
  j excellent thermal and electrical conductors.

Evaluating

19 Cans that contain soup, dog food or vegetables are made mostly of steel, yet are often called tins. Propose a reason why.

20 Graphite is carbon (a non-metal) but it conducts electricity like a metal. Use this information to propose a reason why carbon could be classified as a metalloid instead of a non-metal.

21 Propose what would be the base metal in a ferrous alloy. (Use the element symbols of metals to help you.)

MODULE 2.2

Practical investigations

1 • Making steel stronger

Processing & Analysing

Heating changes the properties of steel because it changes the size of its crystals.

Purpose

To determine which treatment makes steel tougher.

Timing

45 minutes

Materials

- four steel hairpins
- steel wool

- Bunsen burner, bench mat and matches
- wooden peg
- beaker, tub or sink filled with cold water
- pliers (optional)

SAFETY

The hairpin will get red-hot so use a peg at all times to hold it. Water may spit when the hot pin is dropped into it so wear safety glasses at all times.

Procedure

1 Copy the table from the Results section into your workbook.

2 Count the number of times you can bend a hairpin before it snaps. One bend is counted as opening the hairpin up then closing it again. Enter the number in your table.

3 Hold another hairpin with the peg and heat the bend of the pin in a blue Bunsen burner flame until it is red-hot (Figure 2.2.11). Allow it to cool on the bench mat. This process is known as normalising or annealing.

4 Heat another hairpin in the same way, then cool it rapidly by dropping it into a beaker of water. This process is known as **quenching**.

5 Repeat step 4 with the remaining hairpin, then polish the bend with steel wool. Reheat the bend of the pin, removing the pin occasionally to check whether the bend has gone blue. Once it has, remove the pin from the flame and allow it to cool on the mat. This process is known as **tempering**.

6 Bend each of the pins as before, counting the number of times you can bend the pin before it breaks. Record your counts in the results table.

Results

Record all your observations in a table like the one below.

Testing the properties of steel

Treatment	Number of bends needed to break pin	Did the treatment make the pin tougher?
no treatment		
normalising/ annealing		
quenching		
tempering		

Review

1 Compare the processes of annealing, quenching and tempering.

2 Which treatment caused your hairpin to become more:

a brittle (easier to snap)

b malleable (more 'bendy' and less likely to snap)?

3 Fast cooling produces small crystals; slow cooling produces bigger ones. Predict which of the treatments produced the biggest crystals.

4 Propose reasons why bigger crystals make steel tougher than small crystals.

5 Blacksmiths repeatedly heat, hammer and cool (quench) steel when making horseshoes. Propose a reason why.

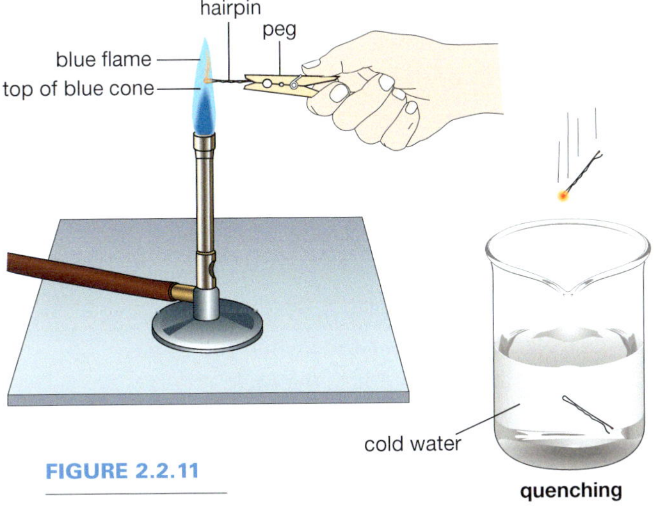

FIGURE 2.2.11

MODULE 2.2

Practical investigations

2 • Making oxygen

Purpose

To prepare and test oxygen gas.

Timing

45 minutes

Materials

- 5 mL hydrogen peroxide solution
- 1 g manganese dioxide pellets

- 1 large test-tube, rubber stopper with opening and glass tube to fit
- hosing to fit glass tube
- 2 test-tubes with stoppers
- test-tube rack
- retort stand, bosshead and clamp
- large container (such as an ice-cream container)
- 10 mL measuring cylinder
- wooden splint
- electronic balance
- rubber gloves

Hydrogen peroxide burns and is toxic. It can explode when heated and may cause fires if in contact with combustible materials. Wear safety glasses, protective clothing and rubber gloves.

Procedure

PART A: PREPARATION OF OXYGEN

1. Use the electronic balance to weigh out approximately 1 g of manganese dioxide pellets.
2. Use the measuring cylinder to carefully measure out 5 mL of hydrogen peroxide.
3. Set up the equipment as shown in Figure 2.2.12.
4. Fill both the two smaller test-tubes with water. Put your thumb over the end on one, upend it and clamp as shown. Put the other one in the test-tube rack for later on.
5. Remove the rubber stopper and drop the manganese dioxide pellets into the large test-tube.
6. Add the hydrogen peroxide and replace the rubber stopper.
7. The inverted test-tube should fill with oxygen gas. Remove the test-tube when full of gas, stopper it and place it in the rack.
8. Fill the other test-tube with oxygen and store it in the rack.

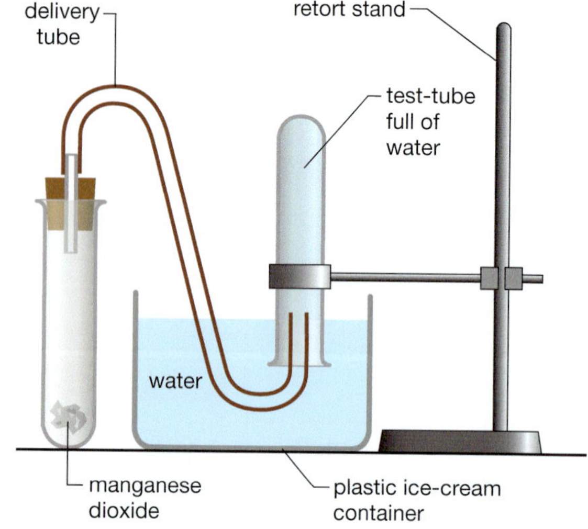

FIGURE 2.2.12

PART B: TESTING OXYGEN

9. Use one tube of collected gas to make as many observations as you can about oxygen. For example, waft the gas towards you and attempt to smell it.
10. Light the wooden splint, allow it to burn for a few seconds and then blow it out. Insert the glowing end of the splint into the second test-tube of oxygen and record what happens.

Results

Record the state, colour and smell of oxygen gas and what it did to the glowing splint.

Review

1. Use your observations to explain why fanning a fire encourages it to burn.
2. The equation for the reaction in this experiment is:

   manganese dioxide + hydrogen peroxide →
   manganese dioxide + water + oxygen gas

   $$MnO_2 + 2H_2O_2 \rightarrow MnO_2 + 2H_2O + O_2$$

   a. Apart from oxygen, what else is produced in this reaction?
   b. Manganese dioxide doesn't actually take part in this reaction but the reaction won't take place without it. Manganese dioxide is a catalyst. Given this information, what do you think is the role of a catalyst?

MODULE 2.2

Practical investigations

• STUDENT DESIGN •

3 • The better conductor

Questioning & Predicting | Evaluating

Purpose

To find out whether wood or graphite is the better conductor of heat and electricity.

Hypothesis

Which substance do you think will conduct heat and electricity better—graphite or wood? Before you go any further with this investigation, write a hypothesis in your workbook.

Timing 60 minutes

Materials

- To be selected by students

SAFETY

A risk assessment is required for this investigation.

Procedure

1 Design an experiment that will test how well wood and graphite conduct heat and electricity (Figure 2.2.13).

FIGURE 2.2.13 Which of these materials is the better conductor?

2 Brainstorm in your group and come up with several different ways to investigate the problem. Select the best procedure and write it in your workbook. Draw a diagram of the equipment you need.

3 Before you start any practical work, assess all risks associated with your procedure. Refer to the SDS of all chemicals used. Construct a risk assessment that outlines these risks and any precautions you need to take to minimise them. Show your teacher your procedure and your risk assessment. If they approve, then collect all the required materials and start work.

See the Activity Book Toolkit to assist with developing a risk assessment.

Hints

- The grey 'lead' in pencils is graphite.
- You will need to construct a simple electric circuit that includes a battery or low-voltage power pack and a light globe.
- Use the STEM and SDI template in your eBook to help you plan and carry out your investigation.

Review

1 a Construct a conclusion for your investigation.

b Assess whether your hypothesis was supported or not.

2 Evaluate your procedure. Pick two other prac groups and evaluate their procedures too, identifying their strengths and weaknesses.

3 The handles of screwdrivers were once made of wood. Use the results of this investigation to propose a reason why.

MODULE

2.3 Acids and bases

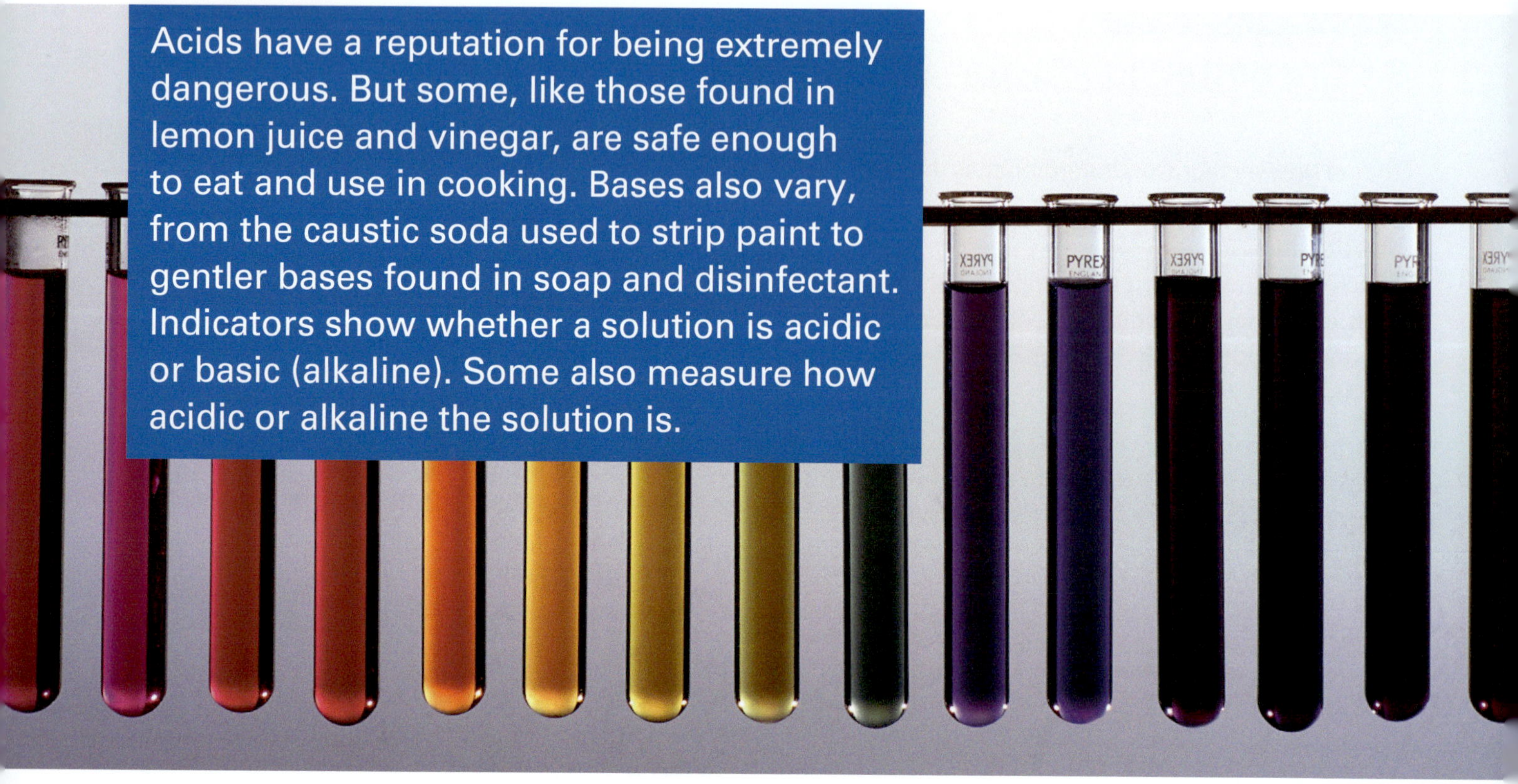

Acids have a reputation for being extremely dangerous. But some, like those found in lemon juice and vinegar, are safe enough to eat and use in cooking. Bases also vary, from the caustic soda used to strip paint to gentler bases found in soap and disinfectant. Indicators show whether a solution is acidic or basic (alkaline). Some also measure how acidic or alkaline the solution is.

Acids

An **acid** is a substance that releases hydrogen ions (H^+) into an aqueous solution (containing water). Examples are the hydrochloric acid that's in your stomach and the acetic acid (ethanoic acid) found in vinegar.

Properties of acids

Acids have similar chemical properties. Acids:

- are corrosive. An acid burn is shown in Figure 2.3.1.
- have a sour taste (think of the taste of vinegar)
- turn blue litmus paper red (Figure 2.3.2)
- react with some metals, releasing hydrogen gas and leaving a salt behind
- conduct electricity
- are neutralised by bases, This means that the acid is made 'safe' by converting it into water and a salt.

Acids are molecular compounds made up of atoms from different elements. For example, a molecule of nitric acid (HNO_3) contains one hydrogen atom, one nitrogen atom and three oxygen atoms. Like nitric acid, all acids have hydrogen atoms in their molecules.

You will almost certainly work with nitric acid in the laboratory, as well as sulfuric acid (H_2SO_4), hydrochloric acid (HCl) and acetic or ethanoic acid (CH_3COOH). These acids are not pure substances but are aqueous solutions of acid mixed with water. When mixed with water, some of the hydrogen atoms in the acid molecule are released to form **hydrogen ions** (H^+).

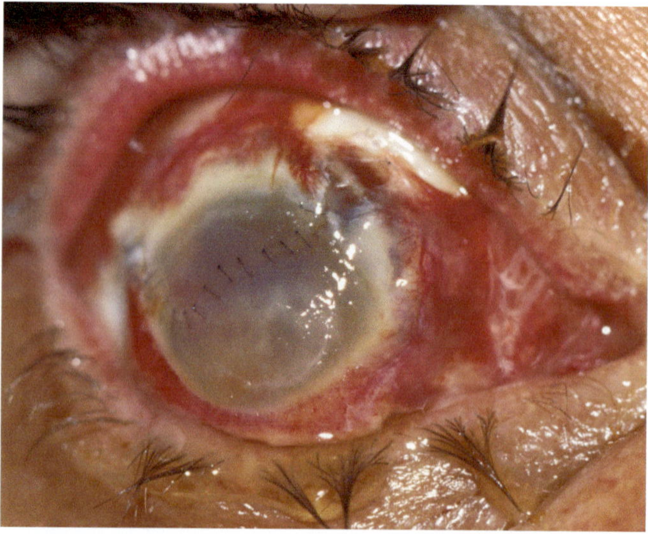

FIGURE 2.3.1 Acid burns can be severe, particularly if the acid is spilt into sensitive tissue such as in the eye.

FIGURE 2.3.2 Acid changes blue litmus paper to red.

The strength of an acid depends on how many hydrogen ions are released. An acid is strong if most of its molecules release hydrogen ions into solution. Nitric acid is an example of a strong acid, as are hydrochloric acid and sulfuric acid. In contrast, an acid is weak if only a few of its molecules release hydrogen ions. An example of a weak acid is vinegar (an aqueous solution of acetic or ethanoic acid).

Figure 2.3.3 compares what solutions of hydrochloric acid and acetic acid would look like if you could see their ions. Other examples of strong and weak acids are listed in Table 2.3.1 on page 66.

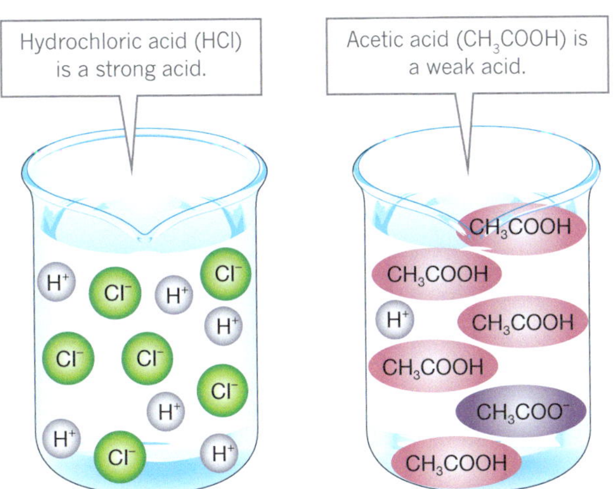

FIGURE 2.3.3 Strong acids such as hydrochloric acid release lots of H^+ ions into solution. Weak acids such as acetic acid (vinegar) release very few H^+ ions.

Bases and alkalis

Ions are not always single 'charged atoms'. Ions can also be charged groups of atoms. This type of ion is known as a **polyatomic** ion (*poly* means 'more than one'). An example is the **hydroxide ion** (OH^-).

A **base** is a substance that releases hydroxide ions (OH^-). You use a weak base every time you use soap or toothpaste. When a base can be dissolved in water, it is also known as an **alkali**. The solution it forms is known as an **alkaline solution**. Bases such as caustic soda can burn you as badly as acids can, and so bases need to be treated with as much care as acids. All bases have similar chemical properties.

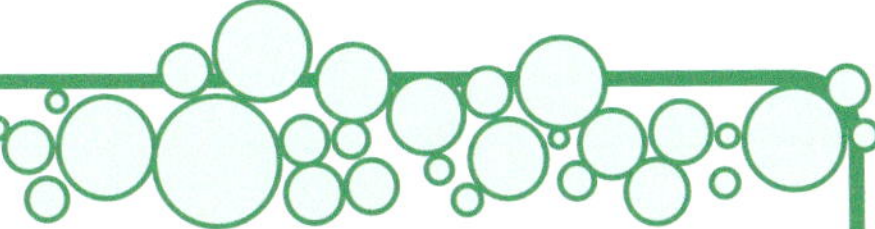

Animal acids and bases

A bite from a bull-ant hurts because the ant injects formic acid (also known as methanoic acid, HCOOH) into a cut made with its pincers. A bee sting also contains methanoic acid. Wasps and jellyfish inject a base. It's a different chemical but it still hurts!

TABLE 2.3.1 Examples of acids

Strong acids		
Acid	**Chemical formula**	**Used for/found in**
hydrochloric	HCl	• cleaning mortar off bricks • your stomach (part of its gastric juices)
nitric	HNO_3	• making fertilisers, dyes and explosives
sulfuric	H_2SO_4	• making other chemicals, dyes, fertilisers, synthetic fibres and plastics

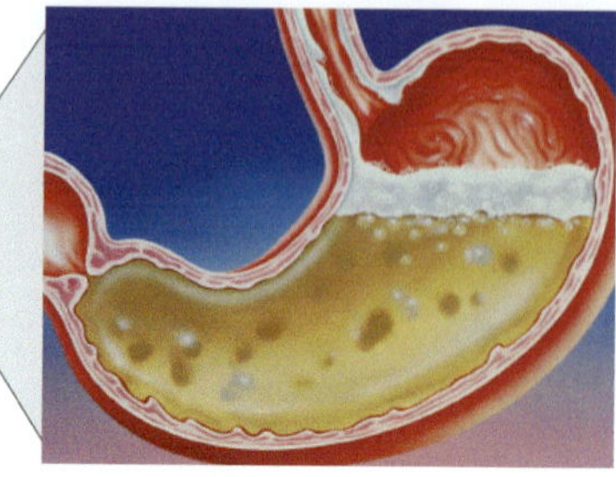

Weak acids		
Acid	**Chemical formula**	**Use for/found in**
ascorbic	$C_6H_8O_6$	• vitamin C
acetylsalicylic	$C_9H_8O_4$	• making aspirin
carbonic	H_2CO_3	• rain water • fizzy soft drinks and beer
citric	$C_6H_8O_7$	• citrus fruits (such as lemons, limes, oranges) • tomatoes
acetic (ethanoic)	CH_3COOH	• vinegar
malic	$C_4H_6O_5$	• apples • most unripe fruits
lactic	$C_3H_6O_3$	• milk, yoghurt • your muscles after heavy exercise, making them hurt
tannic	$C_{76}H_{52}O_{46}$	• wood stains • tea
tartaric	$C_4H_6O_6$	• grapes, bananas

Bases:

- are caustic
- have a soapy, slimy feel
- turn red litmus paper blue (Figure 2.3.4)
- have a bitter taste
- conduct electricity
- are neutralised by acids, producing water and a salt.

Bases form hydroxide ions (OH^-) in solution. Strong bases produce lots of OH^- ions, while weak bases only produce a few. Some strong and weak bases are shown in Table 2.3.2.

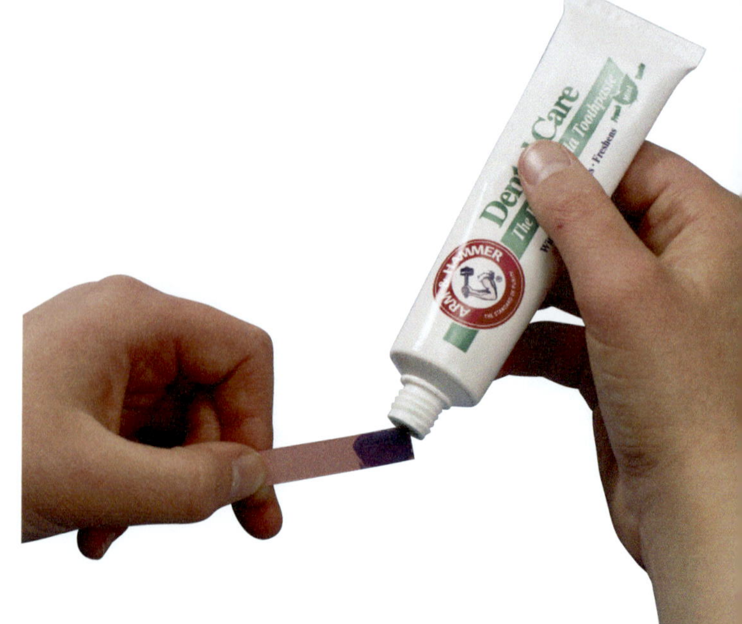

FIGURE 2.3.4 Alkaline solutions turn red litmus paper blue.

TABLE 2.3.2 Examples of bases and alkalis

Strong bases/alkalis		
Base/alkali	**Chemical formula**	**Used for/found in**
calcium hydroxide	$Ca(OH)_2$	• cement, mortar and concrete • stripping hair from hides to form leather • paper production
sodium hydroxide (caustic soda)	NaOH	• producing soap • paint stripper • drain and oven cleaner

Weak bases/alkalis		
Base/alkali	**Chemical formula**	**Used for/found in**
ammonia ammonium hydroxide	NH_3 NH_4OH	• household cleaners
sodium hydrogen carbonate (sodium bicarbonate, bicarbonate of soda or baking soda)	$NaHCO_3$	• baking, to make cakes rise
magnesium hydroxide (milk of magnesia)	$Mg(OH)_2$	• antacids
sodium carbonate	Na_2CO_3	• washing powders

pH

The concentration of hydrogen ions (H^+) in a solution is measured using the **pH** scale. In an acidic solution, there are more hydrogen ions than hydroxide (OH^-) ions. In contrast, an alkaline solution has more hydroxide ions than hydrogen ions.

Pure water is neither an acid nor a base. It is neutral, having equal numbers of hydrogen and hydroxide ions. It has a pH of 7. As Figure 2.3.5 shows, acids have a pH less than 7, while bases and alkaline solutions have a pH greater than 7.

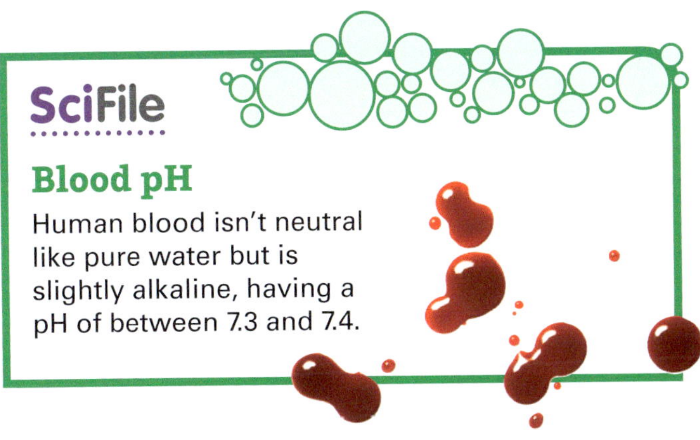

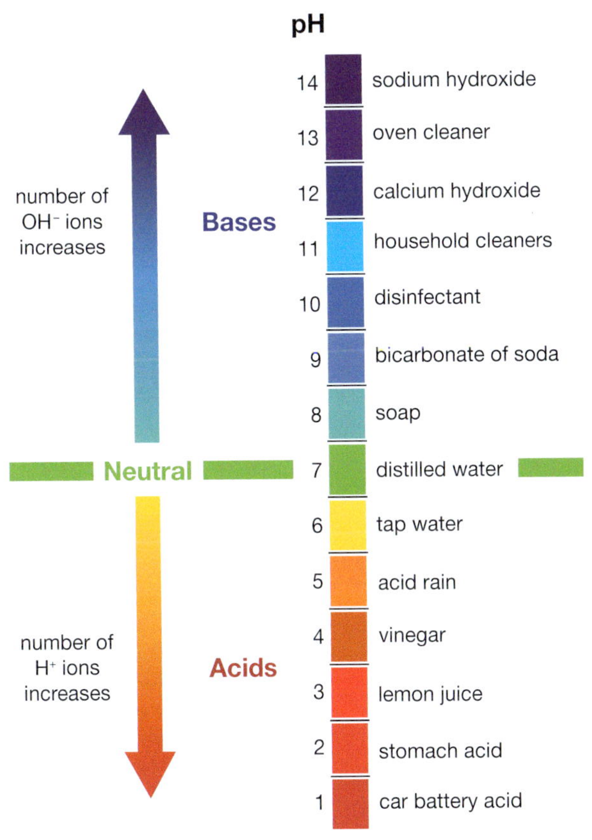

FIGURE 2.3.5 Neutral solutions have a pH of 7. Acidic solutions have a pH less than 7. Alkaline solutions have a pH greater than 7.

Measuring pH

Indicators are chemicals that change colour to show whether a substance is acidic, neutral or basic. A common indicator is **litmus paper,** which turns red when dipped into acids and blue when dipped into a base. While litmus doesn't tell you what the pH of a solution is, other indicators do. As Figure 2.3.6 shows, different indicators change colour at different pH values.

Another way of measuring pH is to use a pH meter. One is being used in Figure 2.3.7.

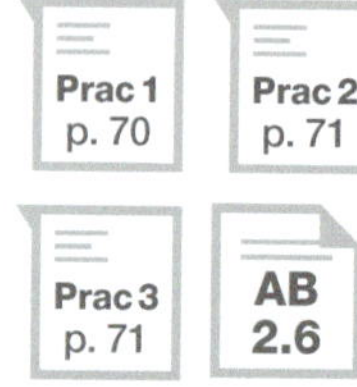

Indicator	Colour of indicator													
	pH													
	1	2	3	4	5	6	7	8	9	10	11	12	13	14
Bromothymol blue														
Litmus														
Methyl orange														
Phenolphthalein				colourless										
Universal indicator														

FIGURE 2.3.6 Different indicators have different colours, allowing pH to be determined accurately.

FIGURE 2.3.7 Pool water pH needs to be regularly monitored to ensure that the water is safe for swimmers.

science 4 fun

Acid or base?

Are the different solutions found around your home acidic or basic/alkaline?

Collect this ...

- samples of various household solutions (such as fruit juices, soft drink, sour and fresh milk, tap water, salad dressing, detergent, shampoo)
- litmus paper (blue and red)
- watch-glass or white tile

Do this ...

1. Pour a little of each solution onto the watch-glass or white tile.
2. Touch one end of a small strip of litmus paper into the solution and then remove it.
3. Record the colour change.

Record this ...

1. Describe what happened.
2. Explain what this tells you about each of the samples you tested.

MODULE 2.3

Review questions

Remembering

1 Define the terms:
  a acid
  b indicator
  c hydrogen ions.

2 What term best describes each of the following?
  a a substance that release OH^- ions
  b a solution of a base dissolved in water
  c an ion with more than one atom.

3 Name the acid that is in:
  a vinegar
  b milk
  c lemons.

4 Name the following acid and bases.
  a CH_3COOH
  b NaOH
  c NH_3

5 List the names and chemical formulas of two strong:
  a acids
  b bases.

6 Name the base that is in:
  a paint stripper
  b cement
  c baking soda.

7 What is the ion formed by the following substances?
  a acids
  b bases.

Understanding

8 Explain why you have a sour taste in your mouth when you vomit.

9 Predict whether litmus paper will turn red or blue when dipped in:
  a washing powder (containing sodium carbonate)
  b orange juice (containing citric acid)
  c lemonade (containing H_2CO_3)
  d cleaner (containing NH_3).

10 Why is universal indicator more useful than litmus?

Applying

11 Identify an example of an ion that is:
  a a single atom that has become charged
  b polyatomic.

12 Use Figure 2.3.6 to identify the colour that the following indicators would be at pH 4.
  a blue litmus
  b phenolphthalein
  c universal indicator.

Analysing

13 Compare the number of H^+ ions in a solution of nitric acid with the number found in ethanoic acid (vinegar) of the same concentration.

14 The most common isotope of hydrogen is ^1_1H. A hydrogen ion H^+ is a hydrogen atom that has lost its single electron. Analyse this information and identify the subatomic particle that makes up a typical hydrogen ion.

Evaluating

15 Heartburn has nothing to do with your heart. It is caused by gastric juices rising from the stomach into the oesophagus. What do you think causes the pain of heartburn?

16 The pH of most public pools is measured using a pH meter, not an indicator. Propose reasons why.

17 Squashed ants have a distinctive smell. What chemical do you think causes the smell?

18 Propose reasons why bricklayers commonly wear gloves when working.

19 Nitric acid is a strong acid but a dilute solution of it might have exactly the same pH as a concentrated solution of vinegar, which is a weak acid. Propose a reason why.

Creating

20 Construct a symbol (that uses no words) to be used on a sticker that would warn people that a bottle contained a concentrated solution of a strong acid like sulfuric acid.

MODULE **2.3**

Practical investigations

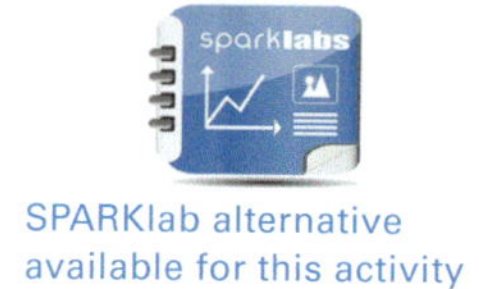

SPARKlab alternative available for this activity

1 • Red cabbage indicator 

Purpose

To make an indicator from red cabbage.

Timing 60 minutes

Most chemicals in this prac are corrosive or caustic, so wear rubber gloves, protective clothing and safety glasses at all times.

Materials

- a few millilitres each of dilute (0.1 M) hydrochloric acid, dilute (0.1 M) sodium hydroxide solution, vinegar, salt solution, distilled water, soft drink and lemon juice
- 1 antacid tablet (such as Alka Seltzer)
- red cabbage leaves (or red flower petals such as carnation, rose or geranium)

- 250 mL beaker
- eyedropper
- hotplate or Bunsen burner, tripod, gauze mat and bench mat
- 8 test-tubes
- test-tube rack

Procedure

PART A: MAKING THE INDICATOR

1 Tear up one or two red cabbage leaves, and place them in the beaker with enough water so that the cabbage is just covered (Figure 2.3.8).

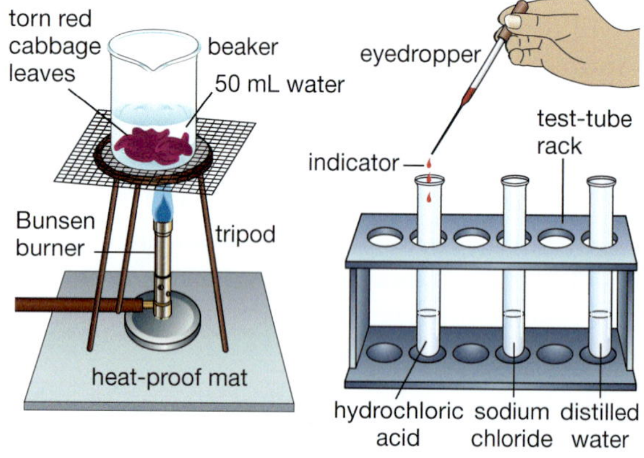

FIGURE 2.3.8

2 Heat the beaker until the water is gently boiling. Continue to boil the water until it has been strongly coloured red by the cabbage leaves.

3 Allow the water to cool and then filter, strain or pick out the cabbage leaves.

PART B: TESTING THE INDICATOR

4 In your workbook, construct a table like the one in the Results section.

5 To the first test tube, use an eyedropper to add about 1 cm of dilute hydrochloric acid.
To the second test tube, add 1 cm of vinegar.
To the third test tube, add 1 cm of distilled water.
To the fourth test tube, add 1 cm of salt solution.
To the fifth test tube, add 1 cm of sodium hydroxide solution.

6 Record what colour the mixture turns in your Results table.

PART C: TESTING UNKNOWNS

7 Add about 1 cm of lemon juice to the sixth test-tube.

8 Add about 1 cm of soft drink to the seventh test-tube.

9 Drop an antacid tablet into the eighth test-tube.

Results

Record your observations in a table like the one below.

Using red cabbage/petal indicator

Test-tube/type of solution	Name of solution	Colour with red cabbage indicator
1 0.1 M strong acid	hydrochloric acid solution	
2 weak acid	vinegar	
3 neutral	distilled water	
4 weak base	salt solution	
5 0.1 M strong base	sodium hydroxide solution	
6 (unknown 1)	lemon juice	
7 (unknown 2)	soft drink	
8 (unknown 3)	antacid	

Review

From their colours, identify which acid or alkaline solution the lemon juice, soft drink and antacid were most similar to.

MODULE 2.3

Practical investigations

2 • Green eggs

Purpose

To use indicators to turn the whites of fried eggs green.

Timing 30 minutes

Materials

- a few millilitres of cooking oil
- 1 raw egg
- red cabbage indicator from Prac 1

The eggs might not be fresh, so do not taste or eat them. Wash your hands thoroughly afterwards.

- small aluminium foil pie dish
- eyedropper
- hotplate or Bunsen burner, bench mat, tripod and gauze mat
- digital camera or mobile phone

Procedure

1. Put a little oil in the aluminium foil pie dish and crack an egg into it. Try to keep the egg yolk intact.
2. Place the pie dish on the hotplate or over the Bunsen burner on a gauze mat and tripod.
3. Gently cook the egg without stirring. As *soon* as the clear liquid part of the egg starts to turn white, use the eyedropper to place a few drops of red cabbage indicator into it.

Results

Use a digital camera or mobile phone to record your observations in both parts of this experiment through photographs or film.

Review

Red cabbage indicator turns red in acid solution, purple in neutral solution and green in basic (alkaline) solution. Identify whether egg white (the material that surrounds the yolk) is acidic, neutral or alkaline.

3 • pH column

Purpose

To construct a series of coloured layers of different pH.

Timing 30 minutes

Materials

- 2 or 3 rice-sized grains of solid sodium carbonate
- 10 mL vinegar
- universal indicator

SAFETY

Sodium carbonate is caustic, so wear rubber gloves, protective clothing and safety glasses at all times.

- 100 mL measuring cylinder
- spatula
- long stirring rod (such as a chopstick)

Procedure

1. Add 90 mL water and 10 mL vinegar to the measuring cylinder.
2. Add a drop of universal indicator.
3. Dissolve 2 or 3 rice-sized grains of solid sodium carbonate (Na_2CO_3) in a small amount of water (maximum 20 mL). Pour into this sodium carbonate solution into the measuring cylinder.
4. Gently add distilled water to the cylinder until it reaches around 90 mL.
5. Gently stir with the stirring rod, trying not to disturb the layers too much.
6. Add a drop of universal indicator.
7. Gently add 10 mL of vinegar to the 90 mL already in the cylinder.
8. Leave the measuring cylinder in a safe place where it will not be disturbed for a few days.

Results

1. After a day, four or five different-coloured layers should be clearly visible. Construct a diagram showing these layers.
2. Identify and label the pH of each band.

Review

1. Describe what happens to the pH as you move towards the top of the measuring cylinder.
2. Explain why the lower layers would be more basic (alkaline) and the top layers more acidic.

CHAPTER 2

Chapter review

Remembering

1 Define the terms:
  - a atom
  - b molecule
  - c crystal lattice
  - d ion
  - e cation
  - f alkali.

2 What are the three types of subatomic particles that make up atoms?

3 Fill in the following statements to show how the atomic number and mass number of an atom are calculated.

  Atomic number = number of ______

  Mass number = number of ______ + ______

4 Name the following chemicals.
  - a CH_3COOH
  - b H_2SO_4
  - c NaOH

5 Name three indicators.

6 What is the pH of pure water?

Understanding

7 Draw a simple diagram that shows the structure of an atom.

8 Name the force in the atom that:
  - a keeps the electrons within the atom
  - b should rip the nucleus apart
  - c keeps the nucleus together.

9 Describe Rutherford's famous experiment and how it contributed to our current understanding of the atomic model.

10 Describe what must happen to an atom to make it:
  - a a cation
  - b an anion.

11 Why are alloys usually better for most purposes than their base metals?

12 Describe why some acids are strong while other acids are weak.

Applying

13 Identify the chemical formulas for these acids and bases.
  - a hydrochloric acid
  - b nitric acid
  - c calcium hydroxide.

Analysing

14 Compare protons, neutrons and electrons, listing their similarities and differences.

15 Classify the following as elements or compounds.
  - a Fe
  - b NaOH
  - c H_3PO_4
  - d O_2

16 A solution was tested with different indicators. The colours they turned were:

  Litmus = purple

  Methyl orange = yellow

  Phenolphthalein = colourless

  Bromothymol blue = blue.
  - a Use this information to identify the pH of the solution.
  - b Classify the solution as acidic, neutral or alkaline.
  - c Predict the colour that universal indicator would turn if it was added to the solution.

17 Compare acids with bases by listing their similarities and differences.

Evaluating

18 Carbon has been known about for over 2000 years. Propose reasons why it was found much earlier than most other non-metals.

19 a Assess whether you can or cannot answer the questions on page 39 at the start of this chapter.
  - b Use this assessment to evaluate how well you understand the material presented in this chapter.

Creating

20 Use the following ten key terms to construct a visual summary of the information presented in this chapter.

metals	carbon
non-metals	hydrogen ion
acids	alloys
diamond	atoms
ions	hydroxide

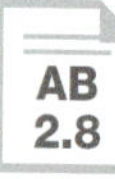

CHAPTER 2 Inquiry skills

Research

1 Processing & Analysing

Use the following search terms to find internet videos on acids, bases and pH: *acid-base video, pH video*.

2 Processing & Analysing

Use the following search terms to find interactive games on the internet: *acid-base games*. One you should try to find is the GEMS Alien Juice Bar Game.

3 Communicating

Research the origin of chemical symbols that don't seem to match their chemical names.

- Search the internet and print out a copy of the periodic table.
- Circle the elements whose symbols do not appear to match their names.
- Research each of these elements to find the origin of its chemical symbol. For example, what language did it come from and what is the name in that language?

Present your findings in a two-column table or spreadsheet.

4 Communicating

Research the life of a scientist who has contributed to our understanding of the atomic model. Find:

- when and where they were born
- information about their childhood and family life
- some information about their education
- their contribution to the atomic model
- any other contributions they made to science.

Present your research as a short biography.

5 Communicating

Some older people are now having the amalgam fillings in their teeth replaced with other materials. Research why.
Present your research as a brochure to give to dental patients.

Thinking scientifically 

1 Acids release hydrogen ions (H^+) into solution. Use this information to identify which of the following substances could *not* be an acid.

A HCOOH
B Fe_2O_3
C H_2CO_3
D $NaHSO_4$

2 pH measures the concentration of hydrogen ions (H^+) in solution. The more concentrated the solution is in H^+ ions, the lower the pH is. An acidic solution has a pH of 5. Water is then added to it. Predict what will happen to the H^+ concentration of the solution.

A It will stay the same.
B It will increase.
C It will decrease.
D It will become the same as water.

3 Predict the pH of the new solution in question 3. It will most likely be:

A 4
B 5
C 6
D 7

4 The diagrams below show four different versions of what could have happened in Rutherford's gold foil and alpha particle experiment. Identify which diagram best represents:

a what he observed in his experiment
b what would have happened if the atoms were solid balls
c if the nucleus and electron cloud were roughly the same size.

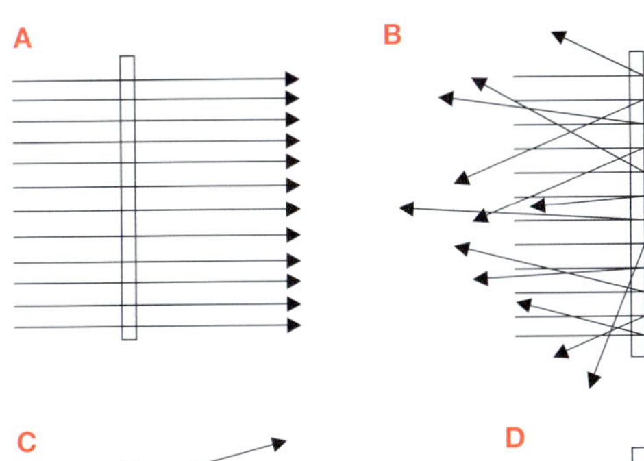

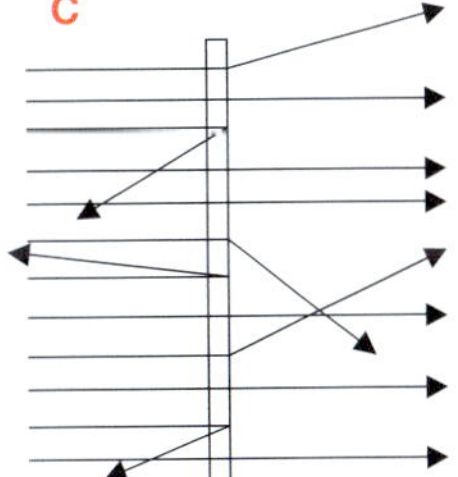

D

CHAPTER 2

Glossary

acid: a substance that releases hydrogen ions into an aqueous solution

alkali: a base that dissolves in water

alkaline solution: a solution made of a base/alkali and water

allotropes: different forms of the same element

alloy: a mixture of a base metal and small amounts of other elements

allotrope

anion: an ion that has more electrons than protons and is negatively charged

atom: the fundamental building block of all materials; it consists of a cluster of protons and neutrons surrounded by a cloud of electrons

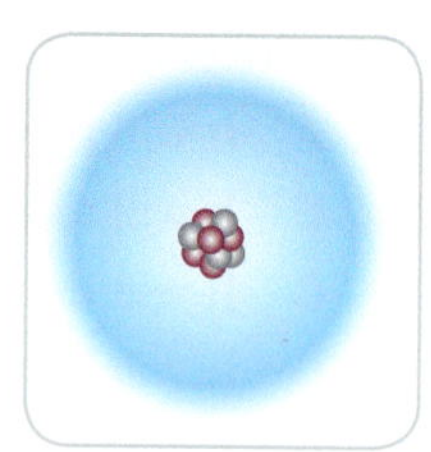

atom

atomic number: the number of protons in a nucleus; the atomic number determines what type of atom it is

atomic symbol: a short-hand notation for describing an atom; it consists of the chemical symbol, atomic number and mass number

base: a substance that releases hydroxide ions

base metal: the main metal in an alloy

brittle: shatters if hit

carat: a scale for measuring the purity of gold

cation: an ion that has fewer electrons than protons and is positively charged

compound: a pure substance that is made up of two or more different types of atom chemically joined

crystal lattice: a grid-like structure of atoms or ions in which each particle is bonded to all of its neighbouring atoms

crystal lattice

ductile: able to be stretched into wires

electron: a small, negatively charged particle; clouds of electrons surround the nucleus of an atom

electron cloud: the region of negative charge surrounding the nucleus, containing the electrons

element: a substance made up of only one type of atom

hydrogen ion: H^+, released by acids

hydroxide ion: OH^-, formed by bases

indicator: a chemical that changes colour to show whether a substance is acidic, neutral or basic

ion: an atom that has gained or lost an electron

litmus paper: a common indicator that turns red in the presence of an acid and blue in the presence of a base

litmus paper

lustrous: shines when polished or freshly cut

malleable: able to be hammered into new shapes

mass number: the number of protons and neutrons in an atom

metalloid: an element that usually displays the properties of a non-metal but conducts electricity like a metal under certain conditions; also known as a semi-metal

molecule: a cluster of atoms that makes up an element or a compound

neutral: having no overall charge

neutron: a particle with no electric charge; it is found in the nucleus of an atom

nucleus: a cluster of neutrons and protons at the centre of an atom

periodic table: a list of all the known 118 elements

periodic table

pH: a scale used to measure the concentration of H^+ ions in a solution

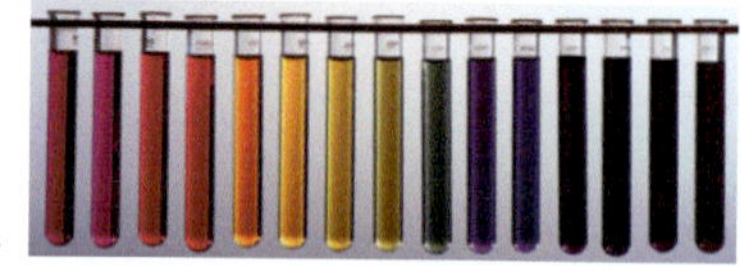

pH

polyatomic: containing more than one atom

proton: a positively charged particle found in the nucleus of an atom

quenching: a process in which a heated metal is cooled rapidly by dropping it into water

stainless steel: a rustless alloy of steel that includes chromium and nickel

steel: an alloy of iron and carbon

subatomic particles: the particles that atoms are made of—protons, neutrons and electrons

tempering: a process in which a metal is heated, cooled rapidly (quenched) and then reheated

CHAPTER 3

Reaction types

Have you ever wondered...

- what causes iron to rust?
- how plants can survive when they don't eat anything?
- why we breathe out carbon dioxide and not oxygen?
- how scientists know the age of fossils?

After completing this chapter you should be able to:

- identify reactants and products in chemical reactions
- model chemical reactions in terms of rearrangement of atoms
- describe observed reactions using word equations
- outline the role of energy in chemical reactions
- outline how conservation of mass can be demonstrated by simple chemical equations
- classify reactions as exothermic and endothermic
- describe the role of oxygen in combustion reactions
- describe how the environment influences our choice of fuels
- compare combustion with other oxidation reactions
- describe how the products of combustion reactions affect the environment
- investigate reactions of acids with metals, bases and carbonates
- evaluate claims relating to antacid tablets
- compare the biological processes in respiration and photosynthesis
- model the structure of isotopes
- describe how alpha and beta particles and gamma radiation are released from unstable atoms
- describe how technology is used in medicine such as the detection and treatment of cancer
- describe the effects on humans of exposure to X-rays
- investigate the work of scientists such as Ernest Rutherford, Pierre Curie and Marie Curie
- *use word or symbol equations to represent chemical reactions.*

AB 3.1

MODULE

3.1 Combustion and corrosion reactions

Chemical reactions happen continually around you. Two important types of chemical reactions are combustion and corrosion. Combustion happens when anything burns or explodes. Corrosion happens when a metal such as copper or an alloy such as steel changes into something else. Similar substances tend to undergo similar chemical reactions. These similarities allow you to predict what might happen if two chemicals are mixed. The similarities become more obvious when chemical reactions are expressed as chemical equations.

science 4 fun

Eating sherbet

An endothermic reaction absorbs energy from its surroundings. What does an endothermic reaction feel like?

YES

Collect this ...

- ½ teaspoon of citric acid
- ¼ teaspoon of baking soda (bicarbonate of soda, $NaHCO_3$)
- 3 teaspoons of icing sugar
- clean mixing bowl, cup or mug
- teaspoon

SAFETY

You should never eat in the laboratory, so only eat sherbet that you have made at home.

Do this ...

1 Add all the ingredients to the small mixing bowl or mug.

2 Use the back of the teaspoon to crush any lumps and to mix everything together.

3 Keep it in a dry place until ready to eat!

Record this ...

1 Describe what happened in your mouth when you ate the sherbet.

2 Explain why you think this happened.

Chemical reactions

In a chemical reaction, new substances form and old ones disappear. **Reactants** are the old substances you started with before the chemical reaction. **Products** are the new substances formed by the chemical reaction.

A chemical equation is a convenient way of showing what happens to different substances in a chemical reaction. A chemical reaction is always written in the form:

reactants → products

A **word equation** is a simple description of what is happening in a reaction. It shows the names of all the chemicals that are reactants and all those that are products. An example of a word equation is:

nitrogen gas + hydrogen gas → ammonia

This word equation shows that nitrogen gas and hydrogen gas reacted to form ammonia. Another, more detailed way of showing what is happening in a reaction is to write a **balanced equation**. The balanced equation for the above reaction between nitrogen and hydrogen gases is:

$$N_2 + 3H_2 \rightarrow 2NH_3$$

A balanced equation shows exactly what is happening in a reaction. The big numbers in front of each substance are called **coefficients**. These numbers show how much of each substance reacted and how much of each reactant was produced. For example, the balanced equation above shows that:

- every single molecule of nitrogen reacts with three molecules of hydrogen
- two molecules of ammonia were formed.

SkillBuilder

Writing word equations

To write a word equation, follow the steps below.

Identify the reactants and products.

As an example, consider the chemical reaction between copper and sulfur dioxide. This reaction forms copper sulfide and oxygen gas. You started with copper and sulfur dioxide, so these are the reactants. Copper sulfide and oxygen gas were produced. These are the products.

Write a word equation.

A word equation is a simple written description of what is happening in the reaction. The reactants are placed on the left side of the arrow and the products on the right.

reactants → products

copper + sulfur dioxide → copper sulfide + hydrogen gas

Worked example

Writing word equations

Problem

A piece of aluminium was dropped into hydrochloric acid. The aluminium dissolved and reacted to form aluminium chloride. As it did so, hydrogen gas bubbled to the surface. Write a word equation for this reaction.

Solution

Thinking: Identify the reactants and products.

Working: Reactants = aluminium, hydrochloric acid

Products = aluminium chloride, hydrogen gas

Thinking: Write a word equation.

Working: reactants → products

aluminium + hydrochloric acid → aluminium chloride + hydrogen gas

Try yourself

Construct word equations for the following reactions.

1. Sodium reacts with iron(II) chloride to form sodium chloride and iron.
2. When propanol is burnt in oxygen gas, carbon dioxide and water are formed.
3. Hydrogen peroxide splits to form water and oxygen gas.

AB 3.2

Energy in reactions

An **endothermic reaction** is a reaction that absorbs heat, taking it from the surroundings and making them feel colder.

In endothermic reactions, the products have more energy than the reactants. Endothermic reactions need energy to proceed and they get their energy from what is around them. An example is what happens in a chemical cold pack that you might use when you have an injury. Packets of ammonium nitrate and water are broken, allowing these substances to mix and react. As they react, they absorb energy from their surroundings, cooling the surroundings down. The sherbet in Figure 3.1.1 acts in a similar way.

While endothermic reactions absorb energy, **exothermic reactions** release energy. In exothermic reactions, the reactants have more energy than the products. During the reaction, energy is released into the surroundings, usually as heat and/or light.

FIGURE 3.1.1 Sherbet leaves your tongue cold because an endothermic reaction absorbs heat from your mouth.

Combustion

Combustion reactions are examples of exothermic reactions. **Combustion** occurs whenever something reacts with oxygen gas (O_2), burning or exploding as it does so. A bushfire is a series of combustion reactions. The chemicals in living plants, dead twigs and leaves burn in oxygen, releasing huge amounts of heat and light energy as they react.

Combustion reactions belong to a type of reaction known as an oxidation reaction—in its simplest form, an oxidation reaction is when an element reacts with oxygen.

SciFile

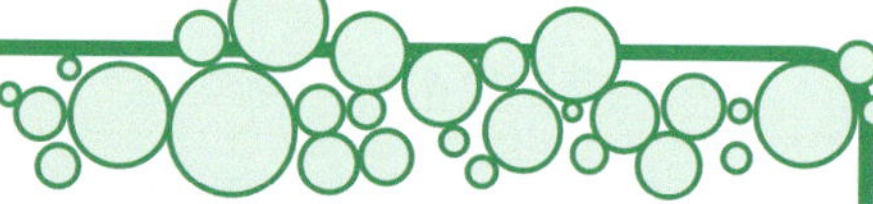

That's shocking!

Explosions generate hot gases that suddenly expand at speeds of up to 8 kilometres per second! These expanding gases form blasts of wind called shockwaves, which can be as deadly as the explosion itself. A shockwave leaves a vacuum at the site of the explosion, and air flowing into this carries rubbish and debris.

Combustion of fossil fuels

Bunsen burners, gas stoves, water heaters and central heating furnaces produce a hot blue flame by burning methane or ethane gas in oxygen (Figure 3.1.2). The reactions are:

methane + oxygen gas → carbon dioxide + water

$$CH_4 + 2O_2 \rightarrow CO_2 + 2H_2O$$

ethane + oxygen gas → carbon dioxide + water

$$2C_2H_6 + 7O_2 \rightarrow 4CO_2 + 6H_2O$$

Petrol is a mixture of highly combustible chemicals called **hydrocarbons**, the most important of which is octane. Octane combusts via the chemical equation:

octane + oxygen gas → carbon dioxide + water

$$2C_8H_{18} + 25O_2 \rightarrow 16CO_2 + 18H_2O$$

Incomplete combustion

The above combustion reactions all need an unlimited supply of oxygen fed into them. These reactions are known as **complete combustion**.

However, oxygen supply is sometimes restricted in some way. This might happen if there is not enough oxygen or the oxygen cannot mix properly with the fuel. This might happen if a fire was started indoors or the substance burning is so dense or tightly packed that oxygen cannot get into it. If the oxygen supply is restricted in some way, then other reactions occur instead. This is known as **incomplete combustion**. Incomplete combustion is still exothermic but does not release as much heat or light energy as complete combustion does.

The reactions below show what happens to methane if oxygen is restricted.

methane + oxygen gas → carbon monoxide + water

$$2CH_4 + 3O_2 \rightarrow 2CO + 4H_2O$$

At the same time another reaction occurs.

methane + oxygen gas → carbon + water

$$CH_4 + O_2 \rightarrow C + 2H_2O$$

Incomplete combustion reactions are 'dirty' because they produce carbon, which is left behind as soot, charcoal or smoke, like that seen in Figure 3.1.3. They also produce the poisonous gas carbon monoxide. In contrast, complete combustion reactions are 'clean'.

FIGURE 3.1.2 A gas stove uses combustion to release its heat (and light).

FIGURE 3.1.3 Smoke and soot are an indication of incomplete combustion.

For example, a sheet of paper burns quickly without much smoke. This is because oxygen can easily get to all parts of the paper, allowing all parts to burn at once. However, crumple the paper into a tight ball and it will burn slowly and produce lots of smoke. This happens because only the outside gets enough oxygen to undergo complete combustion. The supply of oxygen to the inner layers of paper is very limited so much of the ball will undergo incomplete combustion. Hence it will burn slowly and produce lots of smoke.

A Bunsen burner can show both complete and incomplete combustion. If the flow of oxygen to it is good (open airhole), then combustion is complete and the flame is hot, clean and blue. If the air flow is restricted (closed airhole), then combustion is incomplete and a cooler, dirty yellow flame is produced.

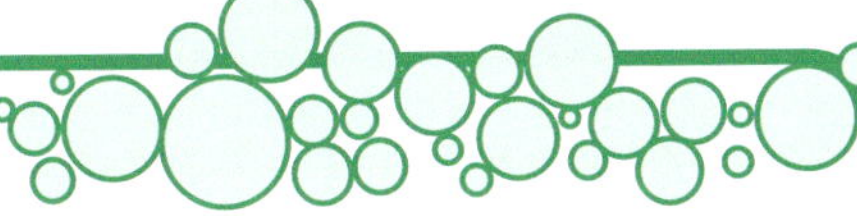

SciFile

Suffocating fires

Fires consume oxygen, so there is less of it to breathe in the region of the fire. During World War II, the German city of Dresden was firebombed. Many of the 25 000 people killed in the attack are thought to have suffocated because of this lack of oxygen.

Pollution and climate change

Water vapour and carbon dioxide are released into the atmosphere whenever fossil fuels such as gas, petrol, oil, coal, diesel and aviation fuel are burnt. Carbon dioxide is a greenhouse gas that traps heat within the atmosphere. Over the past 150 years, we humans have burned huge quantities of fossil fuels to power our cars, ships and aircraft, and to heat their homes and generate electricity. For this reason, the amount of carbon dioxide in the atmosphere has increased to levels that most scientists agree are increasing the atmosphere's average temperature. If this view is correct, then the burning of fossil fuels could be affecting Earth's climate.

If the combustion of these fossil fuels is incomplete, then carbon monoxide and carbon are released.

Carbon adds relatively harmless but dirty soot to the atmosphere. Carbon monoxide gas has no smell, but it is so poisonous that even small amounts of it can kill. Petrol also contains additives that release other poisonous chemicals when burnt. These include oxides of nitrogen and sulfur, both of which can combine with moisture in the air to form smog and acid rain.

Other combustion reactions

A much slower and controlled combustion reaction occurs within the cells of your own body. **Aerobic respiration** combines the sugar **glucose** from the digestion of your food with the oxygen you breathe in. This reaction releases the energy that the cells of your body need. A waste product is carbon dioxide, which you breathe out (Figure 3.1.4).

glucose + oxygen gas → carbon dioxide + water

$$C_6H_{12}O_6 + 6O_2 \rightarrow 6CO_2 + 6H_2O$$

FIGURE 3.1.4 Respiration gives your body the energy it needs. The reaction needs glucose (from your food) and oxygen (breathed in). You breathe out its product, carbon dioxide.

Not all combustion reactions produce carbon dioxide and water vapour. When burnt, magnesium reacts to form magnesium oxide. No other products form.

magnesium + oxygen gas → magnesium oxide

$$2Mg + O_2 \rightarrow 2MgO$$

You can see this reaction happening in Figure 3.1.5.

FIGURE 3.1.5 The light released by the combustion of magnesium is so bright that it can quickly damage your eyes.

Prac 1 p. 84

Prac 2 p. 85

Prac 3 p. 86

Corrosion reactions

Most metals corrode when exposed to water, air or other chemicals. **Corrosion** is a chemical reaction that forms other compounds from these metals.

For example, the iron/steel body of a car slowly reacts with water and oxygen in the air and will corrode until all that is left is a pile of rust.

In a similar way, copper corrodes by reacting with gases in the air to form green **verdigris**, a mixture of copper(II) hydroxide and copper(II) carbonate. The typical green colouring of verdigris is obvious in Figure 3.1.6. The chemical equation is:

copper + water + carbon dioxide + oxygen gas → copper(II) hydroxide + copper(II) carbonate

$$2Cu + H_2O + CO_2 + O_2 \rightarrow Cu(OH)_2 + CuCO_3$$

FIGURE 3.1.6 This copper roof has corroded to form a green coating called verdigris.

Pure silver reacts with sulfur to form a black coating called **tarnish** (silver sulfide). This sulfur comes from hydrogen sulfide in air pollution or from foods such as eggs, fish, onions and pea soup.

silver + hydrogen sulfide → silver sulfide + hydrogen gas

$$2Ag + H_2S \rightarrow Ag_2S + H_2$$

Pure sodium and potassium are such reactive metals that they react with just about anything. Their corrosion is very quick and often explosive because of the hydrogen gas that their reactions produce. Their chemical reactions with water are shown below.

sodium + water → sodium hydroxide + hydrogen gas

$$2Na + 2H_2O \rightarrow 2NaOH + H_2$$

potassium + water → potassium hydroxide + hydrogen gas

$$2K + 2H_2O \rightarrow 2KOH + H_2$$

Rusting

Iron and its alloy, steel, are common and relatively cheap. This makes them the most used metals on Earth. Unfortunately, iron and most types of steel react with air and water to form **rust** (Figure 3.1.7). Rust is known chemically as hydrated iron(III) oxide (chemical formula $Fe_2O_3.H_2O$). Rust is flaky and easy to dislodge. This allows the rusting process to continue into the next layer, progressively making the iron or steel thinner and weaker.

FIGURE 3.1.7 Rust forms when iron is exposed to oxygen and water.

Although an extremely complex reaction, rusting can be summarised by the chemical equation:

iron + oxygen gas + water → hydrated iron(III) oxide

$$4Fe + 3O_2 + 2H_2O \rightarrow 2Fe_2O_3{\cdot}H_2O$$

This equation is often simplified to:

iron + oxygen gas → iron(III) oxide

$$4Fe + 3O_2 \rightarrow 2Fe_2O_3$$

In its simplest form, an oxidation reaction involves an element joining with oxygen in a chemical reaction. For this reason, rusting is considered to be an oxidation reaction.

Corrosion of aluminium

Aluminium is very reactive. The surface metal reacts almost immediately with oxygen in the air, forming a fine layer of dull, grey aluminium oxide (Al_2O_3).

aluminium + oxygen gas → aluminium oxide

$$4Al + 3O_2 \rightarrow 2Al_2O_3$$

Unlike rust, this layer does not flake but acts instead like a tightly bound layer of paint, protecting the aluminium from further corrosion. **Anodising** is a process that deliberately builds up a layer of aluminium oxide to protect the aluminium underneath (Figure 3.1.8).

FIGURE 3.1.8 These cups are made of anodised aluminium. Their surface is a layer of aluminium oxide that was deliberately built up on the surface of the metal and then coloured.

SkillBuilder

Writing balanced equations

Chemists use balanced equations to accurately show what is happening in a chemical reaction. To construct a balanced formula equation, follow the steps below.

Write an unbalanced equation.

Replace the names of each substance in a word equation with their element symbols or chemical formulas. This gives you an unbalanced equation. Consider the reaction:

copper + sulfur dioxide → copper sulfide + oxygen gas

Copper has the symbol Cu and the chemical formula of sulfur dioxide is SO_2, copper sulfide is Cu_2S and oxygen gas is O_2.

Replacing their names with their formulas gives the unbalanced equation:

$$Cu + SO_2 \rightarrow Cu_2S + O_2$$

Identify what elements are unbalanced.

An equation is unbalanced when it has unequal numbers of atoms of a particular element on both sides of the arrow. As Figure 3.1.9 shows, the above unbalanced equation starts with one copper atom, one sulfur atom and two oxygen atoms. However, the equation ends up with two copper atoms, one sulfur atom and two oxygen atoms.

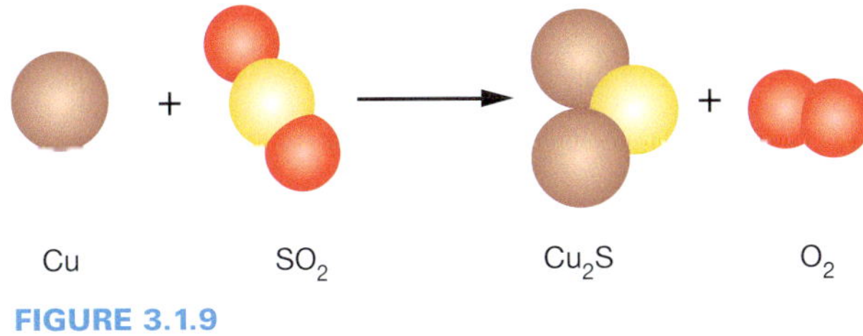

FIGURE 3.1.9

A count of atoms of each element on both sides of the arrow gives:

reactants	→	products
1 × Cu	→	2 × Cu
1 × S	→	1 × S
2 × O	→	2 × O

This suggests that a copper atom appeared from nowhere! That is impossible because atoms never just appear in chemical reactions. Nor do they disappear. They only change in the way they are arranged. This fundamental principle of chemistry is known as the **law of conservation of mass**. To describe accurately the rearrangement that is happening in the reaction, chemical equations need to be balanced.

Balance the equation.

A **balanced equation** has the same numbers of each type of atom on both sides of the arrow. The above reaction would be balanced if the reaction used up two atoms of copper instead of just one. When balancing an equation, you cannot change the small numbers (subscripts) within a chemical formula since that changes the chemical itself (for example, from water H_2O to the bleach hydrogen peroxide H_2O_2). For this reason, you can only change the coefficients of each reactant or product. The coefficients are the big numbers that appear in front of some of the chemicals in a chemical equation. Hence, the correct way of balancing the above equation is:

$$2Cu + SO_2 \rightarrow Cu_2S + O_2$$

Figure 3.1.10 shows what is happening to the atoms in this reaction.

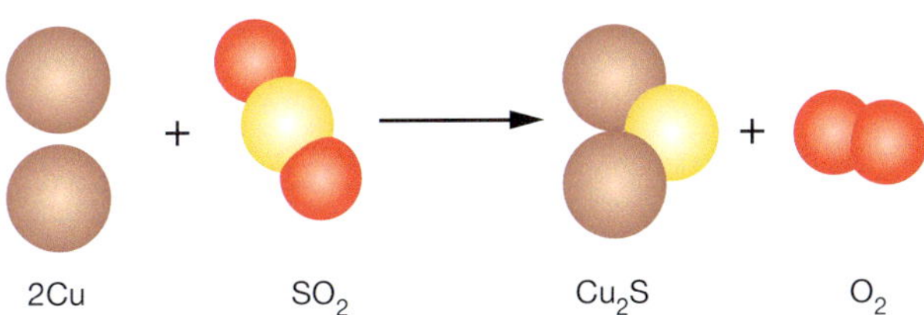

FIGURE 3.1.10 Atoms can't appear from nothing and can't disappear. This is why chemical equations must be balanced.

Check your equation.

Your equation should now be balanced. To make sure that it is, check how many atoms of each element are on either side of the arrow.

reactants	→	products
2 × Cu	→	2 × Cu
1 × S	→	1 × S
2 × O	→	2 × O

Worked example

Writing balanced equations

Problem

Tin (IV) oxide (SnO_2) reacts with hydrogen gas (H_2) to form tin (Sn) and water (H_2O).

Construct a balanced formula equation for this reaction.

Solution

Thinking: Write the unbalanced equation.

Working: $SnO_2 + H_2 \rightarrow Sn + H_2O$

Thinking: Identify which elements are unbalanced.

Working: reactants → products

$1 \times Sn \rightarrow 1 \times Sn$

$2 \times H \rightarrow 2 \times H$

$2 \times O \rightarrow 1 \times O$

Thinking: Balance the equation

Oxygen is unbalanced. By doubling the number of water molecules produced, oxygen becomes balanced. Hence change H_2O to $2H_2O$.

$2H_2 = 2 \times 2$ Hs.

Working: $SnO_2 + H_2 \rightarrow Sn + 2H_2O$

Thinking: Sometimes balancing one element causes another element to become unbalanced. This is what has happened here to hydrogen. The equation now has:

Working: reactants → products

$1 \times Sn \rightarrow 1 \times Sn$

$2 \times H \rightarrow 4 \times H$

$2 \times O \rightarrow 2 \times O$

Thinking: However, doubling the number of H_2 molecules reacting solves the problem:

Working: $SnO_2 + 2H_2 \rightarrow Sn + 2H_2O$

Thinking: Check your equation.

Working: reactants → products

$1 \times Sn \rightarrow 1 \times Sn$

$4 \times H \rightarrow 4 \times H$

$2 \times O \rightarrow 2 \times O$

Try yourself

1 Calcium (Ca) burns in oxygen (O_2) to form calcium oxide (CaO). Construct a balanced formula equation for this reaction.

2 Construct balanced equations by adding coefficients where indicated.

- a $CH_4 + O_2 \rightarrow CO_2 + ...H_2O$
- b $.....AgBr + GaPO_4 \rightarrow Ag_3PO_4 + GaBr_3$
- c $.....NaBr + Cl_2 \rightarrow .....NaCl + Br_2$
- d $SiCl_4 + ...H_2O \rightarrow SiO_2 + ...HCl$
- e $Au_2S_3 + ...H_2 \rightarrow .....Au + ...H_2S$

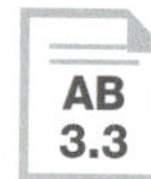

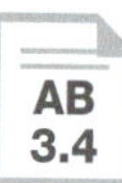

MODULE 3.1

Review questions

Remembering

1 Define the terms:
- a endothermic
- b verdigris
- c anodising
- d tarnish.

2 What term best describes each of the following?
- a a reaction that releases energy
- b a burning reaction
- c a rusting reaction
- d a reaction involving the joining of an element with oxygen.

3 What substance is always required for combustion to occur?

4 What is the word equation and balanced chemical equation for each of the following reactions?
- a combustion of octane
- b corrosion of copper.

Understanding

5 How can a Bunsen burner display both complete and incomplete combustion?

6 Explain why the rusting of iron makes it get thinner and thinner.

7 a Explain what a balanced chemical equation is.
- b The law of conservation of mass states that mass is never created nor is it destroyed. Explain how this law requires all chemical equations to be balanced.

Applying

8 Identify the reactants and products for each of these reactions:
- a iron + sulfur → iron sulfide
- b propane + oxygen → carbon dioxide + water.

9 Identify the products formed when fossil fuels undergo:
- a complete combustion
- b incomplete combustion.

10 The combustion of butene is shown by the following word and balanced formula equations.

butene + oxygen gas → carbon dioxide + water

$$C_4H_8 + 6O_2 \rightarrow 4CO_2 + 4H_2O$$

Identify:
- a the chemical formula for butene
- b reactants for the reaction
- c products of the reaction.

Analysing

11 Classify the following reactions as endothermic or exothermic.
- a A bushfire burns down a forest.
- b A sparkler on a birthday cake is alight.
- c Sherbet cools your mouth.

12 Contrast the corrosion of iron with the corrosion of aluminium.

Evaluating

13 Magnesium is a metal but it burns in oxygen to form magnesium oxide. Assess whether this reaction could also be classified as a corrosion reaction.

Creating

14 Construct word equations for the following combustion reactions.
- a Benzene burns in oxygen gas to produce carbon dioxide and water vapour.
- b Carbon dioxide and water form when hexane burns in oxygen.

15 Zinc corrodes in air by reacting with oxygen gas. It forms a dull, grey substance called zinc oxide. Its balanced formula equation is:

$$2Zn + O_2 \rightarrow 2ZnO$$

- a Use this equation to identify the chemical formula for zinc oxide.
- b Construct a word equation to show what is happening in this reaction.
- c Construct a table to prove that this equation is balanced.

16 Construct word equations for the:
- a complete combustion of pentane
- b incomplete combustion of octane.

MODULE 3.1 Practical investigations

1 • Conservation of mass  

Purpose

To demonstrate that mass is conserved in a chemical reaction.

Timing 60 minutes

Materials

- 150 mL of 0.1 M copper sulfate solution
- 50 mL of 0.1 M sodium hydroxide solution
- 50 mL of 0.1 M sodium carbonate solution
- 50 mL of ammonia solution

- four 250 mL beakers
- marking pen
- electronic balance

All chemicals should be treated as toxic.

Do not taste any chemicals.

Procedure

1. In your workbook, construct a table like the one shown in the Results section.
2. Pour copper sulfate solution into one of the beakers until it reaches the 50 mL mark. Use the marking pen to label this beaker BLUE.
3. To the other beaker add 50 mL of sodium hydroxide solution.
4. Place both beakers on the electronic balance (Figure 3.1.11a) and determine their total mass. Record this mass in a table like the one in the Results section.
5. Carefully pour the copper sulfate into the sodium hydroxide and observe what happens.
6. Place the beaker with the solutions in it and the empty beaker labelled BLUE back on the electronic balance (3.1.11b). Once again determine their total mass. Record their mass.
7. To a clean beaker, add 50 mL of sodium carbonate. Add another 50 mL of copper sulfate to the beaker labelled BLUE. Once again, find their total mass.
8. Carefully pour the copper sulfate solution into the sodium carbonate solution. As before, find the total mass of the full and empty beakers.
9. Repeat steps 6 and 7 but this time use ammonia solution instead of sodium carbonate.

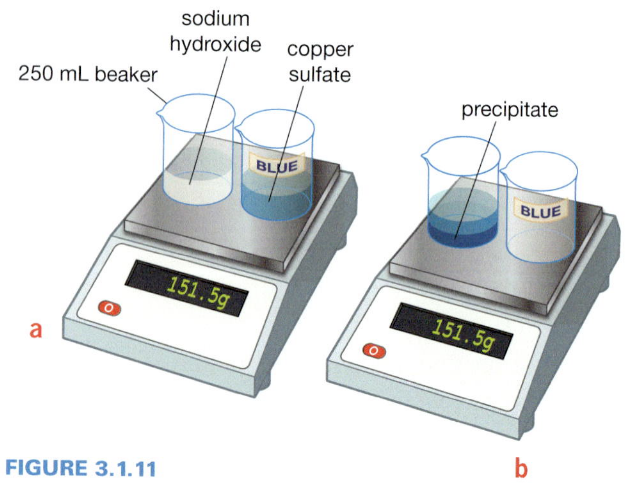

FIGURE 3.1.11

Results

Record all your measurements in your results table.

Mass results

	Total mass before mixing (g)	Total mass after mixing (g)	Observations
sodium hydroxide + copper sulfate			
sodium carbonate + copper sulfate			
ammonia + copper sulfate			

Review

1. What evidence is there that shows that a chemical reaction took place once the solutions were mixed?
2. Compare the total masses before and after mixing.
3. State the law of conservation of mass.
4. Assess whether or not your results support this law.

MODULE 3.1

Practical investigations

2 • Conservation of mass in combustion reactions

Purpose

To show that combustion reactions obey the law of conservation of mass.

Timing 60 minutes

Materials

- clean steel wool
- electronic balance

PART A: COMBUSTION IN THE OPEN AIR

- evaporating dish or watch-glass
- 9 V battery
- electrical leads with stripped ends exposing the wire

PART B: COMBUSTION IN A SEALED CONTAINER

- large Pyrex test-tube
- test-tube tongs
- balloon
- Bunsen burner and bench mat

SAFETY

Sparks will fly in Part A. Make sure they don't hit you or your clothes.

In Part B if sparks enter the balloon it could partly melt or catch fire. Hence, keep it well away from you.

Procedure

PART A: COMBUSTION IN THE OPEN AIR

1 In your workbook, construct a table like the one shown in the Results section.

2 Tease out the strands of a small piece of steel wool (about 1 gram) and place it in an evaporating dish or watch-glass.

3 Use the electronic balance to find the combined mass of the evaporating dish/watch-glass and the steel wool. Record the total mass in a results table like the one shown in the Results section.

4 Attach the leads to the terminals of a 9 V battery. Lower the stripped ends of the leads into the steel wool so that a current can flow through it.

5 Shift the battery around to other parts of the steel wool so that as much of the steel as possible is burnt.

6 When completed, find the mass of the evaporating dish or watch-glass and the burnt steel wool. Record the new mass in the table.

PART B: COMBUSTION IN A SEALED CONTAINER

7 Tease out a strand of steel wool (about 0.5 gram) and slide it into the test-tube.

8 Squeeze all the air out of the balloon and fit it over the opening of the test-tube, as shown in Figure 3.1.12.

9 Find the combined mass of the test-tube, balloon and steel wool and record it in your table.

10 Use tongs to hold the test-tube over a blue Bunsen burner flame. Make sure that you move the test-tube back and forth in the flame and that the flame does not get near the balloon. Stop heating after about 5 minutes. Record what you saw happen in the test-tube and to the balloon.

11 Allow the test-tube to cool completely. When the test-tube is cool, again find the total mass of the test-tube, steel wool and balloon.

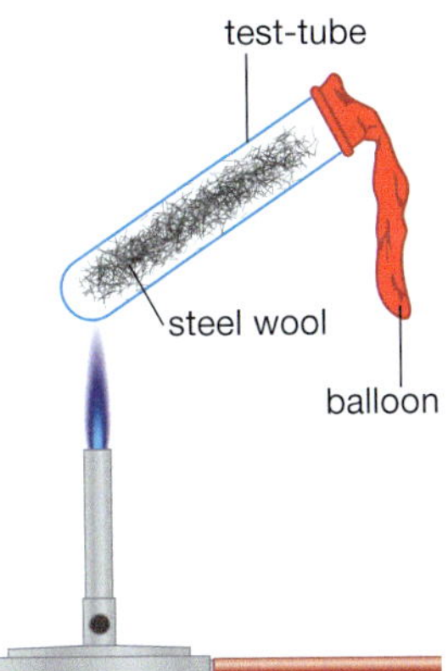

FIGURE 3.1.12

Results

Record all your measurements in your results table.

Mass–combustion results

	Total mass before (g)	Total mass after (g)	Observations
part A			
part B			

Review

1 Apart from something to burn, what other substance is needed for combustion to occur.

2 The total mass in part A most likely increased. Where do you think this increase in mass came from?

3 The total mass at the end of part B was most likely the same or very similar to the total mass at the start. Propose reasons why.

4 Assess which set of results (part A or part B) best demonstrated the law of conservation of mass.

MODULE 3.1

Practical investigations

SPARKlab alternative available for this activity.

3 • Cracker combustion

Purpose

To observe how combustion releases heat energy.

Timing

60 minutes

Materials

- dry biscuits (such as Clix crackers)
- large test-tube/boiling tube
- 10 mL measuring cylinder
- thermometer
- bench mat
- retort stand, 2 × bossheads and clamps
- metal tongs
- Bunsen burner
- electronic balance

SAFETY

An open flame can be dangerous. The room should be well ventilated.

Biscuits containing nuts or sesame seeds should not be used in case of allergies.

Procedure

1 In your workbook, construct a table like the one shown in the Results section.

2 Set up the apparatus as shown in Figure 3.1.13.

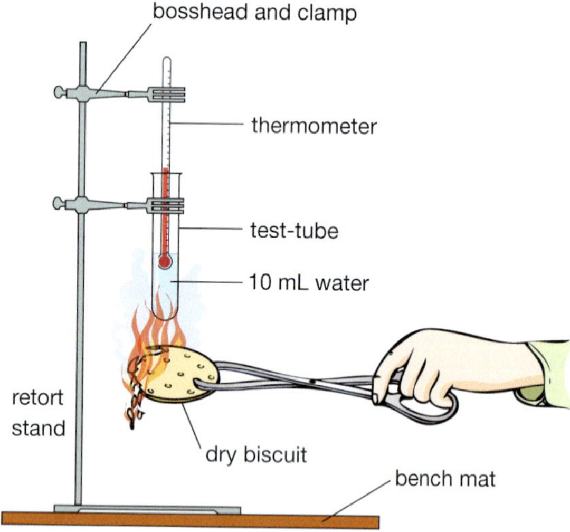

FIGURE 3.1.13

3 Use the measuring cylinder to measure out 10 mL of water. Add it to the test-tube.

4 Measure the temperature of the water in the test-tube. Record this temperature in a table like the one shown in the Results section.

5 Measure the mass of the dry biscuit.

6 Use the tongs to hold the biscuit in a blue Bunsen burner flame. Heat the biscuit until it catches fire.

7 Quickly hold the biscuit under the test-tube. Re-light the biscuit if it goes 'out'.

8 Measure the temperature of the water again once the biscuit has changed completely into ash.

9 Test what happens if the amount or number of biscuits is increased. For example, do two biscuits cause the temperature to rise twice as much?

Results

1 Record all your measurements in your results table or spreadsheet.

Combustion results

	Number of biscuits		
	1	2	3
Mass (g)			
Temperature before (°C)			
Temperature after (°C)			
Temperature rise (°C)			
Energy absorbed (J) = temperature rise x 42			

2 Determine how much energy was absorbed by the water, by multiplying the temperature rise by 42. (It takes 42 J of heat energy to raise the temperature of 10 mL of water by 1°C.) Alternatively, program your spreadsheet to calculate the amount of energy for you.

Review

1 How much energy in joule is needed to raise 10 mL of water 1°C?

2 a Did the biscuit undergo complete or incomplete combustion?

b Justify your answer.

3 The amount of energy absorbed by the water is less than the amount of energy released by the biscuit. Explain why.

4 Construct a conclusion for what happened when the mass or number of biscuits increased.

5 Evaluate your experiment to determine ways in which it could be improved.

MODULE

3.2 Acid reactions

When you have heartburn, taking an antacid tablet brings relief because of a chemical reaction between the acid in your stomach and bases within the tablet. The acid and base neutralise each other, forming harmless salts and water. If the acid had reacted with a metal instead, then hydrogen gas would have formed. With a carbonate, it would have produced carbon dioxide.

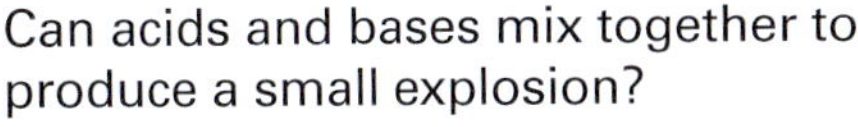

Exploding bags

NO

Can acids and bases mix together to produce a small explosion?

Collect this ...

- baking soda (bicarbonate of soda)
- vinegar
- warm water
- zip-lock plastic bag
- a few squares of toilet paper
- tablespoon
- measuring cup

Do this ...

1. Measure out a tablespoon of baking soda and wrap it up in few sheets of toilet paper. Twist the paper to hold it all in place.
2. Pour ¼ cup of vinegar and ¼ cup of warm water into the zip-lock bag.
3. Zip up the bag but leave a gap big enough for your baking soda parcel to fit through.
4. Go outside and push the parcel in. Quickly zip up the gap and place it on the ground. Stand clear!

Record this ...

1. Describe what happened.
2. Explain why you think this happened.

Acids and metals

Acids can corrode metals, which results in the formation of a salt and hydrogen gas. An example of this type of reaction is shown in Figure 3.2.1.

FIGURE 3.2.1 Iron reacts with hydrochloric acid, forming bubbles of hydrogen gas.

In simple terms, the equation for the reaction between an acid and a metal is:

acid + metal → a salt + hydrogen gas

Examples are:

hydrochloric acid + magnesium → magnesium chloride + hydrogen gas

$$2HCl + Mg \rightarrow MgCl_2 + H_2$$

sulfuric acid + iron → iron(II) sulfate + hydrogen gas

$$H_2SO_4 + Fe \rightarrow FeSO_4 + H_2$$

Salt is normally the term used for sodium chloride (NaCl). However, in these reactions the term is used in a much more general way. Here the term **salt** means any compound formed by when an acid reacts with a metal or a base. Examples of salts are potassium nitrate (KNO_3), nitric acid (HNO_3), magnesium sulfate ($MgSO_4$) and calcium chloride ($CaCl_2$).

Prac 1 p. 94

SkillBuilder

Predicting salts

The name of the salt produced can be predicted by putting the name of the metal in front of the name of the ion formed once the acid loses its hydrogen. For example, nitric acid forms salts called nitrates, hydrochloric acid forms chlorides and sulfuric acid forms sulfates.

Worked example

Predicting salts

Problem

Predict the name of the salt formed when calcium reacts with nitric acid.

Solution

Thinking: Identify the metal.

Working: Calcium is the metal.

Thinking: Identify the ion formed once the acid has lost its hydrogen atom.

Working: Nitric acid forms nitrate ions.

Thinking: Put the two together.

Working: Calcium nitrate

Try yourself

Predict the names of the salts formed when:

a lithium reacts with hydrochloric acid
b iron reacts with nitric acid
c copper reacts with sulfuric acid.

Neutralisation reactions

Acids and bases neutralise each other when mixed. **Neutralisation** reactions take the form:

acid + base → a salt + water

Examples are:

nitric acid + potassium oxide → potassium nitrate + water

$$2HNO_3 + K_2O \rightarrow 2KNO_3 + H_2O$$

sulfuric acid + sodium hydroxide → sodium sulfate + water

$$H_2SO_4 + 2NaOH \rightarrow Na_2SO_4 + 2H_2O$$

Acids and carbonates

When the base is a carbonate, the reaction between it and an acid produces carbon dioxide as well as a salt and water. The reaction takes the form:

acid + carbonate → a salt + water + carbon dioxide

Figure 3.2.2 shows the reaction between the calcium carbonate in chalk and hydrochloric acid. Its equation is:

hydrochloric acid + calcium carbonate → calcium chloride + water + carbon dioxide

$$2HCl + CaCO_3 \rightarrow CaCl_2 + H_2O + CO_2$$

FIGURE 3.2.2 Chalk contains the base calcium carbonate. Here hydrochloric acid reacts with the calcium carbonate to form bubbles of carbon dioxide.

Carbon dioxide is also produced when acids react with hydrogen carbonates. For example, vinegar (ethanoic or acetic acid) and baking soda (sodium hydrogen carbonate) mix and form a froth of carbon dioxide bubbles. You can see this in Figure 3.2.3. Its equation is:

acetic acid + sodium hydrogen carbonate → sodium acetate + water + carbon dioxide

$$CH_3COOH + NaHCO_3 \rightarrow NaCH_3COO + H_2O + CO_2$$

FIGURE 3.2.3 Bicarbonate of soda (sodium hydrogen carbonate) is a base. Vinegar contains acetic acid. When mixed, these chemicals react rapidly, neutralising each other and producing lots of bubbles of carbon dioxide.

FIGURE 3.2.4 The proteins in meat only dissolve in the presence of pepsin and hydrochloric (HCl) acid.

Acid reactions in digestion

Chemical digestion is the process that breaks down large molecules in food into smaller ones that can be absorbed by the body. Chemical digestion happens through a series of chemical reactions, each reaction being sped up by chemicals called **enzymes**. Enzymes in saliva help digest starch, enzymes in your stomach help digest protein and enzymes in your duodenum (the first part of the small intestine) help digest fat and any remaining carbohydrates and protein.

Enzymes are active only when their environment is suitable. For example, the enzymes in saliva are only active when your mouth has a pH of around 7 (neutral). Saliva is slightly alkaline (basic) and so it tends to neutralise acids in foods such as oranges and tomatoes and in drinks such as fruit juices and soft drinks.

An enzyme called pepsin helps digest proteins in your stomach. Pepsin needs a strongly acidic environment to form and to work. This can be seen in Figure 3.2.4. Gastric juice is produced by the lining of your stomach. It contains hydrochloric acid (HCl), giving the stomach a pH of between 0 and 3. This is the ideal environment for pepsin but is too acidic for the enzymes in saliva to keep working. Hence, digestion of starch stops in the stomach and digestion of protein begins. The acid also helps kill potentially harmful microorganisms that may have been in the food you ate.

The part-digested material that emerges from the stomach is far too acidic for the rest of the digestive tract. Hydrochloric acid in this material is neutralised by sodium hydrogen carbonate, which is released by the pancreas if the pH in the duodenum drops below 5. The reaction is:

hydrochloric acid + sodium hydrogen carbonate → sodium chloride + water + carbon dioxide

$$HCl + NaHCO_3 \rightarrow NaCl + H_2O + CO_2$$

The duodenum now has a pH of between 7 and 8, which allows enzymes from the pancreas to digest the remaining starches, proteins and fats. Figure 3.2.5 summarises the role that acids, bases and pH play in digestion.

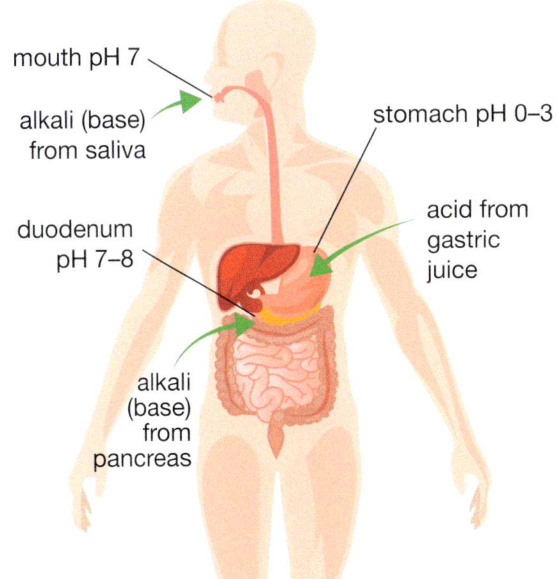

FIGURE 3.2.5 The stomach needs to be strongly acidic for digestion to occur there. The pancreas releases a base to neutralise this acid in the duodenum.

Antacids

Heartburn is caused when hydrochloric acid rises from your stomach into your oesophagus (foodpipe). While your stomach has an acid-resistant lining, your oesophagus does not. This lack of protection means that any acid in your oesophagus will burn it.

Antacids are tablets or liquid medicines that you take to relieve heartburn. Antacids contain bases that neutralise acid rising into the oesophagus and so they bring quick relief. Some antacids float on top of the acidic contents of your stomach. As well as neutralising the acid, these antacids also stop the acid from rising. The effect of antacids is demonstrated in Figure 3.2.6.

The bases in antacid tablets are usually magnesium hydroxide and aluminium hydroxide. The reactions below show how they neutralise hydrochloric acid rising from your stomach.

hydrochloric acid + magnesium hydroxide → magnesium chloride + water

$$2HCl + Mg(OH)_2 \rightarrow MgCl_2 + 2H_2O$$

hydrochloric acid + aluminium hydroxide → aluminium chloride + water

$$3HCl + Al(OH)_3 \rightarrow AlCl_3 + 3H_2O$$

Prac 2 p. 95
Prac 3 p. 96
Prac 4 p. 96
AB 3.5

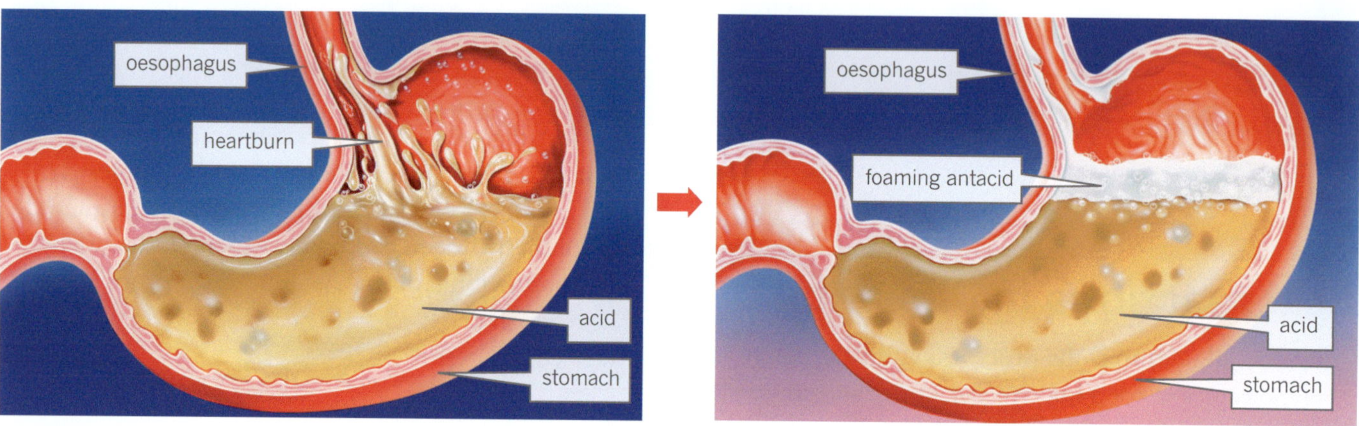

FIGURE 3.2.6 Antacids neutralise stomach acid. The antacid shown here floats on the stomach's acidic contents, stopping it from rising and burning your oesophagus.

SCIENCE AS A HUMAN ENDEAVOUR

Use and influence of science

Acid rain

Rainwater is naturally slightly acidic, with a pH between 5 and 6.

FIGURE 3.2.7 Acid rain has killed many forests in North America and Europe.

Rainwater is slightly acidic because it reacts with some of the carbon dioxide in the air to form carbonic acid.

water + carbon dioxide → carbonic acid

$$H_2O + CO_2 \rightarrow H_2CO_3$$

Pollution can cause rainwater to become even more acidic. The combustion of fossil fuels (particularly coal) releases carbon dioxide, sulfur dioxide (SO_2) and nitrogen oxides (mainly nitrogen dioxide, NO_2) as pollutants into the atmosphere. As a result, rainfall has more carbonic acid in it. It also has strong acids in it formed by the combination of sulfur dioxide and nitrogen dioxide with water vapour in the atmosphere. These form in a series of chemical equations that can be summarised as:

sulfur dioxide + water → sulfurous acid

$$SO_2 + H_2O \rightarrow H_2SO_3$$

sulfur dioxide + water + oxygen → sulfuric acid

$$2SO_2 + 2H_2O + O_2 \rightarrow 2H_2SO_4$$

nitrogen dioxide + water → nitrous acid + nitric acid

$$2NO_2 + H_2O \rightarrow HNO_2 + HNO_3$$

Rainwater dissolves these acids and returns them to Earth as **acid rain**. Depending on the weather, it may return instead as acid dew, fog, sleet, hail or snow. The pH of acid rain is much lower (more acidic) than that of normal rain (pH 6), and has been measured as low as 3.3 (strongly acidic).

The effect on the environment

Although acid rain is not yet a significant concern in Australia, its effects are evident in highly populated and industrialised areas such as the United States, southern Canada, Europe and China.

- Acid rain has killed forests, as shown in Figure 3.2.7. It also removes aluminium from the soil, which then runs into rivers and lakes and percolates downwards into the groundwater. These increased aluminium levels are often enough to kill the plants, animals and fish that use or live in that water.
- Acid rain has killed fish, frogs, water snails, insects and other organisms living in lakes or around lakes. These lakes have increased acidity (decreased pH). As Figure 3.2.8 shows, pH has different effects on different animals. It shows, for example, that frogs can exist in conditions that have a range of pH from 7.0 to 4.0. In contrast, snails only exist in pH 7.0 to 6.0. This means that frogs can survive in more acidic conditions than snails.

pH	7.0	6.5	6.0	5.5	5.0	4.5	4.0
Trout							
Bass							
Perch							
Frogs							
Yabbies							
Snails							

FIGURE 3.2.8 Frogs are least likely to be affected by acid rain. The eggs of most fish will not hatch at pH of less than 5.

SCIENCE AS A HUMAN
ENDEAVOUR

- Acid rain has destroyed buildings and sculptures, like the one in Figure 3.2.9. Calcium carbonate is the major component of the marble and limestone that make up many old buildings and sculptures. Calcium carbonate is a base and so it reacts with the acids in rain, making the rock thinner and more likely to weather, flake and break off. A typical reaction is shown below.

calcium carbonate + sulfuric acid → calcium sulfate + water + carbon dioxide

$$CaCO_3 + H_2SO_4 \rightarrow CaSO_4 + H_2O + CO_2$$

FIGURE 3.2.9 Marble is composed of the base calcium carbonate, which will react with any acid that it comes in contact with. This makes marble statues vulnerable to attack by acid rain.

SciFile

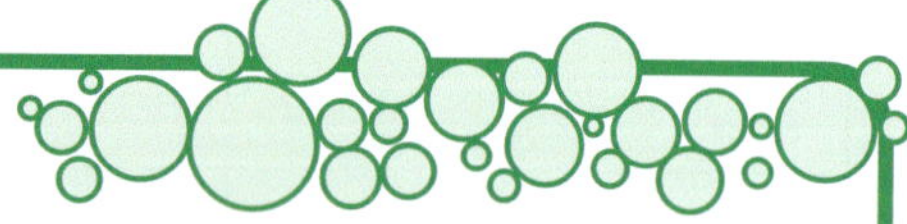

Natural acid rain

Volcanoes spew out huge quantities of sulfur and sulfur-based gases every time they erupt. Lightning produces nitrogen-based gases. These gases react with water vapour in the atmosphere and produce their own acids and acid rain. Studies of ancient ice samples from Antarctic glaciers show that some natural acid rain has always fallen.

Controlling acid rain

Tall chimneys like those in Figure 3.2.10 mean that local areas are less likely to be affected by the chemicals released by them and the acid rain that forms. However, this does not solve the problem. It just pushes the problem onto more distant areas.

Coal-fired power plants are now fitted with 'scrubbers' that remove sulfur-based pollutants from the gases they release. In a similar way, modern cars use catalytic converters to remove nitrogen-based gases from their exhausts.

FIGURE 3.2.10 Pollution can cause acid rain to form. Tall chimneys minimise the damage in the area around them but increase the damage further away.

REVIEW

1 Write word equations to describe how acid rain forms.

2 Use Figure 3.2.8 on page 91 to predict the order in which animals die when the acidity of a lake increases due to acid rain.

3 Propose reasons why forests are more likely to be damaged by acid rain if they are:

  a near cities

  b on the foggy upper slopes of mountains.

AB 3.6

MODULE 3.2

Review questions

Remembering

1 Define the terms:
- a salt
- b chemical digestion.

2 What term best describes each of the following?
- a a tablet you take to neutralise stomach acid and relieve heartburn
- b helper chemicals that speed up a reaction.

3 What are the products of a neutralisation reaction?

4 What gas is formed when an acid reacts with the following?
- a metal
- b carbonate.

5 State the pH:
- a range of the stomach
- b that triggers the release of sodium hydrogen carbonate from the pancreas.

6 What bases do antacids usually contain?

7 Copy the following reactions into your workbook and then fill in the missing gaps to complete them.
- a acid + metal → ...... + ...
- b acid + carbonate → ..... + ... + ...
- c nitric acid + potassium oxide → ..... + ...
- d acetic acid + sodium hydrogen carbonate → ..... + ... + ...
- e H_2SO_4 + Fe → ..... + ...
- f $2HCl + CaCO_3$ → ..... + ... + ...

Understanding

8
- a Explain what causes the pain of heartburn.
- b Explain how antacids work to relieve heartburn.

Applying

9 Use a word equation to explain why the plastic bag in the science4fun on page 87 explodes when the chemicals are allowed to mix.

Analysing

10 Analyse the following word and balanced formula equation.

phosphoric acid + aluminium → aluminium phosphate + hydrogen gas

$$2H_3PO_4 + 2Al \rightarrow 2AlPO_4 + 3H_2$$

Identify:
- a the reactants
- b the products
- c the chemical formula for phosphoric acid
- d the name and chemical formula of the salt produced
- e whether the reaction is an example of an acid–metal reaction, a neutralisation reaction, or an acid–carbonate reaction.

Evaluating

11 Propose reasons why the pH of the stomach varies.

12 What do you think happens to the carbon dioxide formed when acid is neutralised in the duodenum?

Creating

13 Construct word equations for the reactions that happen when:
- a magnesium is added to sulfuric acid
- b aluminium hydroxide neutralises hydrochloric acid
- c hydrochloric acid and sodium carbonate are mixed.

14 Construct word equations for the following acid reactions.
- a Iron filings are sprinkled into a solution of hydrochloric acid. The iron dissolves. In doing so it forms a solution of iron(II) chloride. Bubbles of hydrogen gas rise from the iron filings as they react.
- b Copper reacts differently from other metals in acids. For example, copper dissolves in nitric acid to form copper nitrate and water. Instead of hydrogen or carbon dioxide forming, a poisonous brown gas called nitrogen dioxide is released.

MODULE

3.2 Practical investigations

1 • Making hydrogen

Purpose

To prepare and test hydrogen gas.

Timing

45 minutes

Materials

- 5 cm strip of magnesium ribbon
- 2 M hydrochloric acid

- 1 test-tube, rubber stopper with opening and glass tube to fit
- 2 additional test-tubes
- test-tube rack
- eyedropper
- wax taper

2 M HCl is a strong and concentrated acid.

Wear safety glasses, a laboratory coat or apron and rubber gloves.

Procedure

1. Place the test-tube with the rubber stopper in the test-tube rack.
2. Roll the magnesium ribbon and drop it into the test-tube.
3. Use the eyedropper to add about 2 cm of hydrochloric acid.
4. Immediately turn another test-tube upside down and collect the hydrogen gas produced, as shown in Figure 3.2.11.

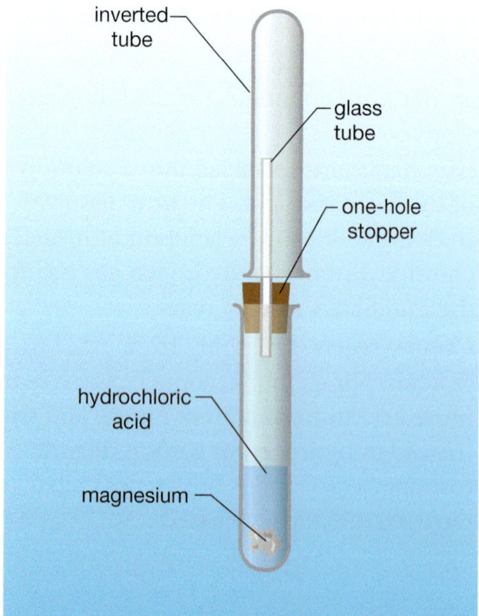

FIGURE 3.2.11

5. After about a minute, remove the upper test-tube, stopper it and stand it upside down in a test-tube rack.
6. Use the remaining test-tube to collect another sample of hydrogen gas. Stopper it and stand it upside down in a test-tube rack.
7. Stop the reaction by carefully adding water to the tube containing the acid.
8. Stand well away from the test-tube rack and light the wax taper.
9. Hold the flame to the mouth of the inverted test-tube of hydrogen gas and remove the stopper.
10. Record what happens.

Results

Record all your observations, including the smell as hydrogen was wafted towards you.

Review

1. Explain how you knew this reaction was producing a gas.
2. Is the reaction happening here an acid–base, acid–metal or neutralisation reaction?
3. Construct a word equation and a balanced formula equation for this reaction.

MODULE 3.2

Practical investigations

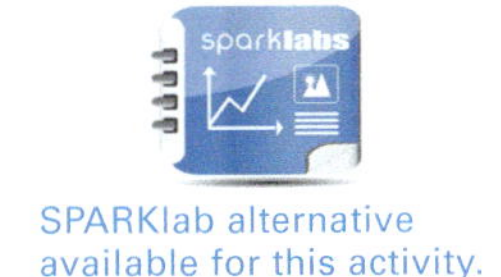

SPARKlab alternative available for this activity.

2 • Evaluating antacids

Antacids contain a base that neutralises some of the excess hydrochloric acid in your stomach.

Purpose

To evaluate how effective different antacids are at reducing acidity.

Timing 45 minutes

Materials

- different antacid tablets
- 0.1 M hydrochloric acid solution
- 250 mL beaker
- liquid universal indicator or pH meter
- eyedropper
- rubber gloves

SAFETY

Hydrochloric acid will burn and irritate open cuts and your eyes so wear rubber gloves and safety glasses at all times.

Do not taste any chemicals.

Procedure

1. Pour about 50 mL of hydrochloric acid into the beaker.
2. Measure the pH of the solution with the pH meter or by putting a couple of drops of the universal indicator in it. Record your observations and pH in a table like the one shown in the Results section.
3. Add an antacid tablet and allow it to dissolve completely.
4. Determine the new pH of the solution with the pH meter or by using the new colour of the solution.
5. Rinse out the beaker and then repeat the experiment with another antacid tablet.

Results

Record all your measurements in a table like this one.

Antacid tablet results

	Tablet 1	Tablet 2	Tablet 3
Colour of universal indicator before antacid was added			
pH of solution before antacid was added			
Colour of universal indicator after antacid was added			
pH of solution after antacid was added			

Review

1. Name the acid that is in your stomach.
2. Explain how this acid can cause the pain known as heartburn.
3. Analyse your results and determine which antacid tablet was most effective.
4. Heartburn has nothing to do with the heart. Why do you think it got this name?

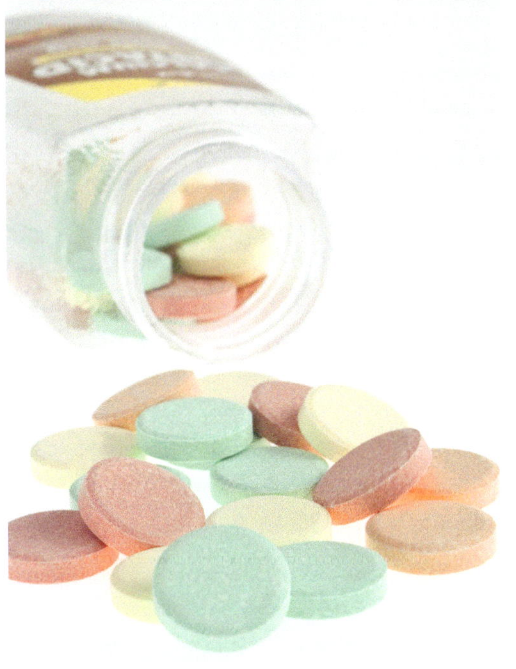

FIGURE 3.2.12 Antacid tablets

MODULE 3.2 Practical investigations

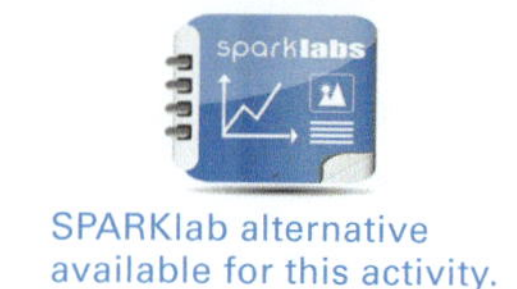

SPARKlab alternative available for this activity.

• STUDENT DESIGN •

3 • Endothermic or exothermic?

Questioning & Predicting Evaluating

Purpose

To determine whether a neutralisation reaction is endothermic or exothermic.

Hypothesis

What do you think a neutralisation reaction is—endothermic or exothermic? Write a hypothesis in your workbook.

Timing 30 minutes to design + 45 minutes

Materials

- vinegar or citric acid
- baking soda (bicarbonate of soda or sodium hydrogen carbonate, $NaHCO_3$)

- materials as selected by students

SAFETY

Vinegar and citric acid are mild irritants, so wear gloves and safety glasses to avoid contact with your skin and eyes. Refer to the SDS for all chemicals when constructing your risk assessment.

Procedure

1 Design an experiment to determine whether the reaction between vinegar or citric acid and baking soda (bicarbonate of soda or sodium hydrogen carbonate, $NaHCO_3$) is endothermic or exothermic.

2 Brainstorm in your group and come up with several different ways to investigate the problem. Select the best procedure and write it in your workbook. Draw a diagram of the equipment you need.

3 Before you start any practical work, prepare a risk assessment. Show your teacher your procedure and risk assessment. If they approve, then collect all the required materials and start work.

See the Activity Book Toolkit to assist with developing a risk assessment.

Use the STEM and SDI template in your eBook to help you plan and carry out your investigation.

Review

1 Contrast endothermic with exothermic reactions.

2 a Classify the reaction that happened here as endothermic or exothermic.
  b Justify your choice.

3 a Construct a conclusion for your investigation.
  b Assess whether your hypothesis was supported or not.

4 Evaluate your procedure. Pick two other prac groups and evaluate their procedures too, identifying their strengths and weaknesses.

• STUDENT DESIGN •

4 • Acid–base rocket

Purpose

To use a neutralisation reaction to propel a model rocket.

Timing 45 minutes

Materials

- vinegar
- baking soda (bicarbonate of soda or sodium hydrogen carbonate, $NaHCO_3$)

- materials as selected by students

SAFETY

Vinegar is a mild irritant, so wear gloves and safety glasses to avoid contact with your skin and eyes. Refer to the SDS for both vinegar and sodium hydrogen carbonate when constructing your risk assessment.

Procedure

1 Search the internet for a design of a rocket that uses a mixture of vinegar and baking soda (bicarbonate of soda or sodium hydrogen carbonate, $NaHCO_3$) to propel it.

2 Print out or save the instructions or video of the rocket you wish to build.

3 Before you start any practical work, prepare a risk assessment.

See the Activity Book Toolkit to assist with developing a risk assessment. Show your teacher your design and your risk assessment. If they approve, then collect all the required materials and start work.

Use the STEM and SDI template in your eBook to help you plan and carry out your investigation.

Review

1 Recall the reaction of vinegar and baking soda by writing its word equation and balanced formula equation.

2 Name the gas that 'powered' your rocket.

3 Use forces to explain how this gas caused your rocket to move.

MODULE

3.3 Reactions of life

Almost all life on Earth depends on two processes called photosynthesis and respiration. Plants use photosynthesis to make glucose, which stores energy for later use. Respiration is needed to release that stored energy. Respiration is also used by animals (including humans) to release energy from the glucose they have absorbed from their food.

Photosynthesis

Photosynthesis is a series of chemical reactions that green plants use to produce a sugar called glucose ($C_6H_{12}O_6$). **Glucose** stores absorbed from sunlight. This energy is stored as chemical energy, which can then be used for all the processes that support life such as growth, repair, reproduction and movement. Glucose can be thought of as a plant's food.

Photosynthesis is a complex series of reactions that can be summarised by the word equation:

$$\text{carbon dioxide} + \text{water} \xrightarrow[\text{chlorophyll}]{\text{sunlight}} \text{glucose} + \text{oxygen}$$

and the balanced equation:

$$6CO_2 + 6H_2O \xrightarrow[\text{chlorophyll}]{\text{sunlight}} C_6H_{12}O_6 + 6O_2$$

In these equations the word *sunlight* above the arrow shows that energy from the Sun is absorbed during this reaction. Therefore, photosynthesis is an example of an endothermic reaction. The equation also shows that the photosynthesis requires a chemical known as **chlorophyll**. Chlorophyll is the chemical that makes plants green. The word *chlorophyll* is written under the arrow to show chlorophyll is necessary for photosynthesis but it is not used up in the reaction.

SciFile

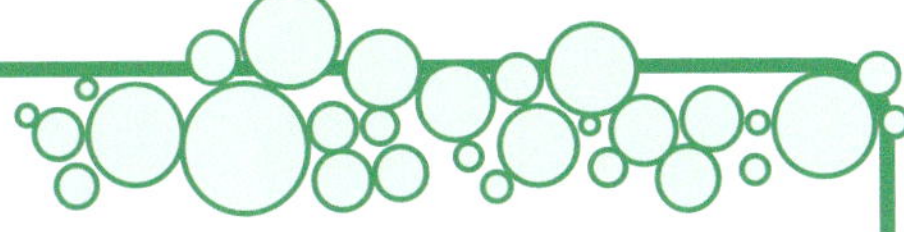

Minty air

The process of photosynthesis was discovered by English chemist, Joseph Priestley (1733–1804). In 1772, Priestley sealed a mouse with some mint under an air tight jar. Without the mint, the mouse would have used up all the oxygen in the jar and eventually suffocated. However, the mint produced enough oxygen via photosynthesis to allow the mouse to survive unharmed.

As long as a plant has enough sunlight, carbon dioxide and water, it can manufacture all the food it needs. Photosynthesis takes place in any part of the plant that is green and exposed to sunlight. Leaves are the most exposed parts of a plant and so most photosynthesis takes place there. To maximise this exposure, most plants have flattened leaves that don't overlap.

Photosynthesis raw materials

Photosynthesis needs carbon dioxide and water as reactants. Photosynthesis also needs sunlight to power its reactions and chlorophyll to make them work.

- Water: Water is carried from the roots up the plant's transport system and into the leaves as shown in Figure 3.3.1.

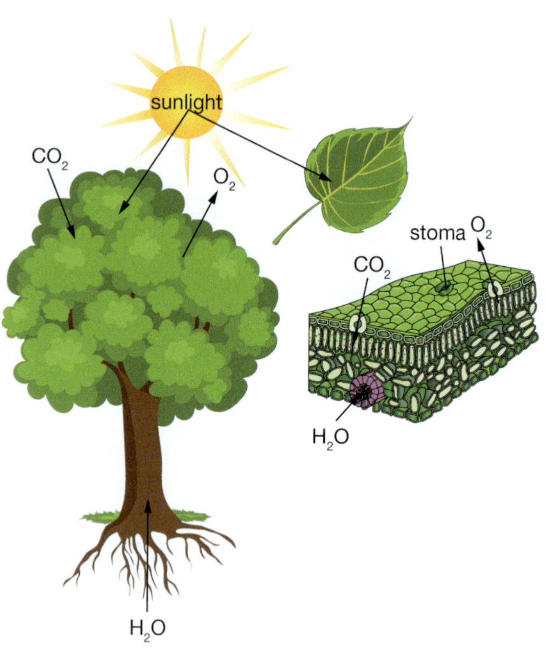

FIGURE 3.3.1 All the raw materials need to be brought together in the chloroplasts of the leaf cells before photosynthesis can take place.

- Carbon dioxide: Carbon dioxide is absorbed from the atmosphere through tiny openings in the leaves called **stomata**. The opening is called a stoma. Typical stomata are shown in Figure 3.3.2. Stomata are also where the oxygen produced by photosynthesis is released back into the atmosphere.
- Chlorophyll and sunlight: Once the water and carbon dioxide are inside the plant's cells, they enter the chloroplasts shown in Figure 3.3.3. **Chloroplasts** contain the chlorophyll necessary for photosynthesis. When sunlight enters the chloroplasts, photosynthesis takes place. The chloroplasts also contain **enzymes**. These are helper chemicals that speed up the reactions of photosynthesis.

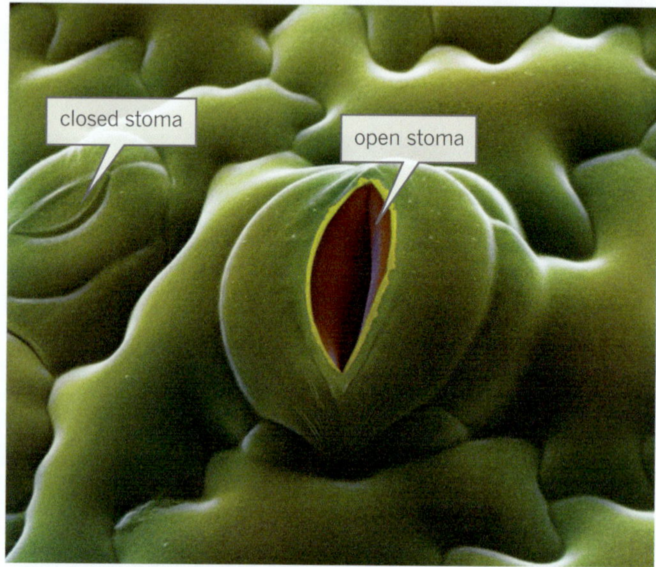

FIGURE 3.3.2 These stomata provide an entry for the carbon dioxide needed for photosynthesis. Stomata also provide an exit for the water vapour and carbon dioxide produced by respiration.

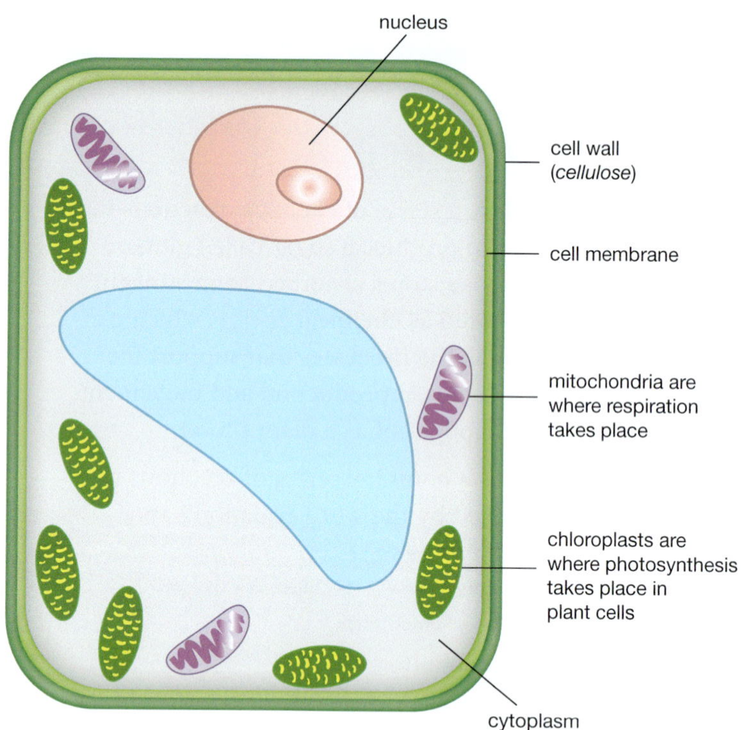

FIGURE 3.3.3 Plant cells contain chloroplasts for photosynthesis and mitochondria for respiration.

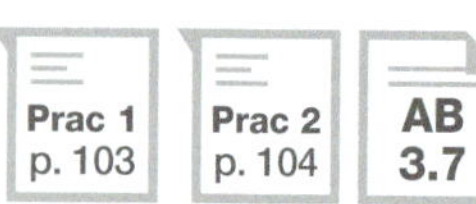

Respiration

While photosynthesis is taking place in the chloroplasts of plant cells, **respiration** is taking place in the **mitochondria**.

In some ways, respiration can be considered the reverse of photosynthesis. During respiration, glucose and oxygen are converted back into carbon dioxide and water. The word equation and balanced chemical equation for this reaction are:

glucose + oxygen → carbon dioxide + water + energy

$$C_6H_{12}O_6 + 6O_2 \rightarrow 6CO_2 + 6H_2O + \text{energy}$$

The equation shows that respiration releases the chemical energy stored in the glucose molecules. Since energy is being released, respiration is an exothermic reaction. The energy released by respiration can then be used for biological processes such as growth, movement repair and reproduction.

Animal cells also have mitochondria to perform respiration; however, unlike plant cells, animal cells do not have chloroplasts. Therefore, animals cannot produce their own glucose. Instead, animals must get the glucose they need from the food they eat.

When you breathe in, you are absorbing oxygen from the air that your cells require for respiration. When you exercise, your body needs more energy from respiration. This requires you to breathe faster and more deeply. You also breathe out strongly to remove the extra carbon dioxide being produced by the reaction. This increased flow of oxygen and carbon dioxide is what is being measured in the cyclist in Figure 3.3.4.

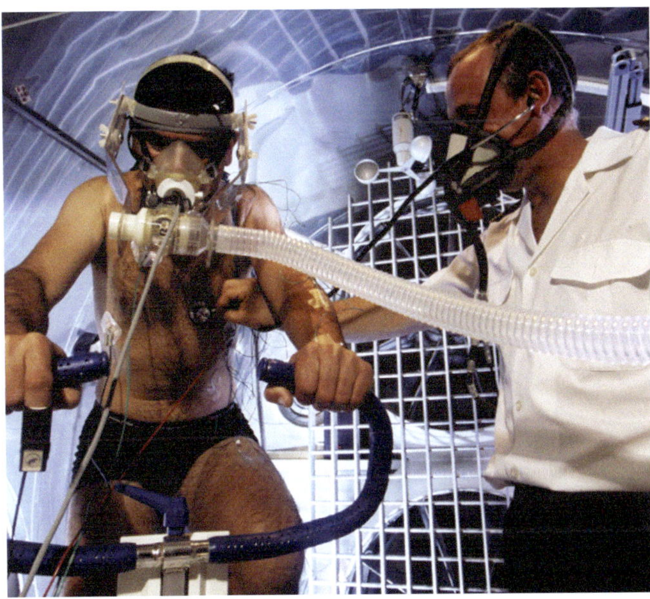

FIGURE 3.3.4 Aerobic respiration needs oxygen. Breathing provides oxygen and gets rid of the carbon dioxide produced.

This type of respiration is known as **aerobic respiration** because oxygen is used to convert the glucose into energy. However, sometimes you will not be able to breathe in enough oxygen to supply all your energy needs. When your oxygen supply is insufficient, some of the cells in your body switch to a process that doesn't require oxygen. This process is known as **anaerobic respiration** and is shown in the chemical equations below.

glucose → lactic acid + energy

$$C_6H_{12}O_6 \rightarrow 2C_3H_6O_3 + \text{energy}$$

Lactic acid builds up in muscles and is what makes them sore after exercising hard.

STEM 4 fun

Colour of chlorophyll

NO

PROBLEM

How can you find out if green chlorophyll is a pure colour or a mixture of colours?

SUPPLIES

- fresh spinach leaf
- coarse sand
- methylated spirits
- mortar and pestle
- plastic cup
- blotting paper

PLAN AND DESIGN Design the solution: what information do you need to solve the problem? Draw a diagram. Make a list of materials you will need and steps you will take.

CREATE Follow your plan. Create your solution to the problem.

IMPROVE What works? What doesn't? How do you know it solves the problem? What could work better? Modify your design to make it better. Test it out.

REFLECTION

1 What field of science did you work in today?
2 If another student was to do this task, what advice would you give?
3 What did you do today that worked well? What didn't work well?

Respiration in plants

Glucose is brought to the cells of a plant by the transport system. Oxygen moves into the plant from the atmosphere through stomata in the leaves.

The products of respiration are carbon dioxide and water. Carbon dioxide diffuses out of the mitochondria into the cytoplasm, a thick liquid sap that fills much of the rest of the cell. Some of the carbon dioxide is taken in by the chloroplasts and used in photosynthesis. The rest leaves the cells and eventually exits the plant through the stomata.

Water is also released by the plant cell and evaporates. It exits the plant through the stomata in the form of water vapour. In hot weather, the stomata close to stop the plant losing too much water.

Prac 3
p. 105

Carbon storage

Plant respiration releases carbon dioxide back into the atmosphere, but in smaller quantities than that used by photosynthesis. This is because the glucose produced by a plant is not used just for respiration. The flow diagram in Figure 3.3.5 shows how glucose is also used to produce other substances.

Specifically, glucose is:

- converted to starch for short-term storage. During the night the starch may be converted back into glucose and then used as a source of energy
- stored long-term as starch in the stems or roots of vegetables such as potatoes and sweet potatoes
- converted into cellulose for the manufacture of plant cell walls
- converted to substances used to make plant oils or proteins (Olive oil is formed this way.)
- converted into a more complex sugar called sucrose ($C_{12}H_{22}O_{11}$) and transported to parts of the plant where photosynthesis cannot occur (such as the roots). Sucrose is the sugar used to sweeten coffee and tea
- used in the process of making vitamins.

In this way, plants store carbon that has been removed from the atmosphere by photosynthesis. Every tree, forest, field and crop can therefore be thought of as a **carbon sink**, effectively trapping the carbon. For this reason, environmentalists are encouraging the preservation of existing forests and the planting of new ones to reduce carbon dioxide in the atmosphere and to reduce the effects of climate change. When trees are chopped down, they cannot take in any more carbon. Rotting returns their carbon to the soil. Burning them releases much of the stored carbon back into the atmosphere as smoke, soot and carbon dioxide.

Respiration in humans

The bloodstream of animals carries glucose and oxygen to the cells. Glucose is one of the products of digestion of food. Oxygen is breathed in, then passes through narrower and narrower tubes until it enters small air sacs in the lungs known as **alveoli**. These sacs are connected to tiny blood vessels that allow oxygen and carbon dioxide to move between the lungs and the bloodstream. You can see this network of tubes in Figure 3.3.6.

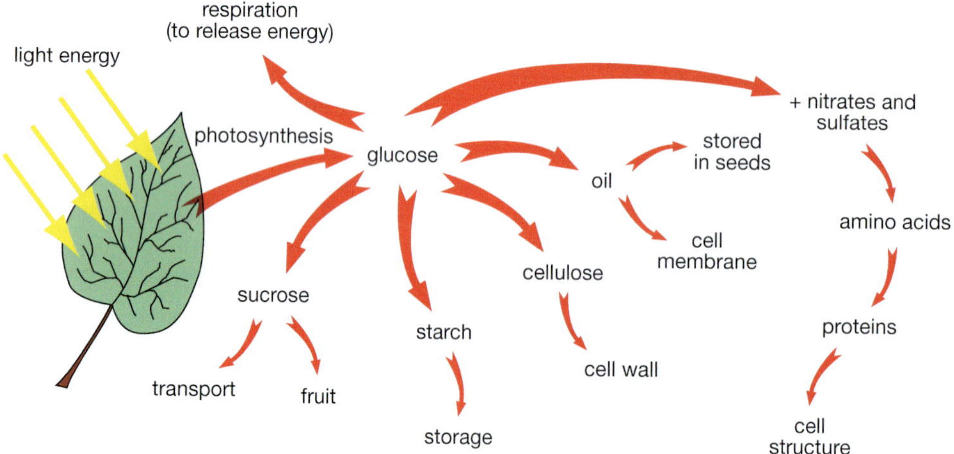

FIGURE 3.3.5 Glucose is not used just for respiration, but also for storing energy, creating fruit and building parts of a plant's cells.

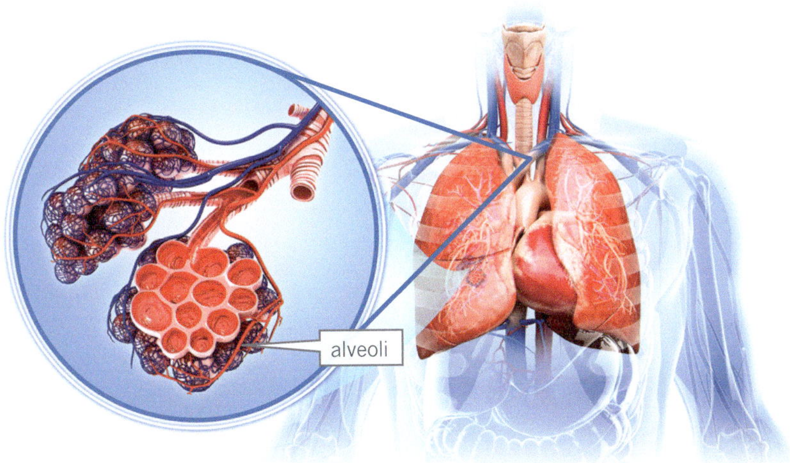

FIGURE 3.3.6 The lungs supply the bloodstream with oxygen so that aerobic respiration can occur in the cells. They also provide an exit for waste carbon dioxide and some water vapour.

The blood also carries away the waste water and carbon dioxide produced by respiration. Some water is re-absorbed into the body. Some is removed by the kidneys to be stored in the bladder and later expelled as urine. Remaining water is breathed out as water vapour. Carbon dioxide moves from the blood into the alveoli in the lungs where it is breathed out.

AB 3.8

Comparing photosynthesis and respiration

The chemical equations for photosynthesis and respiration suggest that the two processes are exact opposites of each other. For example, photosynthesis makes glucose, and respiration uses it. Table 3.3.1 lists the similarities and differences between photosynthesis and respiration.

TABLE 3.3.1 Comparison of photosynthesis and respiration

Photosynthesis	Respiration
makes glucose	uses glucose
uses carbon dioxide	makes carbon dioxide
makes oxygen gas	uses oxygen gas
uses water	makes water
endothermic	exothermic
requires chlorophyll	does not need chlorophyll
occurs only in the chloroplasts of cells of green plants	occurs in the mitochondria of cells of all living things
shuts down at night	happens continuously (day and night)

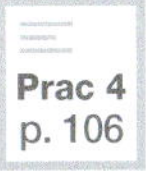
Prac 4 p. 106

SciFile

How do mozzies breathe?

Mammals, birds, reptiles and amphibians have lungs to draw in oxygen and expel carbon dioxide. Amphibians, such as frogs, also exchange gases through their skin. Insects don't have lungs. Instead they have small tubes from the outside of their bodies, feeding the gases directly into and out of their blood. Mosquito larvae (wrigglers) breathe through a tube that sticks out of the water.

MODULE 3.3

Review questions

Remembering

1 Define the terms:
- a photosynthesis
- b anaerobic respiration
- c chloroplast
- d alveoli.

2 What term best describes each of the following?
- a the green chemical found in plants necessary for photosynthesis
- b the part of a cell where respiration takes place
- c small openings in plant leaves that allow gases to move in and out
- d a place where carbon from the atmosphere is trapped, e.g. a tree or crop.

3 Name the following chemicals.
a CO_2 b O_2 c $C_6H_{12}O_6$

4 Apart from energy, list six ways that glucose is used by plants.

5 What chemical causes your muscles to be sore after intense exercise?

Understanding

6 What does the chlorophyll in a leaf help a plant to do?

7 Explain how aerobic respiration releases energy from glucose.

8 Outline how the following get their glucose:
a plants   b animals.

9 Environmentalists sometimes talk about a *carbon sink*. Explain what this term means.

Applying

10 Sucrose is the sugar that is used to sweeten tea and coffee. Sucrose comes from the stems of sugar cane. How does sucrose get in the canes?

11 Equipment was set up as shown in Figure 3.3.7 to gather information about photosynthesis.
- a Identify the gas that would be collected in the test-tube.
- b Demonstrate how the rate of photosynthesis could be determined using this equipment.

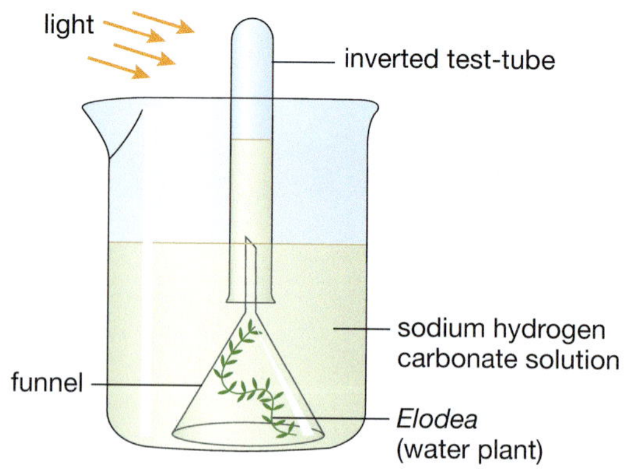

FIGURE 3.3.7

Analysing

12 Compare photosynthesis and aerobic respiration by:
- a writing their word equations
- b writing their balanced formula equations
- c listing the ways in which they are similar
- d listing the ways in which they are different.

13 When exercising hard, some cells use anaerobic respiration to release energy.
- a Compare aerobic and anaerobic respiration by listing the similarities and differences between their:
  - i reactants
  - ii products.
- b Some aerobic respiration still occurs when you are exercising hard. What evidence shows that aerobic respiration still occurs when you are exercising hard?

Evaluating

14 It has been suggested that humans will need to take green plants along with them if they are ever to travel far into space. Propose reasons why.

Creating

15 You get the glucose you need for respiration from the food you eat. The glucose is obtained by digesting proteins, fats and carbohydrates. Starch is a complex carbohydrate that is found in root vegetables such as parsnips, sweet potatoes and potatoes. Create a diagram or flow chart to show how energy from the Sun can be stored as starch in potatoes.

MODULE 3.3

Practical investigations

1 • Looking at stomata  

Purpose

To compare the number of stomata on different leaf surfaces.

Timing 45 minutes

Materials

- leaf cut from a plant
- clear nail varnish

- paintbrush or brush from a bottle of nail varnish
- forceps
- microscope
- slide and coverslip
- small container of water
- dropper

Procedure

1 With the brush, apply a thin coat of nail varnish to a small area on the underside of the leaf (Figure 3.3.8).

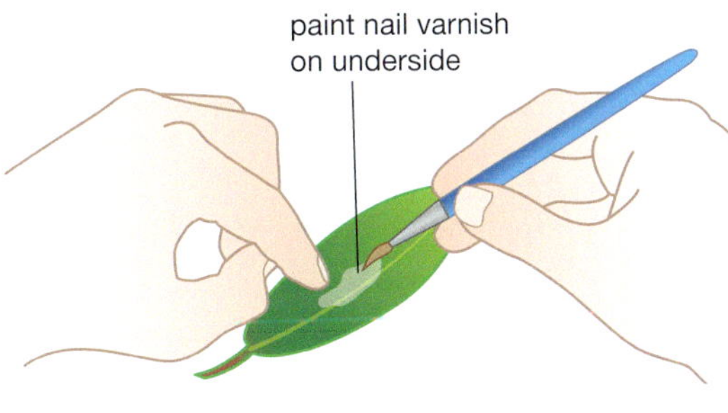

FIGURE 3.3.8

2 Allow the varnish to dry and then use the forceps to gently peel the layer of varnish from the leaf.

3 The nail varnish will have made an exact copy of the leaf surface.

4 Place the film of varnish in a drop of water on the microscope slide and place the coverslip on top, making sure that no air bubbles are trapped.

5 Examine the slide using low power on the microscope.

6 Count the number of stomata in a field of view.

7 Change the microscope to high power and focus on one stoma. Identify the guard cells.

8 Repeat steps 1 to 7 using the upper surface of the leaf.

Results

1 Record the number of stomata you counted in one field of view for both the upper and lower surfaces.

2 Construct a diagram of the stoma and guard cells. Label the parts.

Review

1 Describe the orientation of the leaf you studied when it was on the plant. Did it lie horizontally or hang vertically?

2 a Was whether the stoma you drew open or closed?

  b What evidence led you to this answer?

3 Compare the number of stomata found on the two surfaces.

4 Explain any advantage to the plant of having the observed distribution of stomata.

5 Predict how the number and distribution of stomata could change if:

  a leaves hung vertically rather than horizontally

  b the leaves came from a plant living in a very moist environment

  c the leaves floated on water, like a water lily.

MODULE 3.3

Practical investigations

2 • Testing leaves for starch  

Purpose

To test leaves for the presence of starch.

Timing 45 minutes

Materials

- 1 variegated leaf that has been in sunlight for several hours
- methylated spirits
- iodine solution

- beaker
- hotplate
- test-tube
- test-tube rack
- Petri dish
- tongs
- dropping pipette
- tweezers

SAFETY

Methylated spirits is flammable, so keep it well away from flames and the hotplate.

Wear safety glasses at all times.

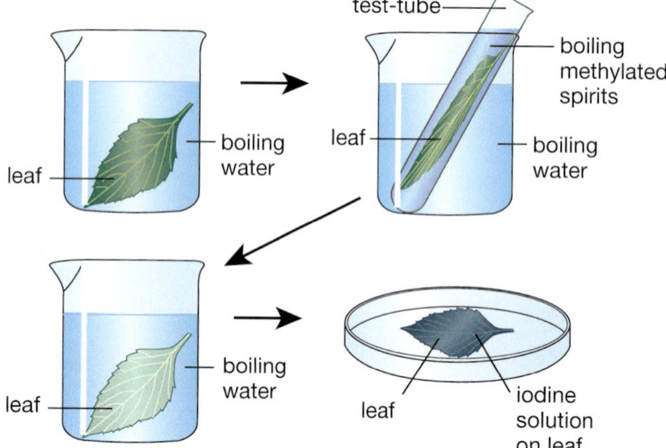

FIGURE 3.3.9

Procedure

1. Sketch the leaf, showing where there is a lot of chlorophyll (the green parts) and where there is little (the paler or cream parts). Keep this sketch for later.
2. Place around 100 mL of water in the beaker and heat it on the hotplate until it boils. Drop the leaf in the water and heat it for about 30 seconds.
3. Use the tweezers to carefully remove the leaf from the water and place it in a test-tube. Keep the water in the beaker boiling.
4. Keeping well away from the hot plate, add methylated spirits to the test-tube so that the leaf is covered (Figure 3.3.9).
5. Place the test-tube into the beaker of boiling water and heat it until the green chlorophyll is removed from the leaf.
6. Using tongs, carefully remove the test-tube from the beaker of boiling water and remove the leaf from the test-tube.
7. Place the leaf back into the beaker of boiling water for a few minutes to soften it.
8. When the leaf is soft, remove it from the water and place it in a Petri dish (Figure 3.3.9).
9. Add 4 drops of iodine solution to the leaf.

Results

Sketch the leaf, showing where starch is present and where it is absent.

Review

1. What did the iodine do when it reacted with starch?
2. Some parts of the leaves contain more starch than other parts. Explain why.
3. Propose a reason why it is safer to use a hot plate in this experiment instead of a Bunsen burner.
4. Why do you think the leaf was boiled in this experiment?
5. a Predict the results you would have obtained if you had tested a leaf picked from the plant at dawn.
   b Justify your prediction.

MODULE
3.3 Practical investigations

3 • Respiring plants  

Purpose

To investigate respiration in plants.

Timing 60 minutes + 30 minutes 1 to 2 days later

Materials

- 9 pieces of water plant such as *Elodea*
- bromothymol blue indicator

SAFETY

Wear safety glasses at all times.

- 9 test-tubes
- 3 test-tube racks
- drinking straw
- 50 mL beaker

Procedure

1. In your workbook, construct a table like the one shown in the Results section. Give your table a title.
2. Two-thirds fill each test-tube and the beaker with water. Add a few drops of bromothymol blue to give the solution an obvious blue colour.
3. Using the drinking straw, bubble exhaled air through the water in the beaker.
4. Observe and record the colour change of the indicator.
5. Label the test-tube racks: ‘Dark’, ‘Low light’, ‘Bright light’.
6. Add the same amount of water weed to each test-tube (Figure 3.3.10).

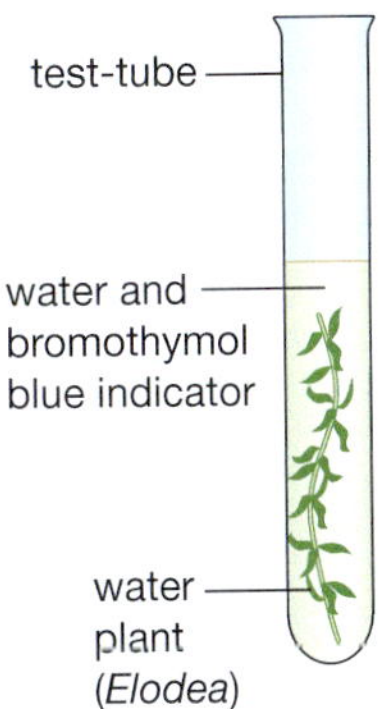

FIGURE 3.3.10

7. Place three test-tubes into each of the test-tube racks.
8. Place the test-tube racks in an appropriate place.
   - dark—a cupboard
   - low light—a corner of the classroom away from the windows
   - bright light—a window sill, but make sure the water does not get hot.
9. Leave the test-tubes for 24–48 hours.
10. Observe and record any changes in the colour of the indicator. Record your observations in a table like that shown in the Results section.

Results

1. Describe what happened when you bubbled exhaled air through the water containing bromothymol blue.
2. Record your observations in your results table.

Treatment	Indicator colour	
	Start	After 24 hours
dark		
low light		
bright light		

3. Bromothymol blue is blue in an alkaline or a neutral solution. It turns green and then yellow in an acidic solution. Carbon dioxide dissolved in water produces a weak acid.

   Use this information to identify whether each solution was alkaline, neutral or acidic.

Review

1. Why did the colour change that occurred when you bubbled exhaled air through the water with bromothymol blue?
2. Describe what happened in each of the test-tubes.
3. Explain why this happened.

MODULE
3.3 Practical investigations

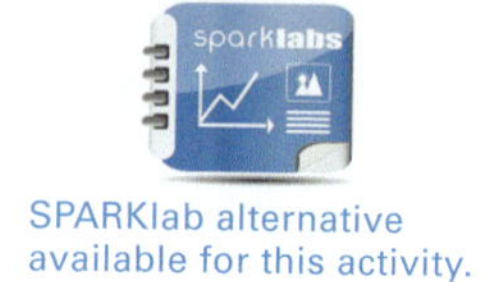

SPARKlab alternative available for this activity.

• STUDENT DESIGN •

4 • Light and photosynthesis

Evaluating

Purpose

To demonstrate that light is needed for photosynthesis to occur.

Timing

60 minutes

Materials

- to be selected by students

Procedure

1. Design an experiment to demonstrate that light is necessary for photosynthesis to occur.
2. Brainstorm in your group and come up with several different ways to investigate the problem. Select the best procedure and write it in your workbook. Draw a diagram of the equipment you need.
3. Before you start any practical work, assess all risks associated with your procedure. Refer to the SDS of all chemicals used. Construct a risk assessment that outlines these risks and any precautions you need to take to minimise them. Show your teacher your procedure and your assessment of its risks. If they approve, then collect your equipment and start work.

   See the Activity Book Toolkit to assist with developing a risk assessment.

Hints

- Think about the product of photosynthesis that you can test easily.
- Consider how you will prevent light reaching plants and how long you will keep the plants in the dark.
- Think about other factors that you will have to control to make this a fair test.
- Use the STEM and SDI template in your eBook to help you plan and carry out your investigation.

Review

1. Recall the word equation and balanced formula equation that summarises the process of photosynthesis.
2. Explain how your observations confirm that light is needed for photosynthesis to occur.
3. Evaluate your procedure. Choose two other prac groups and evaluate their procedures too, identifying their strengths and weaknesses.

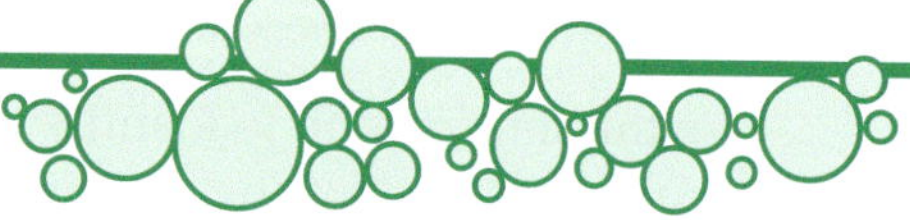

No photosynthesis required!

Sunlight cannot penetrate the ocean beyond 300 metres and so photosynthesis doesn't happen there. However, at the edges of the tectonic plates are vents called 'black smokers'. Chemosynthetic bacteria do not need photosynthesis because they collect energy and nutrients from chemicals dissolved in superheated water erupting from the black smokers. These bacteria provide the basis for a food chain, which includes giant tube worms, spider crabs and mussels.

MODULE

3.4 Nuclear reactions

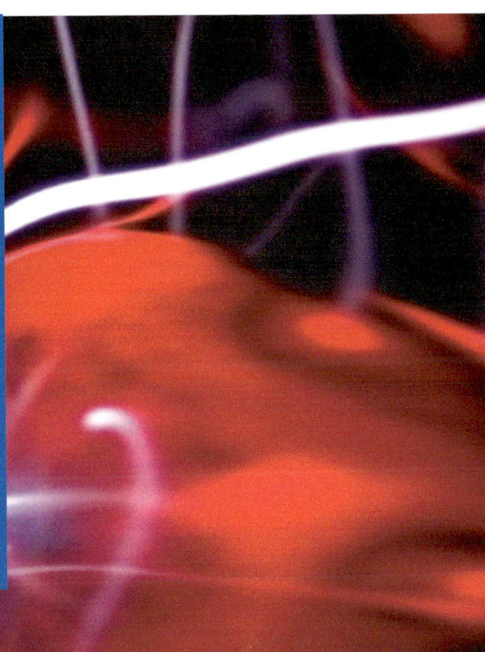

During chemical reactions, the atoms rearrange to form new substances. However, the individual atoms do not change—they stay exactly the same element as they were before. To change an atom from one element to another requires a change at the atom's core—its nucleus. These nuclear changes can release huge amounts of energy and radiation. Understanding these changes has led to the development of nuclear fuels and new medical treatments. However, this technology has also brought the threat of nuclear warfare and the production of nuclear waste.

Nuclear decay

Combustion, corrosion, neutralisation, respiration and photosynthesis are all examples of chemical reactions that convert substances (reactants) into different substances (products). The products often have very different properties from the reactants, but the individual atoms remain the same. To convert an atom into an atom of another element requires a change in the nucleus. This can only be achieved by a **nuclear reaction**.

The nucleus that sits at the centre of an atom is not just standing still. The protons and neutrons are constantly moving and rearranging. In some cases, these rearrangements can cause nuclei to emit (release) high-energy electromagnetic radiation and particles. This process is known as **nuclear decay**. Nuclear decay is a type of nuclear reaction.

Some forms of nuclear decay can cause atoms to change into completely different types of elements. This happens whenever the number of protons in the nucleus changes. For example, some sodium atoms decay by losing a proton. This converts them into neon atoms, as shown in Figure 3.4.1. The process of an atom converting to an atom of another element is known as **transmutation**.

Transmutation cannot happen through a chemical reaction because chemical reactions do not involve changes in the nucleus. Instead, chemical reactions involve the electrons that surround the nuclei of the atoms.

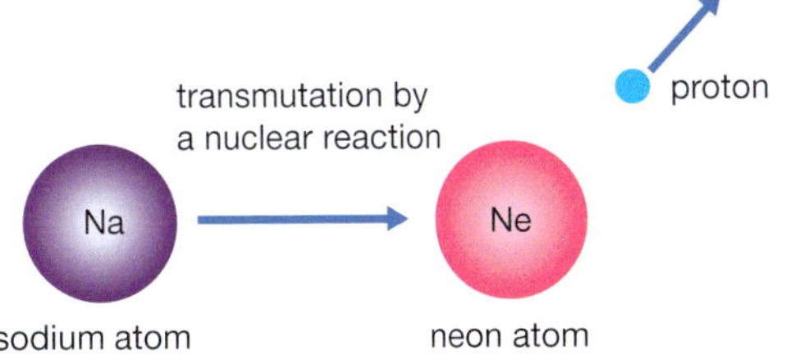

FIGURE 3.4.1 The sodium atom (Na) here is undergoing nuclear decay and changing into a neon atom (Ne). This process is known as transmutation.

Radioisotopes

Isotopes are atoms that have the same number of protons but a different number of neutrons. Hence, isotopes have the same atomic number and belong to the same element but have different mass numbers. In other words, isotopes are different 'versions' of the same element.

Most of the atoms that make up the world around you contain **stable nuclei**. This means that the nuclei will never undergo nuclear decay. However, a tiny fraction of atoms have **unstable nuclei**. These unstable atoms could undergo nuclear decay at any moment. After this happens the atom will usually have a stable nucleus and will not decay any further. These unstable atoms are known as **radioisotopes**—they undergo nuclear decay, making them radioactive.

SciFile

Ambitious alchemists

In the Middle Ages, people known as alchemists tried to turn lead into gold through magic and various chemical reactions. Today, scientists know the alchemists' attempts were pointless and that only a nuclear reaction could convert lead into gold.

Each type of atom may have several isotopes but only some isotopes are **radioactive**. For example, carbon has three naturally occurring isotopes called carbon-12, carbon-13 and carbon-14, as shown in Figure 3.4.2. They are all carbon atoms because they all contain 6 protons and have an atomic number of 6. However, carbon-12 has 6 neutrons (giving it a mass number of 12), carbon-13 has 7 neutrons (mass number 13) and carbon-14 has 8 neutrons (mass number 14). As extra neutrons are added to a nucleus, it becomes unstable. For example, carbon-12 and carbon-13 are stable but carbon-14 is unstable and is therefore a radioisotope.

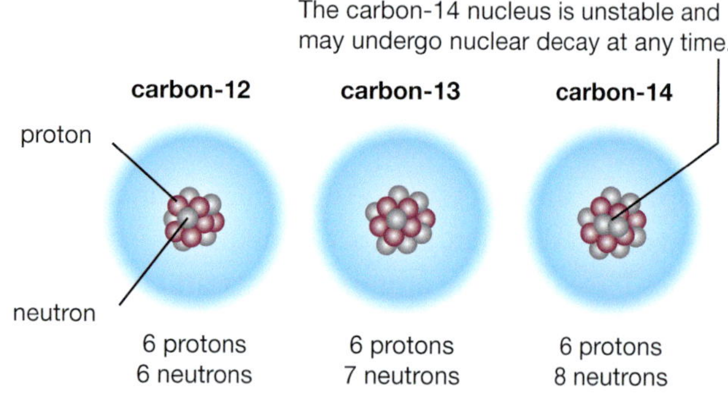

FIGURE 3.4.2 Carbon has three naturally occurring isotopes: carbon-12, carbon-13 and carbon-14. The nucleus of carbon-14 is unstable, so carbon-14 is a radioisotope.

Types of nuclear decay

AB 3.10

There are three types of nuclear decay—alpha decay, beta decay and gamma decay. The three different types of decay are summarised in Table 3.4.1.

Alpha decay

During **alpha decay**, a nucleus ejects an **alpha particle**, which is a cluster of two protons and two neutrons. The alpha particle is given the symbol α. The particle is identical to a helium-4 nucleus, which has the atomic symbol ${}^{4}_{2}He$. An alpha particle is relatively heavy so it moves relatively slowly, at 10% the speed of light.

Alpha decay only occurs in atoms with very heavy nuclei—this is usually where the mass number (protons plus neutrons) is greater than 100. For example, the radioisotope uranium-238 (${}^{238}_{92}U$) undergoes alpha decay as shown in Figure 3.4.3. Through this nuclear reaction, a uranium atom becomes a thorium atom, an entirely different element.

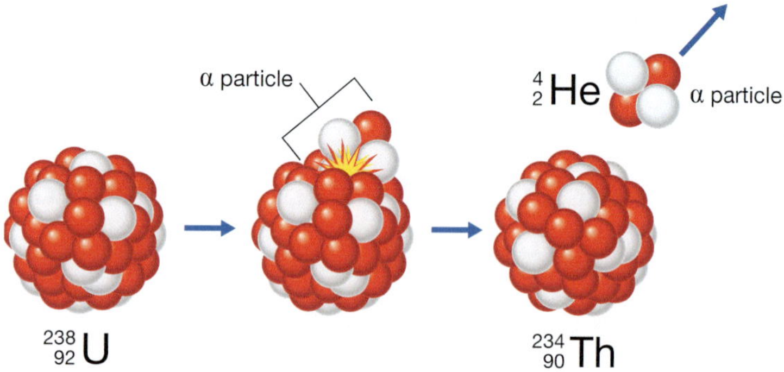

FIGURE 3.4.3 When a uranium-238 nucleus undergoes alpha decay, it becomes a thorium-234 atom. The atom of uranium has changed into an atom of thorium.

Initially, the uranium-238 atom has 92 protons and 146 neutrons. When the uranium-238 ejects an alpha particle, the nucleus loses 2 protons and 2 neutrons. In this way, the atom becomes a thorium-234 atom with 90 protons and 144 neutrons. In other words, the atomic number has decreased by 2 while the mass number has decreased by 4.

Beta decay

Beta decay occurs when the nucleus ejects a **beta particle**, which is given the symbol β. Beta particles are identical to electrons and therefore are very small and have a negative charge. Beta particles are very light and so move at very high speed, about 90% of the speed of light.

When a nucleus undergoes beta decay, a neutron is converted into a proton. This increases the atomic number by one, and so a new element is formed. However, the mass number does not change because the total number of protons and neutrons stays the same.

Carbon-14 undergoes beta decay as shown in Figure 3.4.4. The carbon-14 atom has 6 protons and 8 neutrons. When the atom ejects a beta particle (β), one of the neutrons becomes a proton. This turns the atom into a stable nitrogen-14 atom with 7 protons and 7 neutrons.

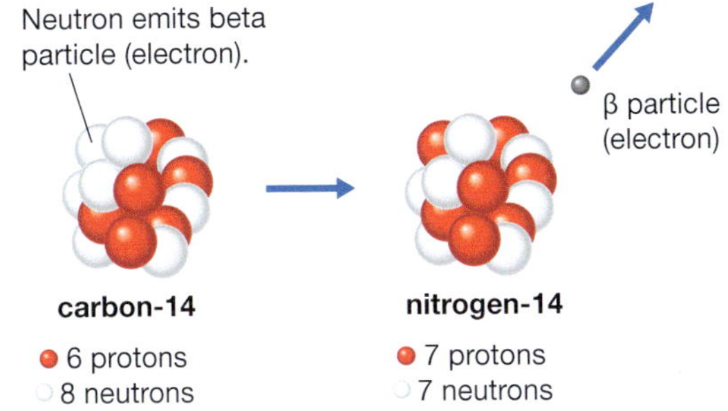

FIGURE 3.4.4 When a carbon-14 nucleus undergoes beta decay, it becomes a nitrogen-14 atom and emits a beta particle.

Gamma decay

Sometimes the protons and neutrons simply rearrange inside the nucleus but do not emit a particle. Instead they emit a form of light and so travel at the speed of light. The radiation emitted is known as a **gamma ray**. This process is known as **gamma decay**. Gamma rays are given the symbol γ. They are like X-rays but are more powerful.

Table 3.4.1 summarises and compares the products formed by alpha, beta and gamma decay.

TABLE 3.4.1 Summary of the products of nuclear decay

	Symbol	Equivalent to	Speed	Charge
alpha particle	α	a helium nucleus	10% the speed of light	+2
beta particle	β	an electron	90% the speed of light	–1
gamma ray	γ	a high-energy x-ray	speed of light	0

Half-life

The rate at which nuclear decay takes place is measured by a radioisotope's **half-life**. The half-life of a radioisotope is the time it takes for half the nuclei to decay. For example, the radioisotope radon-222 decays into polonium-218 with a half-life of 4 days. This means that from 100 radon-222 atoms, 50 would decay over 4 days. Of the remaining 50 nuclei, 25 would decay over the next 4 days, as shown in Figure 3.4.5. And if you waited another 4 days, only 12 or 13 radon-222 atoms would remain.

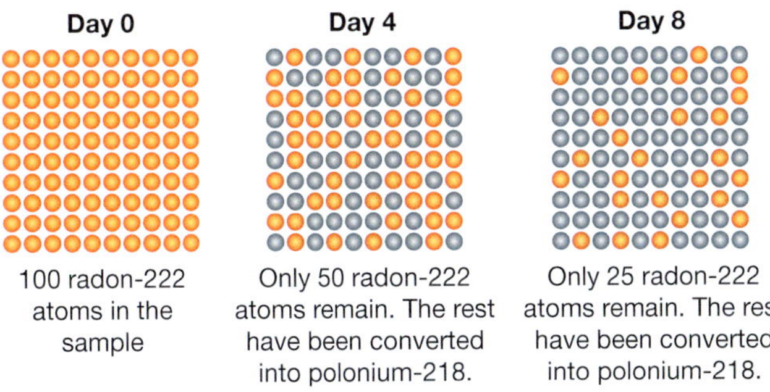

FIGURE 3.4.5 Radon-222 has a half-life of 4 days, so the number of radon atoms halves every 4 days.

The half-life of radioisotopes varies from less than a second to millions of years. Table 3.4.2 lists the half-lives of some common radioisotopes. For example, plutonium-239 has half-life of 24 000 years. It is a waste product of nuclear power plants and its very long half-life poses a significant problem with its disposal. The plutonium cannot be destroyed. Instead it must be buried deep underground until it decays into non-radioactive isotopes.

Prac 1 p. 118

AB 3.9

TABLE 3.4.2 Half-lives of common radioisotopes

Radioisotope	Half-life
gold-200	48 minutes
radon-222	4 days
iodine-131	8 days
cobalt-60	5.3 years
americium-241	460 years
carbon-14	5730 years
plutonium-239	24 000 years
uranium-238	4.5 million years

Carbon dating

The half-life of the radioisotopes carbon-14 can be used by historians and archaeologists to determine the age of fossils like the one in Figure 3.4.6 on page 110. The process is known as **carbon dating**.

The amount of carbon-14 of an organism remains constant throughout its life since carbon-14 breaks down but is replaced by carbon-14 absorbed from the surroundings. When the organism dies, carbon-14 is no longer absorbed. After death, the small amounts of carbon-14 in the organism begin to decay into nitrogen-14 with a half-life of 5730 years. This decay is shown in Figure 3.4.7 on page 110.

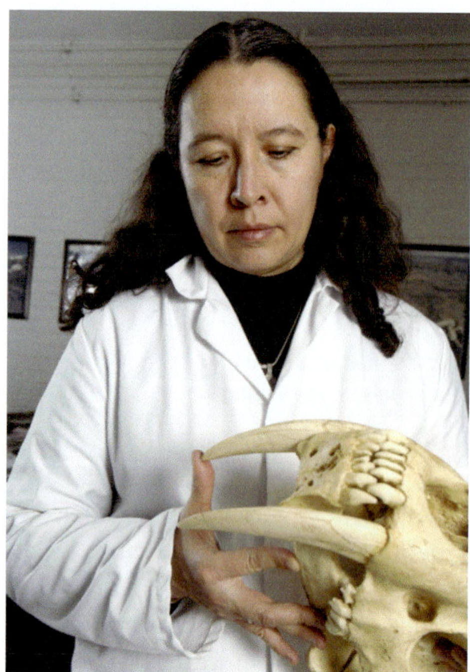

FIGURE 3.4.6 Archaeologists use carbon dating to determine the age of fossils and artefacts. Artefacts are objects made by humans, such as tools.

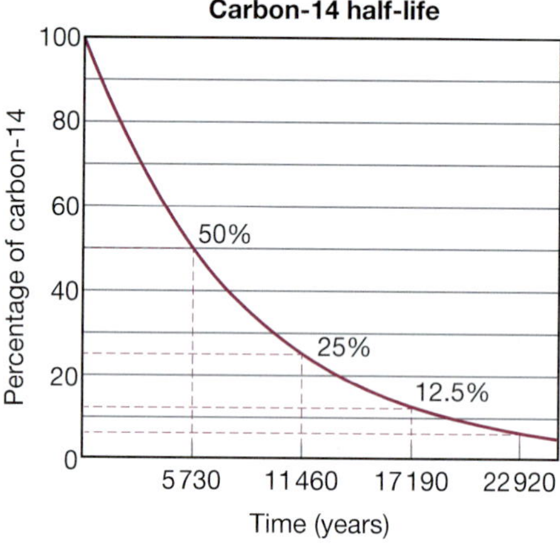

FIGURE 3.4.7 The percentage of carbon-14 atoms in a plant or animal halves every 5730 years after it dies.

By measuring the amount of carbon-14 in fossils and bones, scientists get an accurate idea of when the animal lived. An animal that died 5730 years ago will have half the amount of carbon-14 of an animal living today. An animal that died 11 460 years ago will have a quarter the amount of carbon-14 and so on. Therefore, by measuring the amount of carbon-14 in fossils and bones, scientists can get an accurate idea of when the animal lived. After 50 000 years, so little carbon-14 is left that it no longer gives accurate dates.

Because plants contain carbon, scientists can also use carbon dating to calculate the age of tools, paper and fabrics made from plants.

Nuclear radiation

The term **nuclear radiation** describes any rays or particles emitted (released) by atomic nuclei. The term includes alpha particles, beta particles and gamma rays. Nuclear radiation can be extremely harmful, especially to living organisms. However, it can also be useful in medicine, industrial processes and scientific research.

Biological effects of radiation

Alpha particles, beta particles and gamma rays are particularly damaging to the cells of living organisms. This is because radiation destroys biological molecules and causes unwanted chemical reactions.

Alpha particles, beta particles and gamma rays are all types of **ionising radiation** because they can remove electrons from atoms and molecules, turning them into ions. Exposing cells to ionising radiation can cause them to die or mutate.

Cell death

Cell death occurs when ionising radiation enters the cell and destroys the biological molecules beyond repair. This may result in **radiation burns** or **radiation sickness**.

Radiation burns like the one in Figure 3.4.8 are caused by short exposure to a very large amount of ionising radiation. The radiation damages the cells on the surface of skin or other organs, causing redness and blistering. However, the side effects are not immediately obvious. It may take 1 or 2 days for itching and redness to appear and then 1–3 weeks before burns and blisters appear.

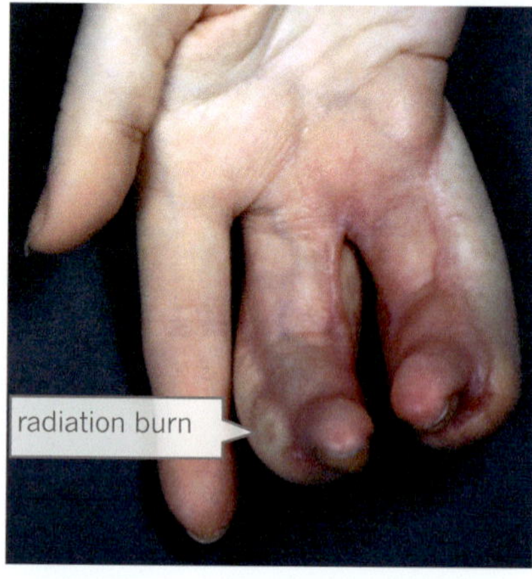

FIGURE 3.4.8 Radiation burns can be just as severe as burns caused by a fire.

Radiation sickness may result from exposure to a large amount of radiation in a short amount of time, or a lower amount of radiation over a longer period of time. The symptoms include nausea, vomiting, fever, hair loss and diarrhoea. The symptoms may not appear immediately but will appear more quickly if the person has absorbed a larger amount of radiation.

Cell mutation

Cell **mutation** occurs when the ionising radiation damages DNA inside the cell without causing the cell to die. The DNA inside a cell contains all the genetic information that tells the cell how to grow and function properly. If the DNA is damaged, then the cell is reprogrammed and may cause the cell to develop into a cancer. An example is the skin cancer in Figure 3.4.9. A cell mutation can be caused by even a small amount of radiation. However, the likelihood of cell mutation increases as the exposure to ionising radiation increases.

When the ionising radiation causes a mutation in sperm or ova (egg cells), the offspring of the organism may be affected. This is known as genetic or inherited mutation.

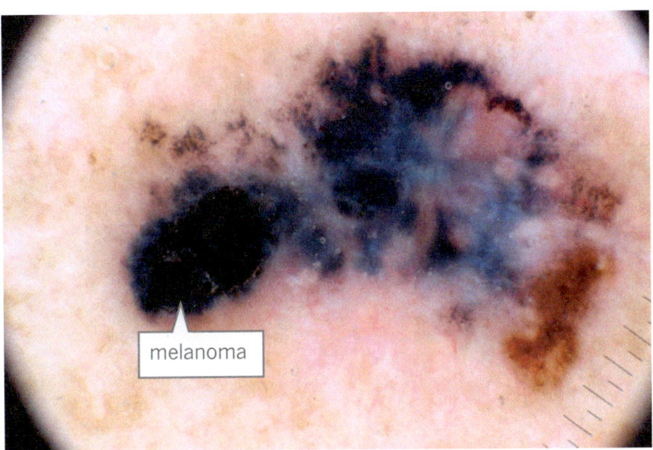

FIGURE 3.4.9 If ionising radiation damages the DNA in cells, it can cause them to turn into cancers. This malignant melanoma is the most dangerous type of skin cancer.

SciFile

Mutants aren't monsters

The mutations caused by radiation are not the monstrous creatures seen in science fiction movies. Instead, radiation exposure simply increases the frequency of naturally occurring mutations (such as albinism, which causes an absence of colour in the skin and hair). This peacock is an albino.

Different animals experience different levels of inherited mutations. For example, the mutations in the offspring of mice and fruit flies are increased significantly if the parents have been exposed to radiation.

However, in humans it is unclear whether large doses of radiation produce mutations in children. Scientists who studied the survivors of the nuclear bombs dropped on Nagasaki and Hiroshima in Japan in 1945 found that children of the survivors did not show an increase in genetic mutations. In contrast, studies of men who worked with radioactive materials showed that the workers of these cities were more likely to have children with leukaemia (a type of cancer common in children).

Properties of radiation

Alpha, beta and gamma radiation all have different properties that determine how much biological damage they can cause. Table 3.4.3 summarises some of these properties.

TABLE 3.4.3 Summary of nuclear radiation

Radiation	Mass of particles	Speed	Penetration depth	Ionisation ability
alpha radiation	7000 times heavier than a beta particle	10% the speed of light	stopped by dead skin or a layer of paper	20 electrons per α particle
beta radiation	same mass as an electron	90% the speed of light	stopped by a 1 mm sheet of aluminium	1 electron per β particle
gamma radiation	no mass	100% the speed of light	stopped by several centimetres of lead or concrete	1 electron per γ ray

Alpha radiation

Alpha particles are large, heavy and slow compared to beta particles and gamma rays. This makes them 20 times more effective at ionising molecules. However, their large size also means that **alpha radiation** can only travel a few centimetres in air and is easily blocked by a thin sheet of paper or even a layer of dead skin. This is shown in Figure 3.4.10 on page 112. As a result, radioisotopes that emit alpha radiation can be handled relatively safely. However, if isotopes emitting alpha radiation get inside the body, the effects can be fatal. Radioactive gases that emit alpha radiation are particularly dangerous when breathed into the lungs.

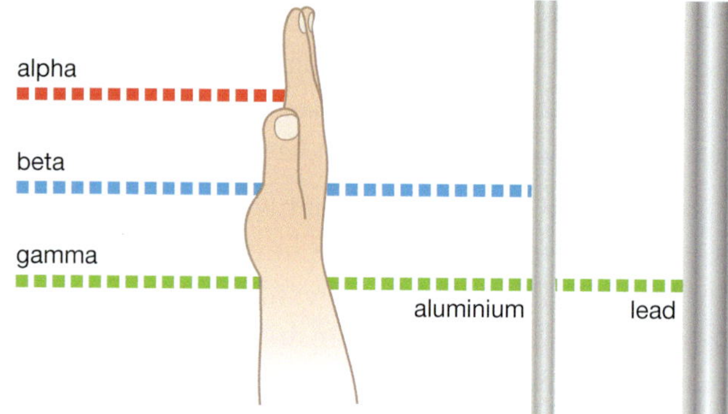

FIGURE 3.4.10 Alpha particles are stopped by a sheet of paper or dead skin. Beta particles are stopped by a 1 mm plate of aluminium. Gamma rays are only stopped by thick lead or concrete.

Beta radiation

The beta particles that make up beta radiation are small and fast. This means that **beta radiation** penetrates the skin more deeply than alpha radiation. As a result, beta radiation is more likely to cause radiation burns to the skin and eyes, like the burns shown in Figure 3.4.11. However, beta radiation can be blocked by a thin plate of aluminium.

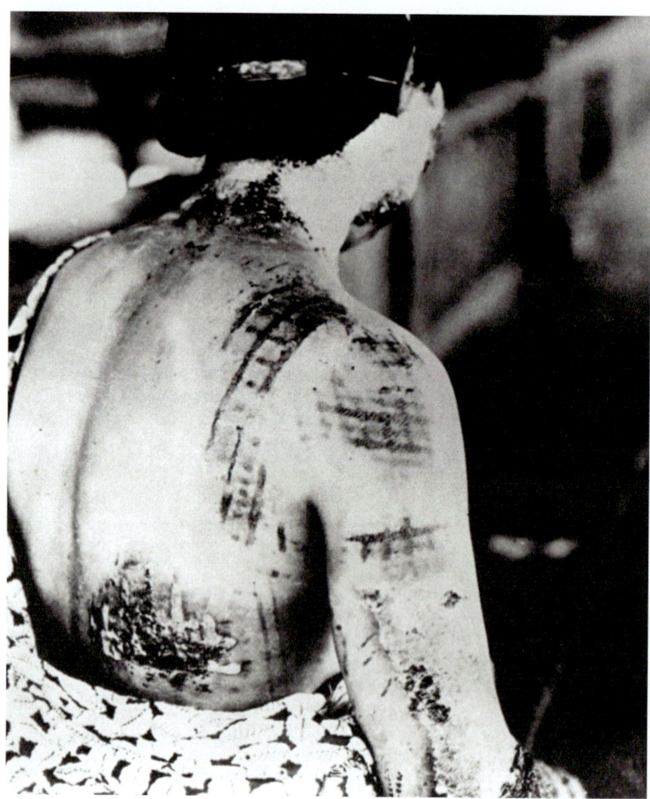

FIGURE 3.4.11 Beta and gamma radiation are the most likely source of radiation burns following a nuclear explosion. This person was burnt by radiation from the atomic bomb dropped on Hiroshima in 1945.

SciFile

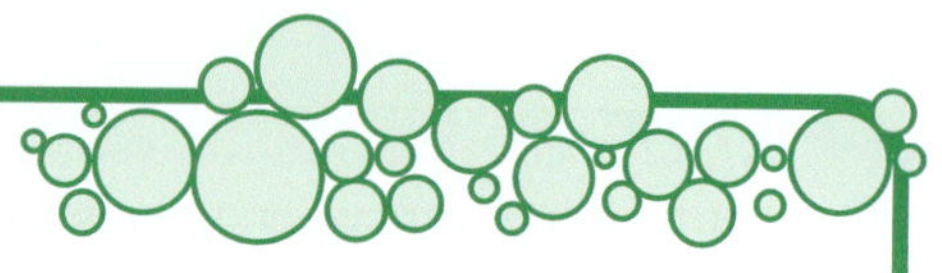

Deadly mines

In the 1940s and 1950s it was discovered that the workers in uranium mines were twice as likely as other people to die of lung cancer. This was due to build-up of the radioactive gas radon-222 in the mines. Radon-222 emits alpha radiation that damages the cells inside the lungs. Today, mines are ventilated properly to prevent radon-222 accumulating.

Gamma radiation

Gamma radiation can travel through skin, bone and aluminium, making it extremely dangerous to humans. Only a thick layer of concrete or lead will block the radiation. This is because gamma radiation is a form of **electromagnetic radiation**, made up of electromagnetic waves rather than particles. This means gamma rays do not have any mass or charge and travel at the speed of light. Other forms of electromagnetic waves include radio waves, microwaves, visible light, ultraviolet light and X-rays. However, only gamma rays, X-rays and certain types of ultraviolet light are powerful enough to ionise molecules and cause cell damage.

Useful radiation

While radiation should be handled with care, it can also be very beneficial if used correctly. Radiation is often used for medical treatments and diagnosis, industrial applications and scientific research.

Medical applications

Although radiation causes cells to become cancerous, it is also one of the most important tools for the treatment of cancers. This type of treatment is known as **radiotherapy**. During radiotherapy, the cancerous tumour is exposed to high concentrations of radiation. This radiation is used to kill the cells in the tumour and stop them multiplying. However, healthy cells may also be damaged. As a result, radiotherapy comes with serious side effects, including skin irritation, ulcers, swelling, nausea, hair loss, heart disease and secondary cancers.

Radioisotopes can also be used for medical diagnosis. For example, radioisotopes can be used to obtain detailed images of the organs inside the body, like the one in Figure 3.4.12. This process is called nuclear imaging. To obtain an image of the internal organs, radioisotopes are injected into the body. These radioisotopes collect in the organs and emit a very low dose of gamma radiation that can be detected outside the body to build up an image of the organs.

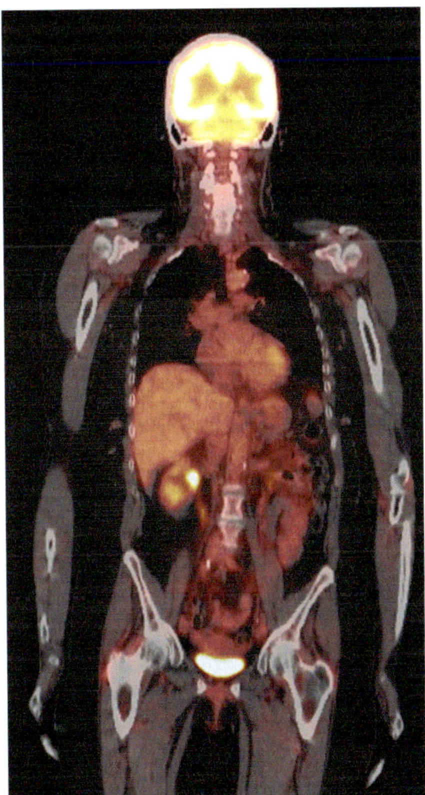

FIGURE 3.4.12 Doctors inject the patient with radioisotopes to obtain images of organs inside the body, like this false-coloured PET scan. Brighter areas show a build-up of radioisotope.

Industrial applications

There is a wide variety of industrial applications for radiation. Radiation is commonly used in the process of sterilisation to kill bacteria in medical equipment and even in food. This means that equipment such as bandages and needles can be sterilised without the need for harmful chemicals. Foods treated with radiation last longer before rotting or going stale.

Radiation can also be used to 'look' inside objects in the same way that X-rays can be used to look inside you. This is useful in exploring for minerals, oil, gas and water. A similar process is also used to determine the thickness of materials such as paper or metal foils, using the technique shown in Figure 3.4.13.

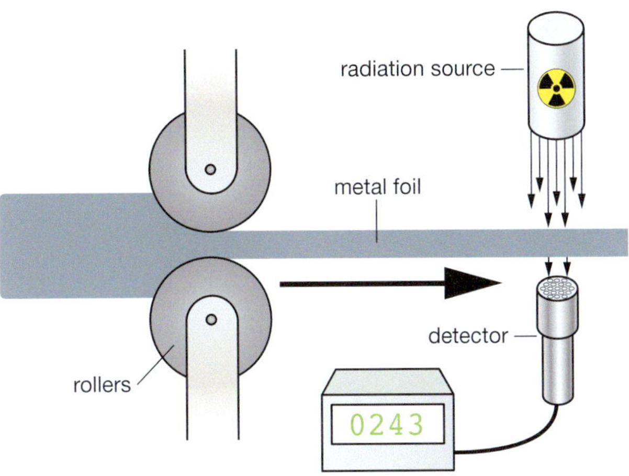

FIGURE 3.4.13 Engineers can accurately measure the thickness of materials by measuring how much radiation can pass through them.

You might even find radiation being used in your home! One type of smoke detector has a small amount of americium-241, which produces alpha radiation. When there is smoke in the air, the alpha particles are blocked and the alarm sounds (Figure 3.4.14).

FIGURE 3.4.14 Some smoke detectors use the radioactive element americium-241 to detect smoke particles in the air.

Working with Science

RADIOGRAPHER

Radiographers, or medical imaging technologists, capture images, such as X-rays, CT (computer tomography) scans and MRIs (magnetic resonance images) (Figure 3.4.15). These help doctors diagnose and treat injuries and disease. Radiographers work closely with radiologists, who interpret the images and diagnose conditions (Figure 3.4.16).

Many aspects of a radiographer's job are highly technical but they also work with and care for patients in explaining medical procedures and providing support. For this reason, they require good communication skills and an attentive and caring nature. They also require technical skills and knowledge of the operation of advanced medical equipment, radiation science, structures of the human body and the impact of injury and disease on these structures. Many radiographers enjoy the variety that comes with the highly technical aspects of their job while also being able to work with patients to improve their health.

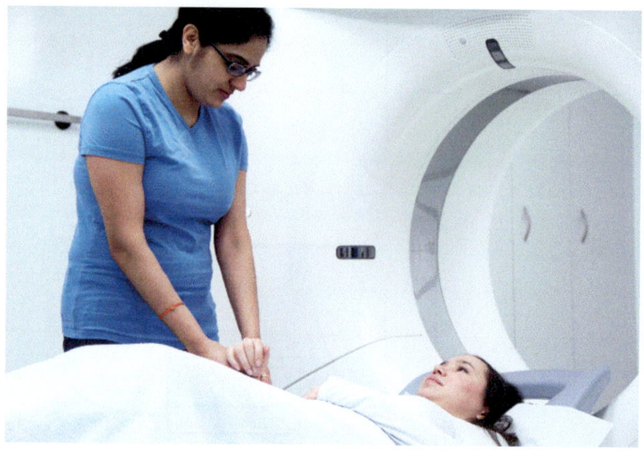

FIGURE 3.4.15 An important part of a radiographer's job is communicating with patients to explain procedures and provide reassurance.

To become a radiographer, you need to complete a degree in Radiation Science or Medical Imaging, such as a Bachelor of Radiography and Medical Imaging or Bachelor of Applied Science (Medical Radiations). There are many opportunities for specialisation in radiography with Graduate Diplomas, Masters or PhD programs. Job opportunities for radiographers have been growing and are expected to continue to rise in the near future. People with these qualifications are highly employable in Australia and overseas, providing many exciting opportunities for travel and professional development.

Review

1 Radiography is used in many medical and non-medical fields. List five uses for radiography.

2 While X-rays and other radiation technologies are beneficial in medicine and many other areas, there are also small risks from radiation exposure. Radiographers are trained in how to manage these risks. What are some ways that radiographers make sure they work safely with radiation?

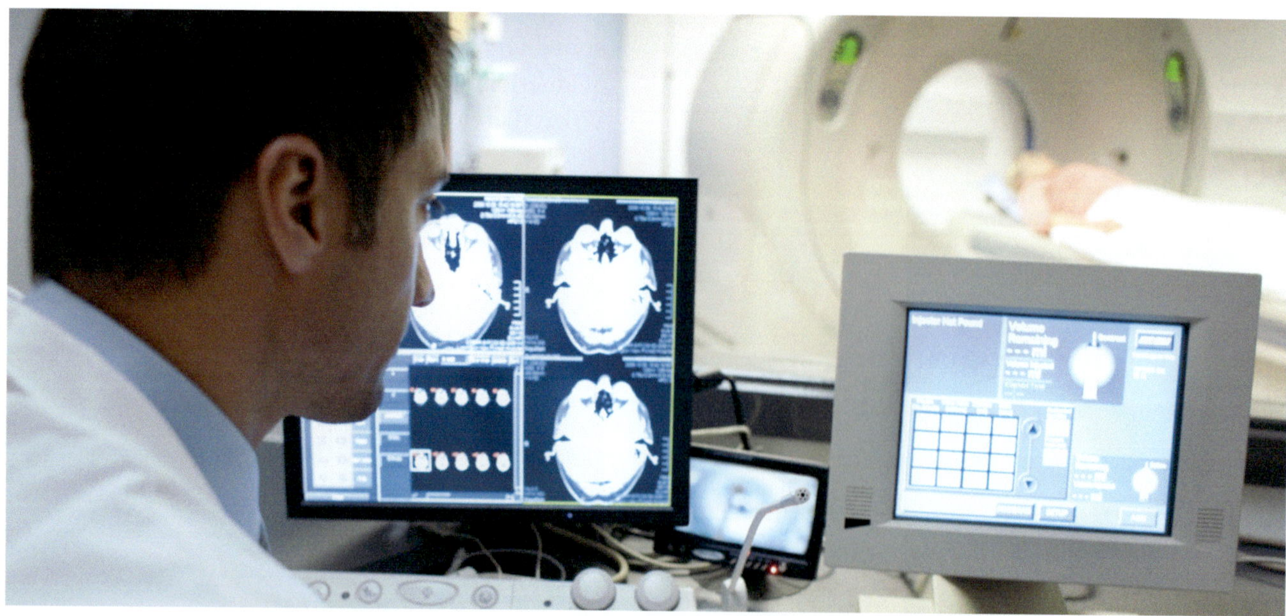

FIGURE 3.4.16 A radiographer examining brain scan images taken with a CT scanner

SCIENCE AS A HUMAN ENDEAVOUR

Use and influence of science

Power of the nucleus

There are two types of nuclear reactions that have changed the face of Earth and caused intense political, environmental and social debate. They are fission and fusion reactions.

FIGURE 3.4.17 Nuclear power provides many countries with much of their electricity.

Fission and fusion

In a **fission** reaction, a large nucleus splits into two almost equally sized pieces. In a **fusion** reaction, two small nuclei come together to form a larger nucleus. Both reactions release huge amounts of energy that can be extremely useful or extremely destructive.

Fission reactions

The most famous fission reaction involves uranium-235. If this radioisotope absorbs a neutron, it forms the highly unstable isotope uranium-236. The uranium-236 then splits into two smaller atoms, krypton-92 and barium-141, releasing a huge amount of energy and three neutrons.

This reaction is used to power naval vessels such as aircraft carriers and submarines and to supply 15% of the world's electricity demand (Figure 3.4.17). This reaction is shown in Figure 3.4.18. However, the reaction creates radioactive waste that cannot be destroyed and must be stored deep underground.

Fission reactions are also used in nuclear weapons. The extreme explosive power of a nuclear reaction can flatten an entire city. This power was demonstrated tragically during World War II when atomic bombs were dropped on the Japanese cities of Hiroshima and Nagasaki, shown in Figure 3.4.19. The bomb blasts instantly killed about 100 000 people in each city and nearly the same number of people died as a result of radiation exposure.

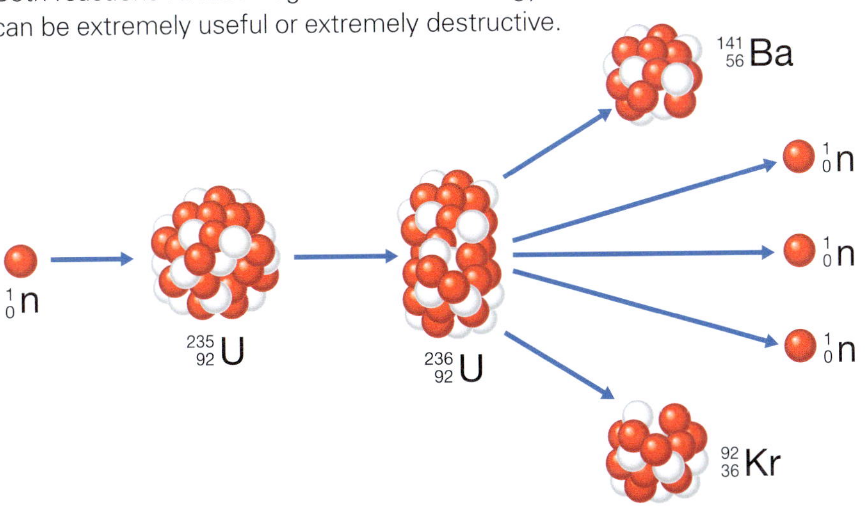

FIGURE 3.4.18 When a uranium-236 nucleus undergoes a fission reaction, it splits into krypton-92 and barium-141 nuclei.

FIGURE 3.4.19 Explosion of a nuclear bomb over Nagasaki in 1945

SciFile

The Fukushima disaster

Japan's first-hand experience of the devastating effects of nuclear weapons has made the people of Japan understandably wary about the use of nuclear power. Their fears were realised in March 2011 when three nuclear reactors in Fukushima went into meltdown after the shock of a magnitude 9.0 earthquake and tsunami. This disaster is considered the second-worst nuclear reactor meltdown after the one in Chernobyl, Ukraine, in 1986.

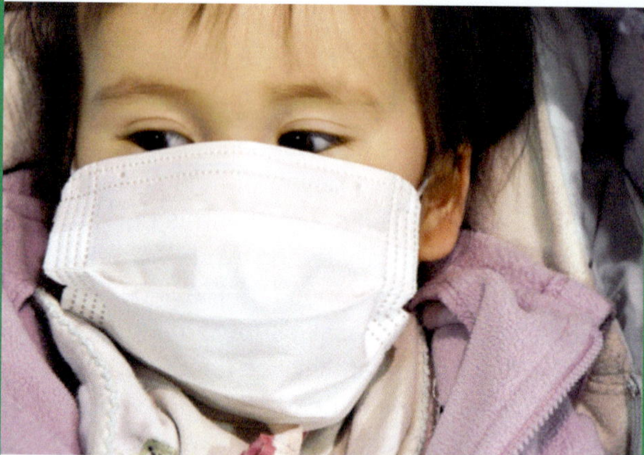

FIGURE 3.4.20 Yunna was a year old when radioactive dust was sent into the air by the Fukushima meltdown.

Fusion reactions

Without fusion reactions, there would be no life on Earth. This is because fusion reactions power the Sun and give us warmth and light. A fusion reaction occurs when two small nuclei combine to form a single nucleus. For example, if two hydrogen-2 nuclei collided, then they might form a helium-4 nucleus, as shown in Figure 3.4.21.

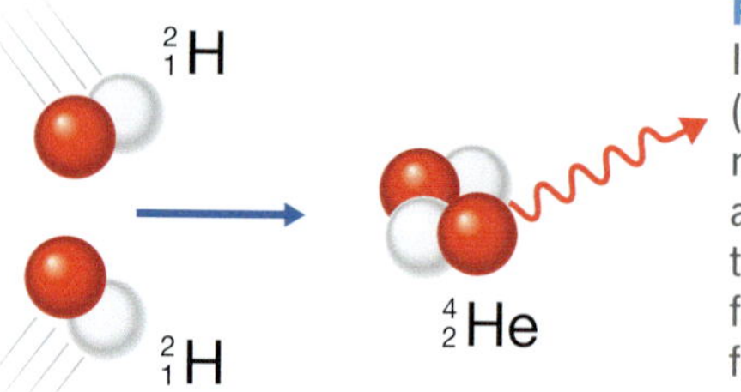

FIGURE 3.4.21 If two hydrogen-2 (deuterium) nuclei collide at high speed, then they may fuse together to form a helium-4 nucleus.

However, the small nuclei strongly repel each other because they both have a positive charge. Therefore, fusion only occurs at extremely high temperatures—over 100 million degrees Celsius. There is no material on Earth that can withstand these temperatures, so scientists must suspend the reaction in mid-air by using a powerful magnetic field like the one in Figure 3.4.22.

FIGURE 3.4.22 Fusion reactions occur at such high temperatures that scientists must ensure the reaction is suspended within strong magnetic fields that produce a 'magnetic bottle'.

If fusion reactions could be controlled, they would provide an extremely powerful and clean source of energy. However, the power of fusion can also be used to create the most destructive weapons on Earth—hydrogen bombs. Fortunately, a hydrogen bomb has never been used in a military attack.

REVIEW

1. What percentage of the world's power supply is provided by nuclear reactions?
2. Why could there be no life on Earth without nuclear fusion?
3. Explain why the number of deaths from a nuclear explosion is not easy to calculate.
4. Discuss whether you would be for or against the use of a small nuclear power generator to power your home.

MODULE

3.4 Review questions

Remembering

1 Define the terms:

- **a** isotope
- **b** radioisotope
- **c** half-life
- **d** mutation.

2 What term best describes each of the following?

- **a** the process of converting an atom into a different type of atom via a nuclear reaction
- **b** a technique to determine the age of fossils by measuring the amount of carbon-14
- **c** a particle ejected from a nucleus that is identical to an electron
- **d** a form of electromagnetic radiation more powerful than X-rays.

3 List three types of radiation in order from least penetrating to most penetrating.

4 What happens to the atomic number and mass number of a nucleus in the following types of decay?

- **a** alpha decay
- **b** beta decay.

5 List four uses of radiation.

6 Gamma rays are just one type of electromagnetic wave. List four others.

Understanding

7 Why are alpha particles, beta particles and gamma rays classified as forms of ionising radiation?

8 Alpha particles are 20 times more effective than beta particles or gamma rays at ionising molecules. Explain why they may be considered the least dangerous nuclear radiation.

9 How are radiation burns and radiation sickness caused?

Applying

10 All oxygen atoms have 8 protons. Three oxygen isotopes are oxygen-16, oxygen-17 and oxygen-18. Use this information to write the atomic symbols for all three isotopes.

Analysing

11 Compare the properties of alpha particles, beta particles and gamma rays.

12 Calculate the atomic number and mass number of the following nuclei after they undergo alpha decay.

- **a** $^{241}_{95}Am$
- **b** $^{240}_{94}Pu$
- **c** $^{210}_{84}Po$

13 Calculate the atomic number and mass number of the following nuclei after they undergo beta decay.

- **a** $^{22}_{11}Na$
- **b** $^{14}_{6}C$
- **c** $^{137}_{55}Cs$

14 The half-life of carbon-14 is 5730 years. Use this to calculate the age of a fossil with a carbon-14 content that is:

- **a** half the normal amount
- **b** one-quarter the normal amount
- **c** one-eighth the normal amount
- **d** one-sixteenth the normal amount.

Evaluating

15 What do you think are the advantages and disadvantages of nuclear power plants? List them in a table.

16 Propose reasons for why radioisotopes that emit alpha radiation are not used for radio-imaging.

Creating

17 Construct a pamphlet for health department workers advising them of the dangers of different types of radiation they may be exposed to in the workplace.

18 Construct a short story describing what you think it would be like to survive a nuclear bomb explosion and the effects of the radiation damage.

MODULE 3.4 Practical investigations

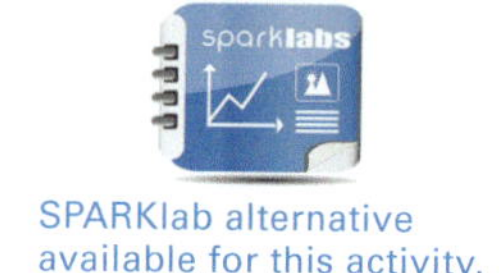

SPARKlab alternative available for this activity.

1 • Half-life  

Purpose

To model radioactive decay and half-life.

Timing 30 minutes

Materials

- a packet of M&Ms® or Skittles or two-sided tokens
- a clean tray or sheet of A3 paper
- a clean jar

SAFETY

It is unsafe to eat food in the laboratory, so do not eat any of the lollies.

Procedure

1. Copy the table from the Results section into your workbook. Alternatively, construct a spreadsheet with similar columns.
2. Count the total number of M&Ms in the packet and put them into the jar.
3. Shake the jar up to mix the lollies around. Pour the jar of M&Ms onto the clean tray or A3 paper.
4. Count how many M&Ms show the letter M facing upwards. Record this number in the table.
5. Place only the M&Ms showing the letter M back into the jar and dispose of the other M&Ms appropriately.
6. Repeat steps 3–5 until there are no M&Ms left in the jar.

Results

1. Record your results in a table like this one, or in your spreadsheet.

M&M count

Number of repeats	1	2	3	4	5
M&Ms showing the letter M					

2. Construct a line graph of the number of M&Ms remaining (those that showed the letter M) versus the number of times the procedure was repeated. Alternatively, program your spreadsheet to plot a graph for you.
3. Compile everyone's results into one table and plot the classroom total of M&Ms remaining with each repeat of the procedure.

Review

1. Describe the shape of the graphs that you produced.
2. The half-life of your M&M sample is the number of throws it took for the number of M&Ms in your sample to reduce to half. Use this definition to determine the half-life of your M&M sample.
3. a Compare your individual results with the class results.
   b Propose which of these results is more reliable.
   c Justify your response.
4. Discuss how this prac models the half-life of a radioactive element.

CHAPTER 3

Chapter review

Remembering

1 Name the following chemicals.
   - a $Fe_2O_3.H_2O$
   - b HCl
   - c $C_6H_{12}O_6$

2 Name the gas produced:
   - a by incomplete combustion, and that is poisonous
   - b when an acid reacts with a metal
   - c when an acid reacts with a hydrogen carbonate.

3 Recall the following reactions by writing their word equations and/or balanced formula equations.
   - a tarnishing of silver
   - b combustion of octane
   - c neutralisation of potassium oxide with nitric acid
   - d photosynthesis.

4 Recall the following reactions by completing their equations.
   - a acid + base → ..... + ...
   - b methane + oxygen → ..... + ...
   - c $CaCO_3 + 2HCl \rightarrow$ ..... + ... + ...
   - d $C_6H_{12}O_6 + 6O_2 \rightarrow$ ..... + ...

5 List three substances required for iron to rust.

6 List three types of ionising radiation.

7 List four ways in which radiation can be useful.

Understanding

8 Burning fossil fuels causes problems for our atmosphere. Describe these problems.

9 Explain why you breathe faster and deeper when exercising.

10 Explain why gamma radiation may be considered the most dangerous type of radiation, even though alpha radiation can ionise atoms more effectively.

11 Explain how ionising radiation causes:
   - a radiation burns and radiation sickness
   - b mutations.

Applying

12 You discover a new element named jellium (Je) that has a half-life of 5 days. Your sample of jellium contains only 256 atoms. Calculate how many jellium atoms there will be after:
   - a 5 days
   - b 10 days
   - c 15 days
   - d 20 days.

Analysing

13 Compare:
   - a combustion reactions with corrosion reactions
   - b the reactions of acids with metals with the reactions of acids with carbonates
   - c photosynthesis with aerobic respiration.

14 Contrast chemical reactions with nuclear reactions.

Evaluating

15 You should never try to break open a smoke detector. Propose reasons why.

16 a Assess whether you can or cannot answer the questions on page 75 at the start of this chapter.
   - b Use this assessment to evaluate how well you understand the material presented in this chapter.

Creating

17 Construct word equations for the following reactions.
   - a Nitric acid neutralises sodium hydroxide to form sodium nitrate and water.
   - b Ethane produces carbon, carbon monoxide gas and water vapour when it burns in a poor supply of oxygen gas.
   - c Lots of hydrogen bubbles rise from calcium as it reacts with water. The end result is a solution of calcium hydroxide.

18 Use the following ten key terms to construct a visual summary of the information presented in this chapter.

chemical reaction
endothermic
combustion
glucose
carbon dioxide
exothermic
photosynthesis
oxidation
oxygen gas
respiration

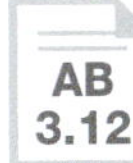

CHAPTER 3

Inquiry skills

Research

1 Processing & Analysing | Communicating

Use the following search terms to find internet videos of different types of chemical reactions: *chemical reaction video, corrosion video, combustion video, acid-base video.*

2 Processing & Analysing

Use the search terms *balancing chemical equations games* to find websites that do the balancing for you.

3  Processing & Analysing | Communicating

Some bacteria in the stomach can damage the stomach lining, causing it to become inflamed. Stomach acid can cause further damage, resulting in an ulcer. Research stomach ulcers and find:

- an image of an ulcer
- the name of the bacterium thought to cause ulcers
- why this bacterium is not killed by gastric juice
- typical symptoms of a stomach ulcer
- treatments for the ulcer, including recommended diet
- details of the Australian scientists who discovered that this bacterium was causing ulcers.

Present your findings as a pamphlet to be provided at doctors' clinics.

4  Evaluating | Communicating

Research anaerobic respiration.

- Find the word equation and balanced formula equation that describes it.
- Contrast anaerobic and aerobic respiration.
- Propose, with reasons, when it is most likely to happen in humans.
- List organisms that use anaerobic respiration to release energy.
- Compare these organisms.

Present your findings as a set of responses to the above tasks.

5 Processing & Analysing

Compare Australia's consumption of power from nuclear power plants by:

- Researching what percentage of Australian power is generated from nuclear power plants.
- Researching and comparing the percentage of power generated from nuclear power plants in five other countries such as the United States, France, Korea, Japan and Russia.

Use a spreadsheet to present your results as a bar or column graph.

6 Evaluating | Communicating

Find out how scientists understanding of radioactivity has changed over time.

- Research the contributions of scientists such as the Curies, Roentgen, Becquerel, Rutherford, Chadwick and Einstein.
- Research new technologies that have appeared as a result of the increased understanding of radioactivity.

Present your research as an annotated timeline.

Thinking scientifically 

1 Fiona dropped a stick of chalk (containing calcium carbonate, $CaCO_3$) into hydrochloric acid (HCl). It fizzed as the reaction produced carbon dioxide gas (CO_2), water (H_2O) and calcium chloride ($CaCl_2$).
Identify the equation for this reaction.

A $CaCl_2 + H_2O + CO_2 \rightarrow CaCO_3 + 2HCl$

B $CaCO_3 + H_2O + CO_2 \rightarrow CaCl_2 + 2HCl$

C $CaCO_3 + 2HCl + H_2O \rightarrow CaCl_2 + CO_2$

D $CaCO_3 + 2HCl \rightarrow CaCl_2 + H_2O + CO_2$

2 Fiona measured the amount of carbon dioxide gas generated as the chalk in question 1 dissolved. She repeated the experiment, but this time she crushed the stick of chalk into a powder. Her results are shown in Table 3.5.1.

CHAPTER 3 Inquiry skills

TABLE 3.5.1 Volume of CO_2 generated by chalk in HCl

Time (s)	Volume of CO_2 generated (cm^3)	
	Stick of chalk	Crushed chalk
0	0	0
15	50	100
30	100	200
45	150	200
60	200	200
105	200	200

From these results, the best conclusion is:

- **A** chalk and hydrochloric acid react together
- **B** chalk always produces 200 cm^3 of CO_2
- **C** crushed chalk reacts faster than a stick of chalk
- **D** hydrochloric acid is dangerous.

3 Figure 3.5.1 shows a graph of the half-life of the carbon-14 radioisotope that is commonly used for carbon dating.

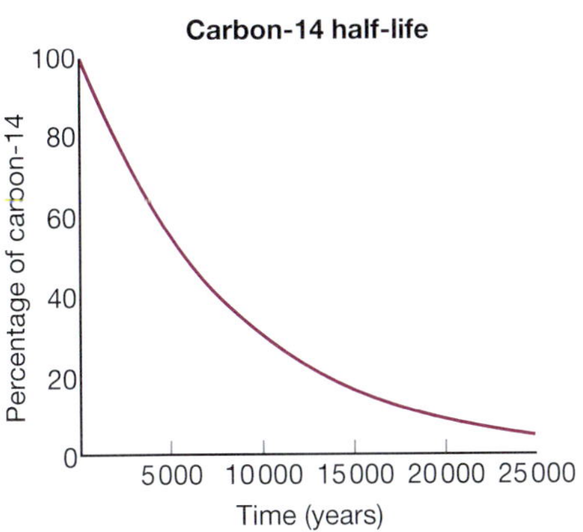

FIGURE 3.5.1

a An archaeologist working in Cairo, Egypt, discovers an old artefact and takes it back to the laboratory for carbon dating. The laboratory results show that there is only 16% of the carbon-14 that would have been found in a similar artifact made today. Which is the best estimate of the age of the artifact?

- **A** 5000 years
- **B** 10 000 years
- **C** 15 000 years
- **D** 20 000 years.

b What percentage of carbon-14 would you expect to be in an artifact that was 10 000 years older?

- **A** 8
- **B** 19
- **C** 32
- **D** 41.

c How old is an artifact that has just 5% of the carbon-14 remaining?

- **A** 2500 years
- **B** 8000 years
- **C** 18 000 years
- **D** 25 000 years.

4 A nuclear power plant worker comes into hospital after having an accident where he was exposed to high levels of radiation approximately 30 minutes before. Initially he seems fine but after an hour he starts to feel nauseated and begins vomiting. He is kept in for observation and after a few days develops diarrhoea.

Use Table 3.5.2 to determine the likely dose of radiation that the worker was exposed to.

TABLE 3.5.2 Symptoms according to level of radiation exposure

Dose (Sv)	Symptoms
0–0.5	no obvious effect
0.5–1.0	vomiting and nausea for about 1 day in 10 to 20% of people; tiredness but no serious disability
1.0–2.0	mild to moderate nausea in 50% of people with occasional vomiting, setting in within 3–6 hours after exposure and lasting several hours to a day
2.0–5.5	nausea in 100% of people; vomiting starting 0.5–6 hours after irradiation and lasting up to 2 days, followed by other symptoms of radiation sickness, e.g. loss of appetite, diarrhoea, minor bleeding
5.5–10	severe nausea and vomiting within 15–30 minutes, lasting up to 2 days, followed by severe symptoms of radiation sickness, e.g. loss of appetite, diarrhoea, minor bleeding
10–20	immediate nausea, diarrhoea and bleeding
> 20	immediate disorientation and coma; onset within seconds to minutes

- **A** 0.5–1.0 Sv
- **B** 1.0–2.0 Sv
- **C** 2.0–5.5 Sv
- **D** 5.5–10 Sv.

CHAPTER 3
Glossary

acid rain: rain that has acids such as nitric acid and sulfuric acid dissolved in it

aerobic respiration: respiration that uses oxygen as a reactant to convert glucose into energy

alpha decay: a nuclear reaction in which a nucleus ejects an alpha particle

alpha particle: a particle made up of two protons and two neutrons, making it identical to a helium nucleus

alpha radiation: a form of ionising radiation made up of alpha particles

alveoli: microscopic air sacs in the lungs that allow oxygen to enter the bloodstream and carbon dioxide to be removed from the bloodstream

anaerobic respiration: respiration that does not require oxygen to convert glucose into energy

anodising: a way of protecting aluminium from corrosion, by deliberately creating a layer of aluminium oxide over it

antacids: tablets or liquid medicines that relieve heartburn

balanced equation: a chemical equation that has the same number of each type of atom on both sides of the arrow

beta decay: a form of nuclear reaction in which a nucleus ejects a beta particle

beta decay

beta particle: a small, negatively charged particle that can be ejected from a nucleus during a nuclear reaction; it is identical to an electron

beta radiation: nuclear radiation that is made up of beta particles

carbon dating: a method for judging the age of artefacts or fossils by analysing the amount of carbon-14 in the fossil

carbon sink: a term used to describe materials that store carbon in their structures; plants and animals can be thought of as carbon sinks

chemical digestion: the process that breaks down large molecules in food into smaller ones that can be absorbed by the body

chlorophyll: the green chemical found in the chloroplasts of green plants that traps the Sun's energy for photosynthesis

chlorophyll

chloroplast: organelle in a plant cell that contains chlorophyll; where photosynthesis takes place

coefficients: the big numbers in front of chemicals in a balanced equation

combustion: a rapid reaction with oxygen that releases energy in the form of heat and/or light

combustion

complete combustion: combustion that occurs when there is plenty of oxygen; it produces carbon dioxide and water vapour

corrosion: the breakdown of metals due to their reaction with other chemicals

electromagnetic radiation: radiation that travels through a vacuum as waves rather than particles

endothermic reaction: a chemical reaction that absorbs energy

enzyme: a chemical that helps a chemical reaction but is not used up in the reaction

exothermic reaction: a chemical reaction that releases energy

fission: a nuclear reaction in which a very large nucleus splits into two smaller nuclei of similar mass number

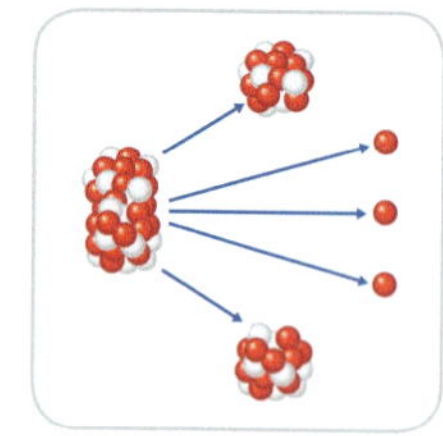
fission

fusion: a nuclear reaction in which two small nuclei come together to form one larger nucleus

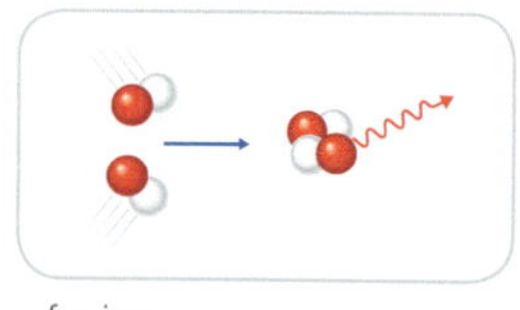
fusion

gamma decay: nuclear decay that involves the release of gamma rays

gamma radiation: a form of ionising radiation made up of gamma rays

gamma ray: a very high-energy electromagnetic wave that is produced when the protons and neutrons in a nucleus rearrange

glucose: a simple sugar with the chemical formula $C_6H_{12}O_6$; a product of photosynthesis and a reactant in respiration

half-life: the time it takes for half the nuclei to decay

hydrocarbons: highly combustible chemicals; petrol is a mixture of hydrocarbons

CHAPTER 3

Glossary

incomplete combustion: combustion that occurs when oxygen is limited; produces carbon (soot, smoke) and carbon monoxide, and does not release as much heat or light as complete combustion

incomplete combustion

ionising radiation: any form of radiation that has the ability to remove electrons from atoms and molecules

isotopes: atoms that have the same number of protons but a different number of neutrons

law of conservation of mass: atoms are not created or destroyed in a chemical reaction; they can only be rearranged

mitochondria: organelles in plant and animal cells where respiration takes place

mutation

mutation: a change in the DNA of a cell that causes it to change how it works and reproduces

neutralisation: a reaction of an acid with a base, forming a salt and water

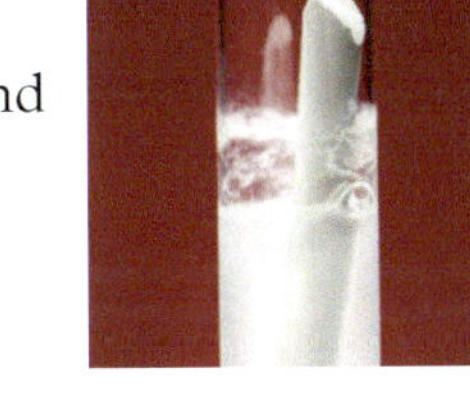
neutralisation

nuclear decay: when a nucleus undergoes a nuclear reaction and emits radiation

nuclear radiation: rays or particles that are emitted by a nucleus during a nuclear reaction

nuclear reaction: a process that causes a nucleus to change, including alpha decay, beta decay, fission and fusion

photosynthesis: endothermic reaction that takes place in green plants; uses energy from sunlight to combine water and carbon dioxide and produce glucose and oxygen gas

products: chemicals produced in a chemical reaction; they are written on the right-hand side of the arrow

radiation burns: redness and blistering on the surface of the skin or other organs caused by intense exposure to ionising radiation

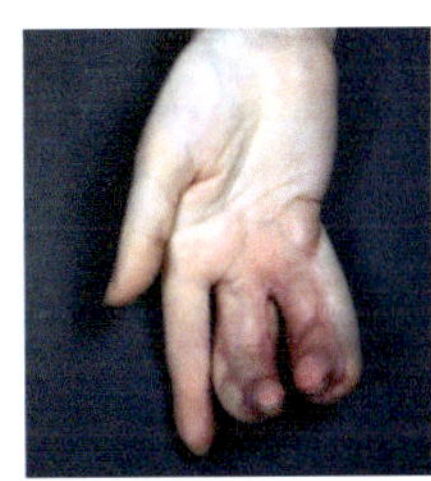
radiation burns

radiation sickness: a condition that results from a large dose of ionising radiation, causing significant cell death; symptoms include nausea, vomiting, fever, hair loss and diarrhoea

radioactive: emitting radiation

radioisotope: an isotope with a nucleus that may undergo a nuclear reaction

radiotherapy: a cancer treatment in which tumours are exposed to high concentrations of radiation

reactants: chemicals that take part in a chemical reaction; they are written on the left-hand side of the arrow

respiration: a series of chemical reactions that releases energy from glucose; it takes place in the mitochondria inside cells

rust: hydrated iron(III) oxide; chemical formula $Fe_2O_3.H_2O$

salt: the term commonly used for sodium chloride, but which covers any compound formed by a metal taking the place of the hydrogen atom in an acid

stable nuclei: nuclei that will never undergo a nuclear reaction

stomata: microscopic holes in the leaves of plants where gases such as oxygen, carbon dioxide and water vapour can enter or exit the plant

tarnish: a black coating of silver sulfide that is produced when silver reacts with sulfur in food or the atmosphere; chemical formula Ag_2S

transmutation: a nuclear reaction that converts one type of atom into a different type of atom

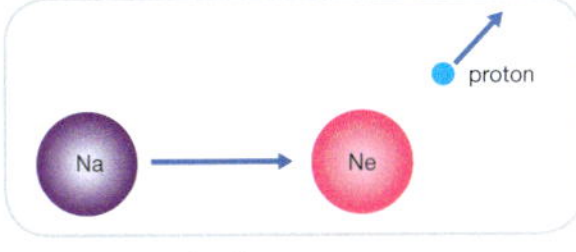

transmutation

unstable nuclei: nuclei that may undergo a nuclear reaction at any time

verdigris: a green coating of copper hydroxide that is produced when copper reacts with moisture, carbon dioxide and oxygen in the atmosphere; chemical formula $Cu(OH)_2$

word equation: simple written description of what is happening in a reaction

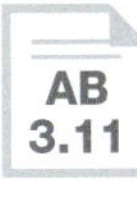

CHAPTER 4

Heat, sound and light

Have you ever wondered...

- why a doona keeps you warmer than just a sheet?
- why tiles feel colder than carpet?
- how musical instruments make different sounds?
- why you cannot see clearly underwater?
- why diamonds sparkle?

LS

After completing this chapter you should be able to:

- discuss how the wave and particle models explain how energy is transferred
- outline how energy moves differently, depending on the material it passes through
- investigate the transfer of heat through convection, conduction and radiation
- use the particle model to explain conduction and convection
- identify safe sound levels for humans and how this affects leisure and the workplace.

This is an extract from the Australian Curriculum 

AB 4.1

MODULE 4.1 Heat

In cold weather, you seek extra jumpers or thicker doonas to keep warm. When it is really hot, you wear less clothing and cool yourself with a fan or by jumping in the pool. Heat is a form of energy that you sense through receptors in your skin. Heat is lost from your skin as you stand in front of a fan and is gained as your body absorbs radiant heat from the flames of a log fire.

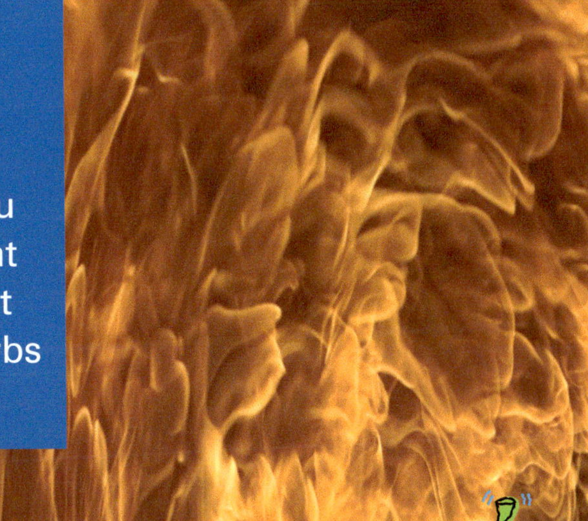

science 4 fun

Burning a balloon!

NO

Can you heat water in a balloon without bursting the balloon?

Collect this ...

- two balloons
- matches
- candle

Do this ...

1 Blow up a balloon and tie its end.
2 Hold a lit candle below the balloon.
3 Observe what happens.
4 Now hold another balloon under a tap and fill it up with water to about the size of a rockmelon.
5 Place a lit candle underneath this second balloon, and again observe what happens.

Record this ...

1 Describe what happened.
2 Explain why you think this happened.

The particle model

Heat is a form of energy that can be transferred through solids, liquids and gases. To understand how this happens, you first need to understand the particle model of matter. In the particle model, atoms are considered as small, hard balls.

In a solid, the particles are closely packed. The particles vibrate on the spot but keep the shape of the substance they form. Particles in a liquid are packed closely together too. The particles vibrate but are also free to move or flow over each other.

The particles of a gas are not bound together at all and are free to move in straight lines until they collide with other gas particles, or the walls of the container in which they are held. Particle model diagrams for a solid, a liquid and a gas are shown in Figure 4.1.1.

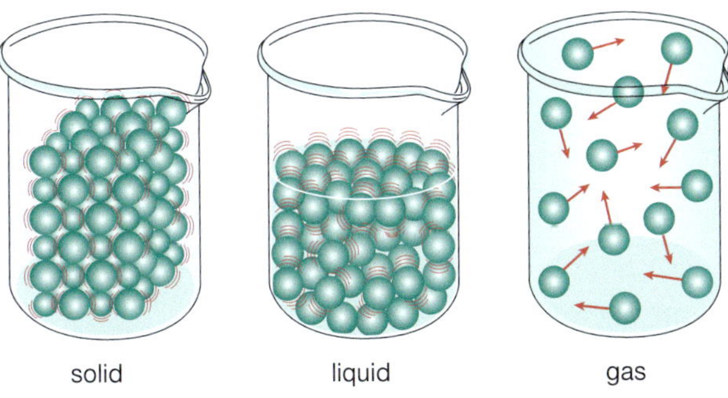

FIGURE 4.1.1 Particle model diagrams for a solid, a liquid and a gas

Heating substances

Heating a substance adds energy to its particles. Some of this energy is stored in the material itself as potential energy. The remaining heat energy increases the kinetic energy of the particles in the material. Kinetic energy is the energy of movement. So if the temperature of a substance increases, then its particles move faster and faster. This spreads the particles further apart and the substance expands. Similarly, particles lose kinetic energy when the temperature decreases. The particles slow down and the substance contracts. Figures 4.1.2 and 4.1.3 show what happens when a solid and a gas is heated or cooled.

When sufficient heat energy is added to a solid or a liquid, the particles break free from each other and the substance changes state and melts or evaporates.

Temperature

Temperature can be measured using a thermometer. A **thermometer** contains a liquid (alcohol or mercury) inside a narrow glass tube. This liquid expands when heated and contracts when cooled. Temperature is read from a scale on the thermometer corresponding to the expansion or contraction of the liquid. Temperature is commonly measured in degrees Celsius (°C).

The Fahrenheit (°F) and kelvin (K) scales are also used to measure temperature. The three scales are compared in Figure 4.1.4.

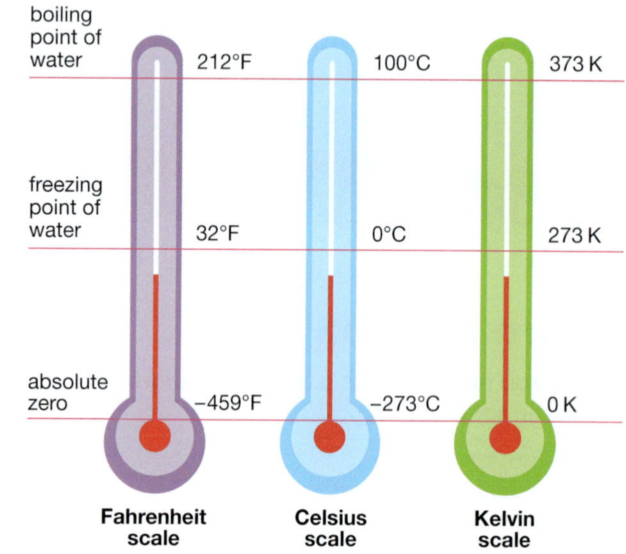

FIGURE 4.1.4 The three scales commonly used to measure temperature

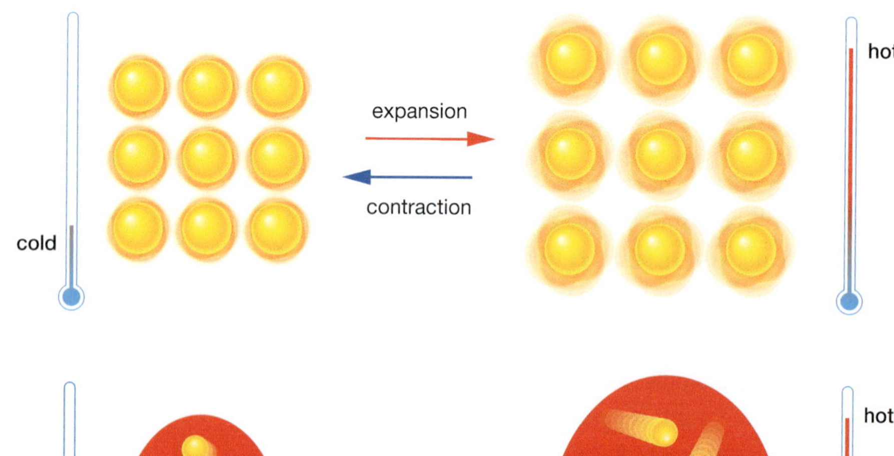

FIGURE 4.1.2 The particles in a solid vibrate more when heated. This causes the solid to expand.

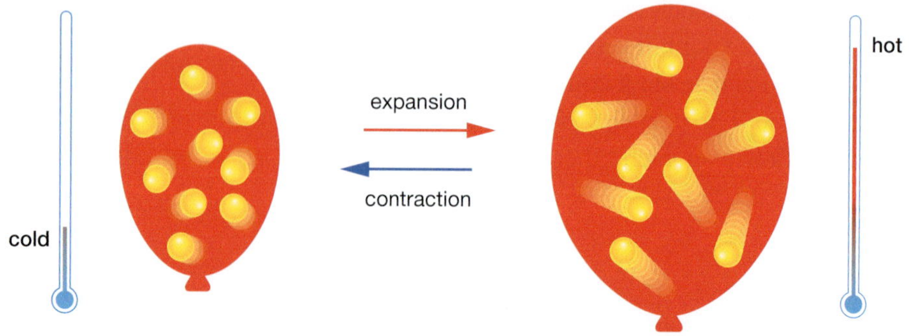

FIGURE 4.1.3 The particles of a gas travel faster when heated. They hit the sides of the container more frequently and with more force. A balloon has flexible walls so this force will cause it to expand.

The **temperature** of a substance is a measure of the average kinetic energy of its particles. The particles of hotter substances move faster than the particles of cooler substances. As the temperature drops, particles lose kinetic energy. Eventually the particles barely move at all. This happens at −273°C, at a point called **absolute zero**.

Thermometers measure temperature but do not measure the heat of a substance. Heat is a form of energy and is a way of describing the total energy of all particles within an object. For example, saucepans A and B in Figure 4.1.5 both contain boiling water. Their temperatures are equal. However, saucepan B contains twice the volume of water and so it has twice the heat energy of saucepan A.

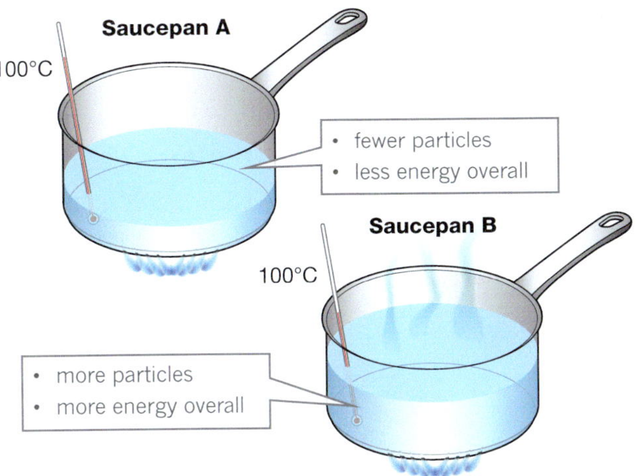

FIGURE 4.1.5 The amount of boiling water in saucepan B is twice that in saucepan A. For this reason, saucepan B has twice the energy.

Heat transfer

Heat flows from areas of higher temperature to areas of lower temperature. The greater the temperature difference, the faster the flow of heat from one object to another. This process of heat transfer can happen in three ways: conduction, convection and radiation.

Conduction

Hold an ice block and your hands get cold. This is because heat flows from your skin into the ice, lowering the temperature of your skin in the process. You know that the ice cube is absorbing this heat because it starts to melt. Heat has flowed from a high temperature (your hands) to a lower temperature (the ice block). This is shown in Figure 4.1.6. Likewise, when you grip a hot cup, heat flows from the cup into your hands warming them up—the cup is at a higher temperature than your hands.

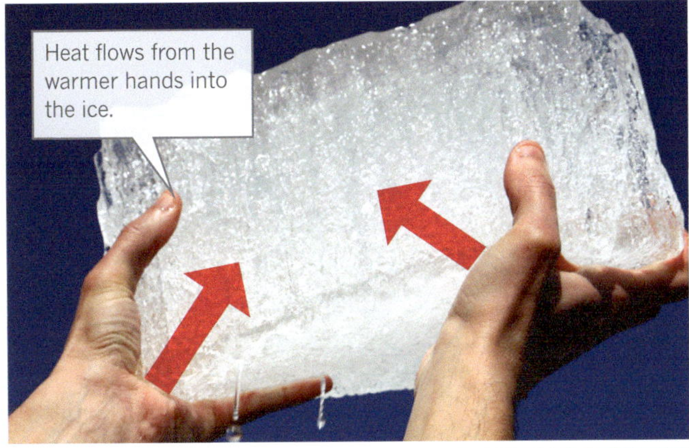

FIGURE 4.1.6 Heat flows from your hands into an ice block, and from a hot cup into your hands.

Hotter substances have faster moving particles than particles in cooler substances. For example, the particles in a cup of hot coffee vibrate rapidly because of its temperature. If a metal spoon is put into the coffee, then the particles of the spoon vibrate faster too. This spreads the heat through the spoon and increases the temperature of the spoon, making it hot to touch. This process of heat transfer by vibrating particles is called **conduction** and is shown in Figure 4.1.7.

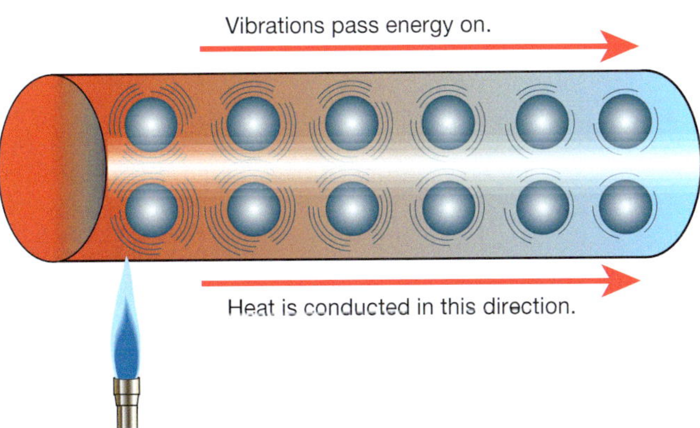

FIGURE 4.1.7 Particles near the flame vibrate more as they absorb heat energy. These vibrations transfer energy to conduct heat along the solid.

Conductors

Some materials conduct heat well while others do not conduct heat at all. A glass of ice-cold lemonade feels much colder than a polystyrene cup of ice-cold lemonade. This is because glass is a better heat conductor than polystyrene. As a result, heat flows from your warm hand into the cooler glass and your hand feels cold. When holding the polystyrene cup, your hand is not losing heat and so it still feels warm.

Substances that transfer heat easily are known as **conductors**. Metals are good conductors of heat. This is why most saucepans are made of stainless steel (Figure 4.1.8). Silver, gold, aluminium and copper are particularly good conductors of heat. Sometimes, saucepans have a copper base to better conduct heat.

FIGURE 4.1.9 Wool fibres and the fluffy polyester, down or cotton filling inside a ski parka trap air and help prevent heat loss from your body.

FIGURE 4.1.8 Metals are good conductors of heat.

Insulators

Plastic, air, cloth, cork, wood and rubber are all very poor conductors of heat, and sometimes can block heat transfer completely. Such substances are known as **insulators**. The handles of a saucepan are usually made from insulating materials to allow you to lift them without burning your hands. An Esky uses insulators such as polystyrene to keep food and drinks cool.

FIGURE 4.1.10 These animals must stay warm in very cold conditions. Polar bears rely on body fat and a thick coat of fur for insulation, and penguins have layers of fat and feathers that they can fluff up to trap more air.

Gases are poor conductors of heat. Air trapped by woollen jumpers and blankets helps to insulate your body from losing heat. Ski parkas, doonas and sleeping bags are filled with cotton, feathers, wool and polyester that also traps air and helps to protect you from the cold (Figure 4.1.9). Similarly, animals like penguins and polar bears that live in cold climates have adaptations that help them stay warm (Figure 4.1.10).

Prac 1 p. 134 | Prac 2 p. 135 | AB 4.2

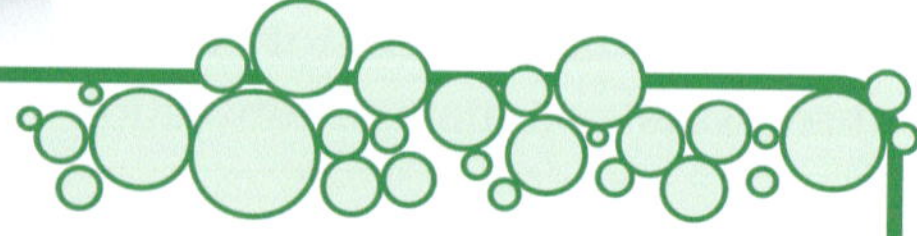

SciFile

Penguins master the cold

Penguins live in some of the coldest conditions on Earth. They huddle together during storms to minimise the surface area of the flock and to minimise the heat loss through conduction to the cold air around them. Most of a penguin's heat should conduct to the ice they stand on, but heat instead conducts from hot blood flowing through arteries passing down their legs to the cold blood flowing through veins returning from their feet.

Convection

As air is heated, its particles gain energy and move further apart. This hot air is less dense than cool air, and so it is pushed upwards by cooler air around it. This method of heat transfer is called **convection**. The air flow it creates is called a convection current. Such a current is shown in Figure 4.1.11, transferring heat from an open fire.

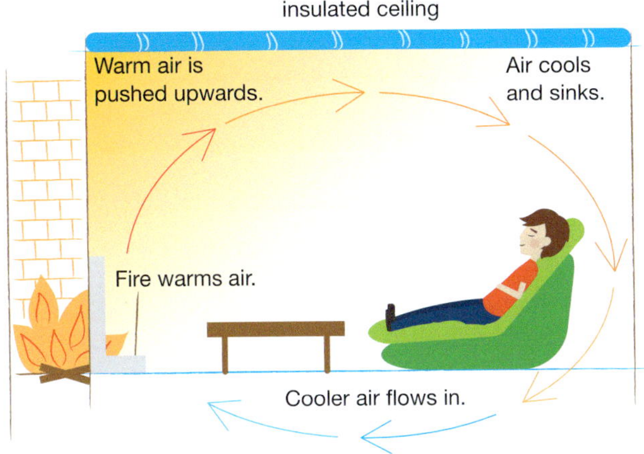

FIGURE 4.1.11 Convection currents gradually spread heat from the open fire through the air in a room.

Heat is transferred by convection in liquids and gases because their particles can move around. Figure 4.1.12 shows how liquid in a saucepan is heated by convection. Convection cannot happen in a solid because the particles can only vibrate and cannot move freely like they do in a liquid or gas.

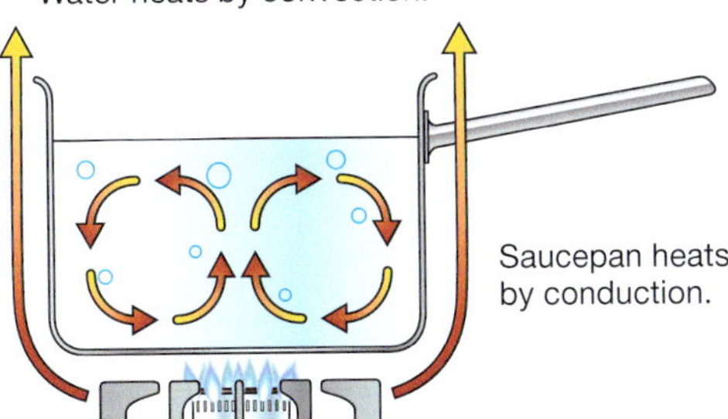

FIGURE 4.1.12 Particles gain heat from the hot base of the saucepan and rise. Cooler liquid sinks down, is heated and the cycle continues.

Convection explains the formation of a sea breeze during the day and a breeze towards the sea at night. This process is shown in Figure 4.1.13. Convection also circulates heat in a hot water system. You can see this in Figure 4.1.14.

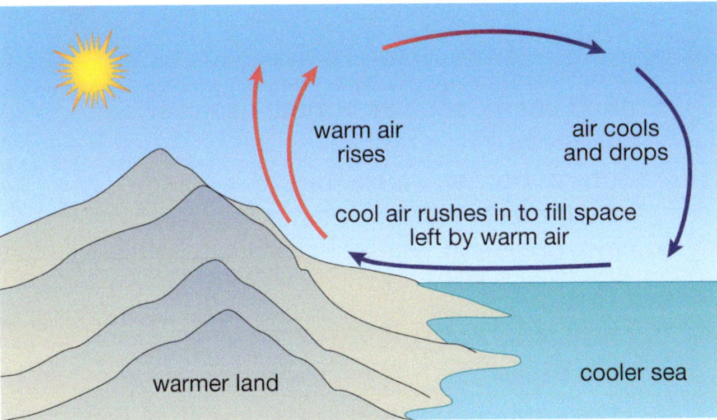

A sea breeze during the day

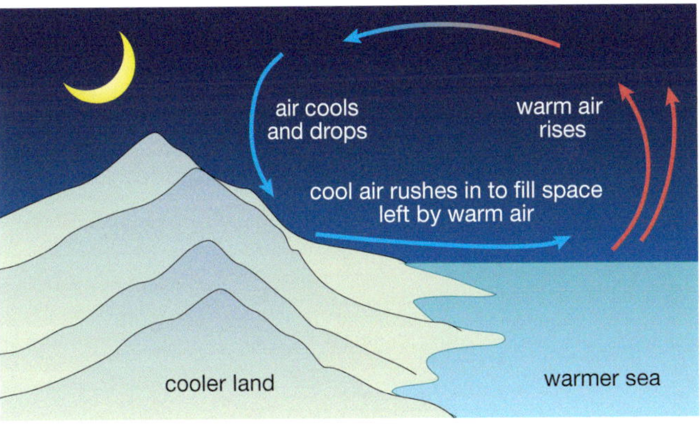

A land breeze at night

FIGURE 4.1.13 In the daytime, land heats up more quickly than the sea. Hot air is pushed upwards by cooler air that flows in towards it, producing a sea breeze. At night, the sea stays warmer for longer than the land and the process is reversed to produce a land breeze.

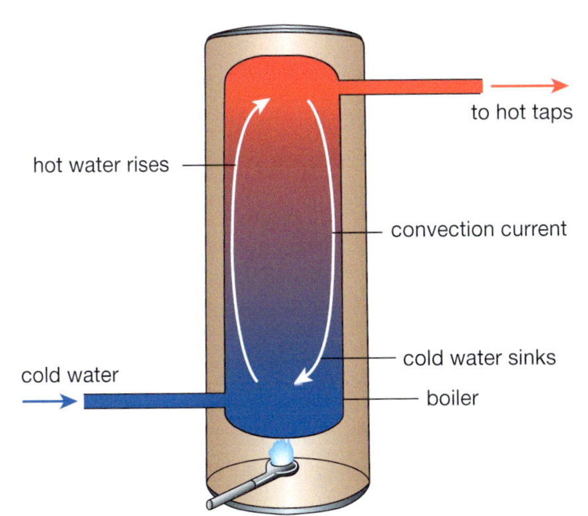

FIGURE 4.1.14 Convection assists in circulation of water in a hot water system.

science 4 fun

Ups and downs!

Can you see convection currents in action?

Collect this ...

- dried beans, such as borlotti beans or chickpeas
- Bunsen burner, gauze mat, tripod and bench mat
- large beaker of water

Do this ...

1 Add dried beans to cover the base of the beaker.

2 Cover the beans with water and then heat the mixture carefully over a Bunsen burner.

3 Turn off the heat after you have observed the behaviour of the beans in the hot water.

Record this ...

1 Describe what happened.

2 Explain why you think this happened.

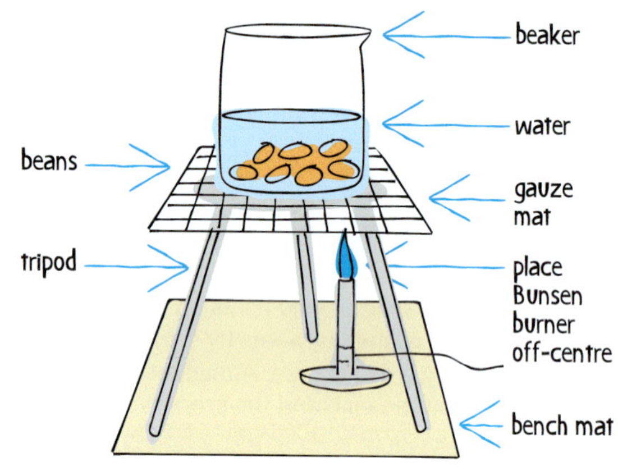

SciFile

Feeling chilly?

Naked mole rats are the only mammals known to not control their body temperature. Their bodies are warmed to the temperature of their burrows, about 30°C.

Radiation

When you go outside and into the sunlight, you can feel the heat from the Sun on your skin (Figure 4.1.15). Heat has travelled through empty space between the Sun and the Earth to reach you. It cannot be transferred by conduction or convection on its journey because there are no particles to vibrate or flow in the vacuum of space. The Sun transfers its heat energy through a process called **radiation**.

Radiation transmits heat as invisible waves that travel at the speed of light, which is around 300 000 km/s. Infrared radiation is heat energy that is transmitted this way. All objects emit (release) some infrared radiation.

FIGURE 4.1.15 Radiation from the Sun travels through the vacuum of space to reach us. It is cooler in the shade because this radiation has been blocked.

The hotter something is, the more heat it radiates. For example, a hot oven radiates more heat than an oven set at a lower temperature. Similarly, the red-hot coals of an open fire radiate such enormous amounts of heat that you cannot sit too close to them (Figure 4.1.16).

FIGURE 4.1.16 You can feel the radiant heat emitted from the glowing coals of an open fire.

When radiated energy hits a surface, the heat may be absorbed into the surface, reflected from the surface or transmitted through the surface.

This is shown in Figure 4.1.17. Often radiation will be partially absorbed, reflected or transmitted according to the material and its colour.

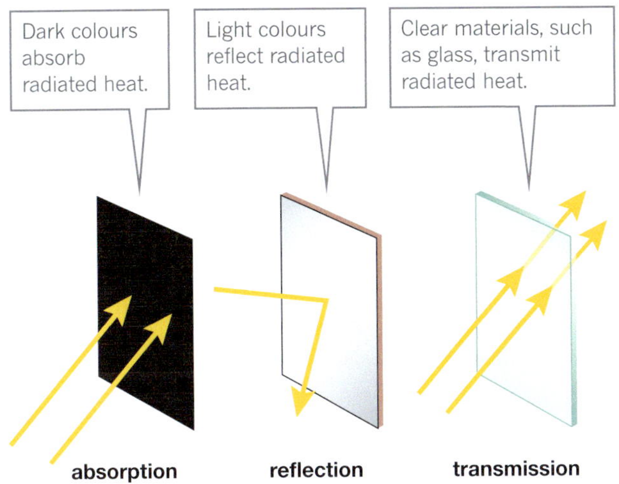

FIGURE 4.1.17 When radiation hits an object, it may be absorbed, reflected or transmitted.

For example, a dark-coloured car heats up more quickly in sunlight than a lighter-coloured car. This happens because dark-coloured objects absorb most of the radiation that fall on them. This means that they absorb much of the heat falling on them from the Sun. Solar pool heaters and hot solar hot water systems use black tubes and collection panels to absorb as much radiant heat from the Sun as possible (Figure 4.1.18). Lighter coloured objects reflect much of the radiation falling on them—most of the heat that falls on them is reflected and so they don't heat up as rapidly.

Prac 3 p. 136 | AB 4.3

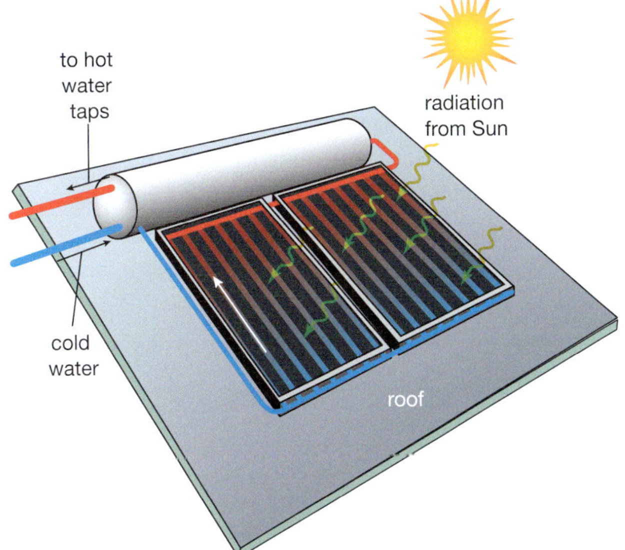

FIGURE 4.1.18 Heat energy radiated by the Sun is absorbed by the cold water within the black collection panels of a solar hot water system.

SciFile

Home insulation

Heat transfer in a home occurs by conduction, convection and radiation. In winter, warm air flows out of the house, and in summer, warm air flows in. Insulation added to the ceiling and walls of a home helps to stop this transfer of heat and makes your home more energy efficient.

STEM 4 fun

Insulated ice cube

PROBLEM

How long can you keep an ice cube frozen?

SUPPLIES

- frozen ice cube
- plastic cup
- assortment of items as possible insulation materials including cotton balls, polystyrene, aluminium foil, rags, petroleum jelly, etc.

PLAN AND DESIGN Design the solution: what information do you need to solve the problem? Draw a diagram. Make a list of materials you will need and steps you will take.

CREATE Follow your plan. Draw your solution to the problem.

IMPROVE What works? What doesn't? How do you know it solves the problem? What could work better? Modify your design to make it better. Test it out.

REFLECTION

1 What field of science did you work in today? Are there other fields where this activity applies?
2 In what career do these activities connect?
3 What did you do today that worked well? What didn't work well?

MODULE

4.1 Review questions

Remembering

1 Define the terms:
   a temperature
   b conduction
   c insulator.

2 What term best describes each of the following?
   a the temperature at which particles barely move at all
   b heat transfer involving particles that are free to move
   c cool air that flows from above a body of water towards land during the daytime.

3 What is the temperature that water freezes on the following scales?
   a Fahrenheit
   b Celsius
   c kelvin.

4 Fill in the following statements with the term that makes them true.
   a Heat always flows from an object of *higher/lower* temperature to one of *lower/higher* temperature.
   b Insulators are *good/poor* conductors of heat.
   c Gases are *good/poor* conductors of heat.
   d On a warm day, a house is warmer upstairs because of *conduction/convection* currents.

5 In the science4fun on page 130, the beans moved about the beaker because of the transfer of heat. Was the heat transfer an example of conduction, convection or radiation?

Understanding

6 Group the following objects as either conductors of heat or insulators.
   a a gold wedding ring
   b a polystyrene cup
   c a metal seatbelt buckle
   d an aluminium fence
   e thermal underwear.

7 Describe how the motion of particles in a solid change as the solid is heated.

8 A wetsuit traps a thin layer of water between the wearer and the neoprene fabric of the suit.
   a State whether water is a good or poor conductor of heat.
   b How does the wetsuit keep the wearer warm?

9 You lose a lot of heat from your head. For most people, their hair protects them from losing too much heat from their heads. Why is hair an effective insulator?

10 You walk barefoot on carpet in the living room of your house and your feet feel warm, yet when you walk into the bathroom and stand on the ceramic tiles your feet feel cold. The carpet and tiles are at the same temperature. Explain why the carpet and the tiles feel so different.

Applying

11 Use the particle model to explain why the following expand when heated.
   a a solid metal rod
   b a balloon.

12 Heat transfer can occur by conduction, convection or radiation. Identify the main method of heat transfer in each situation below.
   a Your feet get hot when you walk on sand at the beach.
   b Your back feels warm when you sit in the sun.
   c You boil water in an electric kettle.
   d You feel cold when you dive into a swimming pool.
   e You feel warm air as you walk into a school disco held in a hall.

Analysing

13 What is the difference between heat and temperature?

14 Two identical bathtubs are filled to the same level with water. The particles in bathtub A move with greater speed than the particles in bathtub B. Analyse this situation to answer the following.
   a Which bathtub will have warmer water?
   b Which bathtub will have more heat energy?
   c As the water cools, each bath loses heat energy. List three places this heat energy could go.

MODULE

4.1 Review questions

15 Water absorbs a large amount of heat energy for a relatively small rise in temperature compared to other substances. Use this information to analyse why in the science4fun on page 125 the balloon does not burn when filled with water.

Evaluating

16 On a hot day, you have a choice of travelling in a red car, a white car or a black car, all of the same model. All have been parked in the sunlight for three hours.

- **a** Which car would you choose?
- **b** Justify your choice.

17 Pyrex glass expands less than ordinary glass when heated. Use this information to propose a reason why Pyrex is used in cooking instead of ordinary glass.

18 Figure 4.1.19 shows the experimental set-up for a radiation experiment. The same-sized black and white cardboard squares are attached to two thermometers close to an incandescent globe.

- **a** What do you think the student is trying to test in this experiment?
- **b** State three variables that must be controlled to ensure a fair test.
- **c** Predict which thermometer will show the highest reading after 5 minutes.
- **d** Discuss reasons for your answer to part c.

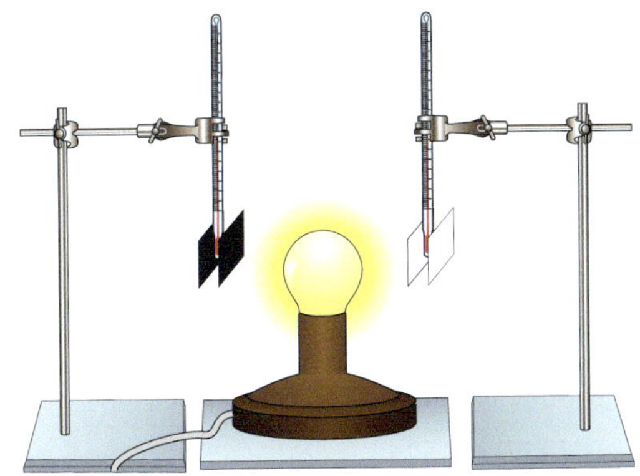

FIGURE 4.1.19

19 Sonja and Marcos are testing the insulating properties of cup A and cup B in their science laboratory. After testing, they plot the graph shown in Figure 4.1.20.

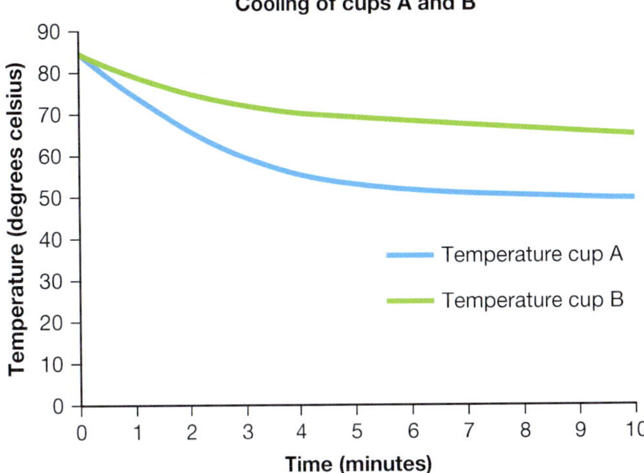

FIGURE 4.1.20

- **a** What procedure do you think the students used in their experiment?
- **b** What is the independent variable being tested?
- **c** What is the dependent variable that Sonja and Marcos are measuring in this task?
- **d** To be a fair test, which variables would Sonja and Marcos need to control in this experiment?
- **e** What is the initial temperature of the contents of cup A and cup B?
- **f** Describe how the temperature of this liquid varied over the 10 minutes in cup A and cup B.
- **g** Assess which cup is the better insulator.

Creating

20 Create a short story in which you imagine that you are one of the beans that was heated in the science4fun on page 130. Describe how your temperature and movement change as the beaker is heated. For at least four stages of your motion, add illustrations showing where you are positioned inside the beaker.

21 **a** Construct a diagram of a new type of suit that will keep you warm in cold conditions.

- **b** On your diagram, label what the suit is made from and how it keeps the heat in.

MODULE **4.1** # Practical investigations

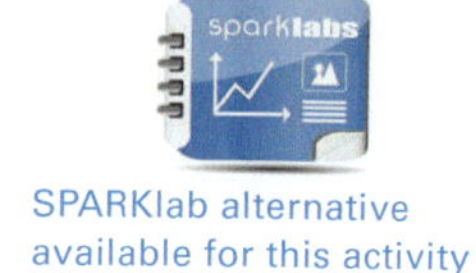

SPARKlab alternative available for this activity

1 • Comparing materials

Purpose

To compare how well plastic, wood and metal conduct heat.

Hypothesis

Which do you think will conduct heat better—a plastic spoon, a wooden stick or a metal spoon? Before you go any further with this investigation, write a hypothesis in your workbook.

Timing

30 minutes

Materials

- butter
- very hot water (from a kettle)
- plastic spoon
- metal spoon
- wooden icy-pole stick
- 250 mL beaker
- small beads or similar
- stopwatch
- ruler

SAFETY
Handle hot water with care.

Procedure

1. In your workbook, construct a results table like the one in the Results section.
2. Put a dob of butter near the top of the handle of the plastic spoon.
3. Push a bead onto the dob of butter.
4. Repeat steps 1 and 2 for the metal spoon and icy-pole stick. Make sure the beads are placed at equal heights and that the same amount of butter is used each time.
5. Carefully place the spoons and the icy-pole stick into very hot water in the beaker as shown in Figure 4.1.21.
6. Time how long each bead takes to fall off, and record your results in a table like the one shown in the Results section.

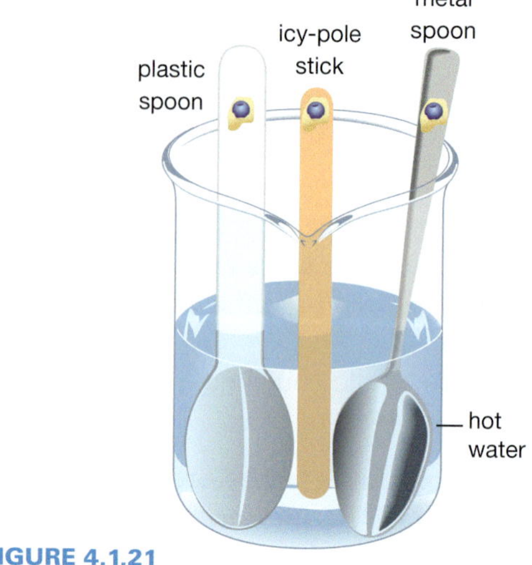

FIGURE 4.1.21

Results

Record all results in your results table.

Time taken to drop (seconds)

Bead on plastic spoon	Bead on icy-pole stick	Bead on metal spoon

Review

1. Which of the materials used was the best conductor of heat?
2. Assess whether or not your hypothesis was correct.
3. Which material was the best insulator?
4. Explain why it was important to place the beads at the same height, and use the same amount of butter.
5. The thermal conductivity of a material is a measure of how well the material conducts heat. It has the unit watts per metre kelvin (W/m/K). The higher this value is, the better the substance conducts heat.

   Use your results to identify which of the following thermal conductivities belong to the plastic spoon, the metal spoon and the wood icy-pole stick.

   0.17 16.0 0.19

MODULE 4.1

Practical investigations

2 • Testing insulators

Questioning & Predicting | Evaluating

Purpose

To test how effective different materials are in insulating heat.

Hypothesis

Which of the materials that you have available for this prac do you think will keep water in a can the warmest? Before you go any further with this investigation, write a hypothesis in your workbook.

Timing 60 minutes

Materials

- a range of insulating materials, such as newspaper strips, cloth, cotton wool, foam, polystyrene beads, foam packing bullets, fibreglass insulation, carpet scraps
- thermometer or temperature probe
- cardboard box
- hot water
- beaker
- stopwatch or clock
- empty soft-drink cans

SAFETY
Handle hot water with care.

You may use a temperature probe to gather temperature data.

Procedure

1 In your workbook, construct a table like that shown below. Alternatively, construct a spreadsheet with similar columns to those in the table.

Water temperature

Time (minutes)	Water temperature (°C)		
	Can with no insulating materials (air only)	Can with insulating material A	Can with insulating material B
0			
2			
4			
6			
8			
10			

2 Carefully measure 200 mL of hot water using a beaker, and pour this into your can.

3 Place the can inside the box.

4 Record the initial (starting) temperature of the water, and then measure and record the temperature every 2 minutes for 10 minutes. Record all your measurements in a table like that shown below.

5 Repeat, using water at the same initial temperature and packing one of the insulating materials into the space between the can and the box, as shown in Figure 4.1.22.

6 Repeat the process, using a second insulating material.

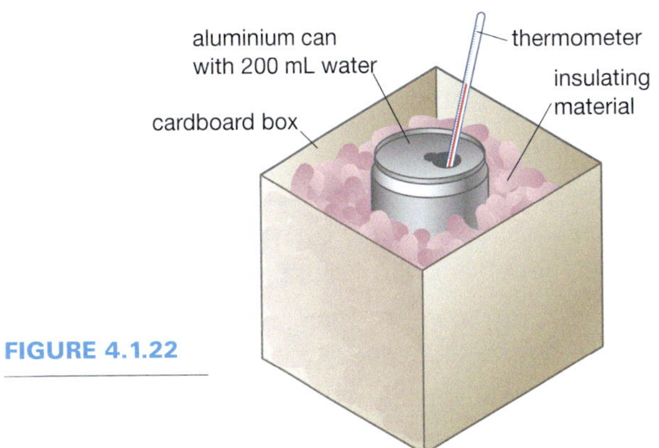

FIGURE 4.1.22

Results

1 In your results table, insert the names of two insulating materials you are going to test.

2 Record all measurements in your table or spreadsheet.

3 Construct a line graph of temperature versus time for each sample tested, to show the temperature drop over time. Alternatively, use data-logging equipment to produce a graph.

Review

1 a Construct a conclusion about which material was the best insulator.

b Assess whether your hypothesis was supported or not.

2 Why was it important to test one can with no insulating materials?

3 Identify any sources of error in your experiment.

4 Outline any improvements that could be made to the design of the experiment.

MODULE
4.1 Practical investigations

• STUDENT DESIGN •

3 • Comparing heat radiation

Planning & Conducting | Evaluating

Purpose

To compare how silver, white and black cans radiate heat.

Hypothesis

Which colour aluminium can do you think will radiate more heat over time—silver, white or black? Before you go any further with this investigation, write a hypothesis in your workbook.

Timing 45 minutes

Materials

- silver, white and black aluminium cans
- thermometer or temperature probe

SAFETY

A risk assessment is required for this investigation.
Be careful when handling hot liquids.

Procedure

1 Design an investigation to compare the amount of heat that is radiated over time from silver, white and black aluminium cans.

2 Brainstorm in your group and come up with several different ways to investigate the problem. Select the best procedure and write it in your workbook. Draw a diagram of the equipment you need.

3 Before you start any practical work, assess all risks associated with your procedure. Construct a risk assessment that outlines these risks and any precautions you need to take to minimise them. Show your teacher your procedure and your risk assessment. If they approve, then collect all the required materials and start work.

See Activity Book Toolkit to assist with developing a risk assessment.

Hints

- Make sure:
  - your cans all contain the same amount of water
  - the water is at the same starting temperature.
- Use the STEM and SDI template in your eBook to help you plan and carry out your investigation.

Results

1 Write a report on your findings.

2 Construct a line graph to display your results.

Review

1 Evaluate your procedure. Pick two other prac groups and evaluate their procedures too, identifying their strengths and weaknesses.

2 a Construct a conclusion for your investigation.

b Assess whether your hypothesis was supported or not.

MODULE

4.2 Sound

Indigenous Australians developed the didgeridoo thousands of years ago. The player blows air into the didgeridoo while vibrating his lips to produce a low rumbling sound. In addition to the didgeridoo, wind instruments like trumpets, flutes and trombones rely on vibrating air to make sounds. Other instruments, like violins, pianos and guitars, produce sound using strings that vibrate.

science 4 fun

Straw clarinets

YES

How does changing the length of a flute or a clarinet change the sound it produces?

Collect this ...

- straw
- pair of scissors

Do this ...

1 Squash the end of the straw and cut it to make a point.
2 Blow into the pointy end of the straw.
3 Try different positions until you get a buzzing sound.
4 As you are making the sound, have a partner carefully cut the other end of your straw.
5 Keep cutting, making the straw shorter and shorter.

Record this ...

1 Describe how the sound changed as the straw got shorter.
2 Explain how you think the straw produced a sound.

Sound waves

Sound is produced when something vibrates, moving back and forth very quickly. Table 4.2.1 shows some common sounds and the objects that vibrate to produce them. When something vibrates, it passes the vibrations into its surroundings, e.g. air. These vibrations create regions of space in which the air particles are bunched together and regions in which they are more spread out. The bunched up areas are called **compressions** and the spread-out areas are called **rarefactions**. Both areas are shown in Figure 4.2.1 on page 138. A **sound wave** is the movement of alternating compressions and rarefactions. Sound waves travel away from the source of a sound, in the same way that ripples of water move outwards when a stone is dropped into a pond.

TABLE 4.2.1 Common sounds and their sources

Sound	Vibrating source
speech	folds of skin (called vocal cords)
drum	drum skin
piano	string inside piano (when you strike a key, the string is struck by a hammer)
saxophone	reed inside the mouthpiece and therefore the air inside the saxophone
car stereo system	speaker cone
a bell ringing	metal casing of the bell (when struck)

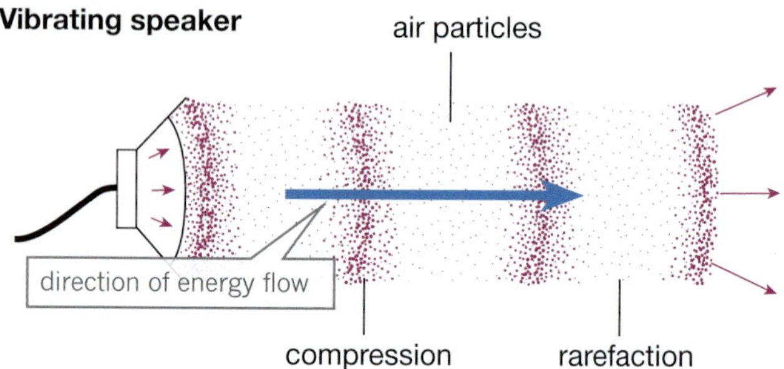

FIGURE 4.2.1 A vibrating speaker produces regions in which air particles are squashed close together (compressions) and regions in which air particles are spaced further apart (rarefactions). The energy moves through air as a sound wave.

Sound relies upon vibrating particles. This means that sound can pass through solids, liquids and gases but not through a vacuum where there are no particles. Hence, sound can pass through railway tracks, water in a swimming pool and air but not through space.

Types of waves

A wave carries energy from one point to another. This can happen in two ways. The energy carried by waves at a beach moves horizontally, but the particles making up the wave move in a vertical direction. This is shown in Figure 4.2.2. This vertical movement explains why a boat or a seagull floating on the sea bob up and down as a wave travels to the shore. This type of wave is called a **transverse wave**. Radiated heat energy is transferred as a type of transverse wave.

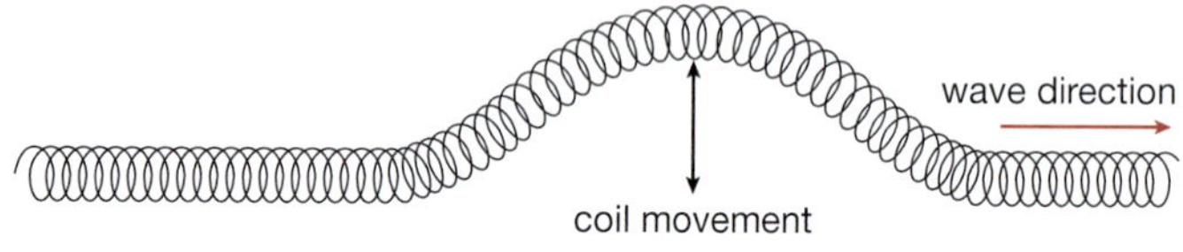

FIGURE 4.2.2 In the transverse wave shown here, the particles of the wave vibrate up and down whilst the wave travels forward.

A sound wave differs from a transverse wave. In a sound wave, the particles that make up the wave move back and forth in the same direction as the wave is travelling. This type of wave is called a **longitudinal wave**, or compression wave. This type of wave is shown in Figure 4.2.3.

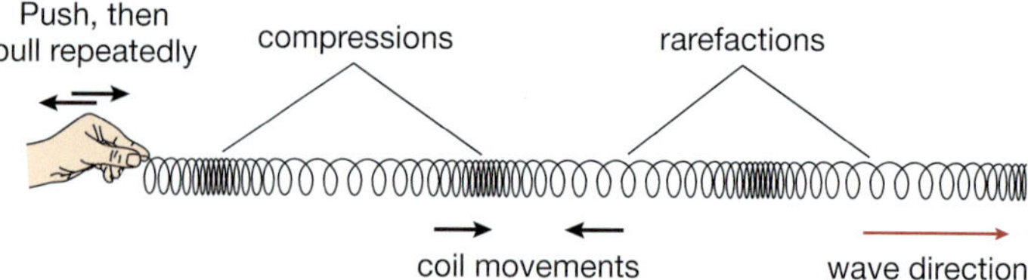

FIGURE 4.2.3 In a longitudinal wave, particles move in the same direction that the wave is moving.

When a guitar string is plucked, it vibrates and a transverse wave moves along the string. This vibrating string then sets up a longitudinal sound wave in the surrounding air particles. This is the wave that carries the sound to our ear. The differences between these waves can be seen in Figure 4.2.4.

Prac 1
p. 145

Transmission of sound

Sound energy is transmitted through a material as longitudinal waves. The particles of the material vibrate as the sound energy flows through it. If you have ever been to a concert or stood near an aircraft that is taking off, then you will have physically felt the energy that can be transmitted by sound waves.

The speed that sound travels through a material depends on the qualities of the material. In general, if a sound wave hits a dense material, made of lots of particles closely packed together, then the compressions and rarefactions will travel quickly. As a result, sound travels faster through solids than through liquids, and faster through liquids than through gases.

Table 4.2.2 shows that sound travels much faster through glass or steel than through air, the temperature of the material also affects the speed of sound transmission.

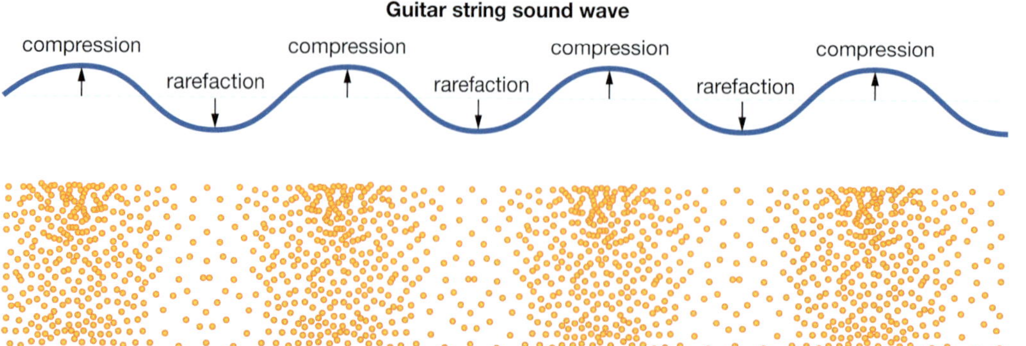

FIGURE 4.2.4 The top or crest of the transverse wave of a vibrating guitar string correlates to the compression of the particles of the longitudinal wave and the bottom or trough of the transverse wave correlates to the rarefaction.

The particles of warmer materials vibrate faster than particles in cooler materials. As a result, sound travels faster through warm air than through cool air and faster through warm water than through cold water.

TABLE 4.2.2 Speed of sound in various materials

Material	Speed of sound (metres/second)
air (at 0°C)	331
air (at 18°C)	342
water	1440
wood	4500
steel	5100
glass	5200

Reflection and absorption of sound

You can usually hear people in the next room if they are noisy. This happens because sound passes through thin walls and is transmitted short distances through most materials.

Hard surfaces, such as concrete or bathroom tiles, reflect sound waves. This reflected sound is heard as an **echo**. The time difference between sending and receiving sound waves can be measured. This difference can be used to calculate the depth of objects under water, using a technique called sonar (sound navigation and ranging), shown in Figure 4.2.5.

In this process, a ship sends a sound wave into the water. The sound wave bounces off any hard surfaces in the water such as fish. We can calculate the depth of objects under water by measuring the time a sound wave takes to bounce off them and return to the ship/surface.

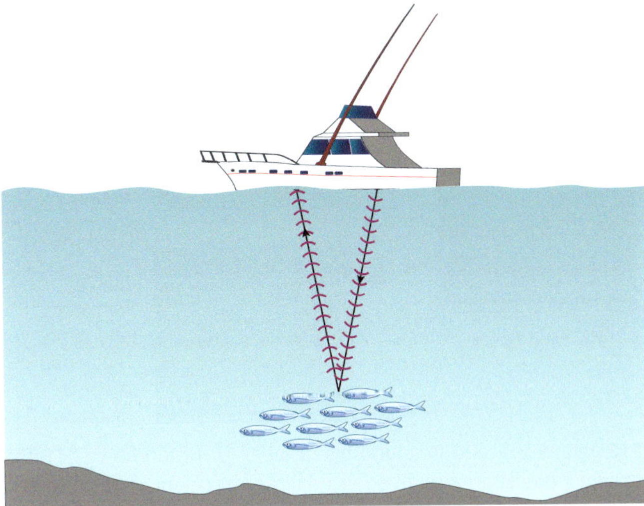

FIGURE 4.2.5 Sonar can be used to determine the depth of objects in the sea, such as this school of fish.

SciFile

Lightning speed

You will see the spectacular effects of a fireworks display before you hear the explosions. This is because light travels much faster than sound.

A bare and empty room has multiple hard surfaces which reflect sound with little if any absorption (Figure 4.2.6). This causes any sound to bounce around as a series of echoes, allowing the sound to be heard for a considerable time. The length of time a sound can be heard for is known as **reverberation**.

Soft materials such as curtain fabric, carpet and cushions absorb sound and convert it into heat. This reduces the reverberation, or length of time a sound is heard.

FIGURE 4.2.6 Different materials give a room different reverberation times.

SciFile

Boom!

Fighter jets regularly travel at supersonic speeds—faster than the speed of sound. As the jet catches up to and then overtakes the sound waves it has produced, a very loud 'sonic boom' is heard. Sonic booms can smash windows and damage the hearing of humans, birds and other animals. The sound of this jet breaking the sound barrier has compressed water vapour in the air to form an instantaneous cloud.

Sound absorption like this is needed in concert halls, so that there is no overlap between the sounds being performed and their echoes, which would otherwise distort what you hear. Figure 4.2.7 compares how well some materials absorb sound.

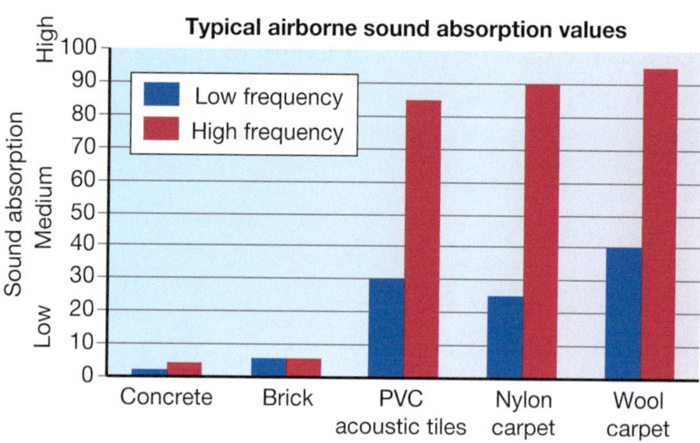

FIGURE 4.2.7 A comparison of the sound absorption levels of different materials

Frequency and pitch

A dog has a low-pitched growl, whereas a bird chirps with a high-pitched sound. The different sounds can be compared by analysing their sound waves using an oscilloscope (also known as a CRO), as shown in Figure 4.2.8.

A source that vibrates rapidly produces sound of a higher pitch, than one that vibrates more slowly. The number of vibrations a sound makes each second is called the **frequency** of a wave. High frequency sound waves have a higher pitch, and low frequency waves have a low pitch. Frequency is measured in **hertz (Hz)**. Sound waves also have a wavelength.

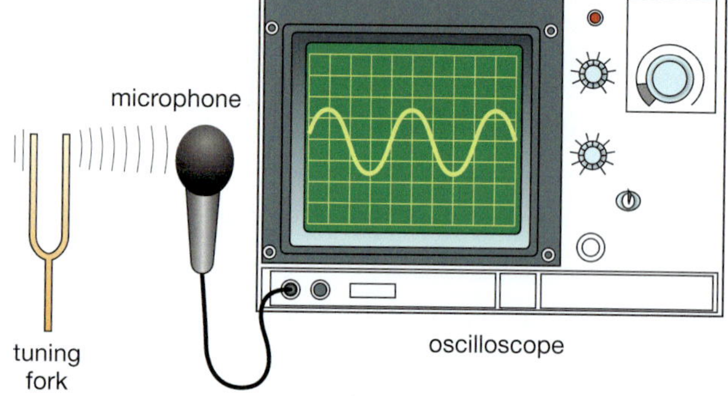

FIGURE 4.2.8 An oscilloscope converts sound waves into electrical signals that can be viewed on a screen.

The **wavelength** of a sound is the distance between two peaks that are next to each other. It is measured in metres (m). A loud sound has a sound wave with steep peaks, also known as greater amplitude. A soft sound has a sound wave with smaller, less steep peaks, also known as lower amplitude. Figure 4.2.9 shows that graphs of louder sounds have larger peaks.

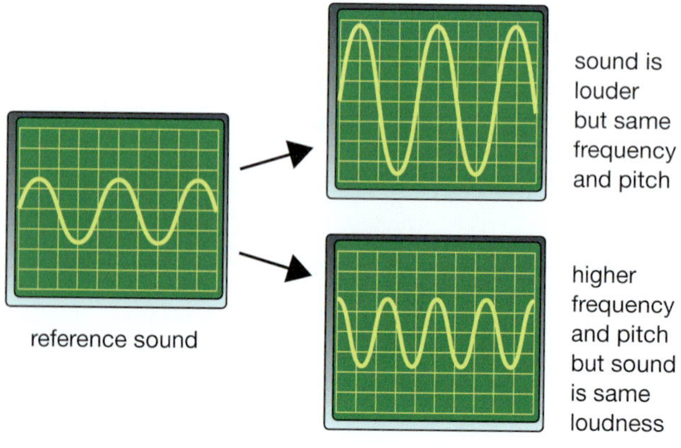

FIGURE 4.2.9 A loud sound has a taller graph on an oscilloscope than a quiet sound. Higher-frequency sound waves have a shorter wavelength.

SciFile

The Doppler effect

Have you ever heard the wail of an ambulance siren rushing past? When an ambulance travels towards you, the sound waves of the siren bunch up. This makes the sound of its siren higher in pitch. As the ambulance moves away, its sound waves are spread further apart and the sound is lower in pitch. This change in pitch is called the Doppler effect, named after the Austrian physicist Christian Doppler who first described it in 1842.

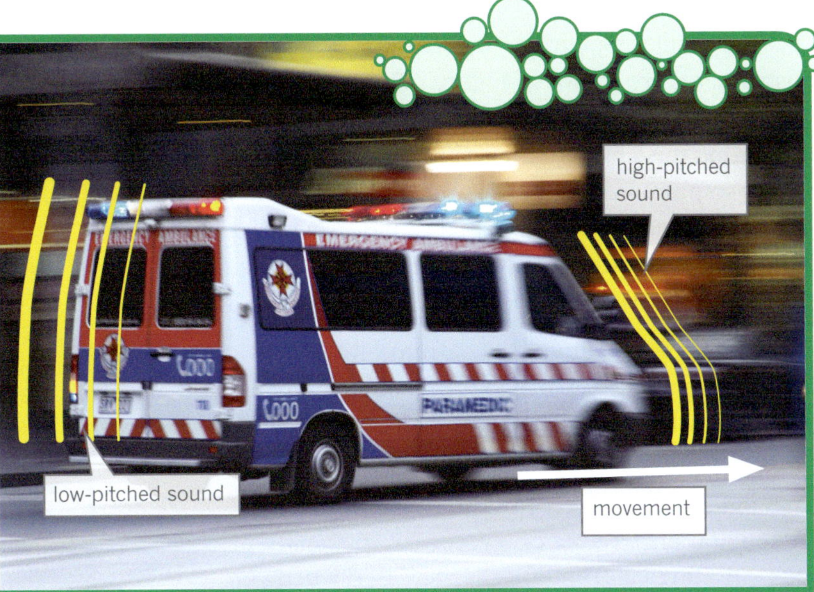

Louder sounds have greater amplitude than softer sounds. The higher the frequency of a wave, the more closely the wave is bunched together and the shorter its wavelength.

When we get older, we lose our ability to hear higher frequencies of sound. Young people can typically hear a range of frequencies up to 20 000 Hz, yet most people over 65 years cannot hear frequencies above 5000 Hz. Hence, some mobile phone ringtones cannot be heard by older adults. As we age, more of the tiny hair cells in our inner ear become damaged or destroyed. This happens most easily to the hair cells that we use to hear high frequency sound and once they are destroyed these cells cannot be repaired.

Many animals, such as dogs and cats, can hear sound frequencies that are outside our human range of hearing. Ultrasound is the name given to sound waves with frequencies above our hearing range. Bats emit ultrasound squeaks with frequencies up to 200 000 Hz, which reflect off surfaces around them and are used by bats to avoid obstacles and to locate food. Elephants can hear a range of frequencies lower than our own hearing range. These frequencies are called infrasound.

Prac 2 p. 145
Prac 3 p. 146

Computer analysis of the way ultrasound reflects from living tissue can be used to create an image, like the one shown in Figure 4.2.10.

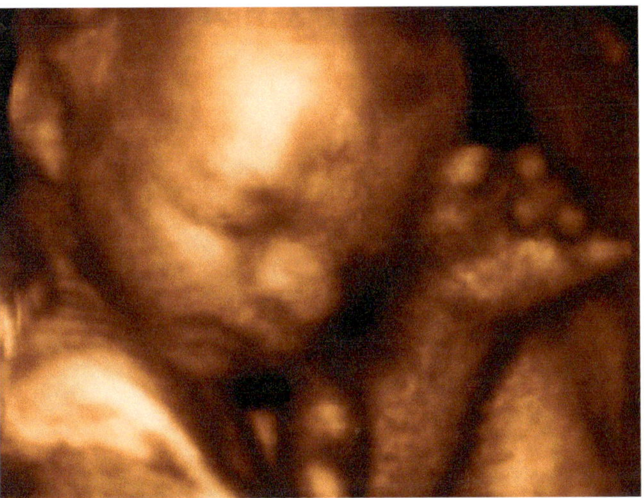

FIGURE 4.2.10 Ultrasound waves pass easily through fluids and soft tissue but are reflected from other layers within the body. Echoes of these waves are detected and analysed by computer to create the image, such as this 3D image of a human foetus.

Musical instruments

All musical instruments produce sounds by vibrations (Figure 4.2.11). They do this in different ways and produce sounds of differing characteristic qualities. These differences can be compared by playing the sound into a microphone attached to an oscilloscope.

FIGURE 4.2.11 Each of these musical instruments uses vibration to create its sound. Guitars use vibrating strings, trumpets use vibration from the player's lips and air inside the trumpet, and drums use a vibrating skin.

Typical oscilloscope traces from the sound of four instruments are shown in Figure 4.2.12. On a guitar, a violin and a piano, vibrations are produced by strings. Changing the length of the string alters the frequency of the sound produced. Longer strings vibrate more slowly, producing lower-pitched sound than shorter strings. When you press a string against the neck of a guitar, you shorten the effective length of that string.

In percussion instruments like drums, the skin stretched over the top of the drum vibrates when you hit it. In instruments like the triangle or the cymbal, the instrument itself vibrates.

In wind instruments, a column of air vibrates. When you play a flute or a recorder, the length of this vibrating column of air is increased when you cover holes along the tube, and shortened when you leave the holes open. A longer vibrating column of air produces a lower pitched sound than a shorter vibrating air column.

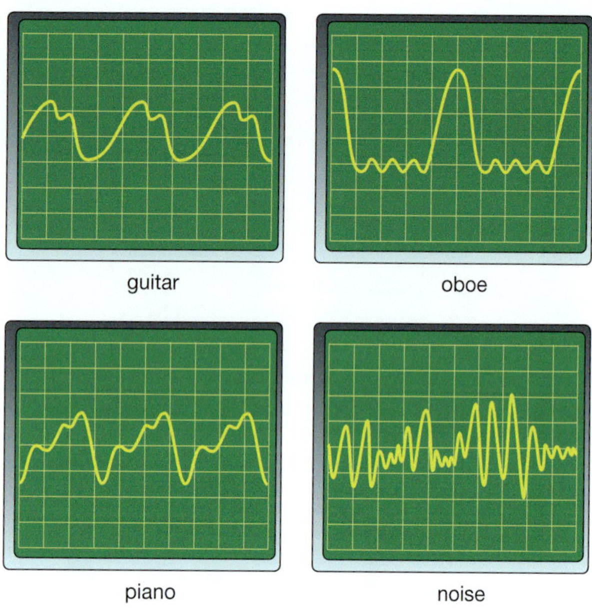

FIGURE 4.2.12 Musical notes produce a smooth, repeating pattern on an oscilloscope. Different instruments produce different characteristic sounds. Background noise shows as an uneven mixture of waves.

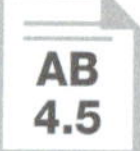

Working with Science

AUDIO ENGINEER

Audio or sound engineers are involved in the technical and mechanical aspects of mixing, recording, editing and producing sound and music. They use their skills in a wide variety of industries, including music, film, television and radio. Audio engineers work both in and out of the studio. They are responsible for setting up audio equipment and performing sound checks and live sound mixing at concerts, sports games and theatre productions (Figure 4.2.13). Audio engineers also design, develop and manufacture audio software and equipment, such as microphones and sound boards. More specialised audio engineers are involved in researching the science behind sound and audio technology. There are many different areas for audio engineers to specialise in, such as video game audio design, live sound engineering and recording engineering.

To become an audio engineer, you will need to complete an audio engineering and sound production course at TAFE or university. Courses are available from certificate level (for example, Certificate IV in Sound Production) to degree level (for example, Bachelor of Sound and Music Design). Because audio engineering involves a lot of technical skills, an interest in maths, engineering and technology is important. Hands-on experience at a studio, community radio station or production company is a great way to gain valuable skills and learn how the industry works. Audio engineering offers an interesting and diverse career pathway and there are opportunities in a wide range of fields for people with these skills in Australia.

FIGURE 4.2.13 Audio engineers can work in a variety of roles, from recording and mixing music in a studio, to sound checks and mixing at live concerts.

Review

What kinds of recorded sounds do you hear every day that might have been produced by an audio engineer?

MODULE 4.2

Review questions

Remembering

1 Define the terms:
   a compression
   b transverse wave
   c frequency.

2 What term best describes each of the following?
   a movement of alternating compressions and rarefactions
   b energy carried by the wave moves in the same direction as the wave particles
   c distance a particle in a wave moves from its rest position.

3 What is the unit used to measure the following quantities?
   a frequency
   b wavelength
   c speed of sound.

4 What is the speed of sound in air at 18 degrees Celsius?

5 What is the source of vibration in a flute?

6 Which of the following statements are true and which are false?
   a Sound is produced by vibrations.
   b Regions of high air pressure are called rarefactions.
   c A sound wave can travel in a vacuum.
   d Waves at the beach are called transverse waves.

Understanding

7 Explain why sound travels faster through solids than through liquids.

8 a What is meant by the term *reverberation*?
   b Explain why an empty room is full of echoes.
   c Predict what happens to those echoes as furniture, curtains and carpet are moved into the room.

9 Predict which of the straws in the science4fun on page 137 would produce:
   a the highest pitched sound
   b sound with the longest wavelength
   c sound of the highest frequency.

10 Group the following surfaces into those that would reflect sound waves and those that would absorb sound waves:
   - marble tiles in a bathroom
   - polished floorboards in a school hall
   - carpet on the floor of a cinema
   - a glass window
   - a lounge room with a velvet couch and a deep shaggy carpet.

Applying

11 Figure 4.2.14 illustrates three traces of sounds on an oscilloscope.

Identify the:
   a sound with the highest frequency
   b sound with the lowest frequency
   c loudest sound.

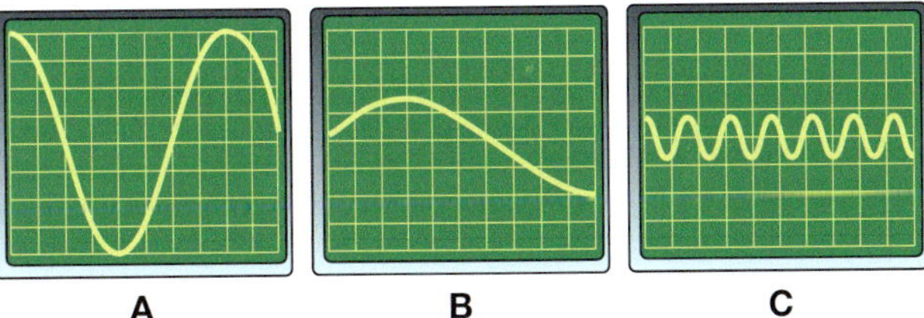

FIGURE 4.2.14

Analysing

12 Analyse the oscilloscope patterns of the sound from a guitar, oboe and piano in Figure 4.2.12 on page 142.
   a Which instrument was being played the loudest?
   b Which instrument had the lowest pitch?
   c Compare the patterns of musical instruments with the shape of the trace that is shown for noise.

MODULE

4.2 Review questions

13 For the ocean wave shown in Figure 4.2.15, determine the:

a wavelength of the wave

b amplitude of the wave.

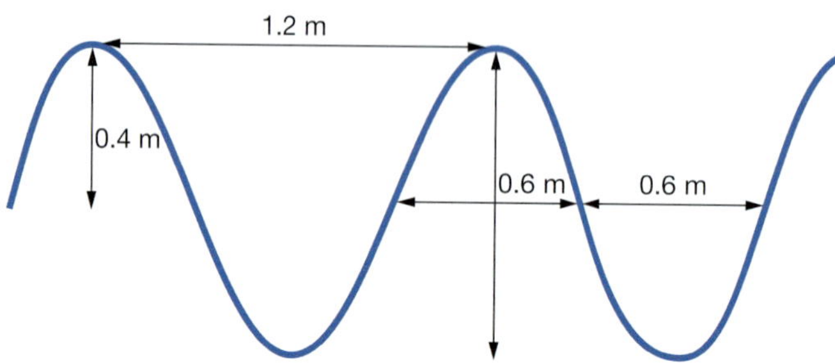

FIGURE 4.2.15

14 Refer to the illustration of the air particles in a sound wave shown in Figure 4.2.16 to complete the following questions.

a Determine the wavelength of this sound wave.

b What would be the distance shown as *x* in the diagram?

c Do the points marked A and B show a compression or a rarefaction?

d Describe the difference in air pressure that would exist at points A and B.

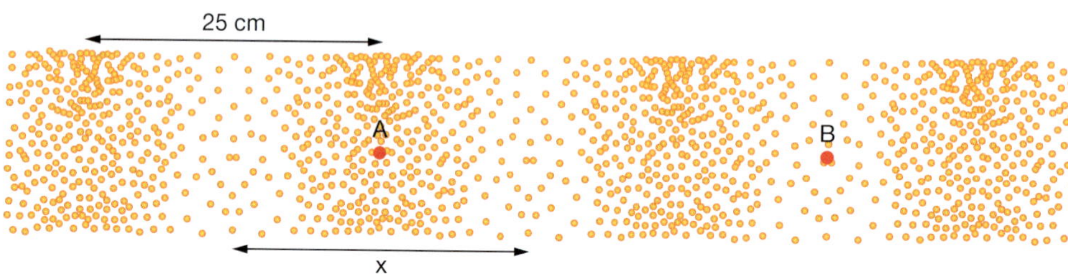

FIGURE 4.2.16

15 Analyse Figure 4.2.7 on page 140 to complete the following.

a Which material listed best absorbs high frequency sounds?

b i Which material reflects the highest proportion of sounds?

ii Propose a reason why this material does not absorb much sound.

c Explain what happens to the sound energy absorbed by materials.

16 Voice recognition uses computer programs to identify speech. How do you think these programs work?

Creating

17 Construct a diagram to contrast ultrasound with infrasound.

MODULE 4.2 Practical investigations

1 • Spring waves

Planning & Conducting | Evaluating

Purpose

To model transverse and longitudinal waves.

Timing 30 minutes

Materials

- slinky spring

Procedure

1. Hold one end of the slinky. A partner holds the other end. Take care not to overstretch the slinky.
2. Move your end of the slinky up and down at a regular speed, as shown in Figure 4.2.17a.
3. Sketch a side view of the waves you made. These are transverse waves, like water waves.
4. Try to alter how you produce the waves so that they are bunched closer together, with greater frequency.
5. You and your partner hold each end still.
6. Tap one end of the spring horizontally so that a pulse travels through the spring, as shown in Figure 4.2.17b.
7. Try to draw what is happening here from a side view. This is a longitudinal wave, like a sound wave.
8. Try to alter how you produce these waves to increase their frequency.

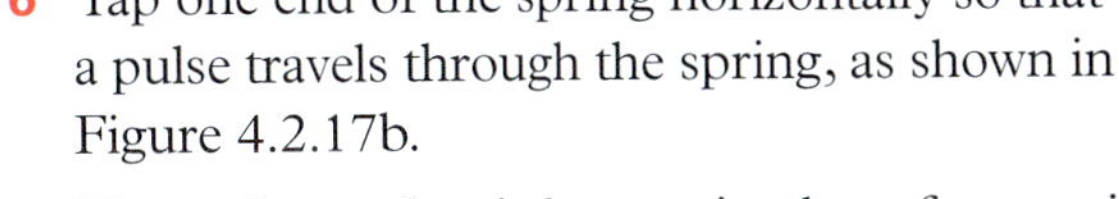

Review

1. Describe how the slinky moves when a:
   - a transverse wave passes through
   - b longitudinal wave is transmitted.
2. How were you able to increase the frequency of the following waves?
   - a transverse waves
   - b longitudinal waves.
3. Which of the two waves you produced is more like a sound wave?

a

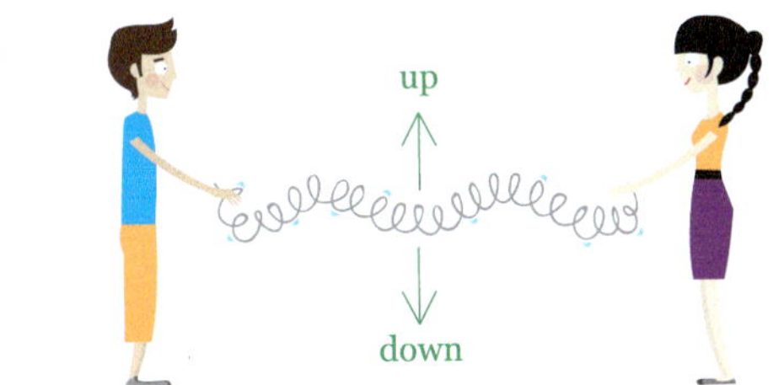

b

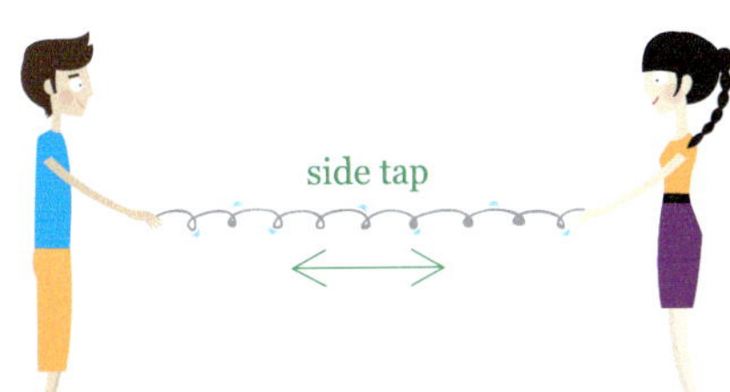

FIGURE 4.2.17

2 • Good vibrations

Purpose

To investigate the differences in producing high-pitched and low-pitched sounds.

Timing 45 minutes

Materials

- water
- 5 beakers (or glasses) of the same size
- pen or a chopstick
- ruler

Procedure

PART A

1. Hold the ruler over the edge of a bench and flick it so it vibrates.
2. Listen to how the pitch changes as you reduce the length of ruler vibrating.

PART B

3. Line up the beakers (or glasses) on your bench.
4. Fill each beaker to a different depth with water.
5. Carefully tap the glass of each using the pen, chopstick or other object, and listen to the variation in pitch.

Review

1. Did the longer or shorter length of vibrating ruler make the highest pitched sound?
2. Construct a graph that shows the difference between sound waves produced.
3. Describe the pitch of sound produced by the beaker with the least amount of water.
4. How do you think that the length of the ruler and the depth of water in the beaker are related to the pitch of sound they produce?
5. Do you think the vibrations would be faster or slower when producing higher pitched sounds?
6. Try and recreate a well-known tune, such as ‘Happy birthday’.

MODULE 4.2

Practical investigations

3 • Producing sound

Purpose

To see and hear the vibrations that produce a sound wave.

Timing 30 minutes

Materials

- water

- selection of tuning forks
- rubber mallet or rubber stopper
- 100 mL beaker
- wooden bench top or sounding box

Procedure

1 Strike a tuning fork with a rubber mallet or on a soft surface such as a rubber stopper or a book.

2 Place the ends of the tuning fork carefully towards your ear and see if you can hear a sound.

3 Strike the tuning fork again and this time hold it on a bench top (Figure 4.2.18) or position it in a hollow wooden sounding box.

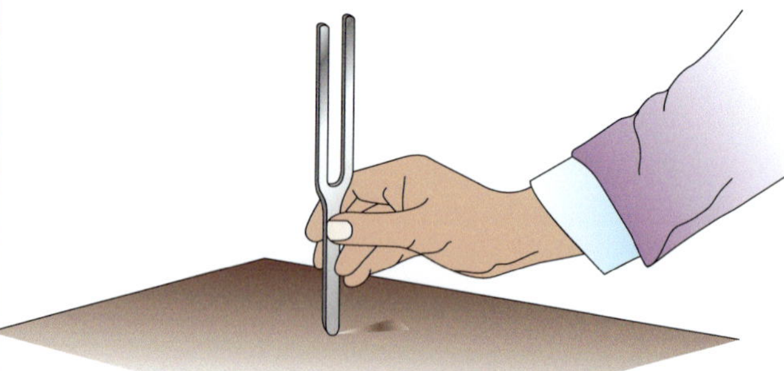

FIGURE 4.2.18

4 Repeat the previous step using a selection of tuning forks that are designed to produce sound of differing frequencies.

5 Half fill a beaker with water.

6 Strike the tuning fork and insert its ends into the water as shown in Figure 4.2.19. Observe the water.

7 Repeat the previous step but record what happens using a phone or video camera.

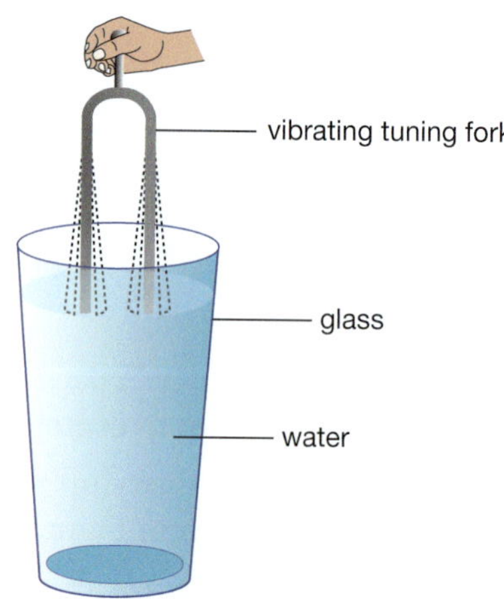

FIGURE 4.2.19

8 Extension:
Place two identical tuning forks in two sounding boxes (Figure 4.2.20). Strike the first tuning fork and carefully observe to see if the second tuning fork vibrates without being struck. This phenomenon is called resonance.

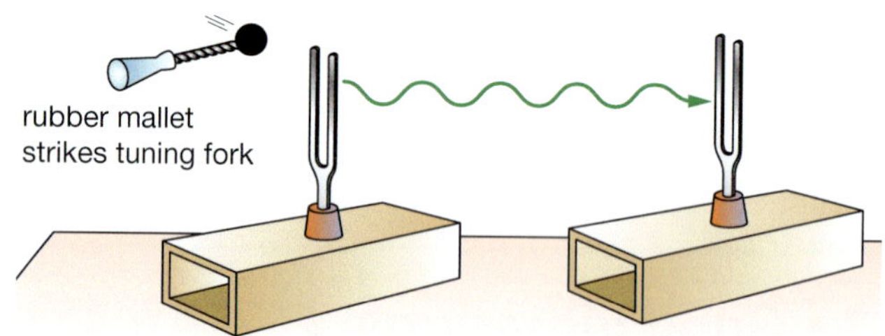

FIGURE 4.2.20

Review

1 a Describe the difference you observed in the sound produced before and after the tuning fork was placed on a bench top or sounding box.

b Propose why the sound changes in this case.

2 a Describe the effect of placing the tuning fork into the water.

b How can this provide evidence that the ends of the tuning fork are vibrating?

3 Compare the pitch of the tuning fork with the frequency in hertz (Hz) that is inscribed at its base. Which tuning fork would produce a higher pitched sound: 256 Hz or 512 Hz?

MODULE 4.2 Practical investigations

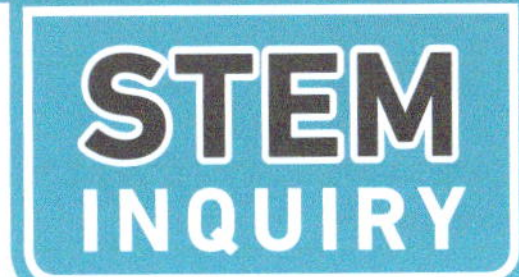

Energy efficient house

Planning & Conducting | Evaluating

Background

You are employed by a construction firm. This firm is known for developing house designs that are sustainable, energy efficient and have spaces suited to specific needs, such as a wine cellar/cool room and a cinema room/music studio.

A client is looking for a house design that uses the best materials and concepts. The client is a musician and environmentalist. Her house will be located in an area where summer and winter weather conditions are extreme.

Problem

Your job is to provide information to the construction firm about energy efficient designs, materials and specific plans for specialty rooms. Using research then trialling materials for heat-proofing or sound-proofing, you will present the firm with the best designs, materials and concepts for this client (Figure 4.2.21).

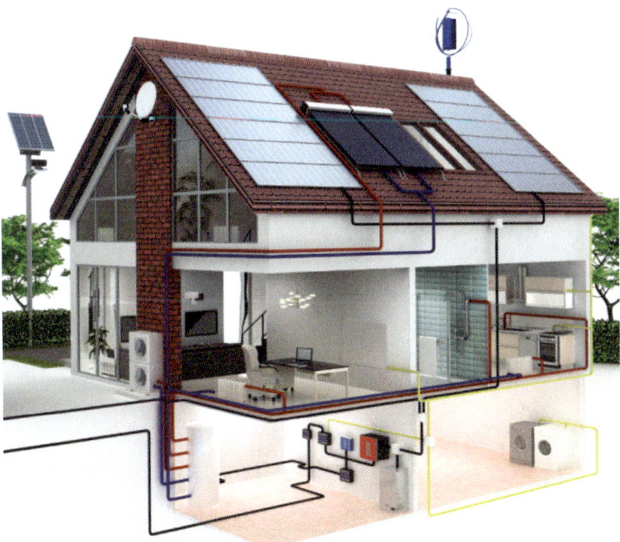

FIGURE 4.2.21 An energy efficient house with ideas for features to research

Engineering design process

The engineering design process is outlined in Figure 4.2.22.

- Identify the purpose.
- Identify the independent, dependent and controlled variables and only change one variable at a time.
- Based on your purpose and the controls and variables, write a hypothesis for this experiment.
- Before you commence your investigation you must conduct a risk assessment and write down safety measures that you will follow to keep yourself and other students safe.
  See Activity Book Toolkit to assist with developing a risk assessment.
- Summarise your experiment in a scientific report including the Purpose, Hypothesis, Materials, Procedure, Risk Assessment, Results (including data presented in tables and/or graphs), Discussion and Conclusion.

Engineering design process

Identify problem

Brainstorm solutions

Design and build prototype

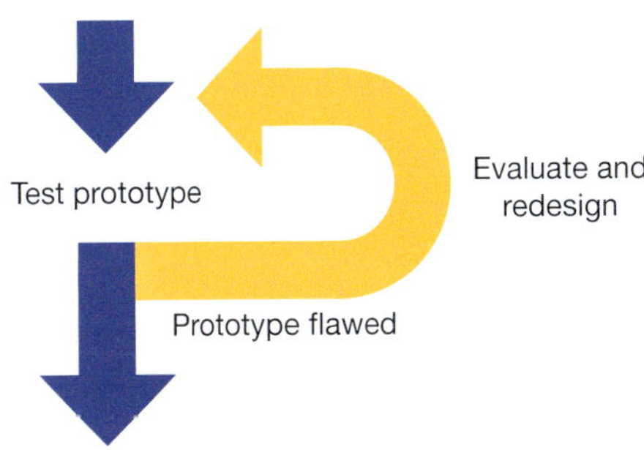

Communicate successful solution

FIGURE 4.2.22 Understanding the engineering design process. Engineering involves the application of science to design, construction and maintenance of structures, machines and devices.

MODULE 4.2

Practical investigations

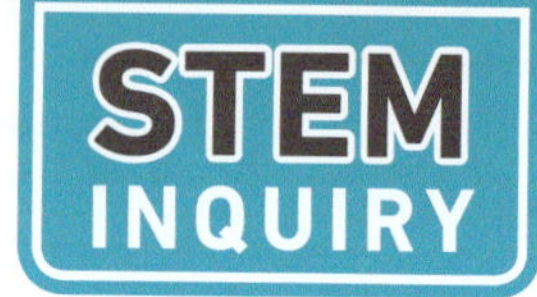

Procedure

1 Your **first stage** is to research and thoroughly understand all concepts involved in the construction of an energy efficient house.

- Brainstorm features or ideas that may contribute to energy efficiency.
- Research and record notes.
- Collate this information for your planning of prototypes.

2 Your **second stage** is to choose which room you want to design.

- Option A: design and build a model/prototype room that needs to stay at a consistent cool temperature at all times, regardless of external temperatures. This is the client's wine cellar or cool room.
- Option B: design and build a model/prototype room where people can play music or movies as loud as they like without disrupting the rest of the house. This is the client's cinema room or music studio.

3 Your **third stage** is to test your prototype room, identify any weaknesses and improve on your design.

- Test your model room and record results.
- Redesign or enhance the model. Test it again and repeat this process until it is the best design.

4 The **fourth stage** is to write a report for your client. Your report should include the following:

- paragraph outlining the key information gained
- explanation of how effective your model is in heat-proofing or sound-proofing
- summary of your results which prove this
- diagram of your design
- list of materials used
- brief statement of why you recommend this concept to your client.

Materials

- cotton wool
- polystyrene
- balsa wood or other model construction material
- carpet
- vinyl
- egg cartons
- aluminium foil
- wooden icy-pole sticks
- cardboard
- glue, masking tape
- heater/heat lamp
- scissors, box knife
- other materials for heat/sound-proofing
- ice-cubes to test thermal insulation
- phone/speaker to test sound-proofing. Download a sound level app to find out the sound readings in decibels.
- poster paper
- other materials to make a poster

Hints

Consider the following points in your investigation.

- What are you trying to do? What is the problem? How will you demonstrate you have solved it?
- Starting with a basic house design, what factors could be altered to make it more energy efficient?
- What are the properties of materials used for thermal- or sound-proofing?
- Investigate and suggest how you could test how the materials you have chosen to protect against different environmental conditions.
- What data needs to be collected and presented as evidence that your house design is efficient? Do you have a control?
- Use the STEM and SDI template in your eBook to help you plan and carry out your investigation.

MODULE

4.3 Light

Scientists have long debated what light is and how it travels. The Sun is a natural source of light, and shadows form when its light is blocked. Light is a form of energy called electromagnetic radiation. X-rays, infrared radiation, ultraviolet light, microwaves and radio waves are other types of electromagnetic radiation.

science 4 fun

Image finder

NO

How far behind a mirror is an image located?

Collect this ...

- 2 birthday candles
- 2 dobs of Blu Tack®
- sheet of glass
- sheet of paper

Do this ...

1 Draw a line down the middle of the sheet of paper.
2 Hold the glass so that it stands along this line.
3 Position a lit birthday candle with Blu Tack in front of the glass, as shown in the diagram.
4 Carefully look into the glass at the image formed.
5 Move an unlit candle behind the glass so that the image of the flame rests on its wick.
6 Measure the distance from each candle to the glass.

Record this ...

1 Describe what happened.
2 Explain why you think this happened.

Properties of light

Radiated heat travels through space from the Sun to Earth as infrared radiation. Light also travels through space from the Sun to Earth, lighting up our day. Both radiated heat and light are examples of electromagnetic radiation. They travel as a wave known as an electromagnetic wave. As Figure 4.3.1 shows, this wave has a complex structure. Like a sound wave, an electromagnetic wave has a specific frequency and wavelength. But unlike a sound wave, an electromagnetic wave does not require a material to transmit through. It can pass through a vacuum.

For example, infrared radiation and light from the Sun travel through the vacuum of empty space to reach Earth. A sound wave in air may travel at around 340 m/s, whereas a light wave travels at an incredible 300 000 km/s.

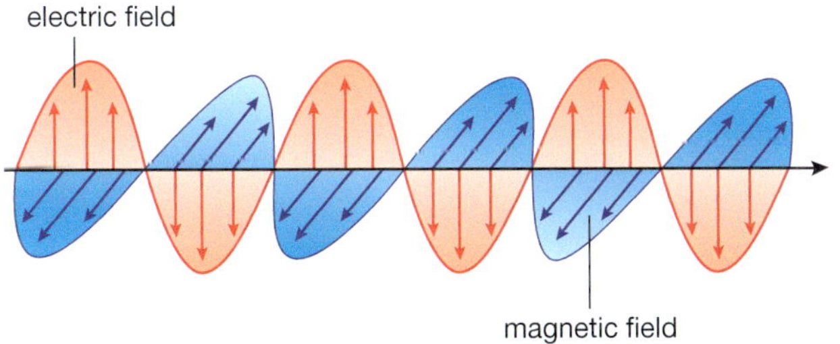

FIGURE 4.3.1 Light travels as an electromagnetic wave made of alternating electric and magnetic fields.

An object that releases or emits light is said to be **luminous**. However, most objects do not produce their own light. They are non-luminous. You see most objects, such as the Moon, because light bounces off them and then into your eyes. This process is called **reflection** and is shown in Figure 4.3.2.

FIGURE 4.3.2 This girl can see the tree and the butterfly because light from the Sun is reflected from them.

The surfaces of most objects are quite rough when viewed up close. These surfaces reflect or scatter light in many directions, and do not form an image. This is called diffuse reflection and is shown in Figure 4.3.4.

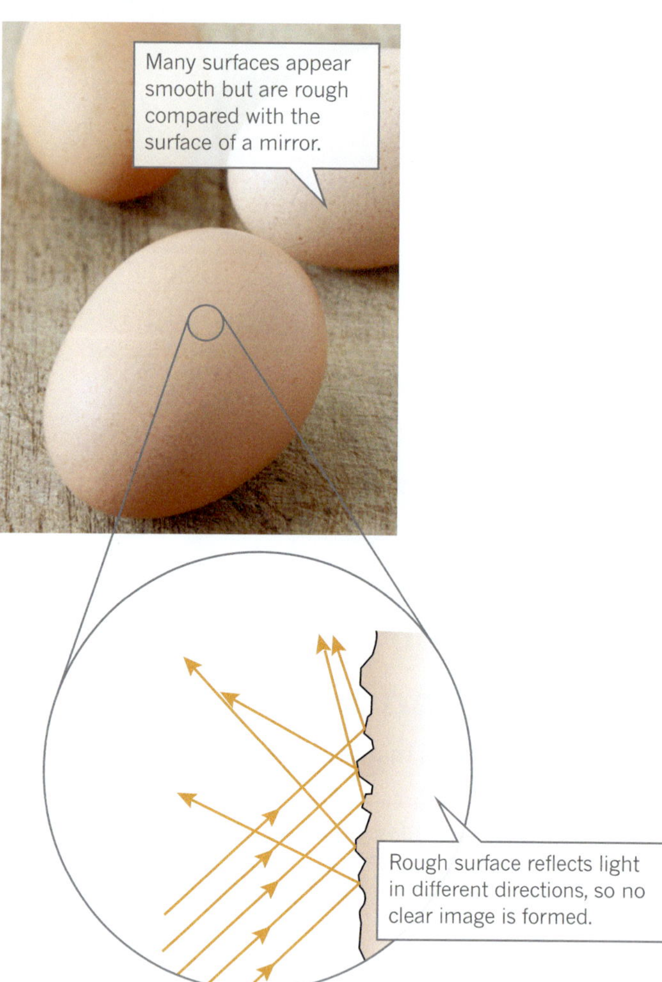

FIGURE 4.3.4 Diffuse reflection occurs from rough surfaces and no clear image is formed.

Diffuse and regular reflection

When light reflects off a very smooth surface such as a mirror or a window, it undergoes regular reflection. This produces a clear image. As Figure 4.3.3 shows, all the light rays are reflected in the same direction.

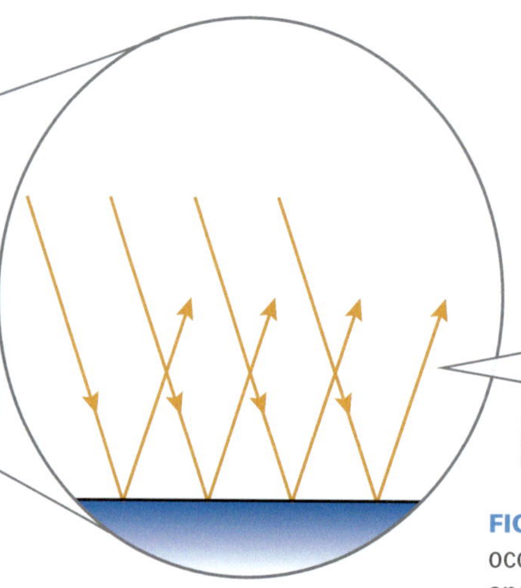

FIGURE 4.3.3 Regular reflection occurs from very smooth surfaces and forms clear, sharp images.

STEM 4 fun

Mirror mirror

YES

PROBLEM

Can you make a mirror out of transparent materials?

SUPPLIES

- clear cellophane or cling film

PLAN AND DESIGN Design the solution, what information do you need to solve the problem? Draw a diagram. Make a list of materials you will need and steps you will take.

CREATE Follow your plan. Create your solution to the problem.

IMPROVE What works? What doesn't? How do you know it solves the problem? What could work better? Modify your design to make it better. Test it out.

REFLECTION

1 What field of science did you work in? Are there other fields where this activity applies?
2 If another student was to do this task, what advice would you give?
3 What did you do today that worked well? What didn't work well?

The law of reflection

When a billiard ball on a pool table is hit so that its path is at a right angle (90°) to the edge of the table, the ball bounces straight back along the same path. When the ball is hit at a different angle to the cushion, it bounces off the cushion at the same angle. In the same way, an incoming ray of light is reflected off a mirror at the same angle. This is called the **law of reflection** and is shown in Figure 4.3.5.

The incoming ray is known as the **incident ray** and the ray that bounces off the mirror is the **reflected ray**. A dotted imaginary line, called the **normal**, is shown at right angles to the surface of the mirror. This is used to measure the **angle of incidence** (shown as i) and the **angle of reflection** (r).

According to the law of reflection:

$$\text{angle of incidence} = \text{angle of reflection}$$
$$i = r$$

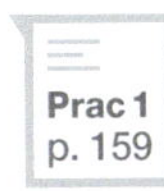
Prac 1
p. 159

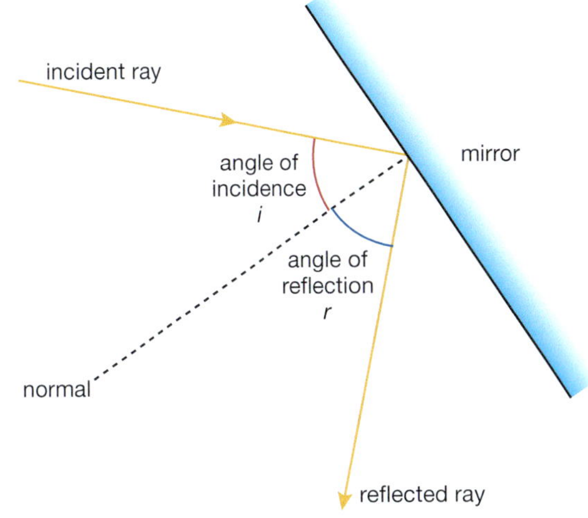

FIGURE 4.3.5 The law of reflection states that the angle of incidence of a light ray is equal to the angle of reflection.

Plane mirrors

A flat mirror is also called a **plane mirror**. When you stand in front of a plane mirror in a fitting room when trying on clothes, your image is the same size as you. Your image appears to be behind the mirror the same distance as you are in front of it. Your image is identical to you in every way except that it is reversed sideways: your right side appears in the mirror as your left, and vice versa, as the example in Figure 4.3.6 shows. This reversal is called **lateral inversion**.

FIGURE 4.3.6 Although this girl is brushing her teeth with her right hand, her image appears to hold the toothbrush in the left hand. The image is laterally inverted (flipped sideways).

Forming an image

When a plane mirror produces an image of an object the image looks as though it is really positioned inside or behind the mirror. Figure 4.3.7 shows why this happens. Some light from the candle flame hits the mirror and is reflected towards your eyes. You see the light from the flame of the candle as though the light travelled in a straight line from a point inside the mirror. The candle shown on the right of the figure does not really exist. It is called a virtual image, a visual illusion created by your brain when you look into a mirror. The image you see in the mirror appears to be positioned as far inside the mirror as the object is positioned outside the mirror. Many light rays are reflected from the actual candle on the left and are used to build up the image of the candle seen in the mirror. Just two rays are shown in Figure 4.3.7, one from the top and one from the bottom of the candle.

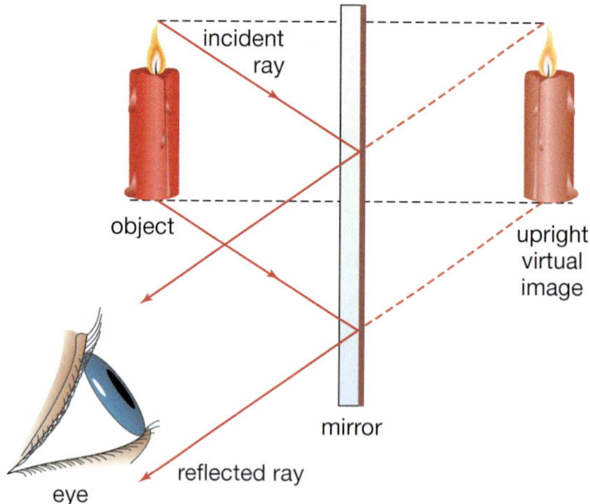

FIGURE 4.3.7 A plane mirror produces a virtual, upright image that is the same size as the object but laterally inverted.

As a result, the candle appears to be inside the mirror. This is called a **virtual image**, because the rays of light do not really meet to produce it.

An image seen in a plane mirror:

- is upright (usual way up)
- is the same size as the object
- is laterally inverted (flipped sideways)
- is virtual
- appears to be located as far inside the mirror as the object is in front.

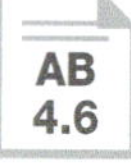

STEM 4 fun

Reflection toy

YES

PROBLEM

Make a toy that uses reflection.

SUPPLIES

paper, cardboard, toilet roll core tube, scissors, tape, glue, BluTack®, straws, cotton, mirror sheets (as mini mirrors), colour marker pens, stickers, anything colourful

PLAN AND DESIGN Design the solution, what information do you need to solve the problem? Draw a diagram. Make a list of materials you will need and steps you will take.

CREATE Follow your plan. Create your solution to the problem.

IMPROVE What works? What doesn't? How do you know it solves the problem? What could work better? Modify your design to make it better. Test it out.

REFLECTION

1. What field of science did you work in? Are there other fields where this activity applies?
2. In what career do these activities connect?
3. If another student was to do this task, what advice would you give?

Refraction

The straw resting in the glass in Figure 4.3.8 appears to be bent. The straw is not bent at all. It appears this way because light bends as it travels out from the water in the glass to the air. This bending of light is called **refraction**.

FIGURE 4.3.8 This straw appears disjointed at the surface of the water, when viewed through the water in the glass.

Light refracts when it travels from one transparent substance into another. Figure 4.3.9 shows light bending towards the normal as it enters a glass block, and bending away from the normal when it exits it.

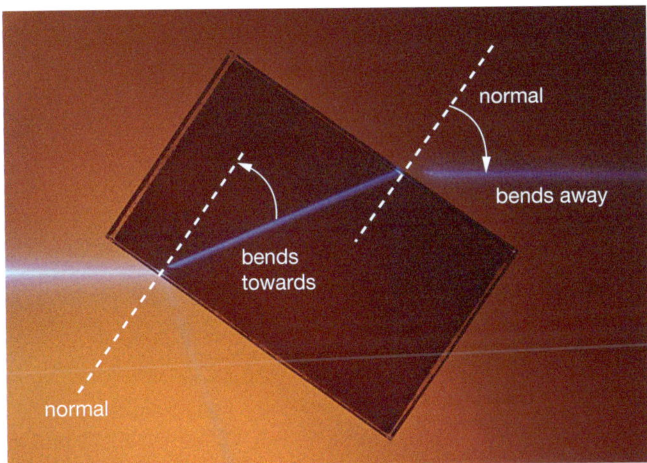

FIGURE 4.3.9 Light bends towards the normal when it enters this glass block and bends away from the normal when leaving.

Why does refraction occur?

Light travels at different speeds through different substances. The differences in its speed result in different amounts of bending, or refraction, as light passes from one substance into another. The **refractive index** is a measure of how easily light travels through a substance. The smaller the refractive index of a material, the faster light will travel through this material.

When light travels through the air and enters a glass block, it slows down and bends. The light bends in the direction of the normal. A car that meets a flooded section of road behaves in a similar way. The first wheel entering the water slows down while the other continues at the same speed, causing the car to slow down on one side and swerve inwards, towards the normal. This is what is happening in Figure 4.3.10. As light leaves a glass block, it speeds up and bends away from the normal.

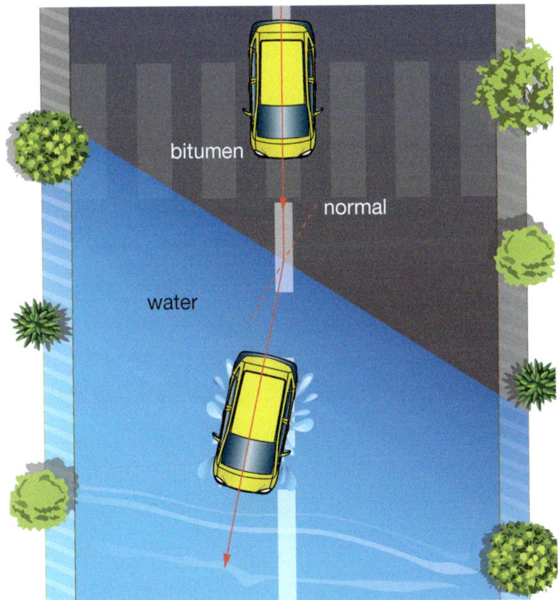

FIGURE 4.3.10 A car that enters a pool of water on a road slows down and its path bends towards the normal.

When a substance causes light to bend a lot, it is said to have a high refractive index. Different substances have different refractive indexes. Air has the lowest refractive index of 1.00 because light travels fastest through air. Light travels more slowly through glass than air, so glass has a higher refractive index than air. Likewise, light travels more slowly in water, diamond and Perspex than through air and so it bends as it enters them. The higher a refractive index, the more light bends when it travels through the substance. Table 4.3.1 shows the speed of light in a few common substances, and the refractive index of each.

TABLE 4.3.1 Speeds of light in different media

Medium	Speed of light (km/s)	Refractive index
air	300 000	1.00
ice	231 000	1.31
water	226 000	1.33
Perspex	200 000	1.49
glass	197 000	1.52
diamond	124 000	2.42

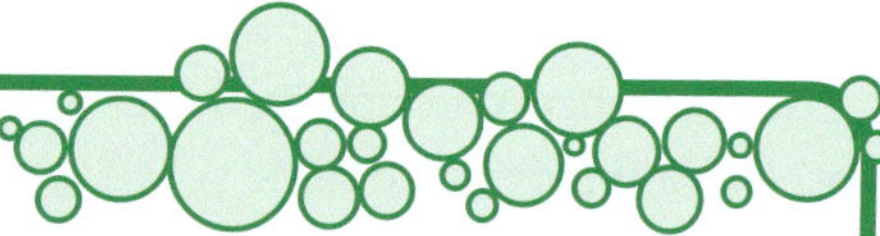

SciFile

Forensic refraction

Investigators can link pieces of broken glass to a window pane smashed at a crime scene if the refractive indices of both samples of glass match. Fragments of glass from car headlights left at a hit-and-run accident can be used to identify the model of the car they came from, and eventually its driver.

The angle an incoming ray of light makes with the normal is called the angle of incidence, *i*. The angle that the refracted ray makes with the normal is called the **angle of refraction**, *r*. Figure 4.3.11 shows that light bends towards the normal when entering a substance of higher refractive index, and bends away from the normal when entering a substance of lower refractive index. Light entering another substance head on is not bent, but continues straight through.

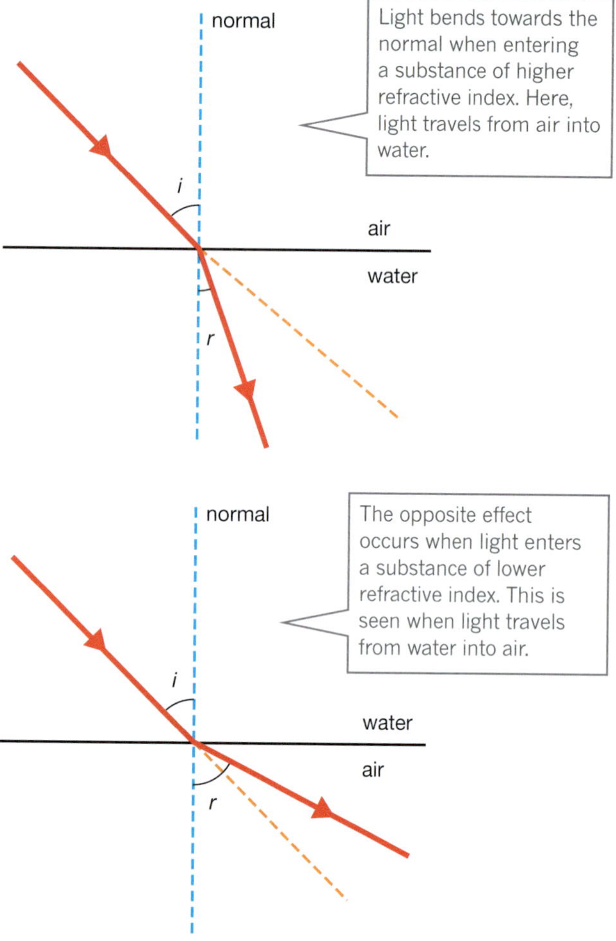

FIGURE 4.3.11 The way light refracts depends on the refractive indices of the substance it is coming from and the substance it is entering.

Depth illusions

When someone is standing in a swimming pool, their legs look shorter than normal. Similarly, rocks on the bottom of a stream always look as though they are in shallower water than they actually are. These depth illusions, like that seen in Figure 4.3.12, occur because light from an object under water is bent away from the normal when it leaves the water surface into air. When you look at this refracted light, your brain traces the light reaching your eyes back in a straight-line path. This makes the object appear to be positioned closer to the surface.

Prac 2 p. 160

AB 4.7

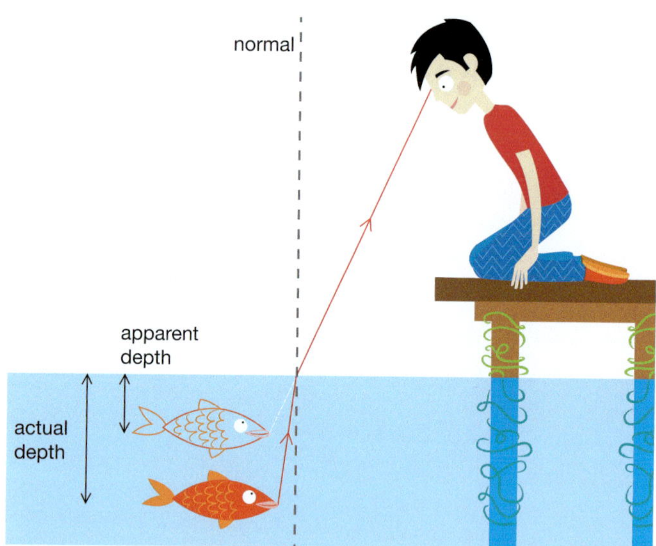

FIGURE 4.3.12 The apparent depth of a fish is less than its true depth in the water. For this reason, hunters using spears to catch fish aim the spear slightly below where they see the fish in the water.

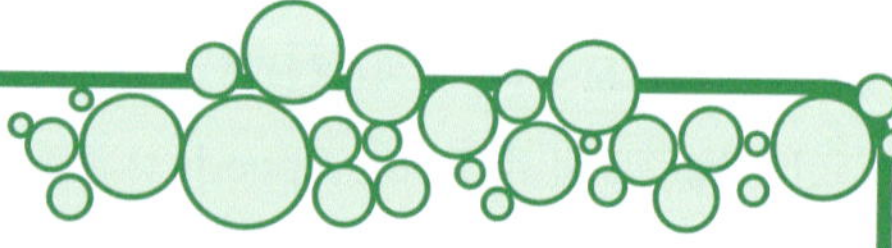

SciFile

Underwater vision

When you are under water looking at objects without goggles or a mask, they appear blurry. Light reaching the corneas of your eyes from water doesn't bend as much as it would if it had come from air. As a result, you can't focus properly. If you wear goggles or a mask, light enters your eyes from air rather than from water. This makes everything clearer!

Total internal reflection

When light enters a substance of lower refractive index, such as from glass into air, it is refracted away from the normal. Figure 4.3.13 shows what happens to this ray as the angle of incidence increases. At an angle of incidence called the **critical angle**, light is refracted so far from the normal that it runs along the boundary of the two substances. For any angles of incidence greater than this, there is no refracted ray.

science 4 fun

The reappearing coin

Can you make a coin appear before your eyes?

Collect this ...

- coin
- opaque cup (one that is not see-through) or evaporating basin
- pencil or pen
- glass of water
- rectangular glass block

Do this ...

1. Put the coin in the cup.
2. Move back until the coin is just out of sight.
3. Have your partner pour water into the cup from the beaker.
4. Can you see the coin again?
5. Now place the rectangular glass block on top of this page of your textbook.
6. Study the print through the block and compare this to when the block is removed.

Record this ...

1. Describe what happened when you viewed the coin and when you looked through the glass block.
2. Explain why you think these things happened, using diagrams to assist your response.

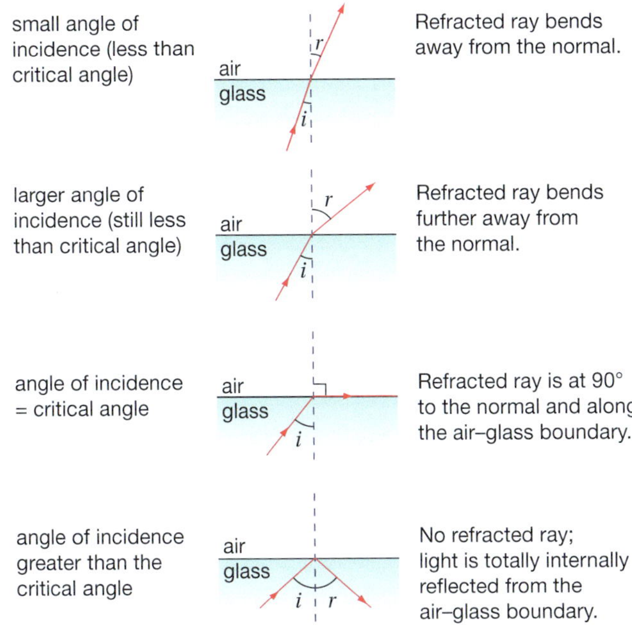

FIGURE 4.3.13 What light does at the boundary between the air and glass depends on its angle of incidence.

Light is reflected from the boundary as though it was a mirror. This is called **total internal reflection**.

Total internal reflection of light explains why diamonds sparkle (Figure 4.3.14). It also explains the upside-down shark seen in Figure 4.3.15 on page 156.

Prac 3 p. 161
AB 4.8

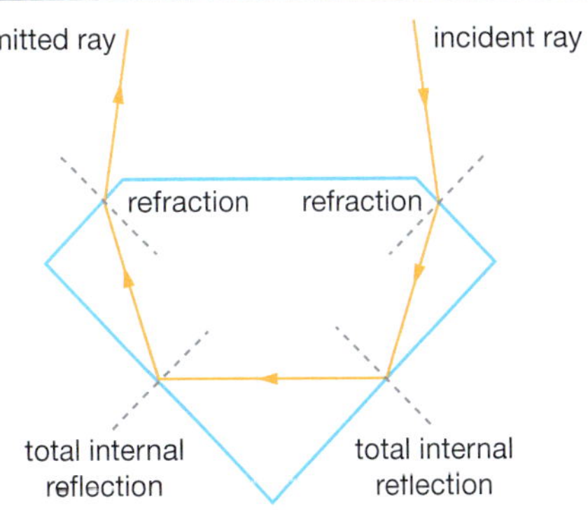

FIGURE 4.3.14 Diamond has a high refractive index and a low critical angle of about 23°. The back of a diamond is cut so that light strikes at an angle larger than the critical angle. This causes light to reflect twice before it emerges, making the diamond look more brilliant.

FIGURE 4.3.15 Light from this shark has been totally internally reflected at the surface of the water. This produces the illusion of a shark swimming upside down above the real one.

Lenses

A lens is a transparent piece of plastic, glass or even jelly that is shaped to curve outwards or inwards. A lens refracts light and focuses it to form images in the eye and in devices such as cameras, telescopes, binoculars, microscopes, spectacles and contact lenses.

Convex lenses

A lens that bulges outwards is called a **convex lens**. These lenses, such as the one shown in Figure 4.3.16, cause light rays to come together, or converge. If a convex lens is held close to an object, it can be used as a magnifying glass. In this case, it produces an upright and enlarged (magnified) virtual image.

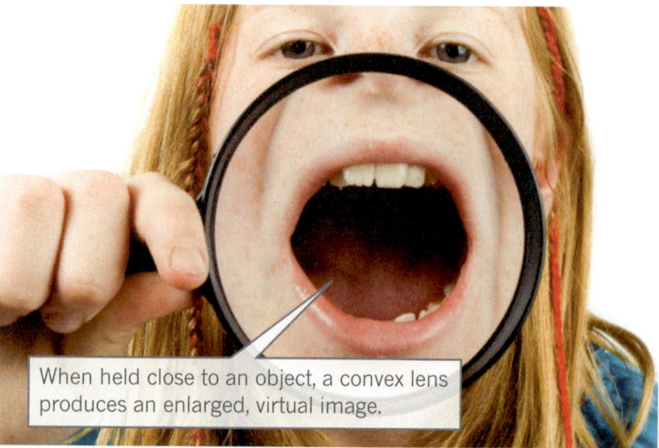

FIGURE 4.3.16 Lenses refract light to produce different types of images. A magnifying glass is a convex lens.

When light reaches a convex lens from a distance, the convex lens can be used to focus the light on a screen to form an image. Figure 4.3.17 shows two convex lenses of different strengths focusing light to a point called the focus of the lens (F). Light is focused to a point on a screen when you see a film at the movies. This is the type of image that is formed by the convex lenses in your eyes. This type of image is called a **real image**.

Concave lenses

A lens that curves inwards is called a **concave lens**. These lenses cause light to diverge, or spread out. They spread parallel light rays as though the rays have come from a point behind the lens, as shown in Figure 4.3.18. A concave lens only produces images that are smaller, upright and virtual. Because this image is virtual, we could not produce the image on a screen because the light does not really cross at this point.

Prac 4 p. 162

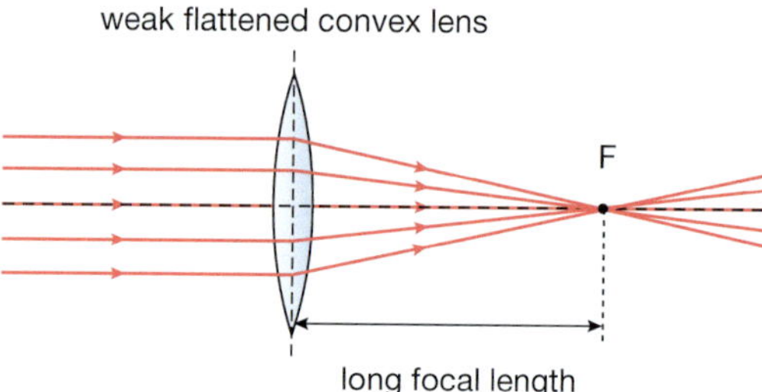

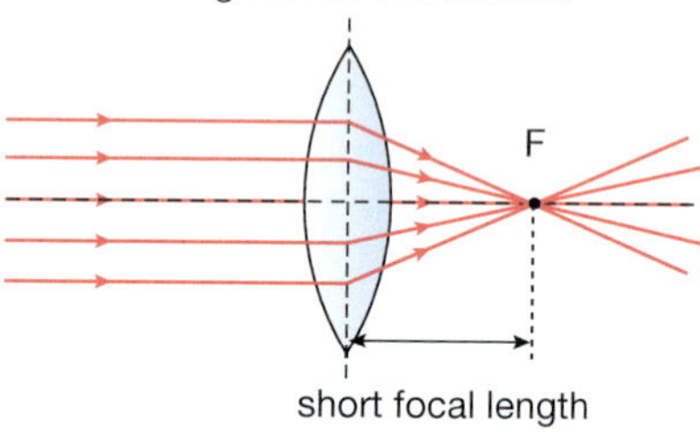

FIGURE 4.3.17 A convex lens brings light rays together at the focus (*F*). Focal length is the distance between the lens and this focus. If a lens has a greater curve, then the lens is stronger and the focal length is shorter.

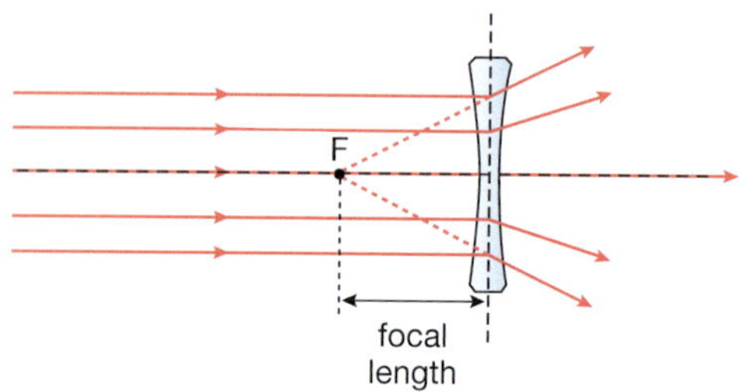

FIGURE 4.3.18 A concave lens spreads light. The focal length is measured by tracing the path of the rays leaving the lens back to the single point that they appear to have come from.

MODULE 4.3 Review questions

Remembering

1 Define the terms:
  a luminous
  b refractive index
  c total internal reflection
  d concave lens.

2 What term best describes each of the following?
  a the angle formed between the incident ray and the normal
  b the type of image formed when light rays do not actually meet
  c the angle of incidence that produces light to be refracted at 90° to the normal
  d a lens that bulges outwards at its centre.

3 a Which letter in Figure 4.3.19 represents each of the following?
    i incident ray
    ii reflected ray
    iii normal
    iv angle of incidence
    v angle of reflection.
  b Name the law that is demonstrated in Figure 4.3.19.

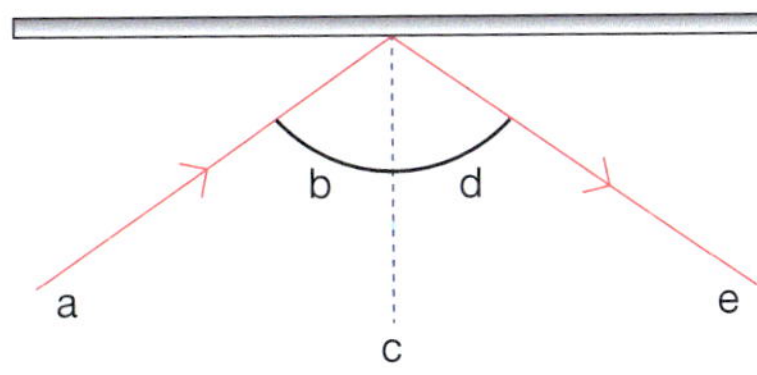

FIGURE 4.3.19

4 List five properties of an image formed by a plane mirror.

5 Match the correct word from the list below to each of the statements that follow. (Note: Some words can be used more than once.)

focus   real   virtual   convex   concave

  a In using a lens, this is the point at which distant light rays meet, or appear to meet.
  b This type of image cannot be produced on a screen or a sheet of paper.
  c This type of lens always produces upright, diminished, virtual images.
  d A convex lens produces this type of image from a distant object.

6 Which of the two alternatives makes each sentence true?
  a Light is refracted *away from/towards* the normal when it passes from glass into air.
  b Light travels at a higher speed through *glass/air*.
  c Total internal reflection can only occur when light attempts to pass from a substance of *lower/higher* refractive index.

Understanding

7 If you were to double the distance of the candle from the glass in the science4fun on page 149, predict where the image would then be located.

8 Why does the boy shown in Figure 4.3.12 on page 154 think the fish is closer to the surface than it really is?

9 What is meant by the term lateral inversion when talking about plane mirrors?

10 Explain why a smooth surface produces a sharp image of an object.

11 a Why does everything look blurry if you open your eyes under water?
  b Describe what you can do to make your underwater vision clear.

12 Explain why the pencils in Figure 4.3.20 seem to be broken in the middle.

FIGURE 4.3.20

MODULE 4.3

Review questions

Applying

13 Identify the property of light that makes the coin reappear in the science4fun on page 155.

14 Use Figure 4.3.11 on page 154 to help you to identify whether light will bend towards or away from the normal in each of the following situations.

Light travels from:

a water to glass **b** diamond to air
c water to Perspex **d** water to ice
e glass to diamond **f** air to Perspex.

15 Figure 4.3.21 shows three rays of light hitting a mirror. Use this diagram to carry out the following tasks.

a Copy the diagram and use a ruler to draw a normal at 90° to the mirror for each of rays A, B and C.

b Extend where each ray will be reflected in front of the mirror.

c Use a protractor to determine which ray has the largest angle of reflection.

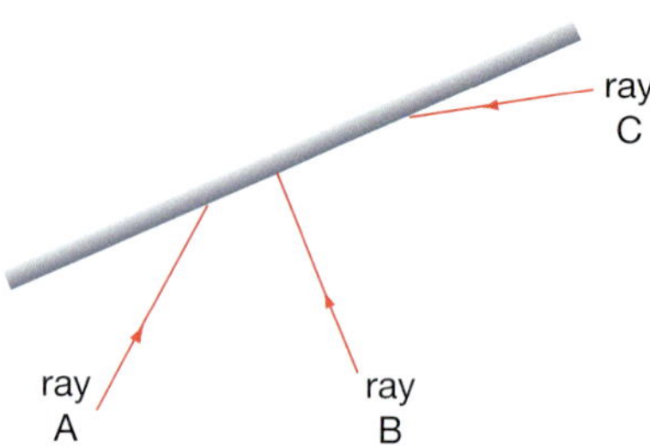

FIGURE 4.3.21

16 The critical angle of light passing from water into air is 48.6°. Construct a diagram to demonstrate how light bends when travelling from water into air at an angle of:

a 60° **b** 48.6° **c** 25°.

17 Copy each diagram in Figure 4.3.22 and draw the normal to the point where light meets each boundary. Construct a ray on each diagram to show the likely path of each ray through each material.

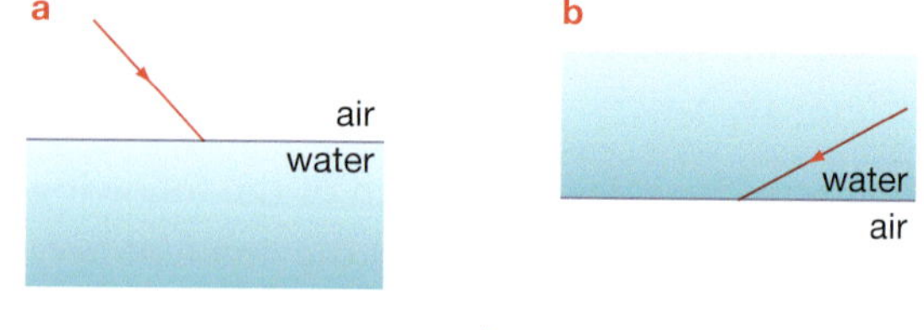

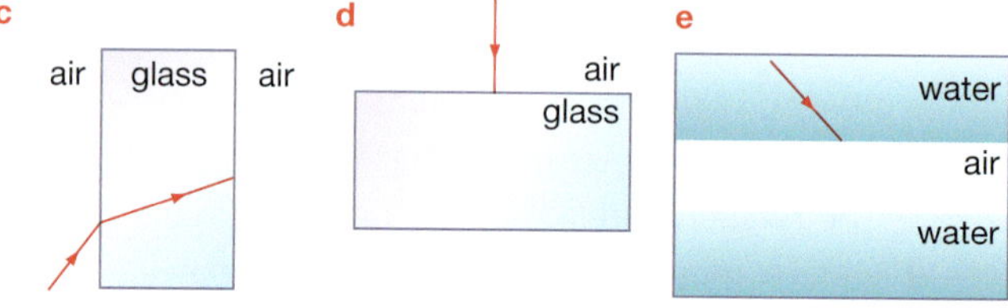

FIGURE 4.3.22

Evaluating

18 **a** Will light entering glass from air speed up or slow down?

b Will light travel faster through diamond or through water?

c Use Table 4.3.1 on page 153 to rank the amount light bends from most to least as it enters the following materials from air: glass, diamond, Perspex, ice, water.

19 Propose a reason why total internal reflection never occurs when light travels into a substance of greater refractive index.

20 A sample of glass is found at the scene of a burglary. How do you think this glass could be used as evidence in a criminal investigation?

Creating

21 Copy the lenses shown in Figure 4.3.23 and construct the path the light rays will take as they pass through each lens.

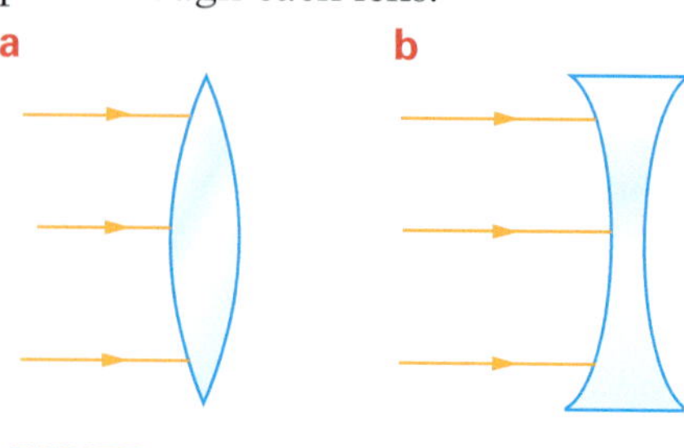

FIGURE 4.3.23

MODULE

4.3 Practical investigations

1 • Law of reflection

Purpose

To verify the law of reflection.

Timing

45 minutes

Materials

- light box and power supply
- sheet of white paper
- plane mirror
- ruler
- pencil
- protractor

Procedure

1 In your workbook, construct a table like the one shown in your Results section.

2 Set up the equipment as shown in Figure 4.3.24, marking the positions of the back of the mirror and the normal line.

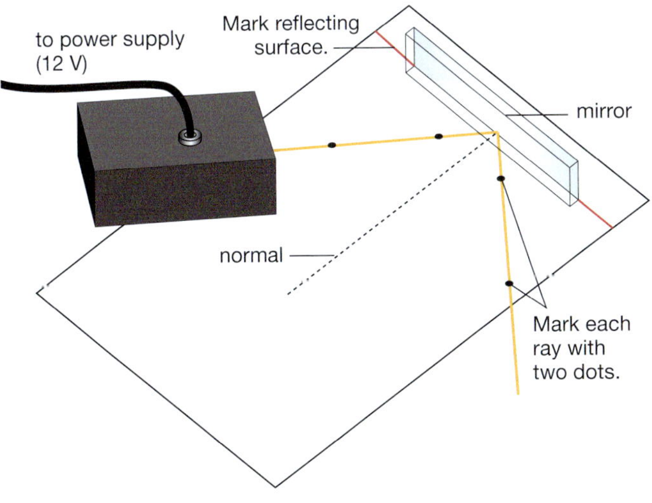

FIGURE 4.3.24

3 Direct a ray of light on an angle towards the centre of the mirror.

4 Mark the position of the incident and reflected rays using two dots, and then use your ruler to draw these rays on the paper. Label these 'ray 1'.

5 Repeat this process for three more rays of different angles.

6 Direct a ray at right angles into the mirror and observe its reflection. Record your observations in a table like the one shown in the Results section.

Results

1 Record all results in your table.

Angles of incidence and reflection

Ray	Angle of incident ray	Angle of reflected ray
1		
2		
3		
4		

2 Select a ray from your diagram and measure the angle that its incident ray (incoming ray) makes to the normal. Enter the angle in the table.

3 For the same ray, measure the angle that the reflected ray makes to the normal. Enter this angle in the table.

4 Complete the table for each ray tested.

Review

1 Assess whether your results support the law of reflection.

2 Describe what happened when you directed light at right angles to the mirror.

3 List at least three examples where you have observed the law of reflection in action (for example, at a cricket match when the ball bounces off the pitch).

MODULE 4.3

Practical investigations

2 • Bending light

Purpose

To observe and measure the refraction of light as it passes through a transparent block.

Timing 45 minutes

Materials

- plastic or glass rectangular block and semicircular block
- light box and single-slit slide
- 12 V power supply
- sheet of white paper
- protractor
- ruler
- pencil

Procedure

1 In your workbook, construct a table like the one shown in your Results section.

2 You will direct a beam of light from air into the glass block and observe its path through the glass. Make two predictions in your workbook by selecting one alternative in each case:

i The angle of refraction will be *greater than/ less than* the angle of incidence for light entering the glass block.

ii The angle of refraction will be *greater than/ less than* the angle of incidence for light leaving the glass block.

3 Trace the outline of the rectangular glass block onto a white piece of paper. Using a ruler, rule a dotted line at right angles from the centre of the top of the block to show the normal (Figure 4.3.25).

FIGURE 4.3.25

4 Place the glass block onto the white paper and direct a single beam of light towards the centre of the block as shown in Figure 4.3.26.

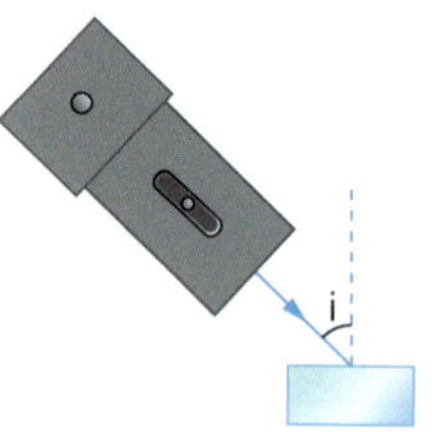

FIGURE 4.3.26

5 Mark a series of dots on the paper to trace the path of light incident on, refracted through and refracted out the other side of the glass block.

6 Remove the glass block and use a ruler to draw the path the light followed on the paper. Add a normal at the point where light left the glass block.

7 Make angle measurements with a protractor to complete a copy of the table shown in the Results section.

Results

Record all measurements in your table.

Angles of incidence and refraction

	Angle of incidence	Angle of refraction
light entering glass block from air		
light leaving glass block into air		

Review

1 State whether light bends towards or away from the normal when it:

a enters the glass block

b leaves the glass block.

2 Compare the size of the angle at which light enters the glass block with the angle at which it leaves.

3 Were your initial predictions correct?

4 Describe whether all the light hitting the glass block was refracted through it, or whether some light followed a different path.

MODULE 4.3

Practical investigations

3 • Total internal reflection

Purpose

To direct a beam of light into a semicircular Perspex block so that it reflects back out again.

Timing 30 minutes

Materials

- a light box and power supply
- protractor
- sheet of white paper
- ruler
- semicircular glass or Perspex block

Procedure

1 Put the semicircular glass block on a piece of white paper and trace around it using a pencil. Rule a dotted line at right angles from the centre of the base of the block.

2 Direct a single beam of light at an angle of about 45° through the curved side of the semicircular block to hit the centre as shown in Figure 4.3.27.

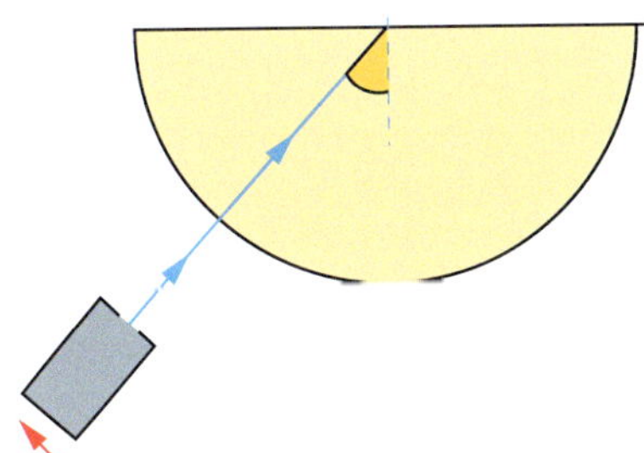

FIGURE 4.3.27

3 You should be able to see a ray of light incident at the centre of the block that is refracted out of the block. Slowly move the light box in an arc away from the centre to increase the angle between the incident ray and the normal. As you do this the refracted ray should get larger and larger, until it exits the block along the straight edge. When this happens, the angle of incidence is the critical angle for the Perspex or glass block. Mark with pencil this angle of incidence.

4 Now increase the angle of incidence and observe what happens to the refracted ray.

5 Remove the block and use a protractor to measure size of the critical angle.

Review

1 Why do you think the ray of light did not bend when it entered the curved surface of the block?

2 Describe what you observed when the angle of incidence was increased by rotating the light box.

3 State the critical angle that you measured for the glass or Perspex block.

4 Describe what you observed happened to the refracted ray for angles of incidence larger than the critical angle.

5 Compare the critical angle you measured with those of your classmates. List possible sources of error in your experiment.

MODULE 4.3

Practical investigations

4 • Comparing curved mirrors and lenses

Planning & Conducting | Evaluating

Mirrors can be curved and can produce real and virtual images in a similar way to lenses. A concave mirror curves inwards like a cave, whereas a convex mirror bulges outwards. Concave mirrors are used in astronomical telescopes.

Purpose

To investigate how light is reflected and refracted from curved mirrors and lenses.

Timing 45 minutes

Materials

- light box with 12 V power supply
- multiple-slit slide
- set of curved mirrors and lenses of different strengths
- pencil
- several sheets of white paper

Procedure

1. Place a multiple-slit slide in the light box and position it on a sheet of white paper.
2. Place a concave mirror in the path of the light rays as shown in Figure 4.3.28.
3. Trace the curved edge of the mirror onto the paper.
4. Use a pencil to mark two dots for each incident and reflected ray on the sheet of paper.
5. Trace the mirror and trace these rays onto the page. Also show the position of the mirror.
6. Repeat the procedure for a concave mirror of different curvature, two convex mirrors, two concave lenses and two convex lenses, all of different thickness. Use a new sheet of paper for each sketch.

Review

1. Describe differences between how the concave and convex mirrors reflected light.
2. How does the reflection of the light beams change when the mirror is more curved?
3. Compare the way light passed through the convex and concave lenses.
4. Describe the effect of using a thicker convex lens compared to a thinner lens.

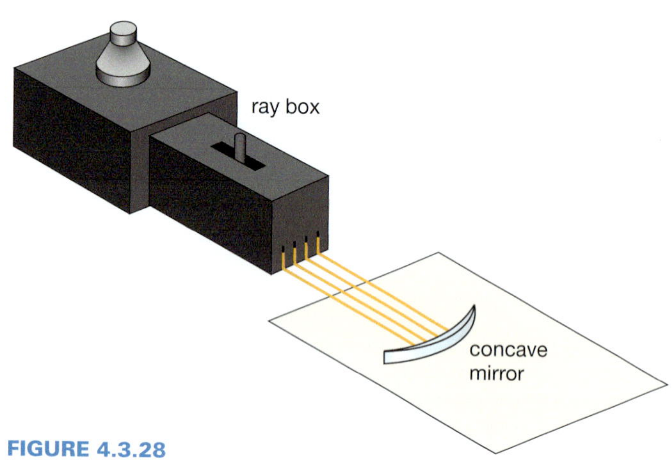

FIGURE 4.3.28

MODULE 4.4

Hearing and seeing

The brain plays an important role in helping us to see and hear. Light reflects from surfaces or travels directly to our eyes. Our eyes process this light to form images which are sent along the optic nerve to our brain. The brain makes sense of this information and provides us with sight. Sound waves travel to our ears making our eardrums vibrate. The amplified vibrations are sent along the hearing nerve to the brain. The brain turns the vibrations into what we hear.

science 4 fun

Fooling your eyes

YES

Light sensitive cells at the back of your eye are able to adjust to how bright or dark your surroundings are. Can you fool these cells?

Collect this ...

- white wall or sheet of paper

Do this ...

1 Position yourself about 30 cm away from the image seen here.
2 Stare at the dot in the centre for about 30 seconds.
3 Look away on to a white wall or piece of paper.
4 What do you see?

Record this ...

1 Describe what happened when you looked at the wall or sheet of paper.
2 Explain why you think this happened.

Hearing sound

To hear a sound, its energy needs to be transmitted through your eardrum and delicate structures of the ears. Your ears collect sound waves as they enter the outer ear and travel through the ear canal. Vibrations of the sound wave strike the eardrum, causing the eardrum to vibrate. This vibration is then passed on to the three tiny bones of the middle ear. The movement of the bones of the middle ear set up vibrations in the fluid of the cochlea which is located in the inner ear. It is in the cochlea that these vibrations are converted into the electrical impulses that your brain interprets as sound. Figure 4.4.1 shows the parts of the ear and how they function.

Over one million Australians have a hearing disability, ranging from mild hearing loss to complete deafness. These problems can be because:

- the ear canal is blocked with wax, preventing the passage of sound waves
- the middle ear is filled with fluid
- the eardrum has been ruptured by an extremely loud noise or as the result of infection
- sensory cells of the ear have been damaged by loud noise

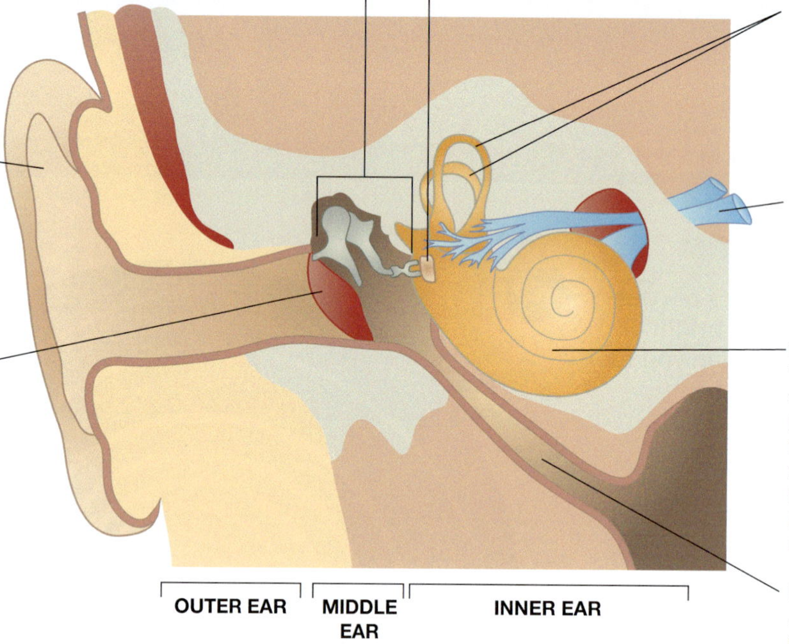

Ossicles
Vibrations of the eardrum pass onto the ossicles, three tiny bones, called the hammer, anvil and stirrup. These bones magnify the vibrations.

Oval window
Separates the middle and inner ear. Vibrations from the stirrup are transmitted to this thin layer of tissue, and continue through to the cochlea.

Semicircular canals
These are filled with fluid and give us our sense of balance. They don't play a role in our hearing but may be affected if we have an infection or ear problem.

Pinna
This is the outer ear. It funnels sound into the ear canal.

Auditory nerve
Electrical impulses travel along this nerve to the brain, which interprets them as sound.

Eardrum
Separates the outer and middle ear. This thin flap of skin stretches tight across the inside of the ear like the skin on a drum. Sound reaching the eardrum makes it vibrate.

Cochlea
This spiral-shaped tube is filled with fluid. Vibrations cause this fluid to move, and are detected by millions of tiny hairs lining the surface of the cochlea. Receptors attached to these hairs convert their movement into electrical impulses.

Eustachian tube
Joins the middle ear to the nose and throat. Air moves into, or out of, the middle ear through this tube to balance the air pressure on the other side of the eardrum.

OUTER EAR | MIDDLE EAR | INNER EAR

AB 4.4

FIGURE 4.4.1 The structures of the ear work together to allow vibrating air to be interpreted by our brain as a sound.

SciFile

Popping ears

As an aircraft takes off, or if you go up a mountain, the air pressure outside your eardrum falls, but the air pressure inside your ear remains the same. This higher pressure air inside your head pushes on your eardrums. When your ears 'pop', the Eustachian tube opens and releases air through your nose and throat, which balances the air pressure between the environment and the air inside your eardrum.

- a defect in the auditory nerve or the tiny hairs of the cochlea prevents sound impulses being transmitted correctly to the brain.

The cochlear implant, or bionic ear, has helped many people with serious inner-ear damage to hear sound for the first time.

The eye

The eye is an incredible organ. Figure 4.4.2 shows the structure of the human eye. Light entering the eye is refracted by the cornea and focused by the convex lens.

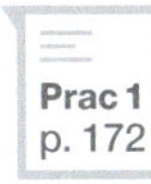
Prac 1
p. 172

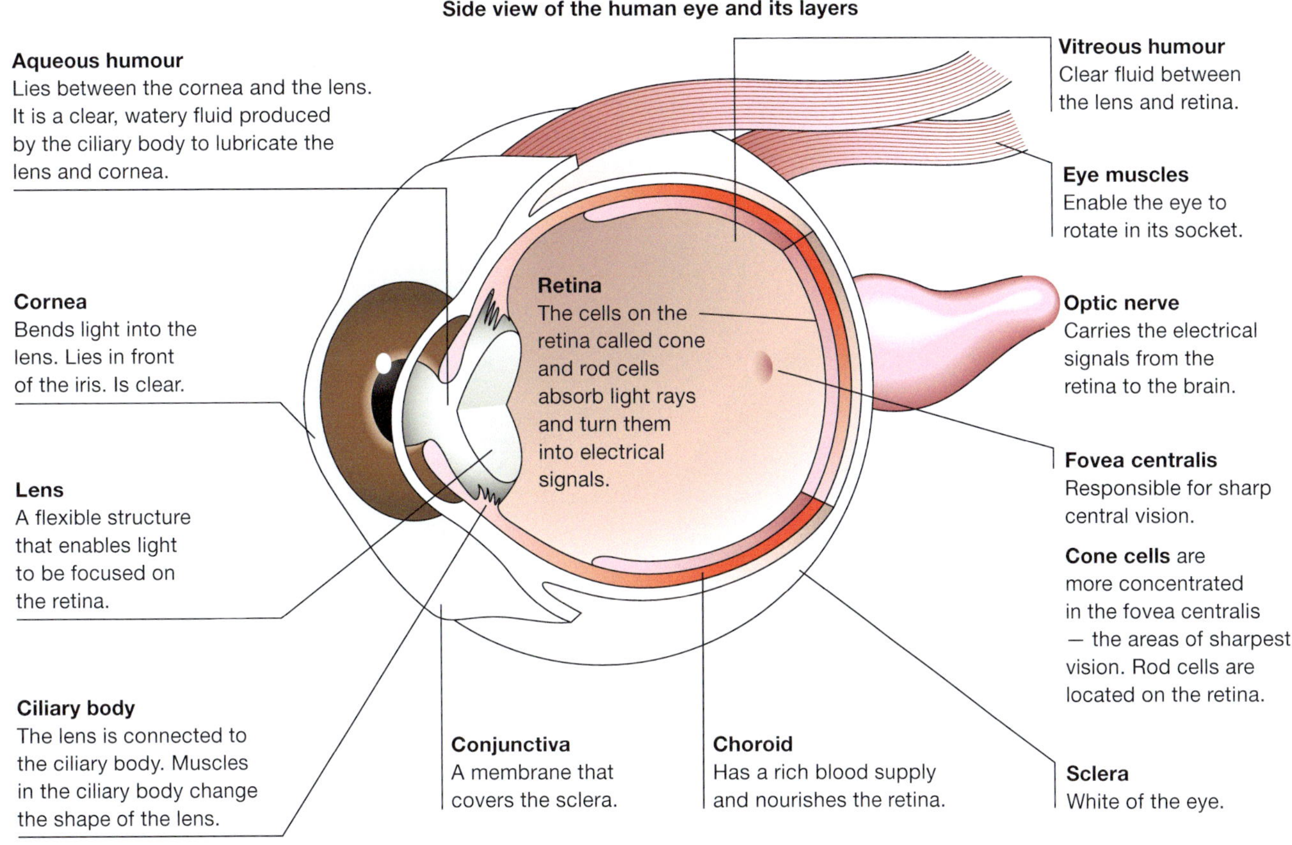

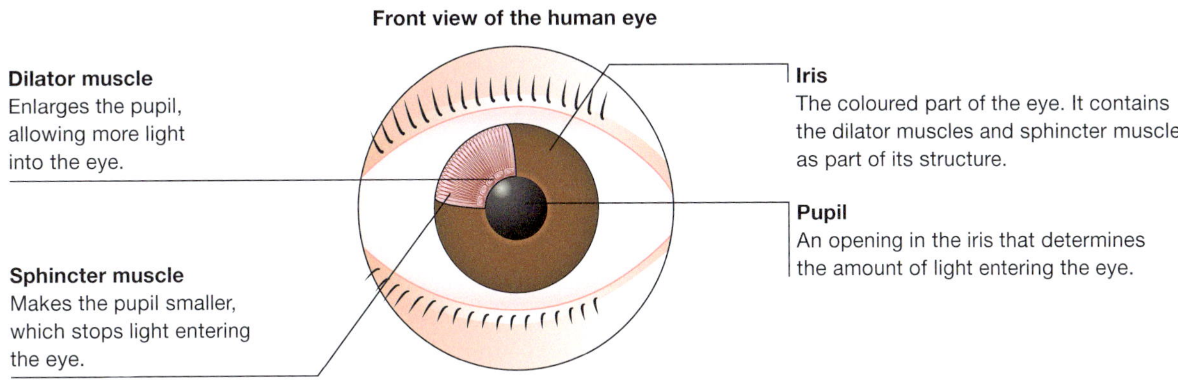

FIGURE 4.4.2 The structures of the eye work together to provide vision.

Figure 4.4.3 shows how incoming light is focused to form a clear, upside-down image on the retina, at the back of the eye. This image is converted into a series of electrical signals, which then travel along the optic nerve to the brain. The brain interprets this information as an image.

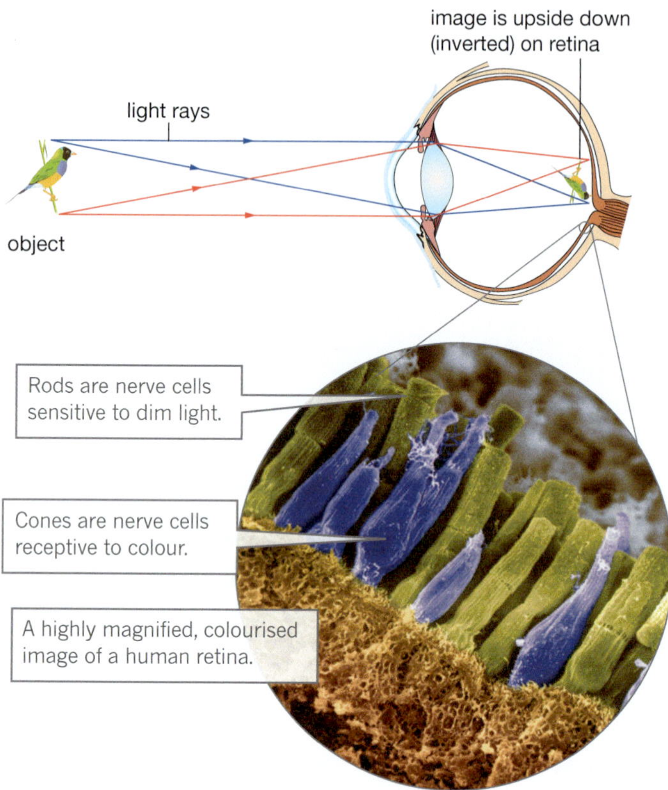

FIGURE 4.4.3 The brain interprets the upside-down image formed on the retina as an upright image. Specialised cells in the retina give information about the colour and shading of what we see.

SciFile

Turning the world upside down

Experiments have been conducted in which people have worn lenses that flip the world upside down. After about a week of bumping into the furniture, they reported that their brain adjusted its view and perceived this view as the right way up. When the glasses were then taken off, everything was upside down again for a while!

Vision problems

The lenses in your eyes focus on objects at different distances by changing focal length. When the muscles attached to the lens contract, the lens stretches, becoming quite flat, and able to focus on distant objects. When these muscles relax, the lens gets much fatter and bends light more, allowing close objects to become focused. This is shown in Figure 4.4.4. The ability of the lens to change shape is called **accommodation**. Unfortunately, as we age, the lenses harden, making accommodation more difficult.

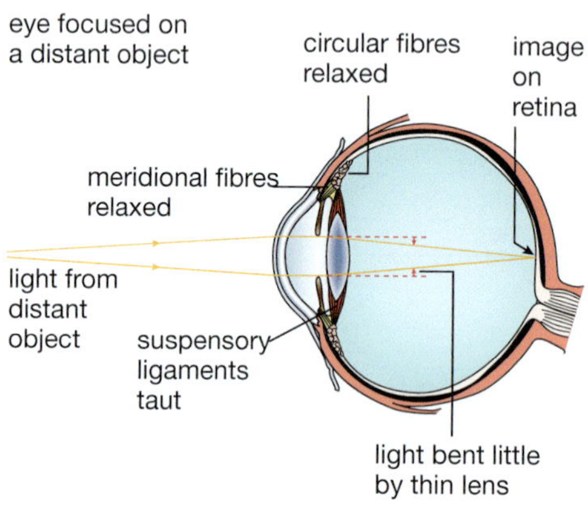

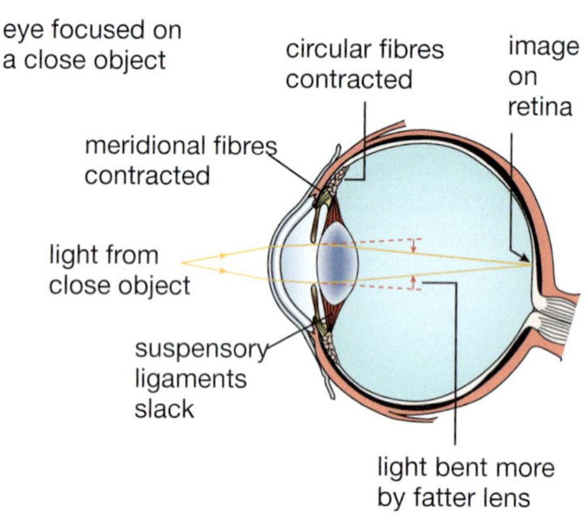

FIGURE 4.4.4 A 40-year-old person's eyesight has just one-quarter of the accommodating ability they had when younger. By age 45 almost everyone needs some form of glasses.

If light is not focused to a point at the retina, then the person will not see a clear image. This commonly leads to **short-sightedness (myopia)** or **long-sightedness (hyperopia)**.

People who are short-sighted can focus on close objects, such as a book, but distant objects, such as children in a playground, are not clear. Figure 4.4.5 shows how a concave lens works to correct short-sightedness.

FIGURE 4.4.5 The eyeball is too long in a person who is short-sighted. A concave lens of appropriate strength can correct this.

People who are long-sighted can see distant objects clearly, but have trouble focusing on close objects. They need to use glasses when reading or doing close work. Figure 4.4.6 shows how a convex lens is used to correct long-sightedness.

Hyperopia (long sight)

close object

convex lens

close object

A long-sighted person's view of the world

FIGURE 4.4.6 This person is long-sighted because their eyeball is slightly too short. A convex lens can correct this.

Bifocals or graded lenses may be used if a person has more than one type of vision problem. These lenses are strongest at the bottom, so a person looks down through this region to read, and looks straight through the lenses to focus on objects further away.

Some people wear contact lenses rather than glasses. These small lenses are worn directly on the cornea of the eye and are made of hard plastic, or from water-absorbing materials. Because contact lenses are in continual contact with the surface of the eye, they must be kept very clean and sterilised regularly. Another treatment for vision problems, rather than wearing glasses, is to undergo laser surgery. Such treatment reshapes the surface of the patient's cornea to alter how it focuses light.

AB 4.9

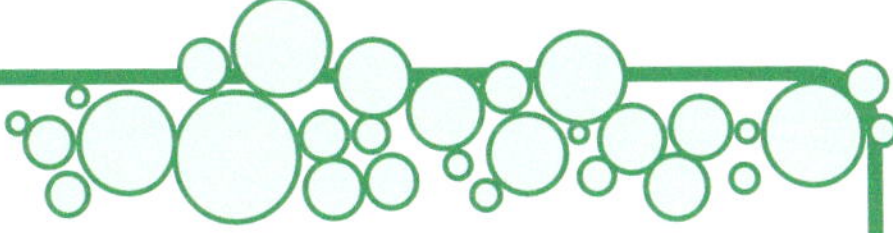

Scifile

Seeing is believing

Surgeons in Britain have restored sight to 42-year-old Martin Jones who lost one eye completely and lost vision in the other in a workplace accident. First, cells were taken from his cheek and were grown across his eye to create a new cornea. Next, surgeons extracted a canine tooth, which was chiselled and hollowed out to hold a man made lens. Then the piece of tooth was implanted into the eye socket.

A hole was cut into this new cornea to allow light to enter the eye and Martin was able to see his wife (whom he met after his accident) for the first time.

SCIENCE AS A HUMAN ENDEAVOUR

Use and influence of science

Workplace hearing protection

Sound waves transfer energy. The greater the amplitude of a sound wave, the louder the sound that is transmitted. The delicate cells in your ears that detect sound need to be protected from excessive noise in the workplace.

FIGURE 4.4.7 Some workplaces, such as the airport where this man works, are noisier than others.

Like the engineer in Figure 4.4.7, many people use noisy machinery at work. Loudness can be measured by a device called a sound level meter. Loudness is measured in **decibels** (unit symbol dB). The decibel scale is shown in Figure 4.4.8.

Exposure to noise levels above 85 dB for long periods can permanently damage your hearing. The degree of damage depends on how loud the noise is, and how long you are exposed to it. People spend much of their day in their workplace, and so exposure to constant loud noise can have a significant effect on their hearing.

Noise destroys delicate sensory cells in the inner ear, called hair cells. These cells detect vibrations and send electrical signals to the brain. Such damage can be seen in Figure 4.4.9. If affected, these cells cannot be replaced.

Repeated exposure to loud noise can also lead to tinnitus, a condition in which a person hears a permanent ringing in their ears. Noise-induced hearing loss is recognised as a major industrial disease in Australia and around the world. It makes hearing higher-frequency sounds more difficult. In turn, this can lead to problems communicating, increased fatigue, stress and anxiety.

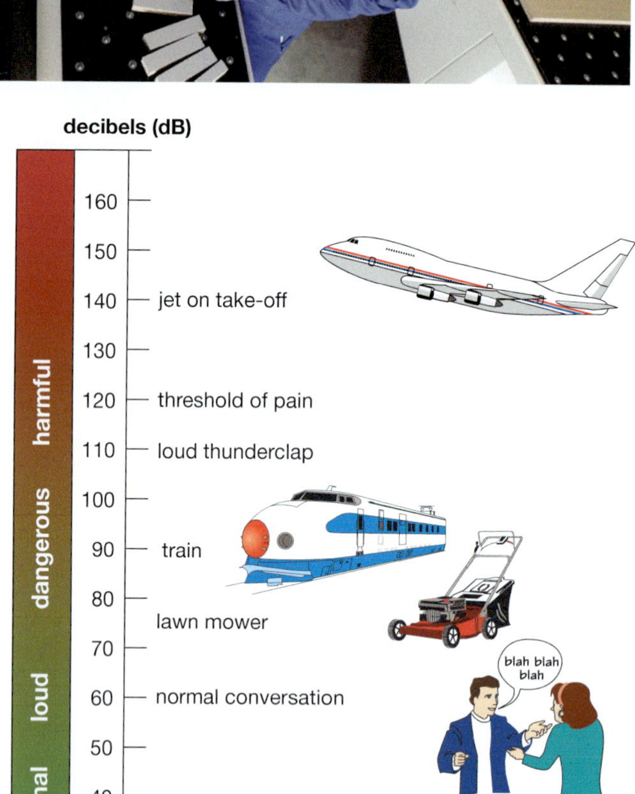

FIGURE 4.4.8 Sound intensity level, or loudness, is measured on the decibel scale.

SCIENCE AS A HUMAN ENDEAVOUR

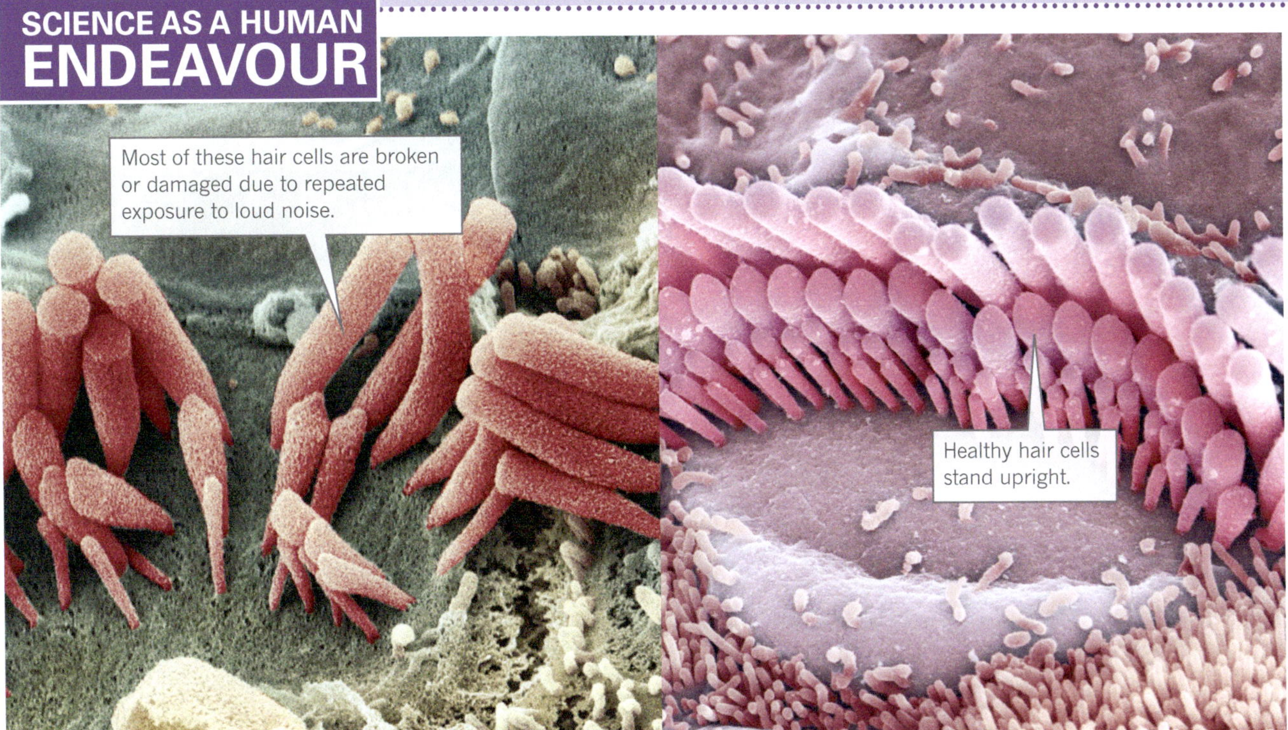

FIGURE 4.4.9 These images show the effect that loud sounds can have upon tiny hair cells in the inner ear.

A code of practice exists for maintaining noise within acceptable levels in workplaces. Workplaces are assessed for their noise levels. Employers operating worksites with high levels of noise are directed to:

- try to reduce noise levels by:
  - replacing outdated, noisy machinery with quieter alternatives where possible
  - ensuring that machinery is regularly maintained
  - reducing metal-to metal contact in machinery by inserting materials to dampen sound
- block noise transmission by:
  - shifting noisy machinery to more remote areas of the workplace
  - fitting sound-absorbing materials to the ceiling or walls
  - using sound-absorbing curtains to screen off an area or machine
- ensure that all areas of loud noise are signposted as hearing protector areas
- ensure that workers are not exposed to sound intensity levels greater than 85 dB averaged over an 8-hour period
- ensure that affected workers wear personal hearing protectors, such as correctly fitted earmuffs or earplugs and that these workers undergo regular hearing tests to monitor their hearing.

REVIEW

1 What happens when the amplitude of a sound wave increases?

2 Use Figure 4.4.9 to estimate the sound level intensity (in dB) of:
  a city traffic
  b two people arguing loudly
  c a chip packet rustling in a cinema
  d a car backfiring.

3 Describe the effect that repeated and prolonged exposure to noise above 85 dB has on a person's ears.

4 List three ways an employer could act to reduce the noise levels around machinery.

5 a Your hearing is usually muffled after a loud concert. Propose a reason why.
  b Members of bands risk hearing damage more than the occasional concert-goer does. Propose a reason why.

6 a What is tinnitus?
  b Explain how it is caused.

7 Portable music devices can produce sounds above 105 dB.
  a Use this information to propose a reason why many doctors believe that tinnitus will become far more common in the future.
  b Assess whether you are at risk of tinnitus in the future due to your current listening habits.

MODULE

4.4 Review questions

Remembering

1 Define the terms:
   a ossicles
   b cochlea
   c pupil
   d accommodation.

2 What term best describes each of the following?
   a the flap of skin separating the outer and middle ear
   b the tube that joins the middle ear to the nose and throat
   c nerve cells that are sensitive to dim light
   d the vision problem of a person with an eyeball that is too long.

3 List the causes of four types of hearing loss.

Understanding

4 a What is the function of the eardrum?
   b Describe the role of the ossicles in the ear.
   c Explain how electrical impulses travel from the ear to the brain.

5 Use one of the two alternatives in each of the following to fill in the sentences to make them true.
   a Light entering the eye is refracted by the lens and the *cornea/retina*.
   b To produce a clear image, light must be focused on the *retina/lens*.
   c The image travels as a series of *light/electrical* signals along the optic nerve to the brain.
   d The aqueous humour is a clear fluid that lies between the cornea and the *retina/lens*.

6 The following table lists the typical focal lengths for different optical devices.

Focal lengths for different optical devices

Object	Focal length (m)
spectacles	1
camera lens	0.05
microscope objective lens	0.004

   a Which object uses the lens with the shortest focal length?
   b Why would this device require the shortest focal length?

7 a Is the image formed on your retina real or virtual?
   b Explain your answer.

8 Describe what happens to the pupils of your eyes when you are:
   a in a dark cinema
   b outside playing in the sun.

9 Tran is short-sighted and forgets to bring his glasses to the cinema. Predict whether you think he would prefer to sit near the front or the back to watch the movie.

10 Explain how a pair of bifocal glasses works.

11 Outline what happens when a person undergoes laser eye surgery.

Analysing

12 Analyse Figure 4.4.1 on page 164.
   a Explain why the shape of the pinna suits the function of this part of the ear.
   b Identify where vibrations are converted to electrical signals.

Evaluating

13 The science4fun, Fooling your eyes, on page 163 is designed to confuse cells at the back of your retina.
   a Which cells do you think become confused?
   b Justify your answer.

14 At a 30-year high school reunion, 29 members of an original class of 34 are wearing glasses. Propose an explanation for this, given that only five of the class wore glasses at school.

15 In young children, the Eustachian tube is narrower and more horizontal than in an adult. Use this information to propose a reason why young children get more ear infections than adults.

MODULE 4.4

Review questions

16 Look at the path of light through the eyeballs A, B and C shown in Figure 4.4.10.

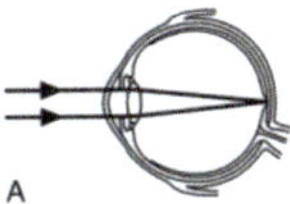
A

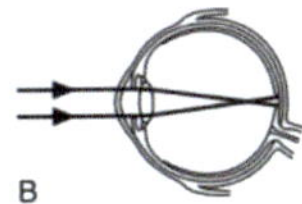
B

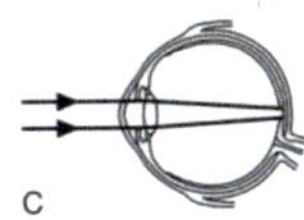
C

FIGURE 4.4.10

a Describe how distant and near objects would appear to a person with the vision associated with eyeballs A, B and C.

b Identify which of the eyeballs would result in normal vision, long-sightedness and short-sightedness.

c Assess which types of corrective lenses would assist the cases of long- sightedness and short-sightedness and match these to the correct eyeball shown.

17 There are no light sensitive cells at the point on your retina where it joins the optic nerve. You are blind in this spot. To identify this blind spot, complete the following steps:

- Hold this page of your textbook (or screen of a device) at arm's length.
- Close your left eye and stare at the cross in Figure 4.4.11 with your right eye.
- Move the page closer to your face while still looking at the cross. Be aware of the dot while you do this.

Finding your blind spot

FIGURE 4.4.11

a Propose a reason why you don't normally observe your blind spots.

b Explain what happened when the dot disappeared.

c If you look at the dot while completing the test, it won't work. Propose why this is the case.

Creating

18 Romeo and Juliet were tested for hearing loss over a range of frequencies. The findings are shown in Table 4.4.1.

TABLE 4.4.1 Hearing test results for Romeo and Juliet

Frequency (Hz)	Hearing loss (decibels)	
	Romeo	Juliet
500	0	5
1000	0	10
2000	0	40
3000	0	45
4000	15	45
5000	15	25

a Construct two line graphs on the same set of axes to display these results. Place frequency (Hz) on the horizontal axis and hearing loss (dB) on the vertical axis.

b Which frequencies did Juliet find it hardest to hear?

c What was Juliet's hearing loss at 1000 Hz?

d At which frequency did Romeo start to experience hearing loss?

19 Construct a flow chart that shows the structures and fluids light travels through on its journey from entering your eye until it reaches your retina.

MODULE 4.4

Practical investigations

1 • Dissection of a bull's eye

Purpose

To examine the structure of a bull's eye and compare it to a human eye.

Timing 45 minutes

Materials

- a bull's eye
- scalpel
- dissecting board
- tweezers
- dissecting scissors
- scrap of newspaper
- gloves

SAFETY

Complete the dissection on a dissecting board. Wear gloves while completing the dissection and when thoroughly cleaning the board after the experiment. Make sure you wash your hands after cleaning up your work area and completing the experiment. Handle scalpels and scissors with care.

Procedure

1 Examine the eyeball. The white part is the sclera and the blue part is the cornea. Carefully cut away fatty tissue from around the eyeball.

2 Locate the optic nerve. This is a bundle of nerves that branch out of the back of the eyeball.

3 Place the eyeball on the dissection board holding it with your thumb (on the cornea) and index and middle fingers above and below the optic nerve. Carefully make a small incision with the scalpel, then using the point of your scissors, carefully cut around the circumference of the eyeball so the front and rear parts separate as shown in Figure 4.4.12. With the first incision the aqueous humour, the liquid that keeps the eyeball firm, oozes out followed by the round glob of vitreous humour that contains the lens.

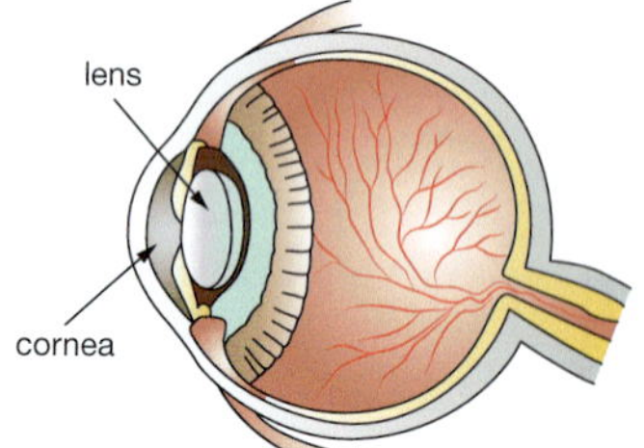

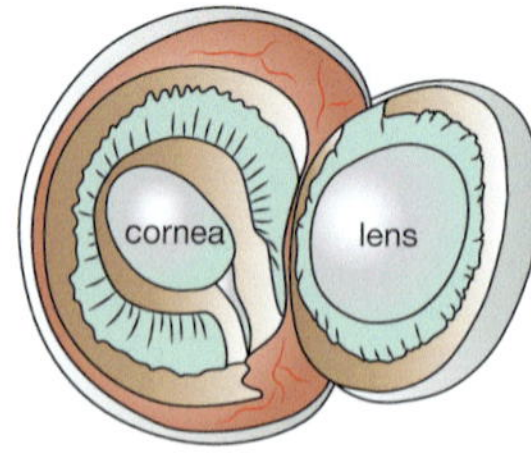

FIGURE 4.4.12 Dissection of a bull's eye

4 Examine the black iris found on the cornea. Incoming light enters the eye through the pupil which is framed by this structure.

5 Carefully remove the jelly-like lens from behind the cornea. The lens is located in the vitreous humor. Although the lens may not be transparent now, it normally would be in a living bull. Carefully hold the lens and look through it to see an image. Also look through the lens onto some newspaper text to see if it acts like a magnifying lens.

6 Examine the pearl-like black inner side of the retina. Can you find the spot where the optic nerve leaves the retina?

7 Make sure all parts of the eye are wrapped in newspaper and disposed of appropriately and your work space is clean. After removing your gloves, wash your hands.

Results

Use a phone or camera to take a series of images of your dissection as you work through each stage.

Review

1 Describe the appearance of the optic nerve and state what type of message travels along this pathway.

2 The eyeball is surrounded by muscles. What do you think these muscles are for?

3 There are two sets of fluid inside the eye. Name these two fluids and describe where each is located in the eyeball.

4 Describe what would happen to the size of the iris when the bull is outside on a sunny day.

5 Describe what would happen to the shape of the lens as the bull focuses on close objects.

6 Identify the type of the lens located in the bull's eye.

7 Describe your observations from looking through the lens at objects and newsprint.

8 Recall the names of the light sensitive cells found in on the retina.

9 Why do you think the part of the retina in which the optic nerve attaches is called the blind spot?

CHAPTER 4

Chapter review

Remembering

1 List the three processes of heat transfer.

2 Name the only process that can transfer heat through the vacuum of space.

3 Does sound travel fastest in a solid, a liquid or a gas?

4 List the parts that make up the middle ear.

5 A beam of light hits a plane mirror at an angle of 45°. What will be the angle of reflection as the light leaves the mirror?

Understanding

6 Explain what causes a land breeze to blow from land over water on a summer evening.

7 Describe what is likely to happen to infrared radiation that hits a:

a black plastic pot plant

b white shade sail over a sandpit

c glass window on a boat.

8 Kim toasts marshmallows on an open fire. Although she can't see any flames, she can still feel heat from the fireplace. Explain why.

9 a How can the pitch of a violin string be changed?

b If a violin string is tightened, predict whether a note plucked would sound higher or lower in pitch.

c Give a reason for your answer above.

10 Predict the sizes of angles x, y and z in Figure 4.5.1.

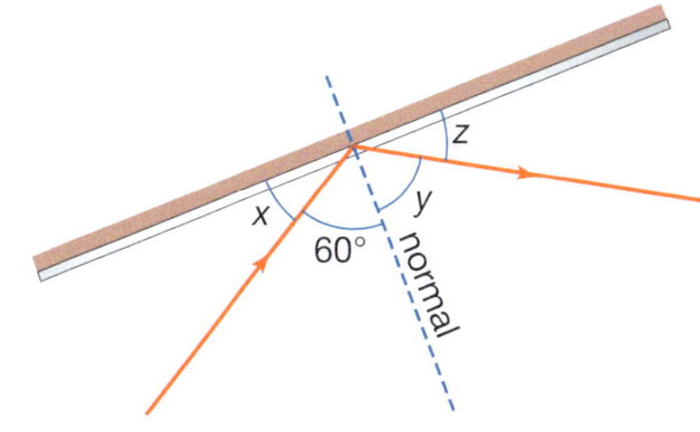

FIGURE 4.5.1

Applying

11 Identify what vibrates in each musical instrument below to produce a sound.

a harp b trumpet c drum.

12 Perspex has a greater refractive index than ice does.

a Identify in which material light travels faster.

b If light travels from ice into Perspex, will it bend towards or away from the normal?

13 A light ray travels through material X and hits the boundary of the transparent material, Y, at an angle of 40° to the surface. It is then refracted into material Y at an angle of 35°. Identify whether X or Y has the greater refractive index.

Evaluating

14 a Assess whether you can or cannot answer the questions on page 124 at the start of this chapter.

b Use this assessment to evaluate how well you understand the material presented in this chapter.

Creating

15 A seagull circling overhead spies a fish below, as shown in Figure 4.5.2. Construct a diagram to show where the fish appears to be when seen by the seagull.

FIGURE 4.5.2

16 Use the following ten key terms to construct a visual summary of the information presented in this chapter.

heat	temperature
conduction	convection
radiation	sound
frequency	wavelength
light	image

AB 4.11

CHAPTER 4

Inquiry skills

Research

1 Processing & Analysing | Evaluating

Three-part inquiry

Select your entry point and complete the relevant parts of this inquiry.

Many people in Australia have a swimming pool in their backyard. However, pool use is limited without heating because the water is too cold, except during the summer. Many people heat their pool to increase its usefulness, which consumes a large amount of energy.

a To keep the heat in the swimming pool and reduce costs, many people use a pool cover, which acts as a thermal blanket. Investigate thermal blankets and their uses. Explain how they reduce heat loss from a pool by conduction, convection and radiation.

The ideal temperature for pool water for swimming is about 26°C but through most of the year the temperature will be much lower than that. In winter, the temperature could be as low as 10°C. To extend the use of their swimming pool many people heat them through a large part of the year.

b Heating a swimming pool requires a large amount of energy. The amount of energy needed will depend on the volume of water and how much it needs to be heated.

- **i** Find out the average amount of water, in litres, needed to fill a domestic swimming pool.
- **ii** Heating water requires large amounts of energy. The energy required to raise the temperature of 1 mL of water by 1°C is approximately 4.19 joules. Using this information, calculate how many joules of energy is required to heat the average swimming pool from 10°C to 26°C. (Remember, 1 litre is equal to 1000 mL.)

c It is quite expensive to heat a pool so once it is heated, many people insulate their pool to retain the heat. To do this they often use a thermal blanket. The blankets used are generally silvery on one side and matt black on the other. Explain why both surfaces are used, which colour will be visible when the pool cover is in place and why.

2 Questioning & Predicting | Communicating

New homes must be constructed to comply with standards to improve energy efficiency. Double-glazed windows and insulation are two methods that improve the energy efficiency of a home. Research these methods to find:

- which process or processes: conduction, convection and/or radiation allow heat to be transferred through a window, allowing heat to escape out in winter and in during summer
- a diagram showing what a double-glazed window looks like
- why a double-glazed window is more effective in insulating a house than a standard window
- how insulation affects the energy efficiency of a home
- images of different types of insulation commonly used in walls and ceilings.

Present your findings as a poster to be displayed at a home show advertising double-glazed windows and insulation.

3 Planning & Conducting 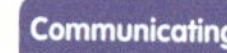

An endotherm is an animal that that can maintain a constant body temperature. For the animal to stay healthy, this body temperature needs to remain fairly constant. Humans and animals have different ways of maintaining their body temperature, including:

- having hair or fur
- shivering when cold
- covering themselves in mud
- panting
- having goose bumps on their skin.

Choose one of these methods and research how it assists in regulating body temperature. Present your findings as a physical model, an animation or as a poster.

Inquiry skills

4 Questioning & Predicting Communicating

Research hearing damage associated with listening to music through headphones and earphones. Find:

- the levels of loudness that are likely to cause long-term hearing damage
- how to reduce the chance of hearing loss associated with headphone use
- why some experts recommend a '60/60' rule: listen to music at 60% of the maximum volume for 60 minutes and then have a break before listening again
- the risks associated with using earphones compared to headphones that fit over the ear
- an estimate of the maximum recommended duration that it is safe to listen to music at maximum volume before causing hearing damage.

Present your research in a brochure or multimedia display to outline key recommendations on how to listen to music through headphones or earphones without risking hearing loss.

5 Questioning & Predicting Evaluating 

Find the depths of the following places and then calculate how long it would take for sonar to return from the bottom of each. Assume the speed of sound in water is 1500 m/s.

- the deepest point on Earth (Mariana Trench)
- the wreck of the *RMS Titanic*
- the shallowest and deepest points in Sydney Harbour.

Present your findings and calculations in written form.

Thinking scientifically 

1 Light reflects from a plane mirror at an angle equal to the angle at which it hits the mirror. Josh directs a ray from a light box onto a plane mirror, as shown in Figure 4.5.3.
The reflected ray in Josh's experiment is:

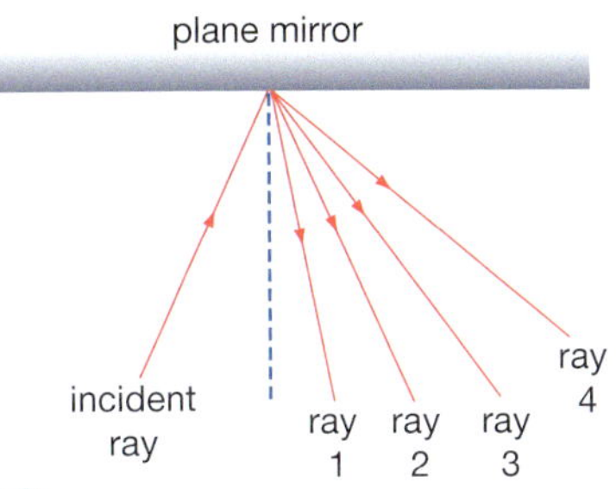

FIGURE 4.5.3

A ray 1
B ray 2
C ray 3
D ray 4.

2 Light refracts (bends) towards the normal when it travels from one medium into another of higher refractive index. It bends away from the normal when travelling into a substance of lower refractive index. Study the ray diagrams in Figure 4.5.4 as light travels between materials 1, 2 and 3.
The materials listed from least to greatest refractive index are:

A 1, 2, 3
B 3, 1, 2
C 2, 1, 3
D 1, 3, 2.

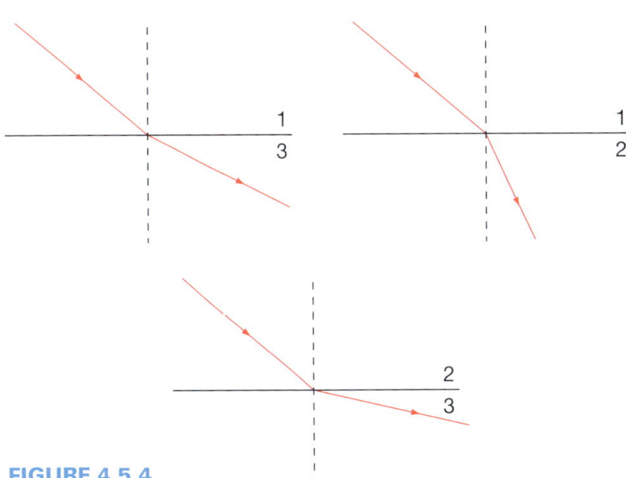

FIGURE 4.5.4

3 In a plane mirror, light is laterally inverted (reflected from left to right). Select the correct image of the house as it appears when reflected through a plane mirror.

A house A
B house B
C house C
D house D.

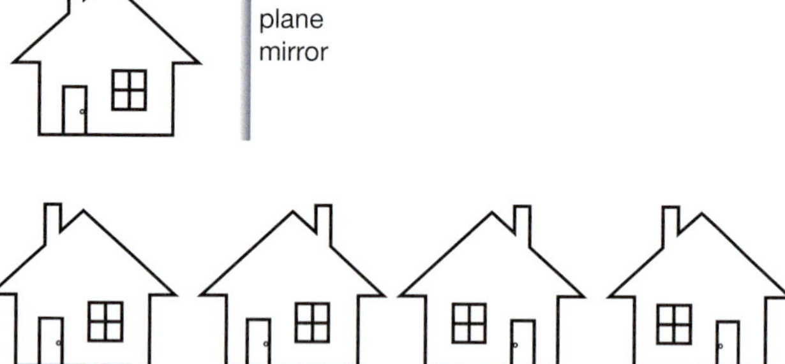

4 A horse has one of the largest pairs of eyes of any animal. The horse has binocular vision and therefore a limited field of view. In this field of view, the horse can easily judge distances between objects. It also has a wide field of monocular vision to the left and right, called its left and right monocular fields. The horse cannot judge distances as effectively between objects it sees in these fields. In a small region behind the horse, it has no vision.

Figure 4.5.5 shows these fields of view, as seen from above the horse.

Identify which of the 1-metre tall objects, L, M, N and O, can be seen by the horse without turning its head.

A only L
B only M and N
C L, M and N
D only L and O.

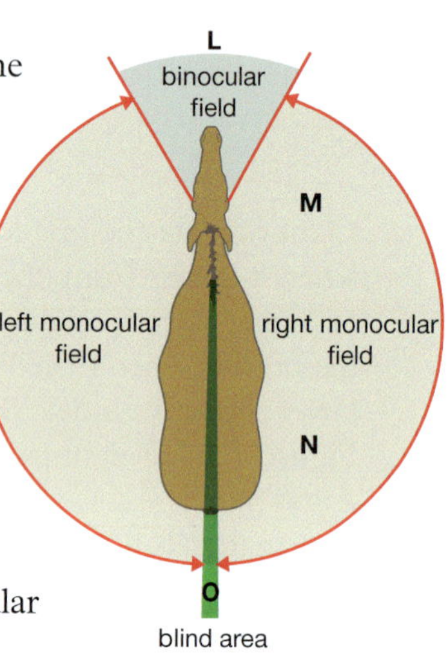

FIGURE 4.5.5

5 Figure 4.5.6 compares the range of hearing of humans and different animals. Use it to answer the following questions.

a Which animal can hear the highest frequency sounds?
b Which animal or animals can hear the lowest frequency?
c The bat and the beluga whale rely on echolocation using ultrasound. Propose why they need to have a higher hearing range than humans.
d Dog whistles are audible to dogs but not to humans. Propose the frequency range of such whistles.
e Would a chicken be able to hear any of the sounds produced by a bat?

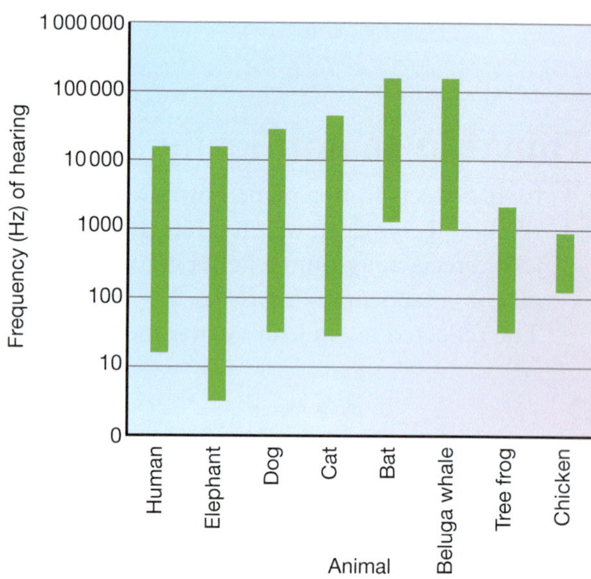

FIGURE 4.5.6 Note that the vertical scale is a logarithmic scale, meaning each division on the scale is ten times as large as the previous division.

CHAPTER 4

Glossary

absolute zero: the lowest possible temperature, –273°C

accommodation: the ability of the lens of the eye to change shape

angle of incidence, *i*: the angle an incoming ray makes with the normal

angle of reflection, *r*: the angle a reflected ray makes with the normal

angle of refraction, *r*: the angle a refracted ray makes with the normal

compression: a region of high pressure in which particles are close together

concave lens: a lens that curves inwards

conduction: a method of heat transfer in which heat is passed by vibration of particles

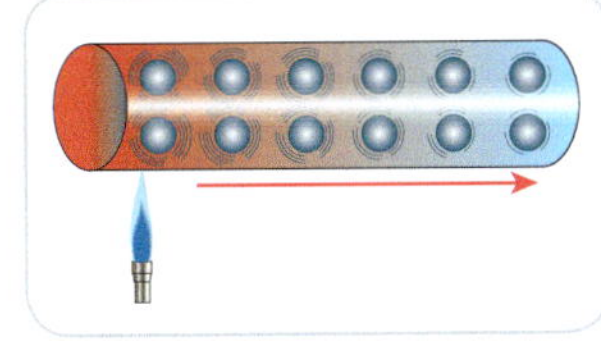

conduction

conductor: a substance that allows heat to flow through it

convection: transfer of heat in a liquid or gas due to less dense, warmer matter rising and denser, cooler matter falling

convex lens: a lens that bulges outwards

critical angle: the angle of incidence of light that produces an angle of refraction of 90°

decibel (dB): unit used to measure loudness

echo: a sound that is reflected and heard a second time

frequency: the number of waves passing a point every second

hertz (Hz): the unit used to measure frequency

incident ray: incoming ray

insulator: a material that does not conduct heat well

lateral inversion: the sideways or left-to-right reversal of an image in a plane mirror

law of reflection: the law stating that light is reflected at the same angle that it is incident, or $i = r$

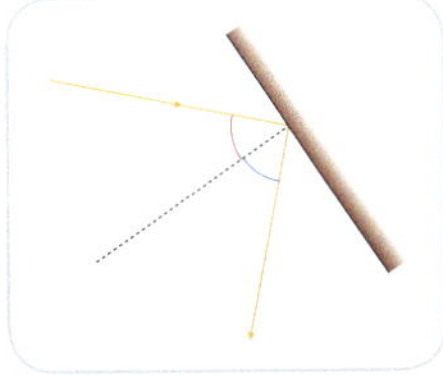

law of reflection

longitudinal wave: a wave in which the vibration is in the same direction as the wave is travelling

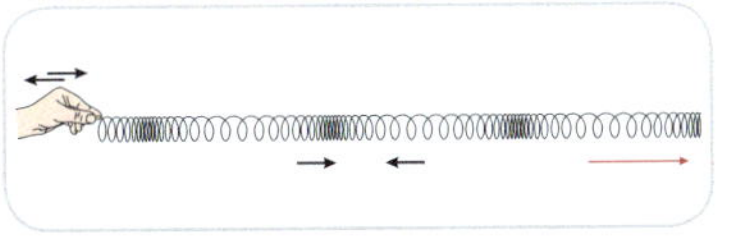

longitudinal wave

long-sightedness (hyperopia): the ability to see distant objects clearly; close objects appear blurry

luminous: an object that releases or emits light

normal: an imaginary line that is drawn at right angles to a surface that light is incident upon

plane mirror: a flat mirror

radiation: movement of heat in the form of electromagnetic waves, which can travel through a vacuum

rarefaction: a region of low pressure in which particles are far apart

real image: an image formed when rays of light actually meet

reflected ray: ray that bounces off

reflection: the bouncing of light off something

refraction: the bending of light as it passes from one substance into another substance

refractive index: a measure of how easily light travels through a substance

reverberation: length of time a sound can be heard

short-sightedness (myopia): the ability to see close objects clearly; distant objects appear blurry

sound wave: regions of high and low pressure originating from a vibrating object and transmitted through a medium

temperature: a measure of the average kinetic energy of particles in a substance that results in how hot or cold the substance is

thermometer: an instrument used to measure temperature

total internal reflection: when light is completely reflected from the boundary of two substances; it occurs when the angle of incidence is greater than the critical angle

transverse wave: a wave in which the vibration is at right angles to the direction the wave is travelling

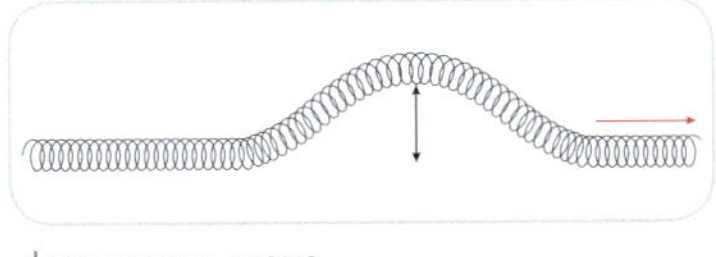

transverse wave

virtual image: an image formed in which the rays of light do not actually meet, but only appear to meet at a point inside the mirror

wavelength: the distance from one peak of a wave to the next

AB 4.10

CHAPTER 5

Electromagnetic radiation

Have you ever wondered ...

- why objects appear different colours?
- how heat gets from the Sun through empty space?
- how night-vision goggles work?
- how your radio works?

LS

After completing this chapter you should be able to:

- *identify situations where waves transfer energy through different mediums*
- *use the wave model to explain different phenomena such as light*
- *describe the properties of waves in terms of the wavelength, frequency and speed*
- describe how electromagnetic radiation is used in radar, medicine, mobile phone communications, and microwave cooking
- *investigate how electromagnetic radiation is used in the detection and treatment of cancer*
- *outline how new mobile communication technologies rely on electromagnetic radiation*
- *describe how science, engineering and technology are used in telecommunication careers.*

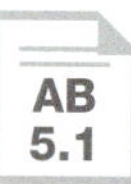

MODULE

5.1 Waves

Sound, light and the heat radiating from a fire transmit their energy via waves. Waves in the sea carry energy with them, as do the waves that shake the land in an earthquake. The Sun and the stars radiate radio waves, microwaves and waves of visible light, infrared radiation, ultraviolet light, X-rays and gamma rays. These different forms of radiation are called electromagnetic radiation, and together form the electromagnetic spectrum.

science 4 fun

Playing with water waves

YES

What do water waves look like?

Collect this ...

- straw
- large bowl half full of water

Do this ...

1. Suck some water from the bowl up into the straw.
2. Quickly take your mouth off the straw and put your thumb on the end of it to keep some of the water trapped inside the straw.
3. Release a small amount of water from the straw by briefly lifting your finger from the end of it. Practise this until you can release the water one droplet at a time. Refill the straw as needed.
4. Wait until the surface of the water in the bowl is still and clear of ripples. Release one droplet of water into the bowl.
5. Release a series of droplets at a constant rate, i.e. with the same time between each.
6. Vary the rate of droplets by increasing and decreasing the time between each droplet.

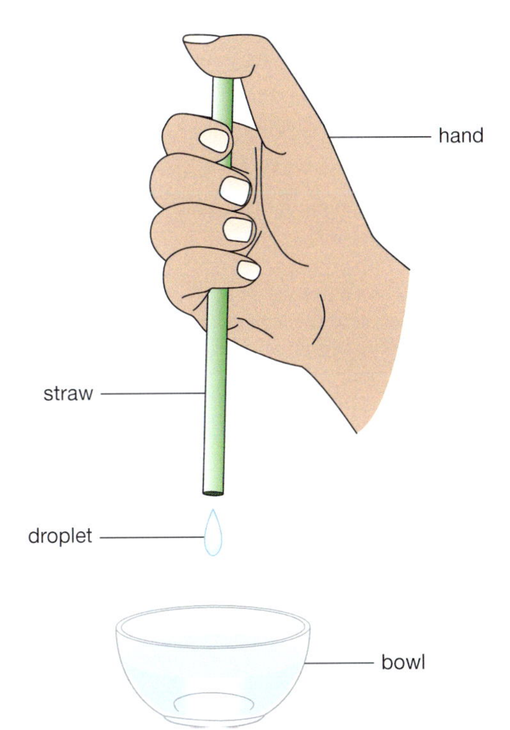

Record this ...

1. Describe the wave patterns formed and how the patterns changed as you changed the rate at which droplets were released.
2. Explain why you think the wave patterns changed.

Wave motion

The transfer of energy without the transfer of matter is called **wave motion**. Figure 5.1.1 shows the wave motion of ripples created from a droplet of water. These ripples travel outwards from the point where the droplet hit the water. The energy of the impact travels outwards, but the actual water particles making up the wave only move up and down.

FIGURE 5.1.1 A water droplet hitting water triggers the outwards spread of circular waves.

There are two types of waves that can transfer energy: transverse and longitudinal waves. The particles of a transverse wave vibrate at right angles to the direction of motion of the wave.

Examples of transverse waves are the waves you see on the ocean and the waves you can flick along a rope, slinky spring or hose. In a longitudinal wave the particles vibrate backwards and forwards in the same direction as the wave motion. Sound waves are longitudinal waves. Both these types of wave are shown in Figure 5.1.2.

Prac 1 p. 187

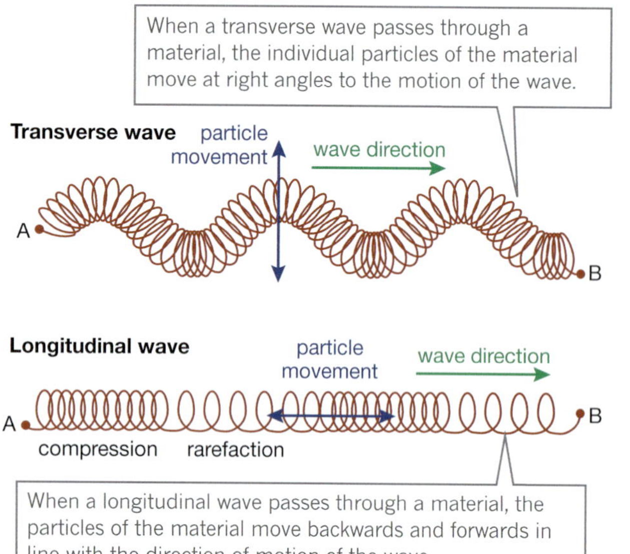

FIGURE 5.1.2 Transverse waves and longitudinal waves both transfer energy but in very different ways.

Wave properties

The number of waves produced each second is called the **frequency** of the wave. Frequency is measured in hertz (Hz), which means cycles (waves) per second.

Wavelength is the distance between two successive waves and is usually measured in metres. The wavelength of some radio waves is several kilometres, whereas the wavelength of visible light is less than one thousandth of a millimetre.

The **amplitude** of a wave is the maximum distance it extends beyond its middle position. Figures 5.1.3 and 5.1.4 show the wavelength and amplitude of transverse and longitudinal waves.

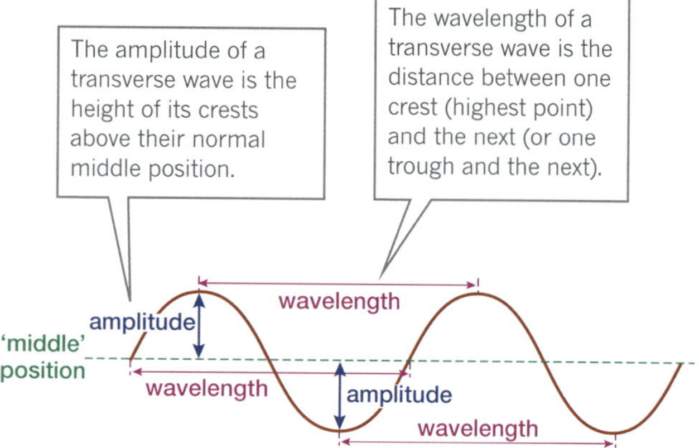

FIGURE 5.1.3 The amplitude and wavelength of a transverse wave

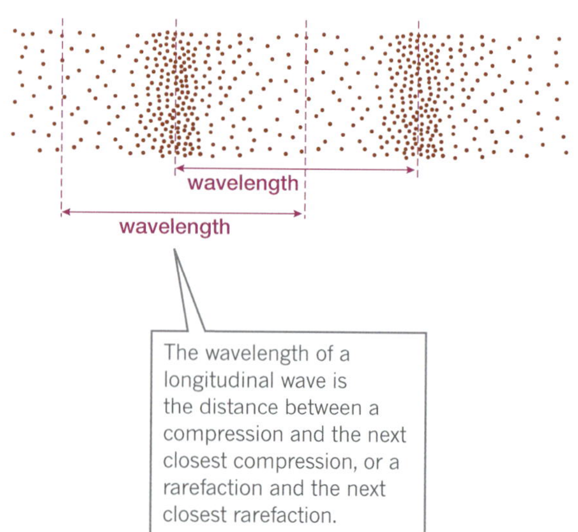

FIGURE 5.1.4 The wavelength of a longitudinal wave

For two waves of the same frequency and wavelength, the larger the amplitude of the wave, the more energy is carried by the wave. This relationship is shown in Figure 5.1.5.

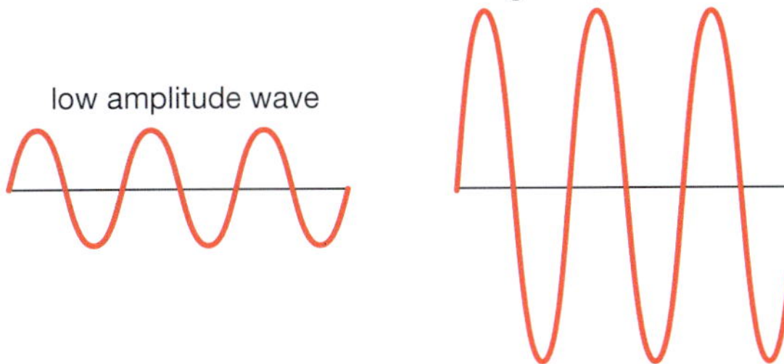

FIGURE 5.1.5 Both these waves have the same frequency and wavelength. The taller wave has the greater amplitude and so it carries more energy.

The wave equation

The speed, wavelength and frequency of a wave are linked by a special relationship. This is called the wave equation. This means that if the frequency of a wave increases, then its wavelength will decrease. Alternatively, if the frequency of a wave decreases, then its wavelength increases. This relationship is shown in Figure 5.1.6.

Wave A

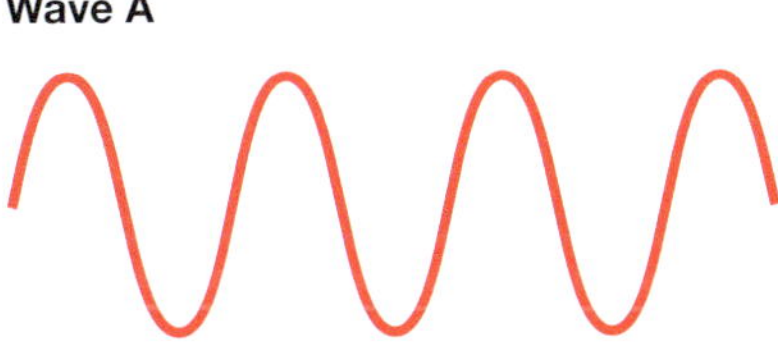

Given that wave A and B are travelling at the same speed.
Wave A has a:
- higher frequency and
- shorter wavelength than wave B.

Wave B

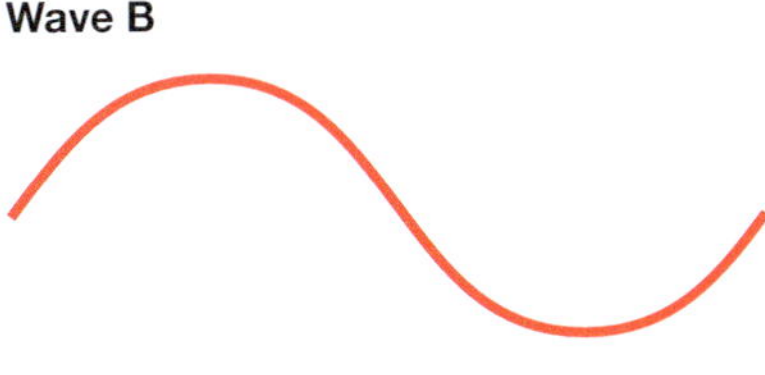

Wave B has a:
- lower frequency and
- longer wavelength than wave A.

FIGURE 5.1.6 Frequency is the number of waves passing every second. The more bunched up the waves, the higher their frequency and the shorter their wavelength.

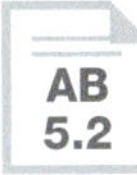

SkillBuilder

The wave equation

The speed, wavelength and frequency of a wave depend upon each other and are linked by a formula called the wave equation:

$$v = f\lambda$$

where: v = speed of wave (m/s)

f = frequency of wave (Hz)

λ = wavelength of wave (m)

(The symbol for wavelength is λ, a letter from the Greek alphabet. The letter is called lambda.)

The equation can be rearranged to calculate frequency:

$$f = \frac{v}{\lambda}$$

It can also be rearranged to calculate wavelength: $\lambda = \frac{v}{f}$

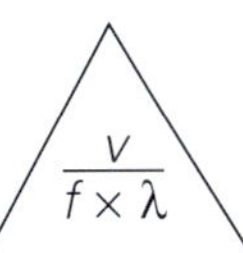

Worked example

The wave equation

Problem

At a beach, a wave hits the shore every 10 seconds. If there is 6 m between successive waves, calculate the speed of the waves.

Solution

Thinking: Determine the frequency of the waves.

Working: $f = \frac{1}{10}$ per second = 0.1 Hz

Thinking: Determine the wavelength of the waves.

Working: $\lambda = 6$ m

Thinking: Refer to the formula triangle to work out which formula to use.

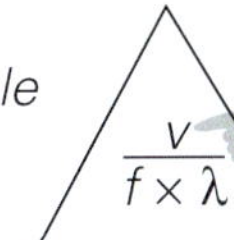

Working: $v = f\lambda$

Thinking: Substitute the values for frequency and wavelength and calculate wave speed.

Working: $v = 0.1 \times 6 = 0.6$ m/s

Try yourself

1 A child playing with a slinky shakes it backwards and forwards 5 times every second. If the waves in the slinky are 30 cm long, calculate the speed of the waves.

2 The speed of sound changes depending on the temperature of the air. On a cold day, a 256 Hz tuning fork produces waves that are 1.25 m long. What is the speed of sound on this day?

3 Gary flicks a hose twice every second to form a wave with a wavelength of 1.5 m. What is the speed of the wave as it travels down the hose?

Musical instruments and the wave equation

The wave equation determines what you hear when playing a musical instrument. For example, when you strike a guitar string, it vibrates to form a transverse wave. More correctly, it forms half a wave as shown in Figure 5.1.7.

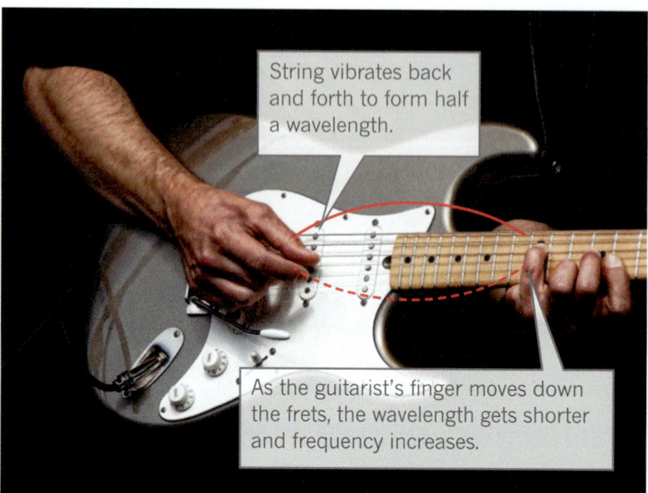

FIGURE 5.1.7 Wavelength decreases as you shorten a string, so the frequency of the note increases.

Whatever note you play on a particular string, the speed of the wave along that string is the same. By moving your finger along the frets you shorten the vibrating length of the string and, therefore, the wavelength of the wave. As the wavelength decreases, the frequency of the note (its pitch) increases.

Place your finger on exactly the same fret but on a different string and you get a note of a different frequency. This is because different strings have different thicknesses, are made from different materials and are under different tensions. This combination of factors alters the speed of the wave that travels down them when struck. The wavelength might be the same but the speed is now different and so its frequency is different too.

Other stringed instruments such as violins and cellos work in the same way. Pianos and harps have many strings of different lengths. Each string vibrates to form a particular wavelength and a particular frequency.

When a woodwind or brass instrument such as a trombone is blown, a longitudinal compression wave is formed in the air inside the instrument. Increasing the length of the trombone causes the wavelength in the tube to increase (Figure 5.1.8).

FIGURE 5.1.8 A person playing the trombone changes the pitch of the note by changing the length of the trombone.

As a result, the frequency of the wave produced decreases and a deeper note is heard. Many wind instruments such as recorders and clarinets do not physically alter the length of their instruments but instead let air escape through holes. When all the holes are closed, the full length of the instrument is used to form the wave.

Speed of sound

The speed of sound waves in air is constant in a particular room at a given time. However, this speed can vary with changes in temperature, air pressure and humidity. For example, at 0°C the speed of sound is around 330 m/s but at 40°C it is over 350 m/s.

FIGURE 5.1.9 When an aircraft flies faster than the speed of sound it creates a shock wave like the one spectacularly shown in this photo.

Opening a hole has the effect of shortening the length of the air column inside the instrument and the wavelength formed in it. As a result, the frequency of sound produced is increased and a higher note is heard.

Electromagnetic radiation

When an electric current flows in a wire, a **magnetic field** is generated around the wire. This field can be detected by placing a magnetic compass near the wire. Similarly, moving a magnet inside a coil of wire generates an **electric field**. This field will cause the electrons in the wire to move producing an electric current. The interaction between electricity and magnetism is called **electromagnetism**.

The Scottish scientist James Clerk Maxwell (1831–79) suggested that a changing electric field could create a changing magnetic field, which would in turn create a changing electric field. These fields would continue to generate each other. He proposed that these fields travelled through space as transverse waves at right angles to each other. This is the structure of an electromagnetic wave. You can see it in Figure 5.1.10.

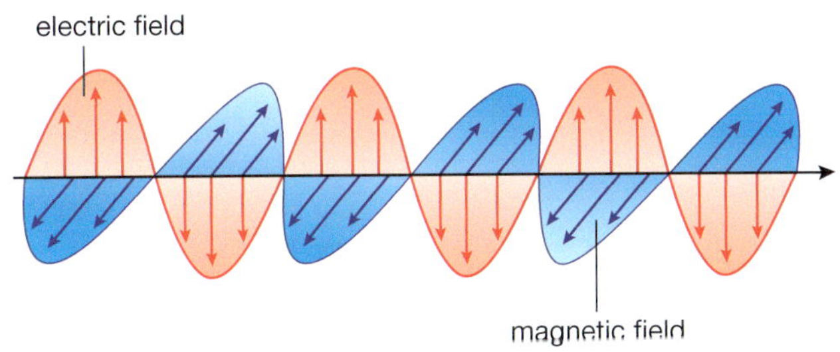

FIGURE 5.1.10 Electromagnetic waves travel as two interconnected electric and magnetic fields moving as transverse waves.

Electromagnetic waves are generated naturally in our upper atmosphere and from stars, including our Sun. Visible light, microwaves and X-rays are examples of **electromagnetic radiation**. These forms of energy travel through space as **electromagnetic waves**.

AB 5.4

The electromagnetic spectrum

The entire range of frequencies of electromagnetic radiation that can be produced is called the **electromagnetic spectrum**. The electromagnetic spectrum is shown in Figure 5.1.11. This ranges from low-frequency radiation, such as radio waves, through to high-frequency gamma radiation. Like sound waves, as the frequency of the electromagnetic waves increases, the wavelength decreases. Electromagnetic waves all travel through empty space at the speed of light, which is approximately 300 000 km/s.

When an electromagnetic wave strikes a substance, it can either pass straight through it or be absorbed by the substance. When electromagnetic radiation is absorbed by a substance, this can cause the substance to heat up or change in some way. For example, sunlight causes the sand on a beach to heat up whereas when sunlight hits a solar panel its energy is converted into an electric current.

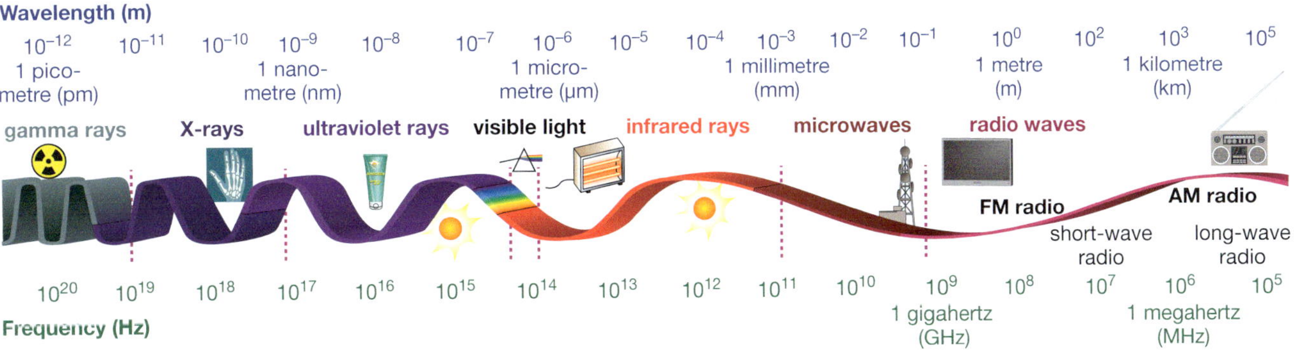

FIGURE 5.1.11 The electromagnetic spectrum shows the complete range of electromagnetic waves. All these waves travel at the speed of light.

SkillBuilder

Scientific notation

Scientific notation is an easy way to handle very large and very small numbers. In scientific notation, numbers are written as a number between 1 and 10 (called the coefficient) multiplied by a power of 10. For example, 10 000 can be written as 1.0×10^4 (or simply 10^4).

The number at the top right of the 10 is called the exponent. For example, in 10^{19}, 19 is the exponent. When the exponent is positive, you can convert the number from scientific notation to decimal form, by moving the decimal point this many places to the right. For example, 1.0×10^7 becomes 10 000 000.

For very small numbers, the exponent is negative. This indicates that to convert the number to a decimal, the decimal point is moved to the left. For example, 1.0×10^{-7} becomes 0.000 000 1.

Figures 5.1.12 and 5.1.13 provide examples of numbers converted to scientific notation.

FIGURE 5.1.12 This reservoir hold 450 GL when full. This is 450 000 000 L or 4.5×10^8 L.

FIGURE 5.1.13 This ant has a mass of 4 mg. This is 0.004 g or 4.0×10^{-3}g.

Worked example

Converting to scientific notation

Problem

State the following in scientific notation:

a 470 000 **b** 0.0006

Solution

a *Thinking: Make the quantity into a number between 1 and 10.*

Working: 470 000 becomes 4.7

Thinking: Determine what factor of 10 it needs to be multiplied by.

Working: $470\ 000 = 4.7 \times 100\ 000 = 4.7 \times 10^5$

b *Thinking: Make the quantity into a number between 1 and 10.*

Working: 0.0006 becomes 6.0

Thinking: Determine what factor of 10 it needs to be multiplied by.

Working: $0.0006 = 6.0 \times 0.0001 = 6.0 \times 10^{-4}$

Try yourself

State the following in scientific notation:

a 21 000 000 000

b 0.000 000 000 009

Worked example

Converting from scientific notation

Problem

State the following in decimal form:

a 1.5×10^3 **b** 7.8×10^{-6}

Solution

a *Thinking: Determine which way the decimal point moves.*

Working: 10^3 indicates that the decimal point moves right 3 places.

$1.5 \times 10^3 = 1.5 \times 1000 = 1500$

b *Thinking: Determine which way the decimal point moves:*

Working: 10^{-6} indicates that the decimal point moves left 6 places.

$7.8 \times 10^{-6} = 7.8 \times 0.000\ 001 = 0.000\ 0078$

Try yourself

State the following in decimal form:

a 3.25×10^4

b 8.13×10^{-2}

MODULE 5.1 # Review questions

Remembering

1 Define the terms:

a electromagnetic spectrum

b wavelength.

2 What term best describes each of the following?

a wave consisting of oscillating electric and magnetic fields

b maximum distance a wave extends from its middle position.

3 A wave is transporting energy through air from right to left. The particles of the air move back and forth in a horizontal direction. What is the type of wave called?

4 Refer to the transverse wave shown in Figure 5.1.14.

State the letter that corresponds to its:

a amplitude

b wavelength.

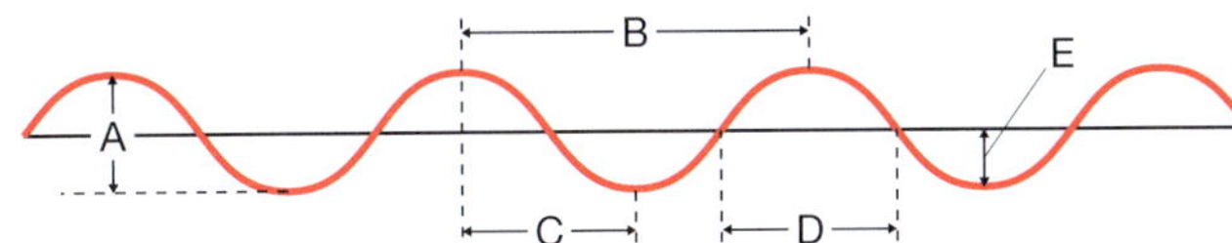

FIGURE 5.1.14

5 Julia and Skye are making waves with a slinky spring that is stretched between them. They halve the frequency of the waves they are producing. What happens to the wavelength produced?

6 What two types of fields interact to produce an electromagnetic wave?

7 List five types of electromagnetic waves.

Understanding

8 Fans in a sporting stadium sometimes move to create a Mexican wave around the stadium. Describe how each person would need to move if they were creating a:

a transverse wave

b longitudinal wave.

9 Atahan strikes a tuning fork of frequency 440 Hz and listens to the sound produced. Hema then strikes another tuning fork and hears a lower-pitched sound. Assuming that the speed of sound in air is constant, describe the difference between the wavelengths of each sound.

10 a Nisha whispers a comment to her friend Sally. Describe the movement of air particles as sound travels from Nisha's lips to Sally's ears.

b Reuben drops a water droplet into a still bowl of water like in the science4fun on page 179. Describe the motion of the water on the surface of bowl as the wave passes across it.

Applying

11 a Rewrite the following in scientific notation:

i 21 000

ii 0.0085

b Rewrite the following in decimal form:

i 3×10^8

ii 6.67×10^{-5}

12 Identify the rarefaction and compression in the sound waves created by the tuning forks in Figure 5.1.15.

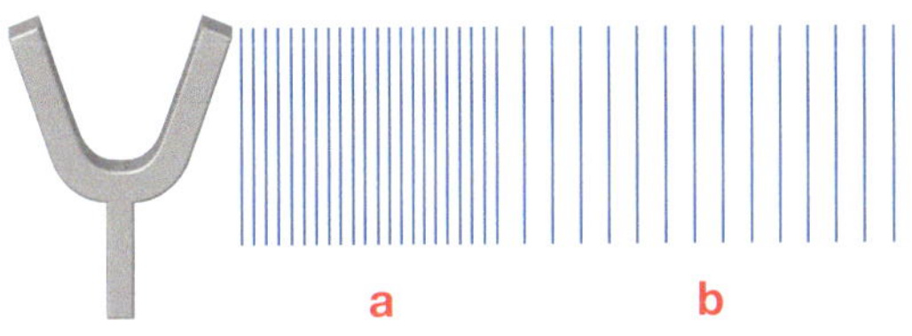

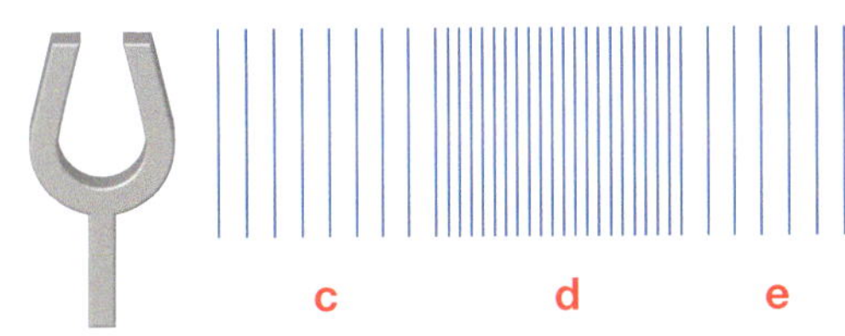

FIGURE 5.1.15

MODULE

5.1 Review questions

13 Look at the transverse wave shown in Figure 5.1.16.
Identify any pairs of letters that are spaced one wavelength apart.

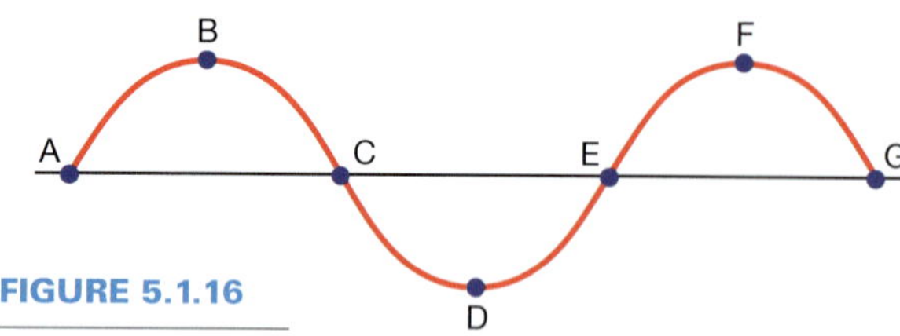

FIGURE 5.1.16

14 a A transverse wave in a string has an amplitude of 5 cm, a frequency of 60 Hz and a distance from crest to the next trough of 15 cm. What is the wavelength of this wave?

b A nervous student paces up and down a corridor waiting to enter an exam room. She paces 10 times in 1 minute. Calculate the frequency of her pacing in hertz.

15 a On a boat, the captain counts 5 waves passing her boat every minute. If these waves have an average wavelength of 120 m, what is the speed of the waves?

b The lowest frequency of sound that most humans can hear is 20 Hz. If the speed of sound is 330 m/s, what is the wavelength of these sound waves?

c Waves travel along a particular guitar string at 425 m/s. If waves have a wavelength of 83 cm, calculate the frequency of the note being played.

16 a Use Figure 5.1.17 to state the wavelength and amplitude of:

i wave A **ii** wave B.

b Which of these waves has the higher frequency?

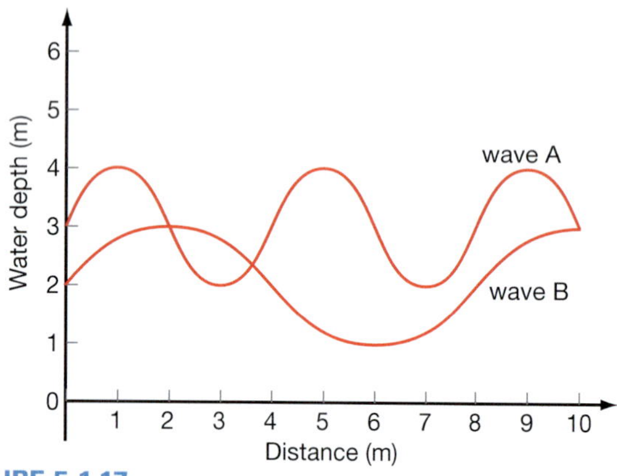

FIGURE 5.1.17

Analysing

17 Sound is a mechanical wave, which is different from an electromagnetic wave. Compare a sound wave with an electromagnetic wave.

18 Use Figure 5.1.11 on page 183 to classify the following frequencies of electromagnetic radiation.

a **i** 1.0×10^{6} Hz

ii 1.0×10^{10} Hz

iii 1.0×10^{14} Hz

iv 1.0×10^{16} Hz

v 1.0×10^{22} Hz

b Which of the frequencies in part a is the highest?

c Which of the frequencies in part a has the shortest wavelength?

Evaluating

19 Ocean waves are continually crashing upon the shore. Why are beaches are not flooded with this continuous supply of incoming water? Why doesn't the middle of the ocean run dry?

Creating

20 Create a poster that explains the difference between longitudinal and transverse waves. Include diagrams of the different types of waves and real-life examples of where they occur and how they are used.

MODULE 5.1 Practical investigations

1 • Water waves  

Purpose

To observe transverse water waves.

Timing 30 minutes

Materials

- ripple tanks or large rectangular plastic tub
- eyedropper
- ruler
- block of wood
- cork
- clock, watch or stopwatch function on smartphone
- camera with video capabilities or smartphone (optional)

SAFETY

Be careful not to splash water near electrical power points.

Procedure

1 Half fill the ripple tank or plastic tub with water.

2 Use the eyedropper to place a droplet of water into the middle of the tub.

3 Carefully watch the waves produced, particularly their shape. For example, are they straight lines, circles or some other shape?

4 Observe the motion of the wave as it spreads out. For example, do the waves 'die out' or do they change shape?

5 Dip the edge of ruler into the water at one end of the tub as shown in Figure 5.1.18 and tap it up and down at regular intervals.

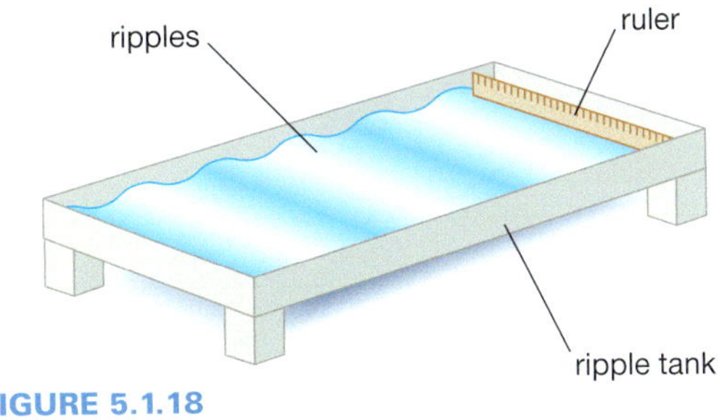

FIGURE 5.1.18

6 As before, carefully watch the waves produced.

7 Count the number of waves passing a point on the tank in 10 seconds.

8 Increase the frequency of waves produced. Observe the effect this has on wavelength.

9 Place a cork in the centre of the tub and repeat step 5. Observe the motion of the cork in the water.

10 Place a block of wood as a barrier in the path of the wave. Observe what happens as waves reach the barrier.

Results

Record your observations by photographing, videoing or drawing sketches of the different types of waves observed.

Review

1 Are water waves transverse or longitudinal? Use the motion of the cork in step 9 to justify your answer.

2 Compare the shape of the waves produced by the eyedropper with the waves produced by the ruler.

3 Calculate the frequency of the waves produced in step 6.

4 Figure 5.1.17 shows the equipment used in this experiment drawn in three dimensions (3D). Construct a scientific diagram that shows it in two dimensions (2D).

MODULE

5.1 Practical investigations

• STUDENT DESIGN •

2 • Wave motion

Questioning & Predicting | Evaluating

Purpose

To use materials to investigate wave motion in rope, string, water or a spring.

Timing 60 minutes

Materials

- Could include a slinky spring, rope, spring attached to a stand and clamp, stringed musical instrument such as a guitar, computer sound recording software (e.g. Audacity)

SAFETY

A risk assessment is required for this investigation.

Procedure

1 Design an experiment that will answer one of the following questions or another related question.
   - How does changing the frequency of a wave affect wavelength?
   - How does changing the tension of a string or spring affect the frequency of a sound wave?
   - How does the changing the thickness or density of a string or spring affect the frequency of a wave travelling through it?
   - How does the frequency of a pendulum (a swinging weight) vary with the size of the hanging mass?

2 Brainstorm in your group and come up with several different ways to investigate the problem. Select the best procedure and write it in your workbook. Draw a diagram of the equipment you need.

3 Write a hypothesis for your investigation.

4 Before you start any practical work, assess all risks associated with your procedure. Construct a risk assessment that outlines these risks and any precautions you need to take to minimise them. Show your teacher your procedure and risk assessment. If they approve, then collect all the required materials and start work.

   See Activity Book Toolkit to assist with developing a risk assessment.

Use the STEM and SDI template in your eBook to help you plan and carry out your investigation.

Results

1 Record your observations in written format, or photograph your wave motion.

2 Describe what you observed when changing variables in your investigation.

Review

1 Construct a conclusion for your investigation.

2 Assess whether your hypothesis was supported or not.

3 Identify any sources of error in your investigation.

4 Evaluate your procedure. Pick two other prac groups and evaluate their procedures too, identifying their strengths and weaknesses.

MODULE

5.2 The visible spectrum

The visible spectrum is the rainbow of colours that combine to form white light. Visible light is just a small band of the frequencies that make up the electromagnetic spectrum. This is the band of electromagnetic radiation that our eyes can detect.

science 4 fun

Polarisation

YES

What happens when you put polarising filters together?

Collect this ...

- two pairs of polarised sunglasses or polarised 3D movie glasses

Do this ...

1. Wear one of the pairs of glasses.
2. Hold the other pair of glasses in front of you so that you can see through a window or to a bright object.
3. Rotate the pair of glasses that you are holding through 90 degrees, i.e. as if the person wearing them was lying on their side.
4. Rotate the pair of glasses that you are holding through another 90 degrees, i.e. as if the person wearing them was standing on their head.
5. If possible, repeat the experiment using combinations of sunglasses and 3D movie glasses.

Record this ...

1. Describe what happened.
2. Explain why you think this happened.

Colour

In 1666, the English scientist Isaac Newton (1642–1727) passed a narrow beam of light through a glass prism. As the light exited the prism, Newton could see the colours of the rainbow, as shown in Figure 5.2.1. Newton realised that white light consists of all of the colours of the **visible spectrum**. He listed the colours making up this spectrum as red, orange, yellow, green, blue, indigo and violet. When all the colours shine at once, they produce white light.

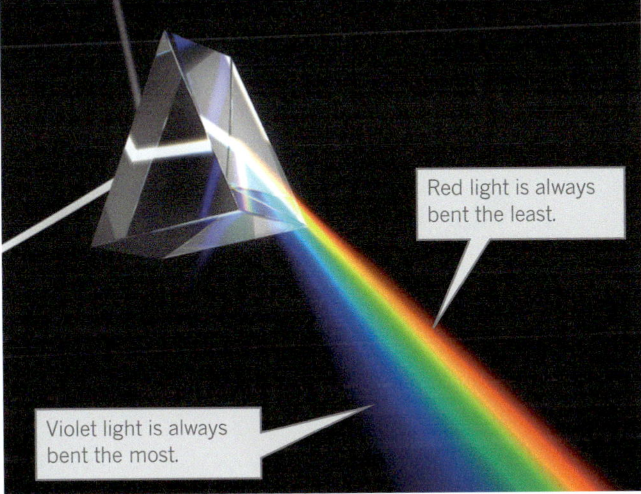

FIGURE 5.2.1 When white light passes through a prism, each individual frequency of light is refracted (bent) by a slightly different amount.

Each colour of light is a wave with a different wavelength and frequency. These are shown in Figure 5.2.2. The wavelengths of visible light are extremely small, ranging from violet light with wavelengths around 400 nm (nanometres), through to red light with wavelengths around 700 nm. To get an idea of how small this is, consider that 1 nm (nanometre) = 0.000 000 009 m = 1.0×10^{-9} m. This means that the wavelengths of visible light are less than one-thousandth of a millimetre long, or about one-hundredth the width of a human hair.

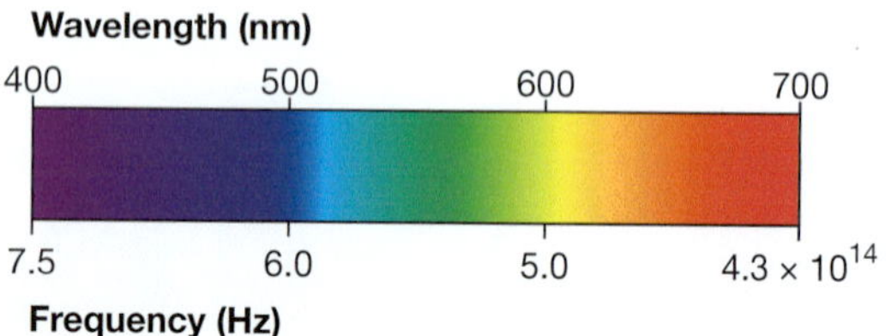

FIGURE 5.2.2 Visible light is a very small portion of the complete electromagnetic spectrum. It is the only part of the spectrum that is visible to our eyes

Seeing in colour

Some apples are red, while others are green. This is because pigments on the surface of the apples determine their colour. Under white light, the apple in Figure 5.2.3 looks red because it reflects red light towards your eyes and absorbs orange, yellow, green, blue, indigo and violet light. In reality, the red apple may reflect a little orange light as well, but this just affects the shade of red that you see. In the same way, a blue shirt reflects blue light (and probably a little green and violet) and absorbs all other colours of light. A white car reflects most of the light and radiant heat that hits it. In comparison, a black car reflects very little light or radiant heat. Most of this radiant heat and light is absorbed. As a result, a black car heats up more rapidly than a white car on a fine day.

FIGURE 5.2.3 A red apple reflects red light and absorbs the other six colours of the visible spectrum.

Objects that are viewed under different coloured lights may look quite different from when they are viewed under white light. For example, compare the four candles in Figure 5.2.4 viewed under white light and then red light.

FIGURE 5.2.4 Coloured candles look very different under different-coloured lights.

Primary colours

White light can be produced by shining all colours together. White light can also be made by using just three colours of the spectrum—red, green and blue. For this reason, these are called the **primary colours** of the spectrum. When you combine light of the primary colours in pairs, the three **secondary colours**—magenta, cyan and yellow—are produced. These combinations are shown in Figure 5.2.5.

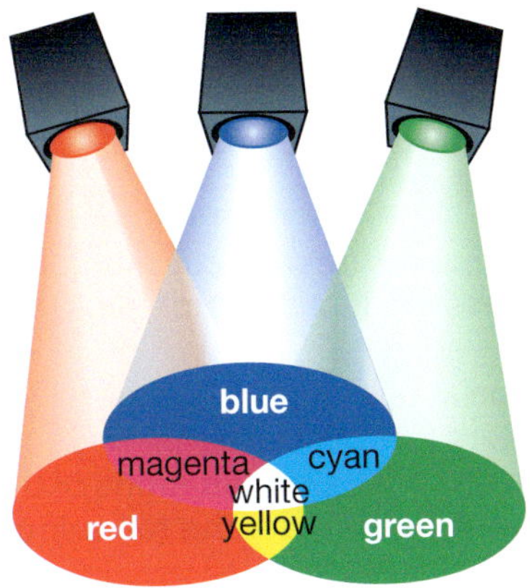

FIGURE 5.2.5
Red + blue light = magenta light
Red + green light = yellow light
Blue + green light = cyan light
Red + blue + green light = white light

Colour filters

A red apple absorbs all colours of the visible spectrum except red light. Similarly, a red piece of cellophane absorbs all colours except red light, which passes straight through. The cellophane acts as a **colour filter**.

A colour filter only allows light of its particular colour to pass through (to be transmitted). Figure 5.2.6 shows the way some combinations of light are transmitted or absorbed by a filter. Coloured filters are used widely in photography and the theatre to provide a range of lighting effects.

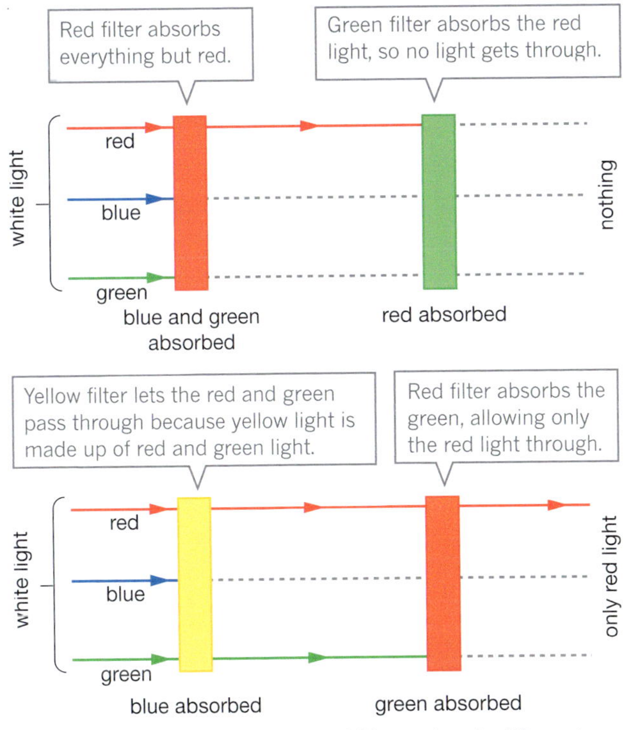

FIGURE 5.2.6 Different coloured filters absorb different colours, and so they affect what you see.

SciFile

Colour-blindness

Your eyes have three types of cells that can detect colour. These cells called cones. Each type of cone cell is sensitive to one of the three primary colours—red, blue or green. Combinations of signals from these cells give a full-colour view of the world. About 4% of people are born with colour-blindness because their cone cells do not work properly.

FIGURE 5.2.7 A person with normal vision will see a particular number in this test. What number can you see?

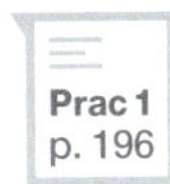
Prac 1 p. 196

Prac 2 p. 197

Colour printing

When all the colours of light are added together, white light is produced. However, if you mixed every colour of paint pigment, then the final mixture would look dark and murky. As more paint pigments are added, more colours are absorbed rather than reflected. This type of colour combination is called subtractive colour mixing.

The three subtractive primary colours are cyan, magenta and yellow. Figure 5.2.8 shows how these three colours can produce all other colours.

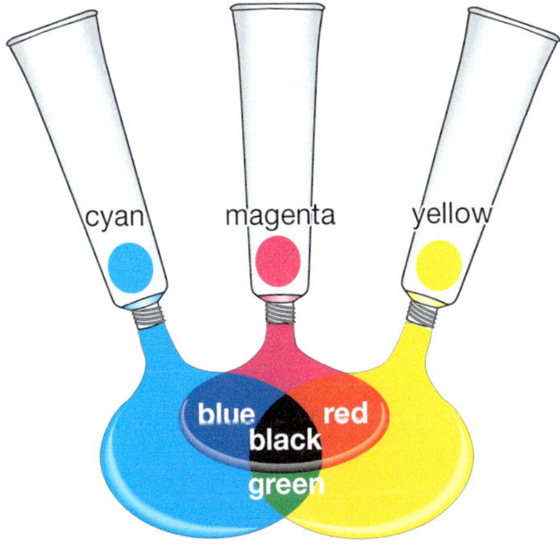

FIGURE 5.2.8 Combinations of the three subtractive colours, cyan, magenta and yellow, can produce every colour of the spectrum.

STEM 4 fun

Coloured apples

PROBLEM

Help a colour-blind person choose the a red, yellow or green apple.

SUPPLIES

- coloured cards, coloured cellophane (or coloured filters from a ray box kit), possibly polaroid sheets
- internet research

PLAN AND DESIGN Design the solution, what information do you need to solve the problem? Draw a diagram. Make a list of materials you will need and steps you will take.

CREATE Follow your plan. Create your solution to the problem.

IMPROVE What works? What doesn't? How do you know it solves the problem? What could work better? Modify your design to make it better. Test it out.

REFLECTION

1 What area of STEM did you work in today?
2 What field of science did you work in? Are there other fields where this activity applies?
3 What did you do today that worked well? What didn't work well?

SciFile

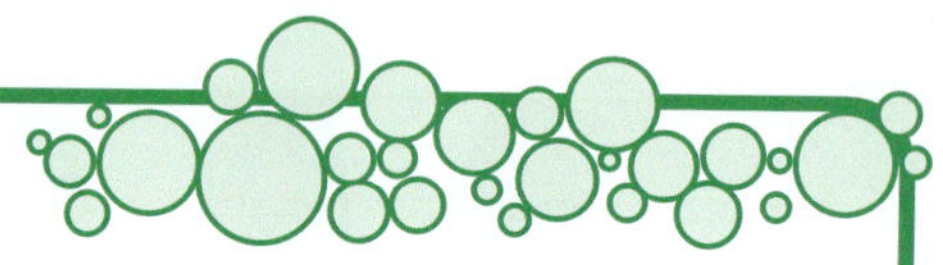

Full colour?

Televisions, video cameras, computers and mobile phones are just some of the devices that use an RGB (red, green, blue) colour model. Their displays consist of many tiny pixels of red, green and blue filters (for LCD screens) or phosphors (for plasma screens). Combinations of the red, green and blue light create the full colour display that you see.

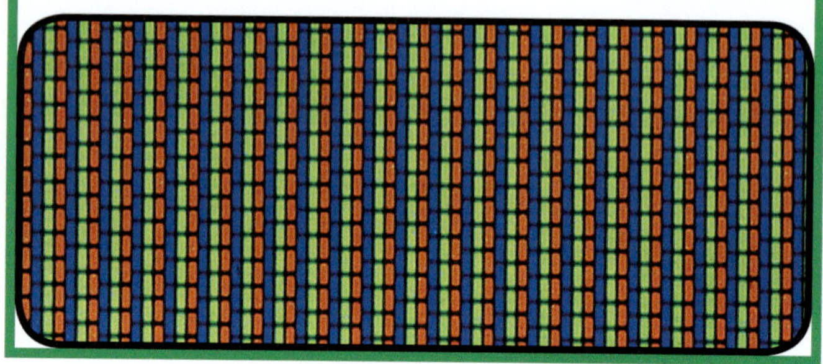

Figure 5.2.9 illustrates the way colour printing operates. Note that in addition to the three subtractive primary colours, black ink is also used in the printing process to increase the contrast of the printed image.

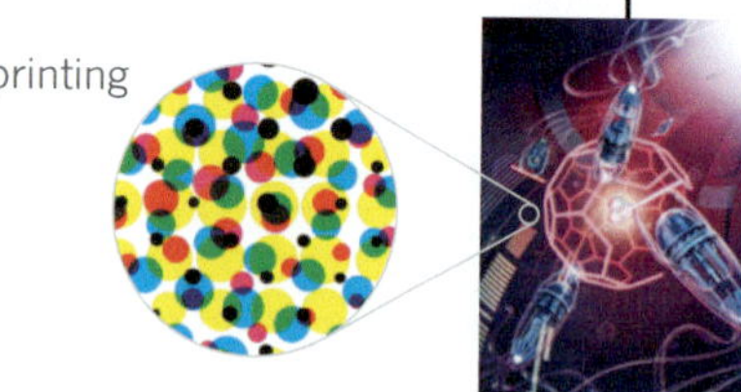

FIGURE 5.2.9 Colour printers produce a full spectrum of printed colour by using only four inks: cyan, yellow, magenta and black.

Polarisation of light

Light is a transverse electromagnetic wave. Unlike a water wave which can only move up and down, a light wave can vibrate in any direction that is perpendicular (at right angles) to the direction of the wave. This means that the light has been **polarised**.

Polarising filters can be used to separate light waves according to their direction of vibration. For example, the vertical polariser shown in Figure 5.2.10 absorbs most of the incident (unpolarised) beam and only transmits light waves that vibrate vertically. The light that passes through the filter is said to be "vertically polarised".

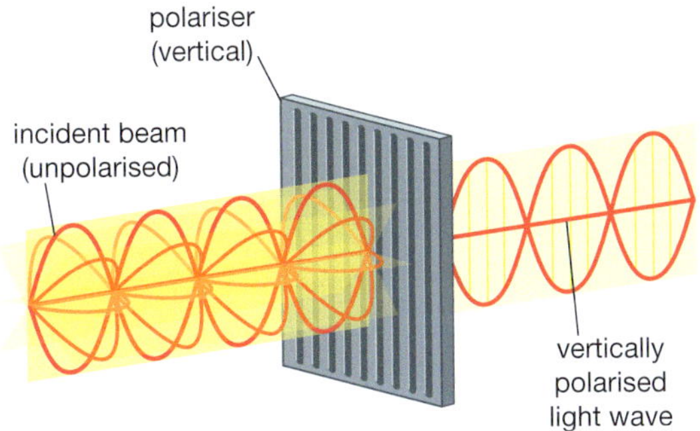

FIGURE 5.2.10 Electromagnetic radiation vibrates at right angles to the direction of travel of the wave. A polariser only allows one plane of vibration to pass through it. The rest of the wave energy is absorbed.

Polarising sunglasses contain polarising filters which absorb much of the incoming light energy, but allow enough light through for you to still see clearly.

You can test your sunglasses to see if they are polarised by holding another pair of polarised sunglasses in front of them. Rotate your sunglasses. If they are polarised, no light will pass through when the two pairs are perpendicular.

Polarisation can be used to create 3D effects in movies. How it is done is shown in Figure 5.2.11.

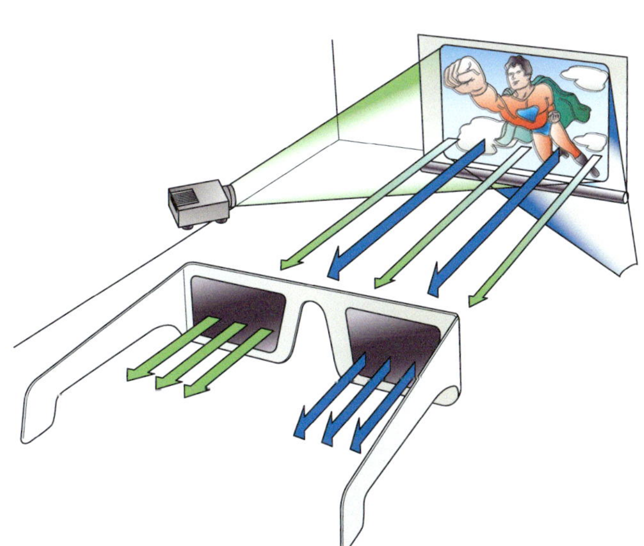

FIGURE 5.2.11 When you view a 3D movie, polarised light is used to show each of your eyes a different set of images. In this example, your left eye would see the horizontally polarised images (shown as green rays) and your right eye would see the vertically polarised images (shown as blue rays).

Dispersion of light

When white light passes through a prism, it splits into the colours of the visible spectrum. This is called **dispersion**. You can see dispersion whenever you notice a rainbow of colours. You will see these colours when you see a rainbow through a spray of water, or look at light reflecting from a soap bubble. You can see this in Figure 5.2.12.

FIGURE 5.2.12 Light reflecting of a soap bubble.

AB 5.3

To see a rainbow, individual droplets of water must be present in the sky and the Sun must be behind you. The water droplets act like tiny prisms. Sunlight enters the water droplets and is dispersed (split) into individual colours. This light is then reflected back out of the water droplet. You only see one colour from each droplet of water. The actual rainbow that you observe is unique to where you are standing because it consists of light that has reflected from many individual water droplets at many different heights in the sky. Figure 5.2.13 shows how dispersion can occur in a raindrop.

FIGURE 5.2.13 Sunlight is dispersed into different colours and is totally internally reflected back out of the water droplet. This produces a rainbow.

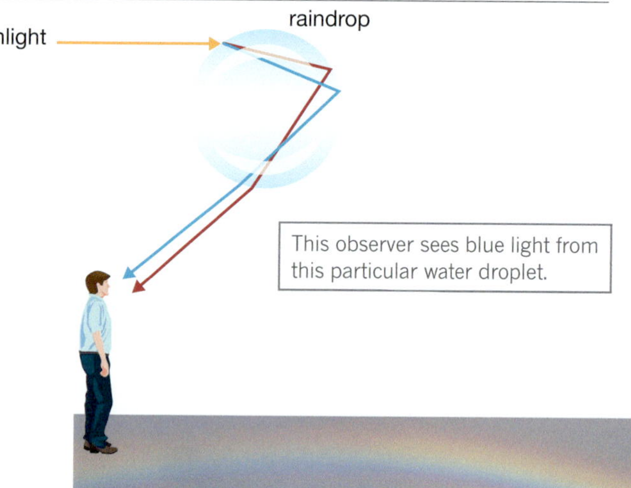

Scattering of light

As sunlight enters Earth's atmosphere, it interacts with air particles and undergoes a process known as **scattering**. If this did not occur, then the sky would appear just as black during the day as it does at night. The amount of scattering depends on the colour and frequency of light being scattered. Light waves of higher frequency are scattered more easily than waves of lower frequency. This means that light from the blue end of the visible spectrum is scattered more than light from the red end of the visible spectrum. As a result, the sky appears blue in the daytime. However, at sunrise and sunset, the Sun is low in the sky which means that light travels through a thicker layer of the atmosphere than in the middle of the day. This situation is shown in Figure 5.2.14. At sunrise and sunset, the blue wavelengths have already scattered and so the sky appears red.

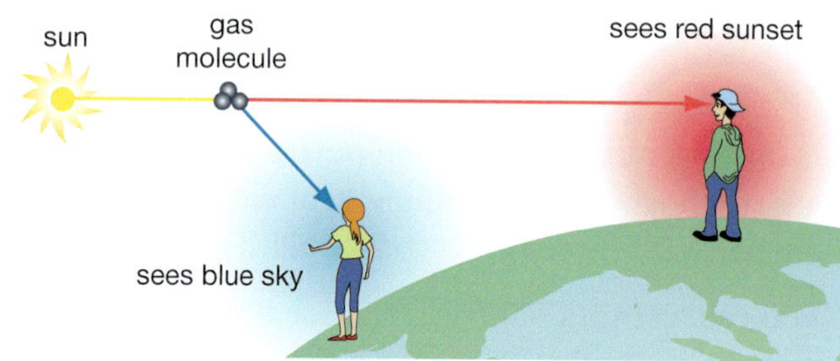

FIGURE 5.2.14 During the day, blue light is scattered and the girl sees a blue sky. At sunrise or sunset, the boy sees a reddish sky because the blue light has been scattered away.

MODULE 5.2 Review questions

Remembering

1 Define the terms:

- **a** visible spectrum
- **b** dispersion.

2 What term best describes each of the following?

- **a** the splitting of light into different colours
- **b** light with waves that are oscillating in a single plane.

3 List the three primary colours of the visible spectrum.

4 List the three secondary colours of the visible spectrum.

Understanding

5 Why does a tree frog viewed in white light look green? Explain your answer in terms of colours of light being reflected and absorbed.

6 Quentin looks at a black bowling ball through a red filter. He expected the ball to look red because this filter allows the transmission of red light. Explain why the ball appears black.

7 Explain why you feel cooler when wearing white clothing in a very warm climate.

8 Explain how polarising sunglasses reduce glare.

9 How can a printer produce full colour images using only a tricolour (three-colour) cartridge and a black ink cartridge?

Applying

10 For each of the cases shown below, identify the final colour that emerges.

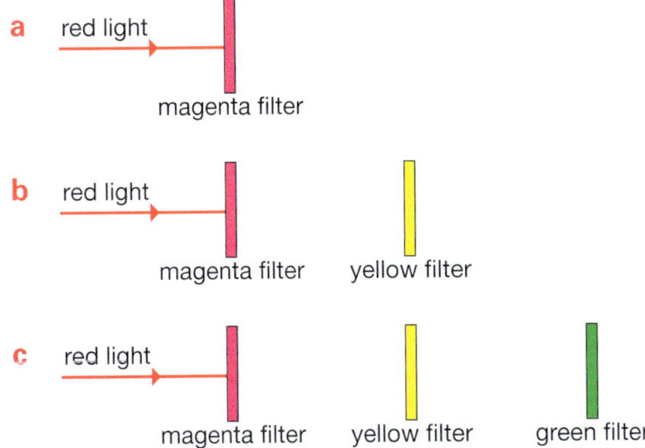

11 Which colours will be reflected and absorbed when white light shines on the objects below?

Key colours reflected and absorbed

Object	Colours reflected	Colours absorbed
red convertible		
yellow banana		
blue jeans		
black bowling ball		
white dove		

12 **a** What colour would a green frog appear under yellow light?

b Identify one colour of light that would make the green frog appear black.

13 Su-Lin and Sofia are dressed as shown in Figure 5.2.15, as they arrive at a night club. What would Su-Lin and Sofia's clothes look like in the nightclub's blue lighting?

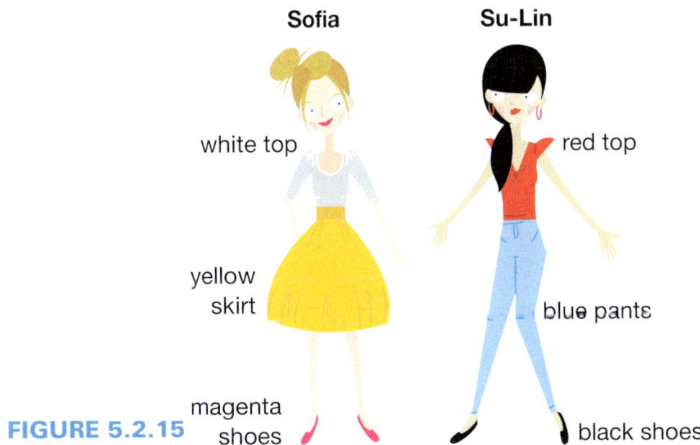

FIGURE 5.2.15

14 Consider the STEM4fun on page 192. Apples come in red, yellow and green. In what other instances would a colour-blind person have difficulty distinguishing the colours?

Evaluating

15 Discuss two properties of light which provide evidence that light travels as a wave.

Creating

16 Design a simple test that you could do to find out whether or not two pairs of sunglasses have polarising filters or not. Consider what outcomes your test would produce if:

- **a** both pairs of sunglasses have polarising filters
- **b** only one pair of sunglasses has polarising filters
- **c** neither pair of sunglasses has polarising filters.

MODULE 5.2

Practical investigations

1 • Combining colour

Purpose

To investigate combinations of coloured light and to explore the behaviour of coloured filters.

Timing 45 minutes

Materials

- light box
- power supply
- set of coloured filters
- set of coloured cards
- sheet of white paper

Procedure

1. Copy tables A and B from the Results section into your workbook. In Table A, predict the colour produced by each mixture of colours.
2. Connect the light box to a power supply and place it on a sheet of white paper.
3. Darken the room as much as possible. Place a red filter and a blue filter in the light box and adjust the mirrored flaps to combine the colours, as shown in Figure 5.2.16.

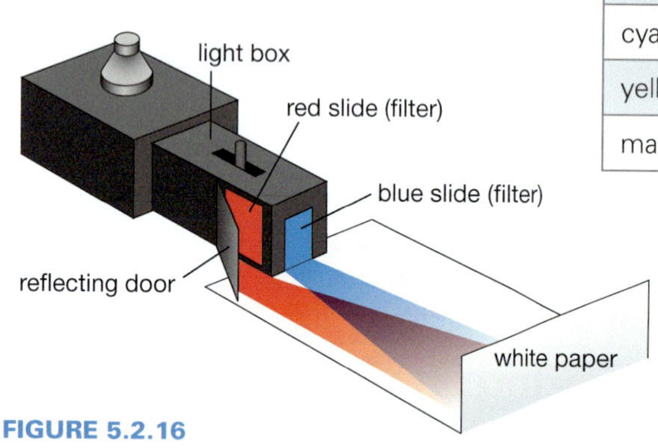

FIGURE 5.2.16

4. Change the filters as necessary to combine the light to complete Table A.
5. Now use one coloured filter at a time and shine light of this colour onto red, blue, green, cyan, yellow and magenta pieces of card. In Table B, record what each card looks like when viewed in each colour of light.

Results

Record all observations in your results tables.

TABLE A Mixing coloured light

First slide	Second slide	Third slide	Predicted colour	Colour produced
red	blue			
red	green			
green	blue			
yellow	cyan			
yellow	magenta			
cyan	magenta			
red	blue	green		
cyan	yellow	magenta		

TABLE B Viewing cards in different coloured light

Colour of slide	Colour of card					
	Red	Blue	Green	Cyan	Yellow	Magenta
red						
blue						
green						
cyan						
yellow						
magenta						

Review

1. List any combinations of colours that produced white light.
2. Discuss whether your results for Table A were as you predicted. Explain any differences.
3. Explain the results you obtained for Table B.
4. Outline ways in which this prac could be improved or extended.
5. Figure 5.2.16 shows the equipment used in this prac drawn in three dimensions (3D). Construct a scientific diagram that shows it in two dimensions (2D).

MODULE 5.2 Practical investigations

• STUDENT DESIGN •

2 • Making a kaleidoscope

Evaluating

A kaleidoscope consists of mirrors in a tube with some brightly coloured objects. A kaleidoscope can be rotated to reveal changing, symmetrical patterns.

Purpose

To construct and test a kaleidoscope.

Timing 90 minutes

Materials

A risk assessment is required for this investigation.

You could use the following or similar materials:

- 3 equal-sized rectangular mirrors (or smooth aluminium foil glued to cardboard)
- sticky tape
- 2 pieces of stiff cardboard
- Petri dish
- cardboard circle (same diameter as Petri dish)
- small coloured objects (such as glitter, sequins, confetti, cut cellophane, buttons and glass beads)
- strong rubber bands
- thin plastic bag or a piece of coloured cellophane

Procedure

1. Design a kaleidoscope that will produce different patterns when the end piece is rotated. Figure 5.2.17 may give you some ideas for your design.
2. Write your procedure in your workbook.
3. Before you start any practical work, assess all risks associated with your procedure. Construct a risk assessment that outlines these risks and any precautions you need to take to minimise them. Show your teacher your procedure and your assessment of its risks. If they approve, then collect all the required materials and start work.

   See Activity Book Toolkit to assist with developing a risk assessment.
4. Once your basic kaleidoscope is built, investigate how the final image is affected by:
   - increasing the number of mirrors used or their orientation
   - using different colours of cellophane over the viewing hole.

Use the STEM and SDI template in your eBook to help you plan and carry out your investigation.

Results

1. Describe the patterns that you saw through the kaleidoscope.
2. Describe the effect of changing the design of the kaleidoscope.

Review

1. Explain how the kaleidoscope creates these images.
2. What improvements do you think you would make to your design of the kaleidoscope if you built another one?

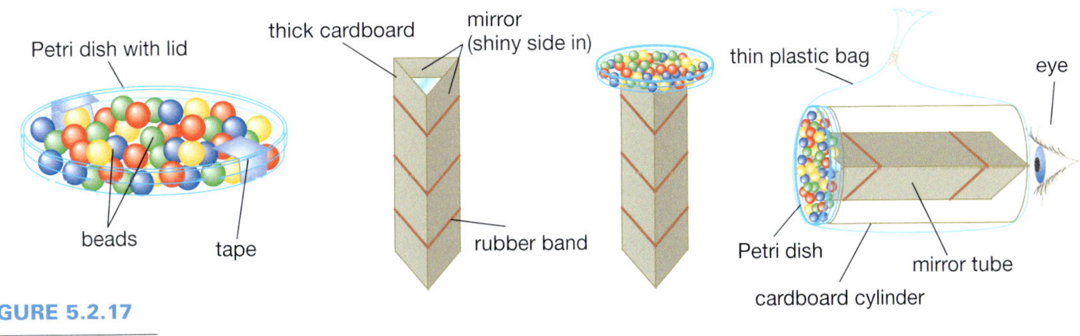

FIGURE 5.2.17

MODULE

5.3 Low-frequency radiation

The electromagnetic spectrum consists of a broad range of types of electromagnetic radiation. Electromagnetic radiation that has a frequency less than that of visible light includes radio waves and microwaves, which form the basis of our modern communication network. Microwave radiation is also used to cook and heat up food, while infrared radiation is the warmth you feel when in front of a fire and when the sun is on your back.

Radio waves

Television and radio networks transmit a signal using **radio waves**. These radio waves are produced by electrons moving backwards and forwards in a transmitting aerial (Figure 5.3.1). Radio waves have the longest wavelengths of all types of electromagnetic radiation. Radio wavelengths range from about a centimetre to many kilometres. Radio waves can travel large distances through the air or empty space. They make electrons in the antenna of your television or radio vibrate, and this can be converted into the sounds or images you see and hear when tuning in.

FIGURE 5.3.1 Radio and television stations broadcast radio waves that are produced by electrons oscillating (moving back and forth).

science 4 fun

Remote signals

YES

Remote controls emit infrared rays, which are detected by sensors on devices such as TV or DVD players.

Collect this …

- TV, CD, DVD or HDD remote control
- its associated player

Do this …

1. Point your remote control directly at your device to turn it on.
2. Vary the direction at which you point the remote control. Which unusual directions will still be effective at turning on the device?

Record this …

1. Describe what happened.
2. Explain why you think this happened.

Different wavelengths of radio waves have different properties and are used for different purposes. These are shown in Figure 5.3.2. The main types of electromagnetic waves used in communication are:

- microwaves (wavelength around 3 cm)
- short radio waves (wavelength around 30 cm)
- long radio waves (wavelengths in the range 3–3000 m).

Microwaves, are useful for 'line-of-sight' communications where the transmitter can point directly at the receiver. Since microwaves can pass straight through Earth's atmosphere, they can also be used for transmissions between Earth and communication satellites.

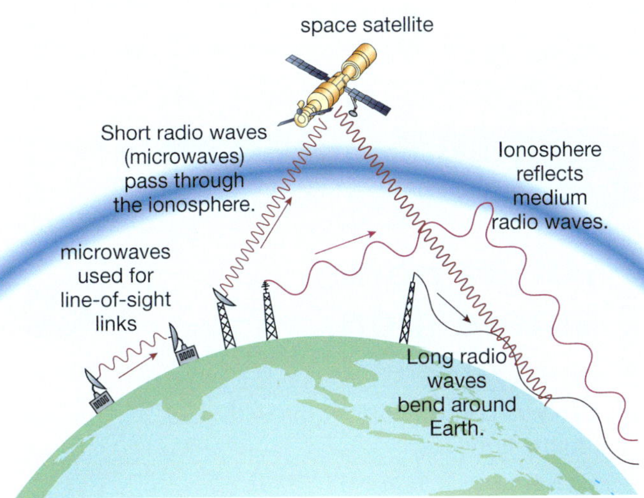

FIGURE 5.3.2 The wavelengths of radio waves can vary from kilometres to tens of centimetres. Long and short radio waves are useful for communication.

In comparison, short-wave radio signals can be transmitted long distances around the globe. This is because they are reflected by the ionosphere, which is a layer of the atmosphere 75–100 km above Earth's surface. If short-wave radio signals are beamed upwards at an angle, they can bounce back and forth between Earth and ionosphere to be detected far away from where the transmitter is located.

Similarly, long radio waves are used for communications because they bend around the Earth's surface when transmitted.

Radio waves are also produced naturally. Objects in space, such as stars, emit radio waves. Because microwaves can pass straight through Earth's atmosphere, they can also be used for transmissions between Earth and communication satellites.

AM and FM radio

Each radio station broadcasts its signals using a carrier wave which has a frequency that is specific to that station. Typically AM signals are transmitted at lower frequencies than FM signals. AM and FM signals also use different methods to encode the music being transmitted.

AM stands for Amplitude Modulation which means the audio signal is used to change (or modulate) the amplitude of the carrier wave. In an FM (Frequency Modulation) system, the audio signal is used to change the frequency of the carrier wave. Figure 5.3.3 compares waves transmitted as AM and as FM.

When you tune your radio receiver to a particular station, you set it to detect the frequency of the carrier wave. Complex circuitry inside your receiver subtracts the carrier wave from the modulated radio waves, leaving only the original audio signal. The full process of radio transmission is summarised in Figure 5.3.4 on page 200.

An FM signal has a wavelength of around 3 metres, whereas an AM signal has wavelengths longer than 100 metres. The longer AM radio waves can bend around large obstacles such as buildings, trees and hills more easily than the smaller FM waves can. This bending around obstacles is called **diffraction**. AM signals travel further than FM signals, but they are of lower quality and are more likely to suffer from interference. You may have noticed this when listening to an AM radio near electrical equipment.

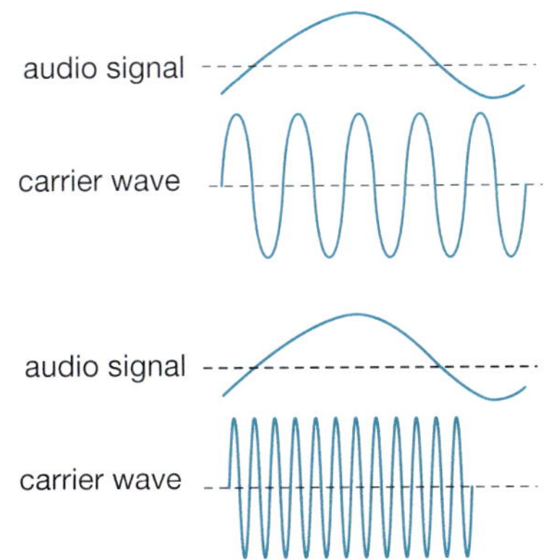

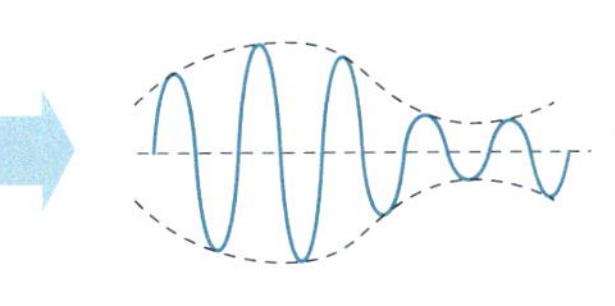

AM wave
The frequency is the same as the carrier wave but the amplitude varies.

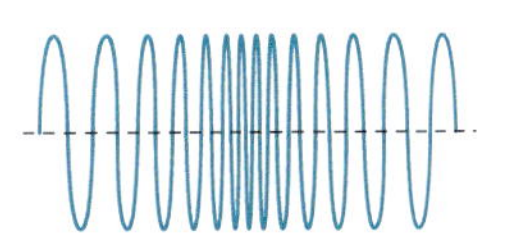

FM wave
The amplitude is the same as the carrier wave but the frequency varies.

FIGURE 5.3.3 AM radio stations transmit waves that are amplitude modulated. In FM, the carrier wave is frequency modulated. Each radio user, such as CB radio, the police or a radio station, operates on its own specific frequency.

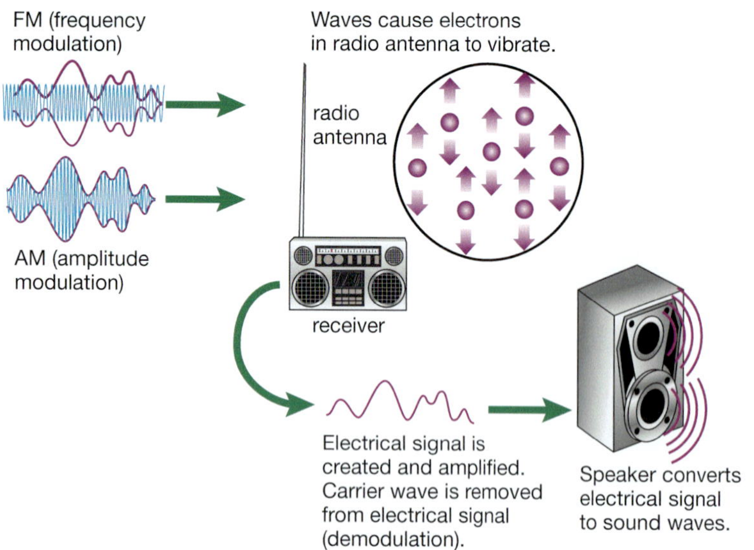

FIGURE 5.3.4 Radio signals detected by a receiver are converted into an electrical signal and then into sound.

FIGURE 5.3.5 The colours in this infrared image indicates temperature – some food is hot (white), some food is cold (black) and the surrounding air is at room temperature (green).

SciFile

Digital radio

Digital radio transmits multiple VHF (very high frequency) signals, filtering these for interference and then recombining the signals. Digitalradio delivers a cleaner, higher-quality sound that is free from crackles.

Microwaves

Microwaves have shorter wavelengths than radio waves and are used in radar and communication systems. Microwaves with wavelengths of about 0.1 mm are used in cooking (Figure 5.3.5). Microwaves are absorbed by water, fats and sugars in food, causing these molecules to vibrate and heat up. Because the heating occurs inside the food without warming the surrounding air, the food cooks quickly but sometimes unevenly. Glass, paper and many plastics don't absorb microwaves, and metal reflects microwaves.

STEM 4 fun

Future home automation

PROBLEM
Can you control your home with your smartphone?

SUPPLIES

- pen and paper

PLAN AND DESIGN Design the solution: what information do you need to solve the problem? Draw a diagram. Make a list of materials you will need and steps you will take.

CREATE Follow your plan. Create your solution to the problem.

IMPROVE What works? What doesn't? How do you know it solves the problem? What could work better? Modify your design to make it better. Test it out.

REFLECTION

1. What area of STEM did you work in today?
2. What field of science did you work in? Are there other fields where this activity applies?
3. What did you do today that worked well? What didn't work well?

Infrared radiation

Heat is transferred from the Sun to Earth as **infrared radiation**. Infrared rays have lower frequencies than red light rays. The prefix 'infra' means 'below'.

All objects emit infrared radiation; the hotter something is, the more infrared radiation it emits. You cannot see this radiation, but can detect its presence as warmth on your skin or by using an infrared camera, as shown in Figure 5.3.5 and Figure 5.3.6.

FIGURE 5.3.6 In this false colour image, different intensities of infrared radiation are assigned different colours. White corresponds to the hottest regions, while the coolest regions appear turquoise.

AB 5.5

Modern communications networks

Modern communications networks carry vast amounts of data from landline telephones, mobile phones, radio, TV and the internet. Data such as data files, voice conversations and satellite images are relayed.

Analogue or digital signals

Some mediums, such as the copper wires connected to landline phones, were originally designed to carry an **analogue signal**. In an analogue signal, the information, such as changes in a speaker's voice, is transmitted as changes in the voltage of the signal. You can see a typical analogue signal in Figure 5.3.7. Analogue signals are limited because they suffer from signal loss and interference as they travel.

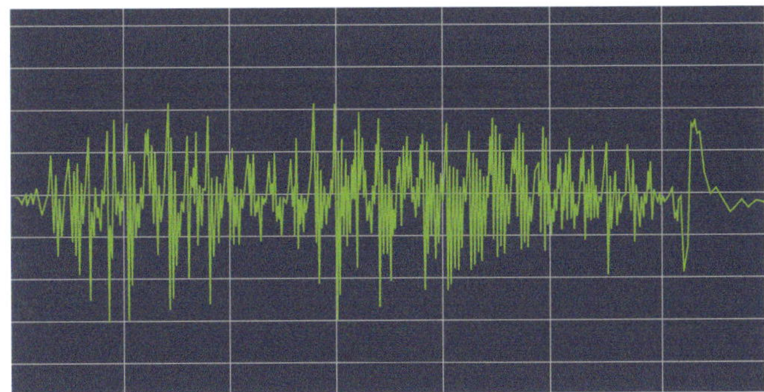

FIGURE 5.3.7 An analogue signal replicates the message it transmits.

More modern communication systems, such as cable television and the internet, rely on a **digital signal**. In a digital system, the information is coded into a series of 'on' or 'off' pulses using a **binary number system**. The advantage of a digital signal is that it can still be read even when there is significant interference. This means that digital signals are clearer and more accurate. Our mobile phone network is digital, as is the music we listen to on CD or smartphones. You can see a typical digital signal in Figure 5.3.8.

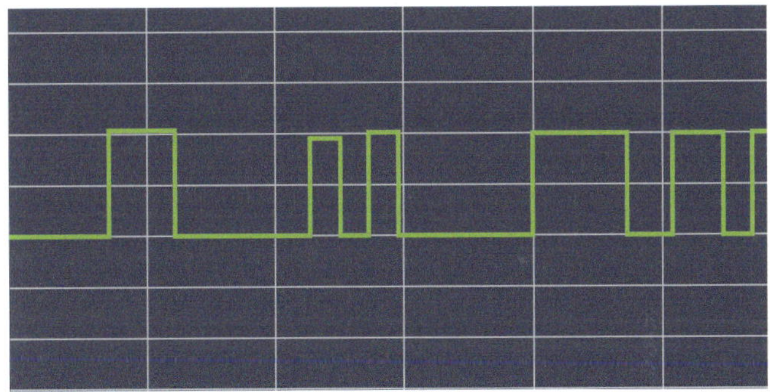

FIGURE 5.3.8 A digital signal

Increasing capacity

Until relatively recently, all telephone signals were sent along copper wires. To limit signal interference, twisted-pair copper wires were used. These consist of a pair of insulated copper wires twisted around each other. A 600-pair copper cable can carry 600 two-way conversations. These cables are inexpensive and do not suffer much signal loss when used over short distances.

SciFile

Binary versus decimal

We are all familiar with the decimal number system that uses the ten digits from 0 to 9 (*deci* = ten). In comparison, the binary number system (*bi-* = two) has only two digits: 0 and 1.

Figure 5.3.9 shows the structure of a coaxial cable. Coaxial cables can carry more data than twisted-pair cables. Australian cities have been linked by underground coaxial cables since the 1960s. Two cables of 50 tubes can carry 2700 two-way conversations. These cables need repeaters positioned every 45 km to reinforce the signal.

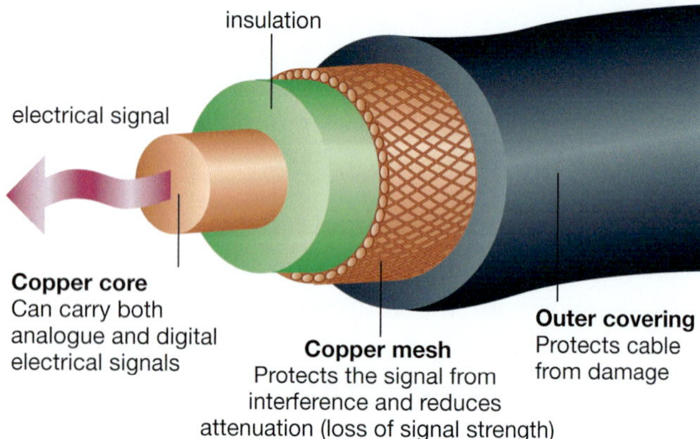

FIGURE 5.3.9 The inner core of this coaxial cable transmits analogue and digital signals. The outer layers protect the signal from interference and loss of strength.

To transmit more data along a copper wire, different signals are sent along the wire at the same time. In analogue systems, this is possible when using carrier waves of different frequencies. At the receiving end, these different frequencies are sorted back into separate signals. This is called frequency division multiplexing (FDM).

Channels of communication

As Australians make more and more use of digital technology, these is an increasing demand for **bandwidth**—that is, the ability to send information more quickly through a single communication channel.

Optical fibre

In Australia from the late 1980s, a new network was laid next to the underground coaxial cable that linked cities. This network was made from **optical fibre**. Optical fibres are thin, flexible tubes made of glass or plastic. A typical optical fibre consists of a central core of pure silica glass surrounded by a layer of less dense glass, called the cladding. This layer is coated with a plastic jacket to minimise interference. The structure of fibre optic cable is shown in Figure 5.3.10. Laser light is used to transmit signals down a fibre optic cable.

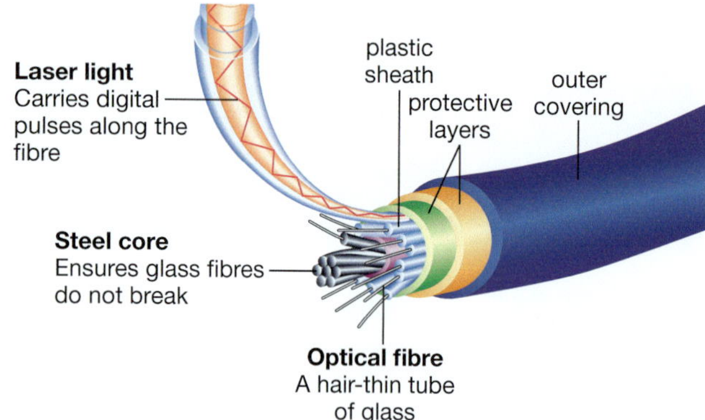

FIGURE 5.3.10 A fibre optic cable is made up of many single optical fibres.

A laser produces a narrow beam of light waves that are said to be **coherent**. Coherent waves are in step with each other—the peaks of all the waves occur at the same place and the troughs of all the waves also occur together. The difference between normal light and laser light is illustrated in Figure 5.3.11.

In an optical fibre transmission system, the signal is coded as a series of binary numbers. This signal is sent down the optical fibre as a series of millions of flashes of laser light. The optical fibre carries the signal very efficiently because the light rays are totally reflected internally inside the cable.

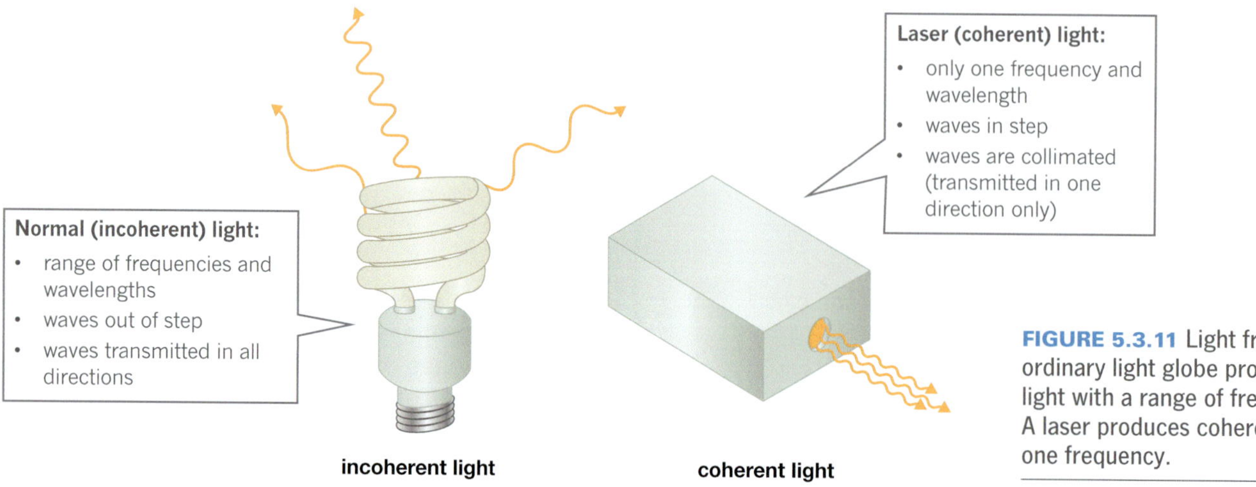

FIGURE 5.3.11 Light from an ordinary light globe produces light with a range of frequencies. A laser produces coherent light of one frequency.

SciFile

Slippery cables

Fibre optic cables manufactured for use in Australia have a unique outer nylon jacket. This jacket protects the cabling from termites, which are unable to grip its slippery outer surface.

An optical receiver at the other end of the fibre converts these pulses back into a digital signal. Much higher frequency signals can be sent along a fibre optic system than can be sent through copper wires. As a result, a fibre optic system has much greater bandwidth than a copper system. A single optical fibre can carry over 30 000 telephone calls. This bandwidth can be expanded by sending signals of different frequency along the fibre at the same time. In addition, optical fibre is lighter and more flexible than copper cable. A signal can be transmitted around 200 km without any signal loss, and it suffers less interference than signals sent along a copper network.

Prac 1
p. 207

Microwave links

Microwave signals are directed in short, straight-line paths. About 2000 phone conversations can be carried using a microwave system, with repeater towers required every 50 km. Microwave satellite links are used for long-range mobile communications and for communications in remote areas.

SciFile

Cell phones

The region around each mobile phone base station is called a cell. Each cell uses a different frequency to transmit its signals. This is why mobile phones are sometimes called 'cell' phones. If you move from one cell to another while you are talking on the phone, your phone should switch from one base station to another. Sometimes this does not work properly and the call will drop out.

Mobile phone networks

When you dial a number on your mobile phone, the phone transmits a digital signal as microwaves through the air. In your local area there will be a number of base stations each with an antenna designed to receive this signal. The base station that receives the strongest signal from your phone sends your call to the mobile phone exchange. The call is then sent by copper wire or optic fibres through a series of exchanges until it gets to the base station near the person you have called. The call is then converted back into a microwave signal and transmitted to the phone of the person you have called. This arrangement is shown in Figure 5.3.12.

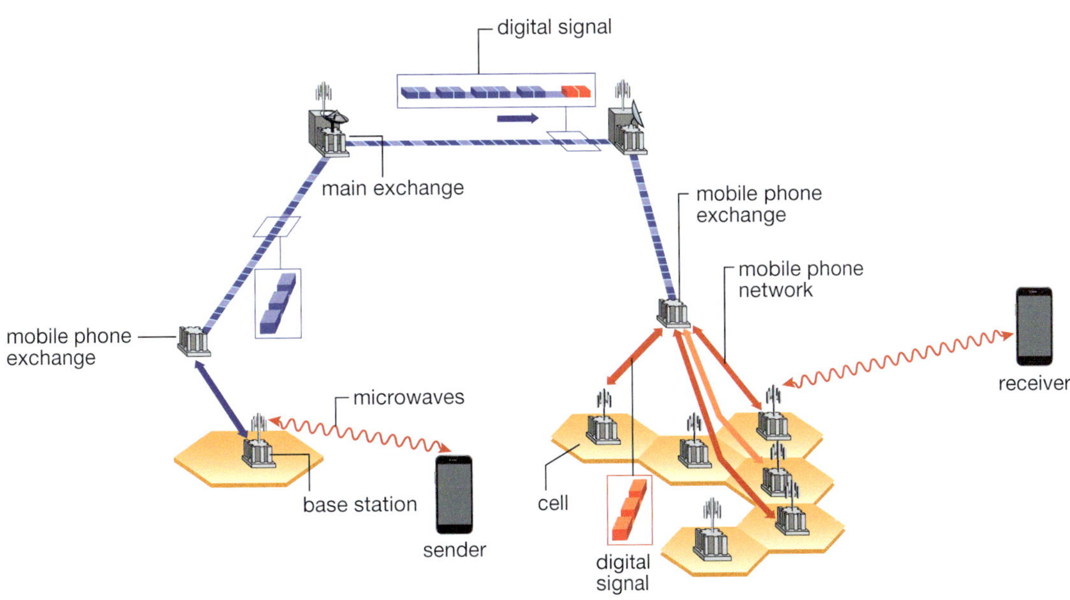

FIGURE 5.3.12 A mobile phone call is transferred by microwaves from the phone of the sender to the base station, then through the wires or optic fibres of the telephone exchange system, and then again by microwaves to the phone of the receiver.

The internet

Using the internet, you can connect with people around the world in an instant. Documents that used to take days to reach a destination can be downloaded in seconds. A *router* is a device that manages the connection between your computer and your internet server. It is this device that is responsible for making sure your message reaches where it is meant to go.

Data to be sent as a downloaded file or email is first split into a 'packet' made up of about 1500 bytes (each byte is a group of eight binary digits). These packets then travel over a 'packet-switching network' in which each individual packet is directed along the best pathway for it to reach its destination.

Scifile

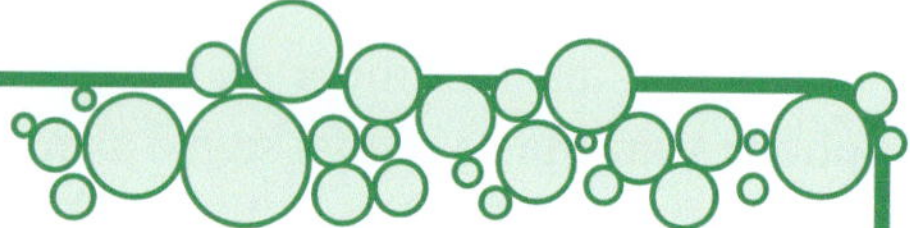

A nibble, a bit or a byte?

A 'byte' is a series of eight binary digits. Each digit (which can either be a 0 or a 1) is called a 'bit'. Sometimes, a series of four binary digits is called a 'nibble' (half a byte).

An 8-digit decimal number can have 100 milllion different values (i.e. between 0 and 99 999 999) because 10^8 = 100 000 000. Similarly, each digital byte can be decoded into a number between 0 and 255 since 2^8 = 256.

Wireless internet networks

Wireless internet is a method of transmitting an internet connection using radio waves. It allows a wi-fi (wireless fidelity) enabled device such as a mobile phone, laptop, tablet, video game console, Bluetooth or MP3 player to connect to the internet when within range of an access point. An example is shown in Figure 5.3.13. Wi-fi devices connect with each other in a similar way to mobile phones. Wi-fi is a wireless alternative for internet access within local area networks (LANs). A wi-fi signal does not have a long range, only about 30 metres indoors and 100 metres outdoors.

FIGURE 5.3.13 This WLAN (wireless local area network) card enables the laptop to link to a wireless broadband internet network. Most modern laptops have a built-in wireless card.

In regions further away from a wireless network that is connected to the internet, the wi-fi device cannot pick up a signal. In such cases, wi-fi is not an alternative to an internet system that operates using coaxial cable or optical fibres.

Working with Science

BROADCAST ENGINEER

Natalie Biady

FIGURE 5.3.14 Natalie Biady works for the ABC as a broadcast engineer.

Broadcast engineers are the people working behind the scenes, making sure that you can receive clear radio and television signals. Natalie Biady is a broadcast engineer who works for the ABC (Figure 5.3.14). Her job involves looking after the television and radio networks across Australia, studio feeds and remote broadcasts. She does a lot of technical work and troubleshooting to ensure that transmissions are maintained or quickly restored when there are network failures. Usually system changes are done in the middle of the night, when there are fewer people watching, so sometimes Natalie has to work during unusual hours.

Natalie completed a Bachelor of Engineering, majoring in telecommunications. There are also Certificate and Diploma courses in media technology and telecommunications that can lead to graduate programs at television and radio stations. With the shift from analogue to digital technology, there is a growing need for broadcast engineers. If you are interested in digital technology and like solving problems, then a job in broadcast technology could be for you. Natalie finds working with telecommunications technology exciting and challenging as there is always something new to learn.

Review

1 Radio and television broadcasting are an important part of our telecommunication system. What other forms of telecommunication are important in our society?

2 List the forms of telecommunication do you use every day.

SCIENCE AS A HUMAN ENDEAVOUR

Use and influence of science

Wi-fi helps CSIRO scientist win top award

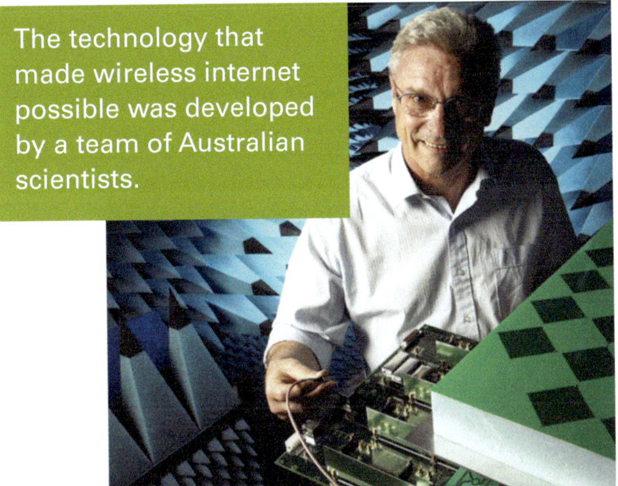

The technology that made wireless internet possible was developed by a team of Australian scientists.

FIGURE 5.3.15 Dr John O'Sullivan

In 2009, almost two decades after pioneering high-speed wireless now used by almost a billion people each day, John O'Sullivan (Figure 5.3.15) won one of Australia's top science awards.

The CSIRO scientist was awarded the prestigious Prime Minister's Prize for Science for 2009 for his WiFi technology now found in millions of laptops, printers and wireless access devices.

One of Australia's most significant scientific breakthroughs, Dr O'Sullivan and his team found a way to speed up wireless networks in 1992—a problem that had previously confused international scientists.

The idea has since generated a large amount of money for the CSIRO: around $205 million.

Dr O'Sullivan was given $300,000 at a gala event in Canberra on Wednesday.

Former Prime Minister Kevin Rudd said the award recognised Dr O'Sullivan's major contribution to astronomy as well as his groundbreaking wi-fi technology.

'While looking for exploding black holes Dr O'Sullivan created a technology that cleaned up intergalactic radio waves,' Mr Rudd said.

'Then in 1992, he and his colleagues at CSIRO realised that the same technology was the key to fast, reliable, wireless networking in the office and home.

'Their patented invention is now built into international standards and into computers, printers, smart phones and other devices used by hundreds of millions of people every day.'

Mr Rudd called it one of the most significant achievements in CSIRO's 83-year history and said it illustrated how scientific research can be used for everyday practical solutions.

Dr O'Sullivan and the CSIRO team beat 22 international labs to solve the 'multipath' problem—or the interference caused by reflected radio waves which slows down network speeds.

They found a way to accelerate them by splitting radio channels apart, essentially turning a one-lane road into a super highway and making wireless about five times faster.

Dr O'Sullivan said the idea was born out of a need to 'cut the wires'.

'And to cut the wires, we needed to make it as fast as the wires,' he told AAP.

Seeing the technology in use in millions of devices around the globe, 'I can't help but feel proud,' he said.

'Even though we thought it had huge potential, I'm just blown away with how many applications there are now.'

The Age 28 October 2009 © 2010 AAP

REVIEW

1. Name the prize that Dr John O'Sullivan was awarded in 2009.
2. Describe what the prize was awarded for.
3. List devices that use this technology.
4. Describe what Dr John O'Sullivan was looking for when he created the new technology.
5. What the world would be like without wi-fi? Make a list of things that you would have to do differently each day if you did not have access to wireless technology.

MODULE

5.3 Review questions

Remembering

1 Define the terms:

a coherent

b digital signal.

2 What term best describes each of the following?

a continuous signal that varies in amplitude or frequency with the information being transmitted

b electromagnetic radiation with frequencies slightly lower than visible light, detected by our skin as heat.

3 Name a natural source of radio waves.

4 What is the typical wavelength of an AM radio signal?

5 What is the number system that is used in digital communications?

6 What are two disadvantages associated with analogue signals compared with digital signals?

7 What is the term that describes the amount of data a communication channel can carry?

8 How far apart are repeater towers positioned in microwave surface links?

9 List four common devices that may be wi-fi enabled.

10 State the indoor and outdoor range of a wi-fi signal.

Understanding

11 Explain why AM radio waves travel further than FM radio waves.

12 Define the the term frequency-division multiplexing.

13 Explain how a signal is transmitted along an optical fibre.

14 Explain the function of a router.

Applying

15 A TV remote control transmits its signal using infrared waves. Sometimes the TV's infrared receiver detects the signal even when the remote control is not pointed directly at it. What property of infrared waves makes this possible?

Analysing

16 Compare normal light with laser light.

17 a Compare frequency modulation (FM) with amplitude modulation (AM).

b Identify the advantages and disadvantages of each technique.

18 Draw a diagram to compare AM with FM radio waves.

19 Compare the number of two-way conversations that can be carried by a typical twisted pair cable, coaxial cable and fibre optic cable.

Evaluating

20 Why do you think food sometimes cook quickly but unevenly in a microwave oven?

21 Why do you think both coaxial cable and optic fibres wrapped in an outer plastic jacket?

Creating

22 Construct a design for a new application of wireless internet technology; for example, the projection of a 3D replica of your body shape into a changing room to allow virtual fitting of clothing while internet shopping.

MODULE 5.3

Practical investigations

1 • Using light 

Purpose

To model how light is totally internally reflected along an optical fibre.

Timing 45 minutes

Materials

- length of optical fibre
- light box
- power pack
- 3 glass prisms

SAFETY

Ensure that the light box has cooled before handling it.

Procedure

1. Connect the light box to the power source and direct a single beam of light at a glass prism.
2. Rotate the prism. Look at what happens to the incident ray as you do this. (The incident ray is the one coming into the prism from the light box.)
3. Find a position in which all the light is reflected from the inside glass surface. This is called total internal reflection.
4. Place a second prism in the path of the reflected beam and rotate it so that the beam is totally internally reflected again by the second prism.
5. Repeat this process once more, using a third prism. You should be able to make the ray of light travel in the opposite direction to its original pathway using these three prisms. Use Figure 5.3.16 as a guide.
6. Sketch the orientation of your three prisms and the reflected rays of light.
7. Curve an optical fibre into a shape that is similar to that of the light pathway that you have just drawn. Direct light along the fibre.
8. Change the shape of the fibre to see if the light can be transmitted through it.

Review

1. Describe what happened to the ray of light as you rotated the first prism.
2. Construct a diagram to show what happened when you tried to shine light through the optical fibre.
3. Discuss what happened when you tried to direct the light through the fibre when bent in different directions.
4. Identify features of an optical fibre that make it a useful channel of communicaton using light.

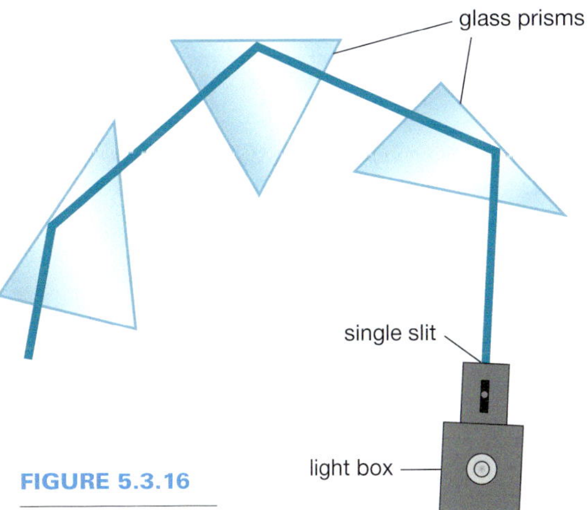

FIGURE 5.3.16

MODULE 5.4

High-frequency radiation

High frequency radiation has shorter wavelengths than visible light: from about one-ten-thousandth of a millimetre to one-hundred-billionth of a metre. Such tiny waves can penetrate cells in your body and damage your DNA. This property of high-energy radiation is used in medical diagnosis, and to kill cancer cells during radiotherapy. However, it is also why exposure to UV radiation from the Sun can cause skin cancer.

UV sunscreen

YES

What sort of radiation does sunscreen protect you from?

Collect this ...

- different brands of sunscreen or moisturiser

Do this ...

Examine the packaging of a bottle of sunscreen to find out:

- its sun protection factor (SPF), e.g. 15, 30+, 50+
- what sort of ultraviolet radiation it protects against, e.g. broad spectrum, UVA, UVB.

Repeat this for as many brands of sunscreen or moisturiser as you can find.

Record this ...

1 Describe what happened by recording your data in a table.
2 Explain which types of sunscreen you think provide the best protection.

Ultraviolet light

Ultraviolet (UV) light is radiation with a higher frequency than violet light ('ultra' means 'beyond'). Sunlight contains UV light in addition to infrared and visible light. Your body needs some exposure to UV light to produce vitamin D. Although you cannot see UV light, it can tan or burn your skin. High exposure to UV light can cause skin cancers such as melanoma. UV light can also cause cataracts in your eyes. Approved sunglasses and sunscreens can offer us some protection from these rays.

The Bureau of Meteorology issues daily UV index forecasts like the one shown in Figure 5.4.1 to help you take precautions to protect yourself against damage from UV radiation.

Prac 1 p. 214

SciFile

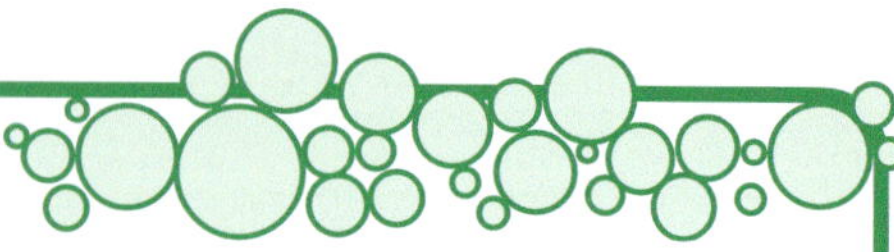

Skin cancer alert

Australia has the highest rate of skin cancer in the world. Each year, 1300 people in Australia die from the disease.

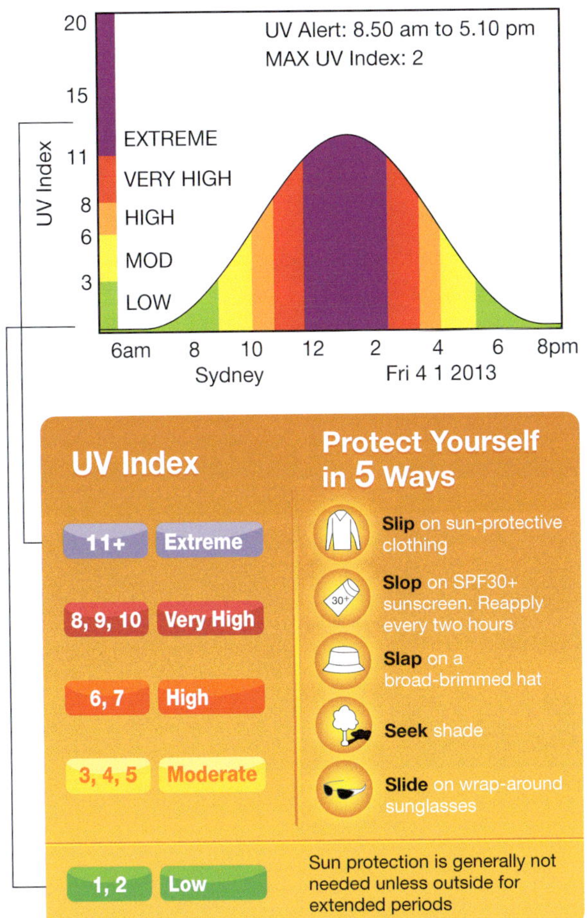

FIGURE 5.4.1 The Bureau of Meteorology issues daily Sunsmart UV alerts for each capital city in Australia.

Some objects, such as the rocks shown in Figure 5.4.2, **fluoresce** (glow) when hit by UV light. This is because they absorb the energy from the UV light and emit visible light. Manufacturers of white paper, teeth whiteners and some laundry powders add fluorescent particles to make these objects appear brighter. UV light is also used to sterilise objects.

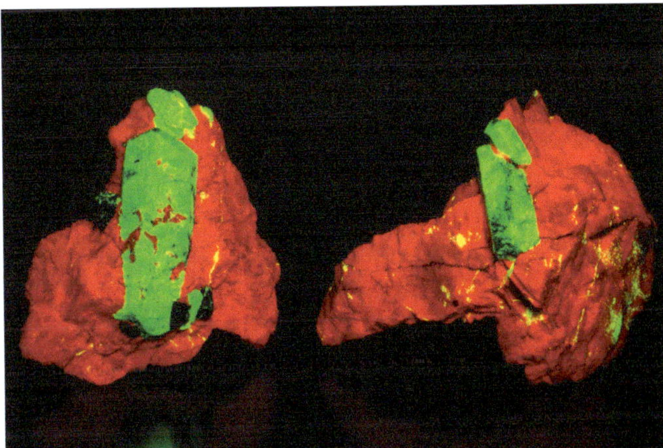

FIGURE 5.4.2 Many rocks, including calcite, gypsum, ruby, talc, opal, quartz and fluorite, are fluorescent and glow under UV light.

SciFile

Glowing notes

In 1966, fake bank notes were widely circulated in Australia. Since then, our bank notes have become extremely difficult to reproduce illegally. They are now made of plastic instead of paper and they now incorporate many complex security features. For example, the serial number of a bank note and a patch below its denomination (value) are printed in fluorescent ink. This causes them to glow when placed under UV light.

X-rays

X-rays have great penetrating power and so are used to investigate the structure of objects and to find flaws in metals. This radiation has such high energy that it can damage cells and tissues, and also affect the genetic material inside cells. X-rays are produced when electrons hit a metal surface. This happens inside an X-ray tube. X-rays are used in radiology, to produce images of bones, like Figure 5.4.3. They are also used in radiotherapy, in which X-rays are targeted at cancer cells to kill them or stop them from multiplying.

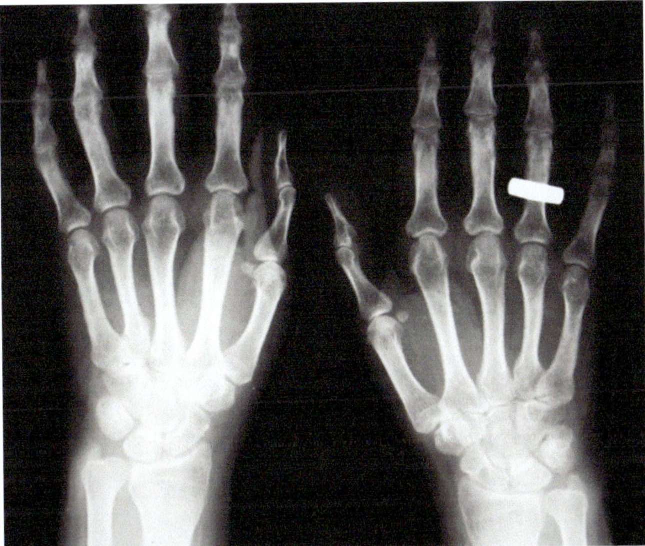

FIGURE 5.4.3 X-rays travel through human flesh, but not through bone. This makes them useful in producing images of the structures inside the body.

When a patient undergoes a computed tomography (CT) scan, the X-ray sources and detectors rotate around the person. Computers then analyse the data from the CT scanner to create images of organs in the body.

Because of the high energy of X-rays, it is important that people who work with them use protective lead shields and monitor their exposure levels. This is done using a personal radiation monitoring device, such as the one shown in Figure 5.4.4. The device is worn for up to three months. The total, or accumulated, radiation dose is then measured. The employer must ensure that this dose remains below a certain value, to protect the worker from possible harm.

FIGURE 5.4.4 A personal radiation monitoring device measures a person's exposure to X-rays,and other forms of harmful radiation.

Figure 5.4.5 shows a baggage X-ray inspection system. This is a machine used to screen luggage for dangerous items. An X-ray generator inside the machine produces X-rays, which are directed at the luggage as it passes through on a conveyor belt. X-rays pass more easily through less-dense items than through items of high density. Sensors behind the luggage detect variation in penetration of X-rays through the sample. Differences in density can be displayed as different colours so that any items of concern can be identified. An example is shown in Figure 5.4.6.

AB 5.6 AB 5.7

FIGURE 5.4.5 X-ray baggage inspection systems are widely used to screen luggage at airports.

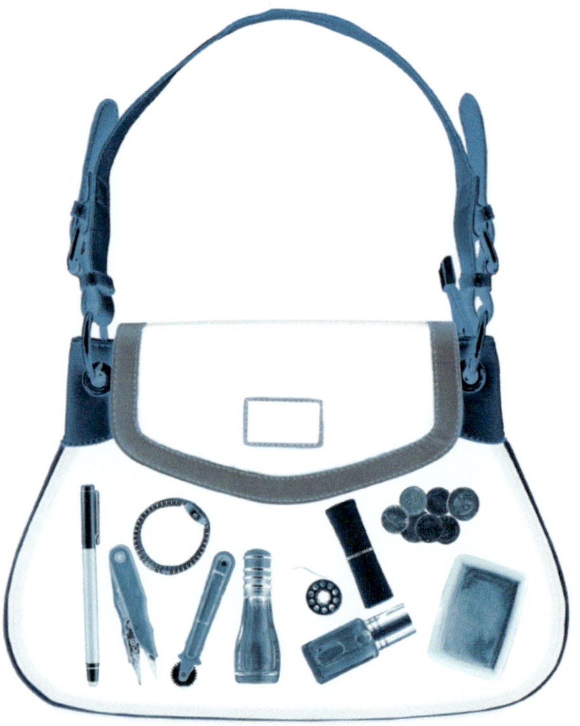

FIGURE 5.4.6 Variations in the density of luggage items mean that X-rays penetrate samples to different extents. Such variations are used to build an image.

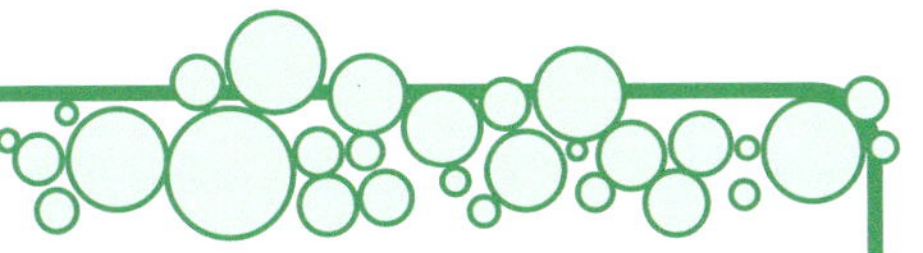

SciFile

Sharp pain!

In January 2004, Patrick Lawler of Denver, USA, visited a dentist, complaining of tooth pain and blurred vision. The dentist found the problem: a 10 cm nail that the construction worker had unknowingly fired through the roof of his mouth 6 days earlier! The nail was safely removed.

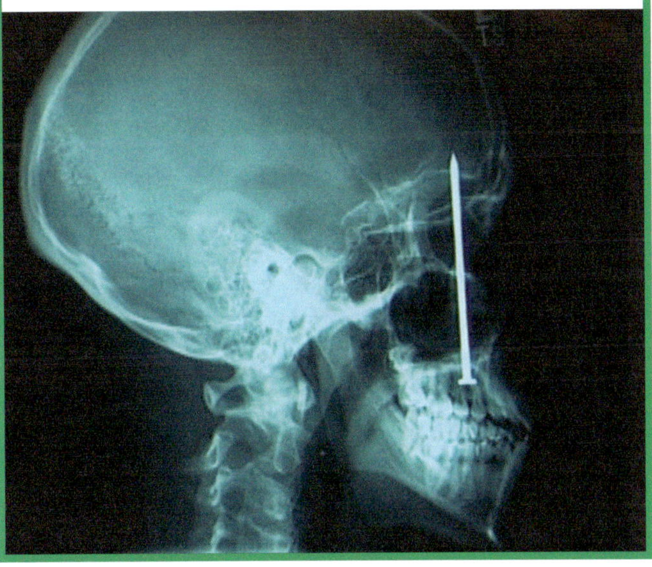

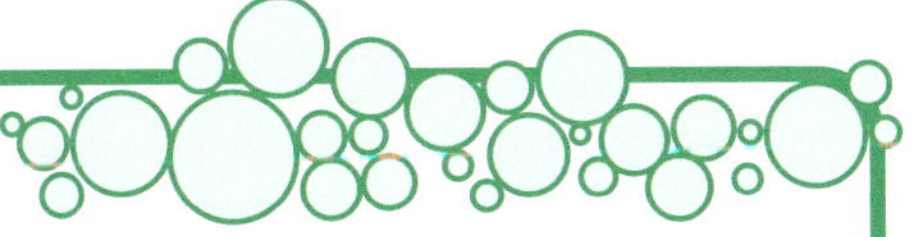

SciFile

Does gamma radiation make you turn into the Incredible Hulk?

The Incredible Hulk is a famous superhero character who has been depicted in numerous comic books and movies. The character first appeared in 1962. In this first story, scientist Dr Bruce Banner becomes the Hulk when he receives a massive dose of gamma radiation. In reality, this would almost certainly result in cancer or death rather than superpowers!

Gamma rays

Gamma rays have wavelengths of about one-hundred-billionth of a metre. Only a thick sheet of lead or a concrete wall will stop them. Gamma rays are produced by radioactive materials, nuclear power plants and nuclear bombs, and can be detected with photographic film or a machine called a Geiger counter. Gamma rays can turn atoms into ions by causing them to release electrons. This ionising ability of gamma rays is used in radiotherapy to target and kill cancer cells in patients.

Gamma rays are also useful in medical diagnosis. In positron emission tomography (PET), a patient is injected with small amounts of a radioactive material. This material emits gamma rays, for a short period of time. The gamma rays are detected by a PET scanner or camera. Computer analysis converts this data into a three-dimensional image. PET scans, like the one shown in Figure 5.4.7, allow doctors to study how parts of a patient are functioning.

AB 5.8

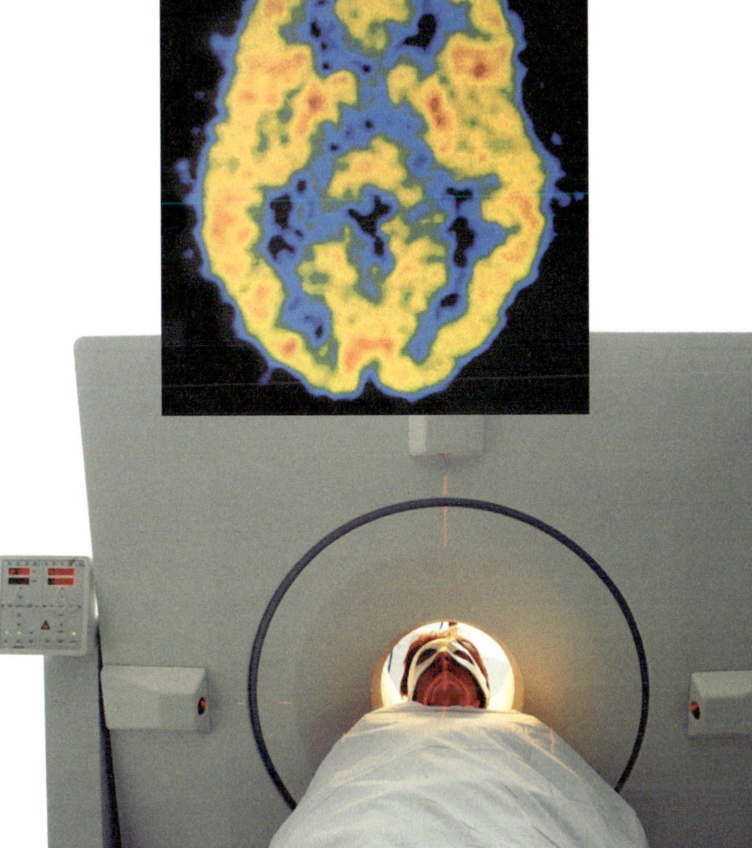

FIGURE 5.4.7 PET scans allow doctors to study how parts of a patient are functioning.

SCIENCE AS A HUMAN ENDEAVOUR

Use and influence of science

Human energy powers medical devices

Your body is constantly in motion, even when you are not aware of it; the beating of your heart, your lungs expanding and contracting, and muscles twitching.

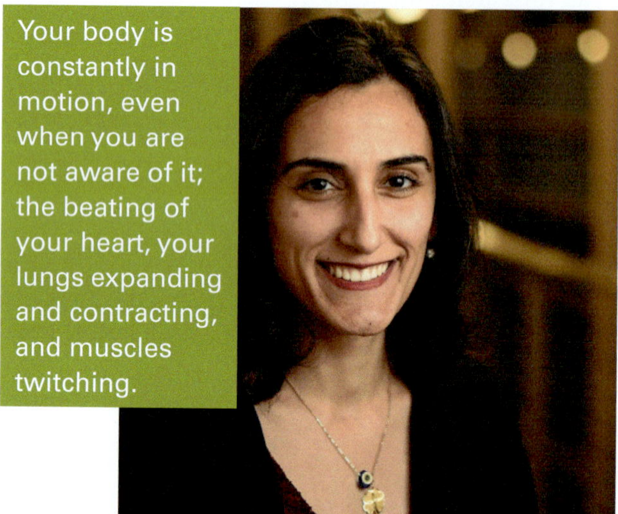

FIGURE 5.4.8 Dr Canan Dagdeviren has designed electronic devices that use piezoelectric materials to convert the movement of our bodies into electricity.

Scientist Canan Dagdeviren (Figure 5.4.8) has designed small electronic devices that can convert the motion of your body into electricity. As a young girl growing up in Turkey, Dagdeviren was inspired by Pierre Curie, his brother, Jacques, and their discovery of the piezoelectric effect: the ability of some materials to generate electricity from movement. In the Curie brothers' case, it was the movement of crystals under pressure that generated electrical sparks. For Dr Dagdeviren, it's the movement of the human body.

Dr Dagdeviren began researching piezoelectric materials, such as zinc oxide, during her PhD and has used them to develop electronic devices with a wide range of applications. Her devices are designed to be worn on the skin or inside the body and are particularly important for the field of medicine. She has developed sensors that can measure blood pressure and detect changes in skin cells that can warn of early signs of skin cancer (Figure 5.4.9). These devices also have energy harvesting capabilities which allow them to power lifesaving technology, such as cardiac pacemakers, that are currently powered by bulky batteries that need to be changed every five to ten years.

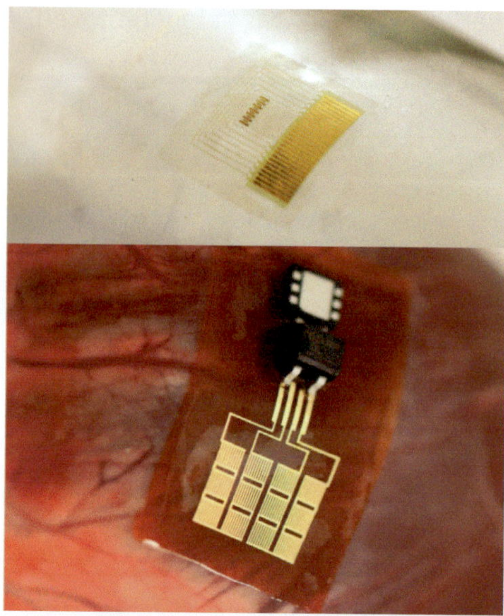

FIGURE 5.4.9 Dr Dagdeviren's skin sensor is made from stretchable materials and can be worn to detect early signs of skin cancer. She has also developed devices that can be worn inside the body and are capable of converting energy from the movement of organs into electricity.

Our current healthcare system is based on check-ups with doctors, tests that give us a snapshot of our health at that time and sometimes treatments are required for conditions that are detected late. Dr Dagdeviren's devices could revolutionise healthcare, allowing us to continually monitor changes in our body and alert us to early warning signs of disease. This information could enable an entirely different approach to medicine, with less invasive treatments and more preventative measures that are based on real-time data from our bodies.

REVIEW

1 Electromagnetic radiation is critical to many medical technologies. List three other important medical devices or procedures that rely on electromagnetic radiation.

2 List some ethical concerns that may arise from continually monitoring and collecting data on people's health.

3 Dr Dagdeviren's devices have many potential uses. List three other uses for this technology.

MODULE 5.4
Review questions

Remembering

1 Define the terms:
  a fluoresce
  b X-ray.

2 What term best describes each of the following?
  a damaging rays emitted in a nuclear explosion
  b electromagnetic radiation with frequencies slightly higher than visible light, can do damage to skin cells.

3 Which vitamin can your body only produce after some exposure to UV light?

4 List five ways you can protect yourself from harmful UV radiation.

5 Name the machine that detects gamma rays.

6 Specify what type of barrier is needed to stop the penetration of gamma radiation.

7 Why can some laundry detergent manufacturers claim that their product will leave clothes 'brilliantly white'?

Understanding

8 Explain why a patient is injected with a short-lived radioactive material before having a PET scan.

9 All baggage at airports is checked by passing it on a conveyor belt through an X-ray machine (Figure 5.4.10). How do X-ray baggage scanning machines work?

FIGURE 5.4.10 An airport X-ray machine and monitor

Applying

10 Which type of radiation has a greater wavelength: infrared or ultraviolet?

11 a What happens to the penetrating ability of electromagnetic radiation as its wavelength decreases?
  b Justify your answer.

12 The Sun emits a range of UV frequencies. UVA is the 'safest', UVC is the most dangerous while UVB is somewhere in between. Identify the form of UV radiation that has the:
  a longest wavelength
  b highest frequency.

Evaluating

13 Some electromagnetic radiation can pass through materials and some cannot. This penetrating power or radiation changes as the frequency of the radiation changes. Use the information in the table below to describe this pattern.

Penetrating power of different types of radiation

Type of radiation	Penetrating power
radio waves	some cannot penetrate Earth's atmosphere
visible light	reflects off most solid objects, except transparent materials like glass and some plastics
ultraviolet (UV) rays	can penetrates the first few layers of human skin cells
X-rays	can pass through soft tissue but not bone
gamma rays	only stopped by several metres of concrete or 40 cm of lead sheeting

14 Full-body X-ray scanning machines have been used to screen passengers before boarding an aircraft. These machines use low-energy ionising X-rays. The amount of radiation delivered to a person being scanned is equivalent to the radiation dose a passenger absorbs in a few minutes of flying. Some people argue that these machines should be banned. Propose arguments for or against the use of these machines.

15 a What is the purpose of radiotherapy?
  b Why do patients undergoing radiotherapy often suffer unwanted side effects to their treatment?

16 Some banknotes and passports include text and images that can only be seen under UV light. How do you think these security measures work?

MODULE

5.4 Practical investigations

• STUDENT DESIGN •

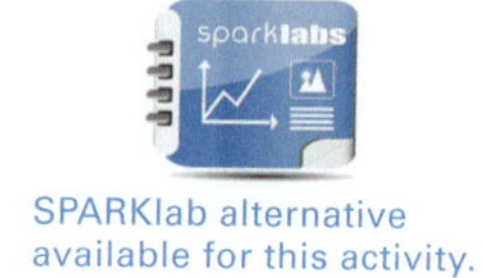

SPARKlab alternative available for this activity.

1 • UV protection

Purpose

To test how effectively different materials block UV radiation.

Timing 60 minutes

Materials

- UV colour-changing beads or a UV sensor
- a range of materials to test, such as Polaroid sunglasses, cellophane of different colours, glass, transparent plastic, shadecloth, various clothing fabrics, and other items

A risk assessment is required for this investigation.

FIGURE 5.4.11 Wearing a wide-brimmed hat, sun-protective clothing and SPF 30+ sunscreen will help to protect this child's delicate skin from harmful UV rays.

Procedure

1. Design a test that will investigate how well different samples block UV radiation.
2. Brainstorm in your group and come up with several different ways to investigate the problem. Select the best procedure and write it in your workbook. Draw a diagram of the equipment you need.
3. Before you start any practical work, assess all risks associated with your procedure. If using chemicals, then refer to their SDS. Construct a risk assessment that outlines these risks and any precautions you need to take to minimise them. Show your teacher your procedure and your risk assessment. If they approve, then collect all the required materials and start work.

   See Activity Book Toolkit to assist with developing a risk assessment.

Use the STEM and SDI template in your eBook to help you plan and carry out your investigation.

Results

Record all your results in a table.

Review

1. List the materials you tested from most effective at blocking UV to least effective.
2. Analyse any sources of error in your experiment.
3. Evaluate your procedure. Pick two other prac groups and evaluate their procedures too, identifying their strengths and weaknesses.

CHAPTER 5

Chapter review

Remembering

1 What type of electromagnetic radiation is emitted by radioactive materials?

2 What is the approximate wavelength of an FM radio wave?

3 Which is more likely to experience static: AM or FM radio waves?

4 Recall which secondary colour is produced when red light and blue light are combined.

5 A hummingbird flaps its wings 120 times in a second. State the frequency of its beating wings.

6 a Define the term *wi-fi*.
  b List three situations in which such technology is useful.

Understanding

7 Describe two uses of microwave radiation.

8 a List three products that may contain fluorescent additives.
  b Explain why these are added to these products.

9 Su-Ann sits on a beach and watches the waves roll in. She calculates that the waves are arriving every 5 seconds, with a frequency of 0.2 Hz. A short while later Su-Ann calculates that the waves are arriving every second.
How has the frequency of the waves changed?

Applying

10 Huong and Callum are relaxing on a beach. Huong is wearing red-tinted sunglasses and Callum wears yellow-tinted ones. Emily jogs past them wearing a white T-shirt, green shorts and a magenta cap. Identify the apparent colours of Emily's clothes to Huong and Callum.

Analysing

11 Compare a transverse wave and a longitudinal wave.

12 Contrast infrared and ultraviolet radiations.

13 Use Figure 5.1.10 on page 183 to classify the electromagnetic waves that have the following wavelengths:
  a 2 cm b 3 km
  c 0.0008 m d 0.000 000 3 m

14 Pigment X reflects mostly orange light with a little red and yellow, but absorbs other colours. Pigment Y reflects mostly green light, with some blue and yellow, but absorbs all other colours. You dye your favourite socks in a mixture of X and Y. What colour will your dyed socks appear?

Evaluating

15 Imagine that optical fibres had never been invented. How would the world be different?

16 a Assess whether you can or cannot answer the questions on page 178 at the start of this chapter.
  b Use this assessment to evaluate how well you understand the material presented in this chapter.

Creating

17 Tony plays a scale on his flute and ends on a note of frequency 220 Hz. The pressure variation of this sound is represented by Figure 5.5.1. Tony then plays another note that has a frequency of 440 Hz and is twice as loud as the first note. Construct a pressure variation diagram showing the differences between these two notes.

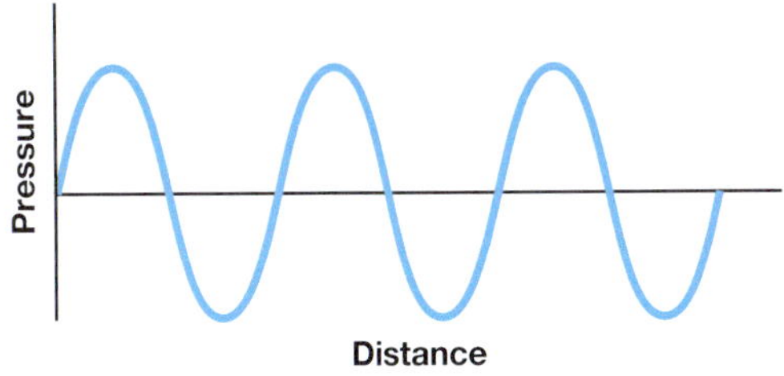

FIGURE 5.5.1

18 Use the following ten key terms to construct a visual summary of the information presented in this chapter.

electromagnetic radiation	microwaves
electromagnetic waves	visible light
electromagnetic spectrum	X-rays
gamma rays	radio waves
infrared radiation	ultraviolet light

AB 5.10

CHAPTER 5

Inquiry skills

Research

1 Processing & Analysing | Communicating

Research the frequency and wavelength of some common musical notes.

- Research to find the frequency of the 'middle C' note on a piano. Express your answer in hertz (Hz).
- Assuming that the speed of sound in air is 340 m/s, use the wave equation to calculate the wavelength of this note. Express your answer in metres.
- Similarly, research to find the frequency of the C notes one octave higher and one octave lower than middle C. Express these answer in metres.
- Compare the wavelengths of these C notes. Is there a mathematical relationship between them?
- The pentatonic scale consists of five notes. Many songs are written using the notes of this scale. Research the frequencies and calculate the wavelengths of these notes. Is there a mathematical relationship between them?

Present your research as a PowerPoint presentation, using diagrams and audio files where possible to illustrate the notes and patterns between them.

2 Processing & Analysing

You can use the wave equation to calculate the wavelength of transmission of radio waves from radio stations. Conduct research to find the frequency of transmission of the radio stations that are popular with your class.

- For AM stations, frequencies will be stated in kilohertz (kHz), where 1 kHz = 1 000 Hz.
- For FM stations, frequencies will be stated in megahertz (MHz), where 1 MHz = 1 000 000 Hz.
- Use the wave equation stated in terms of wavelength: $\lambda = \frac{v}{f}$. The speed of a radio wave is the same as the speed of light. This is 300 000 000 m/s.
- Use speed (in m/s) and frequency (converted to Hz) to calculate the wavelength (in m) of transmission for five radio stations (including at least one in each band: AM and FM).

Present your results in a table.

3 Communicating

Research and construct a timeline to outline the history of the development of one of the following technologies:

- the USB device
- iPods
- MP3 players
- social media
- email
- World Wide Web.

Present your research as a poster. Include in your report a summary of key people who contributed to the development of the technology.

4 Communicating

Heating in a microwave oven can be quite uneven. This has led to people receiving serious burns from 'superheated' liquids. Research to:

- find out what 'superheating' is
- describe how this happens in a microwave oven
- develop a list of precautions to prevent such burns from occurring.

Present your findings as a newspaper ad or plan for a TV commercial warning of the dangers and providing helpful advice.

5 Communicating

A number of techniques of medical diagnosis or treatment are possible using forms of electromagnetic radiation. These include radiology, radiotherapy, CT scans and PET scans. Select one of these techniques and research to find:

- the type of electromagnetic radiation involved
- how this radiation is utilised
- any precautions that need to be taken when using this technique
- conditions the technique is most commonly used to treat or diagnose.

Present your findings as a poster or an information brochure for a radiography or doctor's clinic.

CHAPTER 5

Inquiry skills

Thinking scientifically

1 Sheena is carrying out an experiment using three tin cans. These cans are identical except that one is painted black, one is painted white and one is painted light grey. Sheena's teacher tells her that dark colours are better absorbers of infrared radiation than lighter colours. The air temperature in each can is 20°C at the start of the experiment. The cans are placed near a heater.

Select the likely temperature of the black, silver and white cans after an hour.

A black 30°, light grey 40°, white 50°

B black 50°, light grey 20°, white 40°

C black 50°, light grey 40°, white 30°

D black 30°, light grey 50°, white 40°

2 The unit of measurement for radiation dose is the millisievert (mSv). Table 5.5.1 reveals the effective radiation dose involved in particular medical procedures and diagnostic tests.

TABLE 5.5.1 Effective radiation dose

Procedure	Effective radiation dose (mSv)
CT abdomen and pelvis	10.000
radiography spine	1.500
CT head	2.000
bone densitometry (DEXA)	0.001
mammography	0.700

CT: computed tomography

Which of the following lists the effective radiation dose absorbed by a person in order of increasing dose?

A bone densitometry (DEXA), CT head, CT abdomen and pelvis, radiography spine, mammography

B CT head, CT abdomen and pelvis, radiograph spine, mammography, bone densitometry (DEXA)

C bone densitometry (DEXA), mammography, radiography spine, CT head, CT abdomen and pelvis

D CT abdomen and pelvis, CT head, radiography spine, mammography, bone densitometry (DEXA)

3 Jimmy listens at close range to a trumpet being played. If the final note of the song has a frequency of 440 Hz, calculate the number of sound waves passing Jimmy each second.

A 110

B 440

C 220

D 880

4 Study Figure 5.5.2, which shows the energy and wavelength of various types of electromagnetic radiation.

Identify which list ranks this radiation in order from longest to shortest wavelength.

A X-rays, infrared radiation, microwaves, radio waves

B X-rays, visible light, UV light, infrared radiation

C radio waves, UV light, microwaves, infrared radiation

D radio waves, infrared radiation, UV light, X-rays

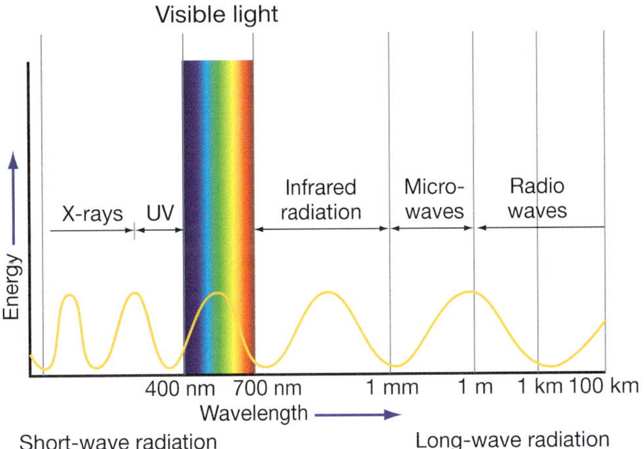

FIGURE 5.5.2 Electromagnetic spectrum

5 Infrared radiation has a band of energies just below those of visible light. Given that red light has a wavelength of approximately 4.0×10^{-7} m, predict which wavelength below would be classed as infrared radiation.

A 2.0×10^{7} m

B 2.0×10^{3} m

C 2.0×10^{-9} m

D 2.0×10^{-5} m

6 A certain type of radiation has a wavelength of 300 nm. This type of radiation is most likely to be classified as:

A infrared radiation

B visible light

C UV

D X-rays

CHAPTER 5

Glossary

amplitude: the maximum distance a wave extends beyond its middle position

analogue signal: a continuous signal that varies in amplitude or frequency with the information being transmitted

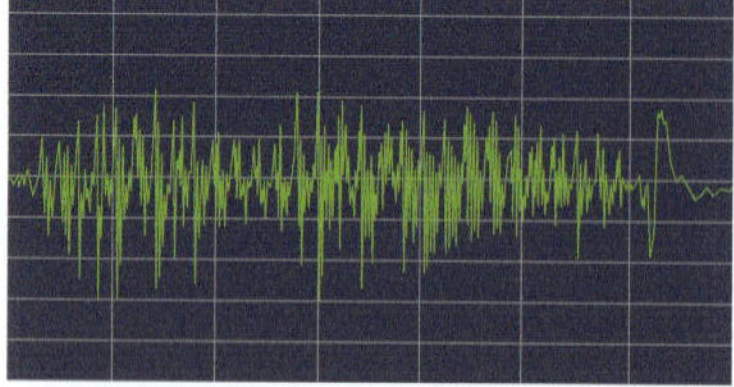

analogue signal

bandwidth: the amount of data that can be transmitted through a communication channel

binary number system: number system consisting only of two digits: 0 and 1

coherent: light waves that are 'in step'

colour filter: a transparent material that allows light of a particular colour to pass through

diffraction: bending of a wave around an obstacle

digital signal: a signal consisting of a series of 'on' or 'off' pulses

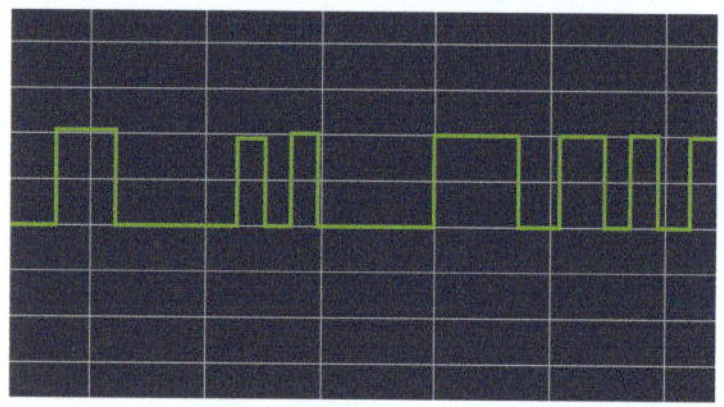

digital signal

dispersion: splitting of white light into separate colours

electric field: a region of electrical influence in which charged particles will move

electromagnetic radiation: electromagnetic waves consisting of oscillating electric and magnetic fields travelling at the speed of light

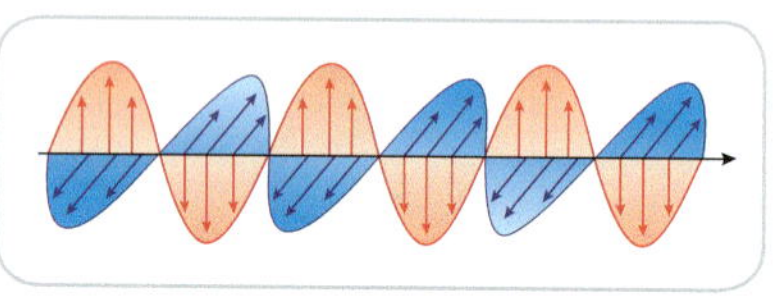

electromagnetic radiation

electromagnetic spectrum: the entire range of frequencies of electromagnetic radiation, from high-frequency gamma rays to low-frequency radio waves

electromagnetic wave: transverse electric and magnetic fields positioned at right angles to each other and travelling through empty space at the speed of light

electromagnetism: the phenomenon of electric and magnetic fields interacting with each other

fluoresce: absorb UV light and emit visible light

frequency: number of waves produced each second and is measured in hertz

gamma rays: extremely high-frequency electromagnetic radiation emitted by radioactive materials

infrared radiation: electromagnetic radiation with wavelengths slightly longer than those of visible light, detected by our skin as heat

magnetic field: a region of magnetic influence in which a magnetic object (like a compass) will move

microwaves: electromagnetic radiation with wavelengths ranging from fractions of a millimetre to tens of centimetres, used in communication and cooking

optical fibre: a narrow tube of glass or plastic used to transmit pulses of light

polarised: electromagnetic radiation that is oscillating in a single plane

primary colours of light: red, green and blue

radio waves: electromagnetic radiation with wavelengths ranging from less than a centimetre to hundreds of kilometres, used in communication

scattering: the interaction of light with particles in the atmosphere, depending on colour and frequency of the light

secondary colours of light: cyan, yellow and magenta

ultraviolet (UV) light: electromagnetic radiation with frequencies just above those of visible light, contained in sunlight

visible spectrum: the range of colours that can be seen by the eye (red, orange, yellow, green, blue, indigo and violet)

wave motion: the transfer of energy without transferring matter

wavelength: distance between two successive waves and is usually measured in metres

wireless internet (wi-fi): a method of transmitting an internet signal using radio waves

X-rays: high-frequency electromagnetic radiation that can penetrate materials

AB 5.9

CHAPTER

6 Electricity

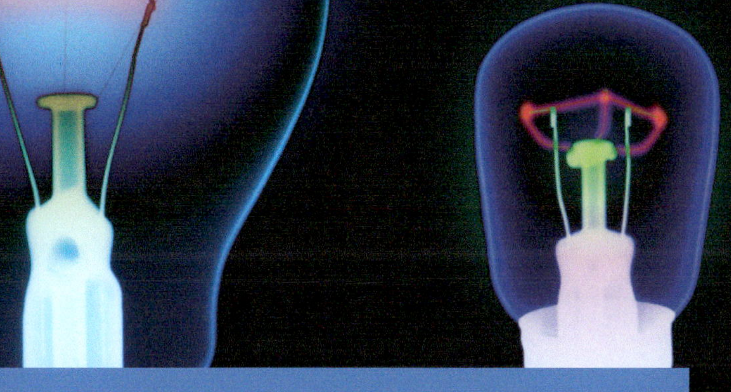

Have you ever wondered...

- how energy is carried around an electric circuit?
- how the circuits in houses are wired up?
- how electricity is generated?
- how electric motors work?

After completing this chapter you should be able to:

- discuss the particle model and how it is used to understand electricity
- investigate factors that affect the transfer of energy through an electric circuit
- *investigate parallel and series circuits*
- *measure the voltage drop and current through circuit components*
- *compare circuit design with household wiring*
- *investigate the properties of components such as LEDs, and temperature and light sensors*
- *explore the use of sensors in robotics and control devices*
- *describe the magnetic field around magnets*
- *outline how the movement of a magnet and a wire can produce electricity*
- *describe how a magnet and a current from a battery can produce movement.*

AB 6.1

MODULE 6.1

Simple circuits

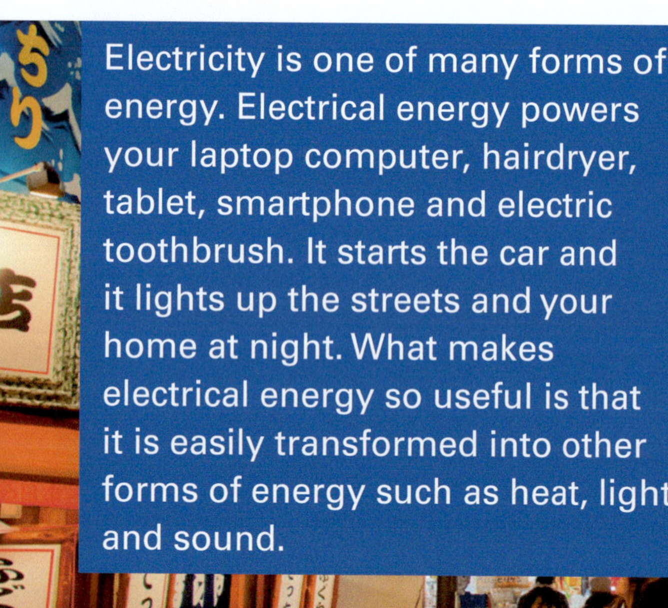

Electricity is one of many forms of energy. Electrical energy powers your laptop computer, hairdryer, tablet, smartphone and electric toothbrush. It starts the car and it lights up the streets and your home at night. What makes electrical energy so useful is that it is easily transformed into other forms of energy such as heat, light and sound.

science 4 fun

Sparks on opening!

YES

In a solid, the particles are strongly bonded, so lots of energy is needed to break them apart. Can you see the energy released when these bonds break?

Collect this ...

- self-stick envelope
- sugar cubes
- pliers or multi-grips

Do this ...

1. Do this activity at night or in a very dark room. Allow your eyes time to get used to the dark before you start.
2. Seal a self-stick envelope. While in the dark, open the seal as quickly as you can (don't tear the envelope, just open its seal).
3. If you're unsure what happened, then seal the envelope and repeat.
4. While still in the dark, crush a sugar cube with pliers or multi-grips.

Record this ...

1. Describe what happened.
2. Explain why you think this happened.

Electric charge

Everything is made of atoms, which consist of **protons**, **neutrons** and **electrons**. As Figure 6.1.1 shows, protons and neutrons are located in a small and dense core called the **nucleus**. Tiny particles known as electrons spin around the nucleus. Protons and electrons are electrically charged: protons carry a positive charge (+) and electrons carry a negative charge (–). Neutrons carry no charge and are said to be **neutral**. Overall, the atom is neutral, because the numbers of protons and electrons are always equal. Their opposite charges balance each other out and so the atom has no overall charge.

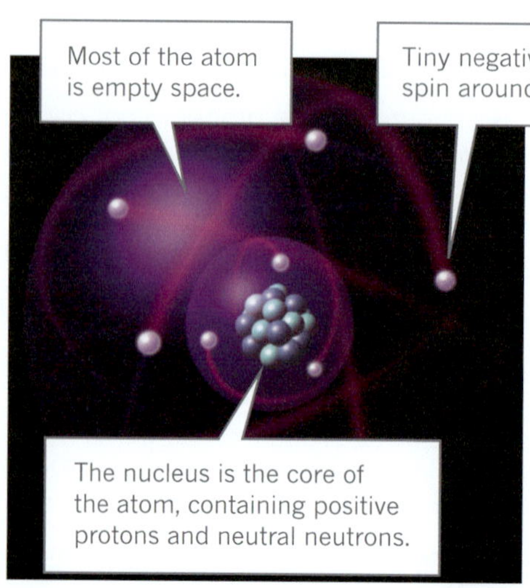

FIGURE 6.1.1 Atoms are neutral but contain charged particles. An ion is a charged atom that forms when there are unequal numbers of protons and electrons.

Sometimes electrons can be knocked off an atom or added to it, giving the atom an overall charge. This 'charged atom' is known as an **ion**. If electrons have been removed, then the ion has more protons than electrons, giving it a positive charge. If an atom has electrons added to it then it forms a negative ion. This is because it has more electrons than protons.

Static electricity

Static electricity is the build-up of electric charge on a surface. This build-up of charge most commonly occurs because the surface has been rubbed against another surface (Figure 6.1.2). Rubbing can cause electrons to be rubbed off one surface, charging it positive (it has lost negative electrons). These electrons are transferred to the other surface, charging it negative (it has gained negative electrons).

Static charge usually leaks away after some time into its surroundings, including the air around it. This returns the materials to their original neutral state. However, if the build-up of charge continues, the electrons may jump across a gap from the negatively charged surface back to the positively charged surface. As they jump back, the electrons release all their energy in one go. This converts the energy into the heat, light, sound and motion (kinetic energy) that you observe as a spark or lightning bolt.

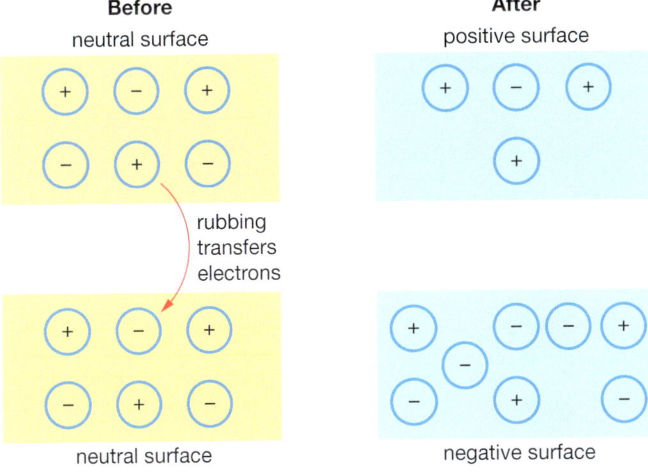

FIGURE 6.1.2 Rubbing can transfer electrons from one surface to another, charging both surfaces.

SciFile

Strike me lucky!

A bolt of lightning:

- is about 5 cm wide
- has a temperature of around 30 000°C (hotter than the surface of the Sun)
- is the main source on Earth of plasma, the fourth state of matter
- could power a 25 W light globe for a year!

Current electricity

The electricity you get from a battery or a power point is not static electricity. It is made up of electrons moving along a wire, like those in Figure 6.1.3. This movement of charge is called an electric **current**. These moving electrons carry energy that is transformed into other forms of energy as the electrons pass through things like light globes (transforming electrical energy into light), heating elements (into heat) and motors (into movement).

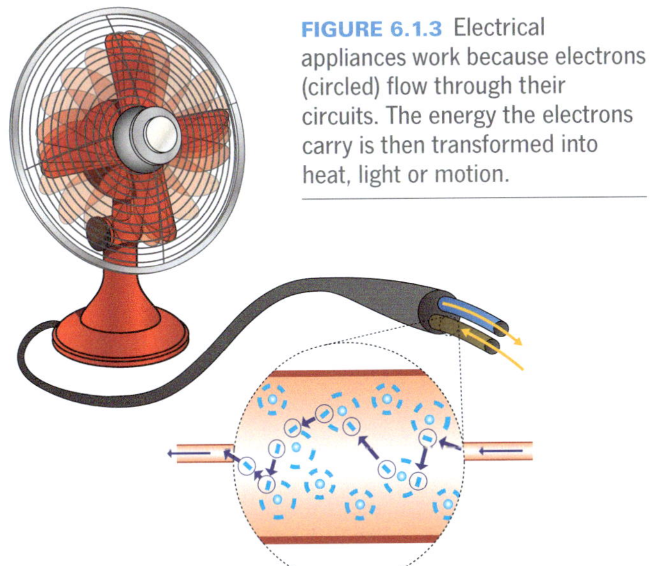

FIGURE 6.1.3 Electrical appliances work because electrons (circled) flow through their circuits. The energy the electrons carry is then transformed into heat, light or motion.

Simple electric circuits

Electrons travel along a path to deliver their energy. This path is called an **electric circuit**. As Figure 6.1.4 shows, an electric circuit needs:

- an energy source, such as a battery, a power point or a generator like the dynamo on a bike. This supplies the electrons in the wire with the energy they require to get them moving around the circuit
- an energy user, such as a light globe, heating element resistor or motor. These devices convert the energy that electrons are delivering to them
- wires to connect everything, making the circuit complete.

Any break in an electric circuit stops the flow of electrons and stops them from delivering their energy. Most electric circuits have switches that deliberately break the circuit, turning it on and off.

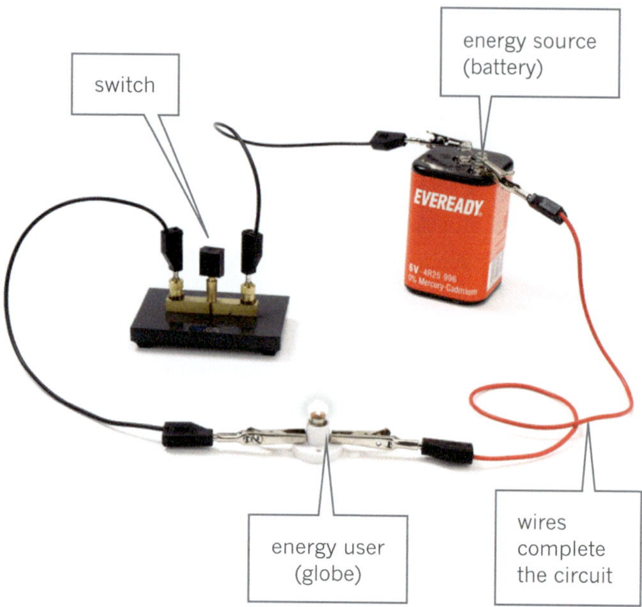

FIGURE 6.1.4 A circuit needs an energy source, an energy user and wires to connect them all. The circuit usually has a switch too.

Circuit components

The different parts of a circuit are known as its **components**. Each component is given a different symbol. This makes diagrams of circuits easier to draw and easier to understand. Some of these components are shown in Figure 6.1.5.

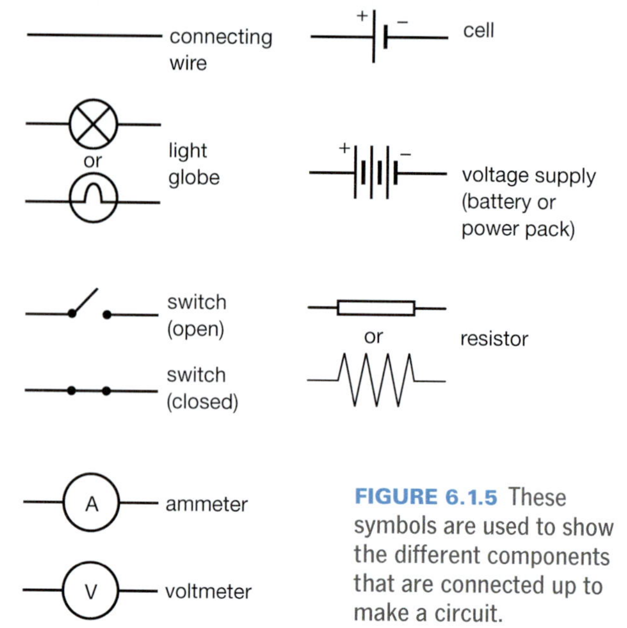

FIGURE 6.1.5 These symbols are used to show the different components that are connected up to make a circuit.

Circuit diagrams

A **circuit diagram** is a simplified and shorthand version of a real circuit. It shows how all the components in the circuit are connected.

A torch is an example of a very simple circuit. Its energy source is a battery and its globe transforms electrical energy into light. The circuit diagram for the torch is shown in Figure 6.1.6. The battery supplies the electrons with energy. As the electrons flow through the globe, they lose almost all their energy, which is transformed into light energy and some heat energy. The electrons then travel back to the battery, where their energy is replenished.

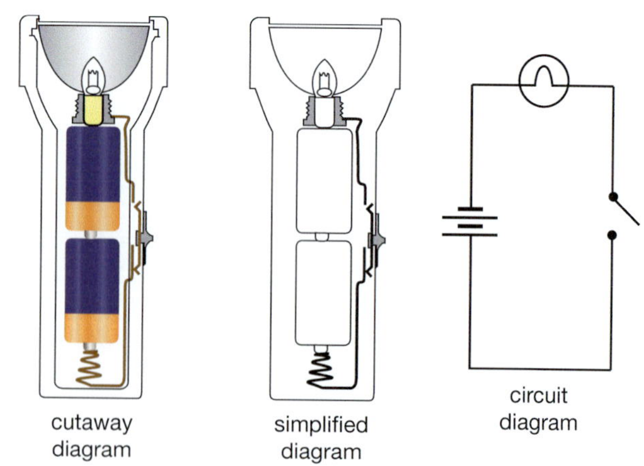

FIGURE 6.1.6 A torch has a battery, a globe, wires and a switch.

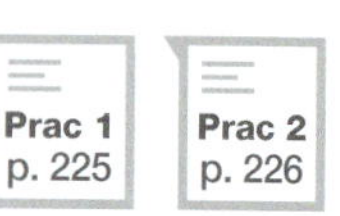
Prac 1 p. 225
Prac 2 p. 226
Prac 3 p. 226

MODULE
6.1 Review questions

Remembering

1 Define the terms:
- a nucleus
- b static electricity
- c current
- d circuit diagram.

2 What term best describes each of the following?
- a having no charge
- b globes, switches, cells and resistors
- c path around which electrons travel.

3 What are the electric charges of each of the following?
- a an atom
- b a proton
- c a neutron
- d an electron.

4 What forms of energy are released when a spark jumps across a gap?

5 Name the charged particles that carry an electric current through a circuit.

6 What does an electric circuit need?

7 List three examples of each of the following components of an electrical circuit.
- a an energy supplier
- b components that use electrical energy.

8 Draw the symbols for the following electric components.
- a globe
- b battery
- c single cell
- d switch.

Understanding

9 Fill in the following statements to make them true. To complete each statement, insert the sign = or > (is greater than) or < (is less than).
- a In an atom, the number of protons ___ the number of electrons.
- b In a positive ion, the number of protons ___ the number of electrons.
- c In a negative ion, the number of protons ___ the number of electrons.

10 Would the particles in Figure 6.1.7 be neutral, positively charged or negatively charged?

a

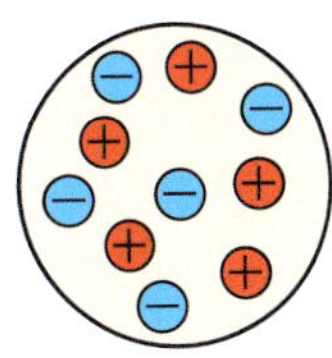

b

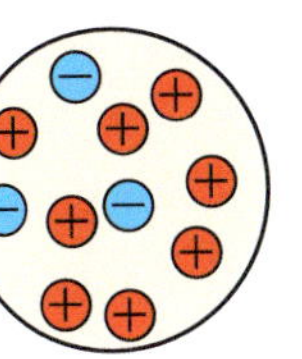

c

FIGURE 6.1.7

11 Draw a simple circuit diagram that shows how a torch works.

Applying

12 Identify whether the following are displays of static electricity or current electricity in action.
- a the spinning motor of a hairdryer
- b lightning
- c a spark felt when you touch a doorknob after walking across carpet
- d the TV when it is on.

Analysing

13
- a Why do you think a desktop computer will not work when the switch is off or its plug is out of the power point?
- b Use current and energy to justify your answer.

14 Contrast static electricity with current electricity.

15 Analyse the circuits shown in Figure 6.1.8 and identify which globe(s) would never light up.

a

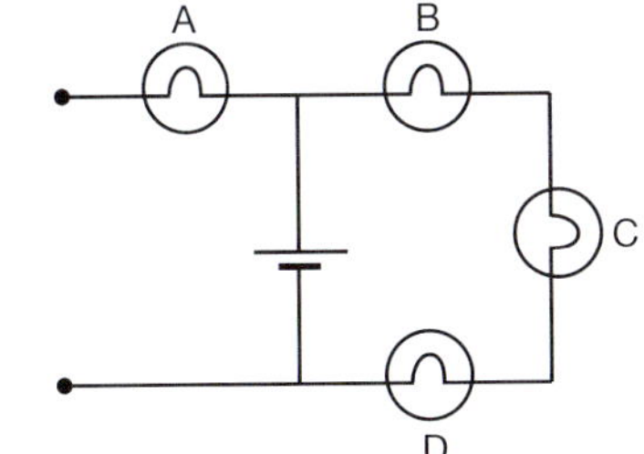

b

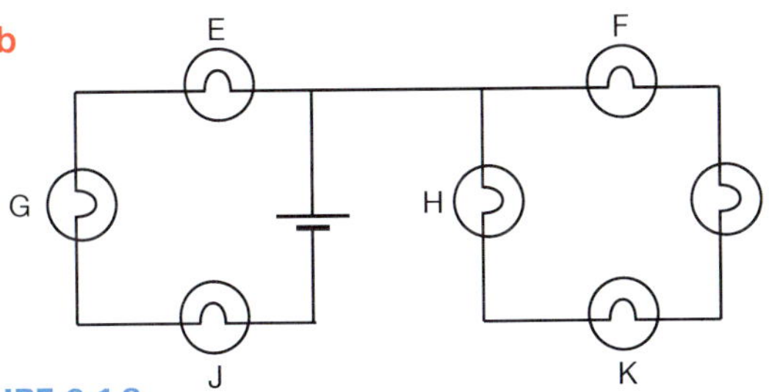

FIGURE 6.1.8

MODULE
6.1 Review questions

Evaluating

16 It is difficult to get all the grains of rice out of the plastic bags they come in, because some always stick to the sides. Propose an explanation for this observation.

17 When you rip open a self-stick envelope or crush a sugar cube in the dark, you should see small flash of light as the bonds break. Why do you think this happens and how do you think it is related to static electricity?

Creating

18 **a** The component symbol for a resistor can be drawn in two different ways. Draw each version.

b A variable resistor is one that can be changed. List examples of devices that would use variable resistors.

c One version of the component symbol for a variable resistor is shown in Figure 6.1.9. Draw a diagram of what you think could be another version of the component symbol for a variable resistor.

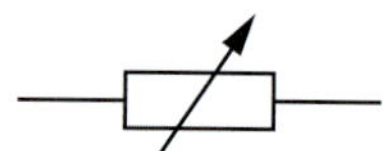

FIGURE 6.1.9

d What do you think the arrow across the variable resistor represents?

19 Construct circuit diagrams for the circuits shown in Figure 6.1.10.

a

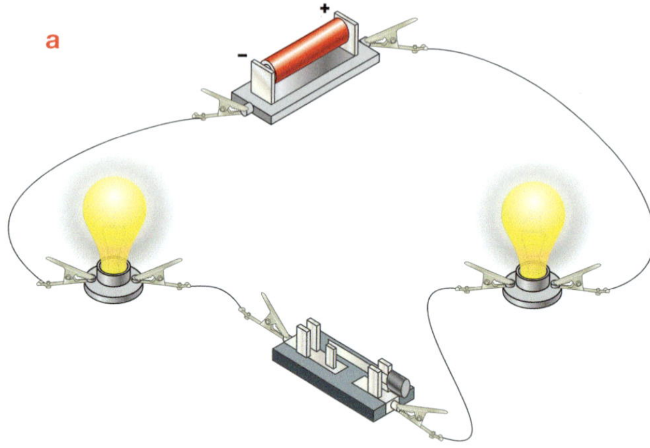

b

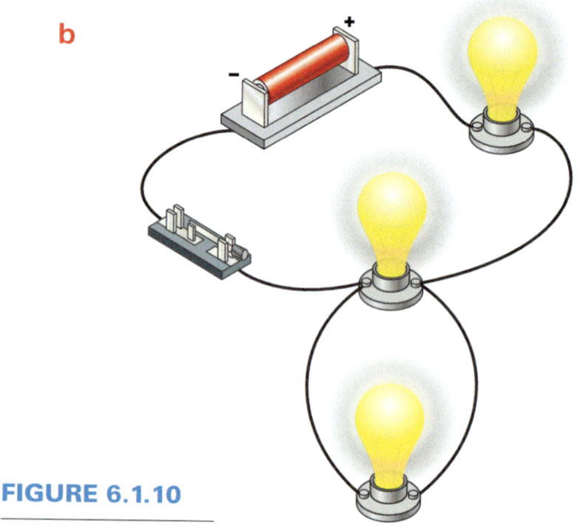

FIGURE 6.1.10

MODULE
6.1 Practical investigations

1 • How steady are you?

Purpose
To construct a simple, fun electrical circuit.

Timing
60 minutes

Materials
- 1 wire coat hanger
- wooden board
- 2 screws
- 1 empty plastic pen casing
- sticky tape
- electrical leads
- alligator clip
- globe
- switch
- battery
- access to pliers or multi-grips
- access to screwdriver

Procedure
1. Cut a length of coat hanger wire and bend it into a twisted shape. Use screws to secure it to the wood so that the wire stands upright.
2. Cut another length of coat hanger wire and bend one end into a loop around the twisted wire. Insert the other end into the plastic body of a used pen. Secure both ends with tape.
3. Connect up the circuit as shown in Figure 6.1.11.
4. Test how steady your hand is by trying to pass the loop along the twisty wire without making the globe light up.

Review
1. Why does the globe only light up when you touch the twisty wire with the loop?
2. Electrical current can pass along a coat hanger wire. Justify this claim.
3. Imagine you touched the loop halfway along the twisty wire. Contrast what is happening to the electrons in each half of the twisty wire when this happens.

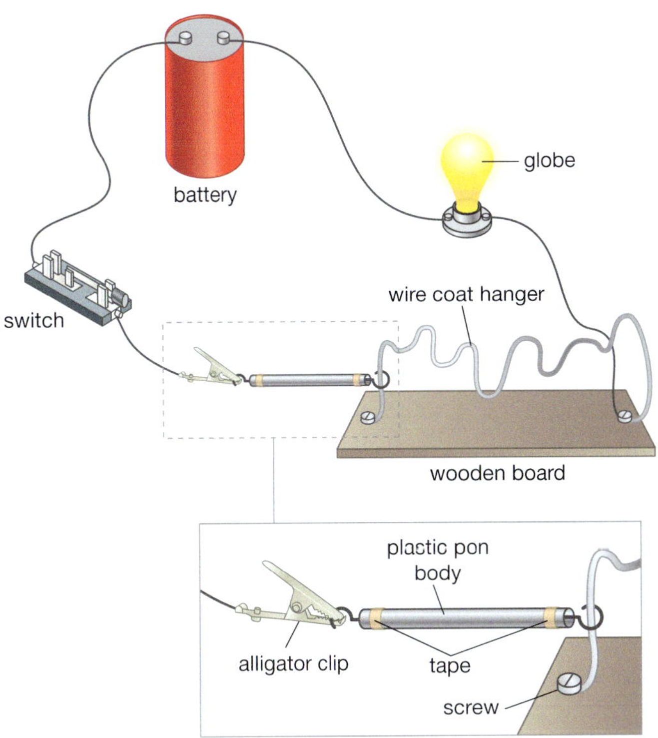

FIGURE 6.1.11

MODULE

6.1 Practical investigations

• STUDENT DESIGN •

2 • Super-sparker

Questioning & Predicting 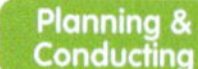

Purpose

To construct a static electricity device called a super-sparker.

Timing 60 minutes

Materials

As selected by students

SAFETY

A risk assessment is required for this investigation. See Activity Book Toolkit to assist with developing a risk assessment.

Be aware that this device produces sparks.

Procedure

1 Search the internet to find out what a super-sparker is and instructions on how to construct one. Copy, print or save the instructions.

2 Before you start any practical work, assess how you are going to build and test your super-sparker. List any risks that it might involve and what you might do to minimise those risks. Show your teacher your planned design, testing procedure and your assessment of its risks. If your teacher approves, then collect all the required materials and start work.

Use the STEM and SDI template in your eBook to help you plan and carry out your investigation.

Results

Record your observations.

Review

1 To charge your super-sparker and produce sparks, you needed to rub something against something else. Identify the two materials that rubbed against each other in this case.

2 The super-sparker didn't spark immediately. It only sparked when you did something to it. Describe what you needed to do.

3 For your super-sparker to discharge and produce a spark, it needs to be earthed. This means that a current needs to be able to travel from it to the ground. Explain how this happened in your design.

• STUDENT DESIGN •

3 • Static electricity race

Purpose

To use static electricity to get a soft-drink can rolling without touching it.

Timing 45 minutes

Materials

- empty soft drink can
- balloon
- flat, smooth table

SAFETY

A risk assessment is required for this investigation. See Activity Book Toolkit to assist with developing a risk assessment.

Procedure

1 Rub the balloon against your hair until the balloon is charged.

2 Place the charged balloon near the can but do not touch it. The charged balloon should attract the can and get it rolling across the table.

3 Once you have got the can rolling, test different ways of getting it to roll even faster.

4 After a few tests, run a 'can race' where you compete against another group. To do this, line up your cans and see which group can attract their cans so that they roll fastest. The winner is the group whose can falls off the end of the table first.

Hints

- Possible variables that might test to get the can rolling faster are:
  - how long you rub the balloon against your hair
  - the distance the balloon is held from the can
  - whether the balloon is held at a constant distance from the can or whether it is 'stabbed' in and out towards the can.
- Use the STEM and SDI template in your eBook to help you plan and carry out your investigation.

Review

1 Describe what made your can roll fastest.

2 Which group had their can fall off the table first? Why do you think their can rolled fastest?

MODULE

6.2 Measuring electricity

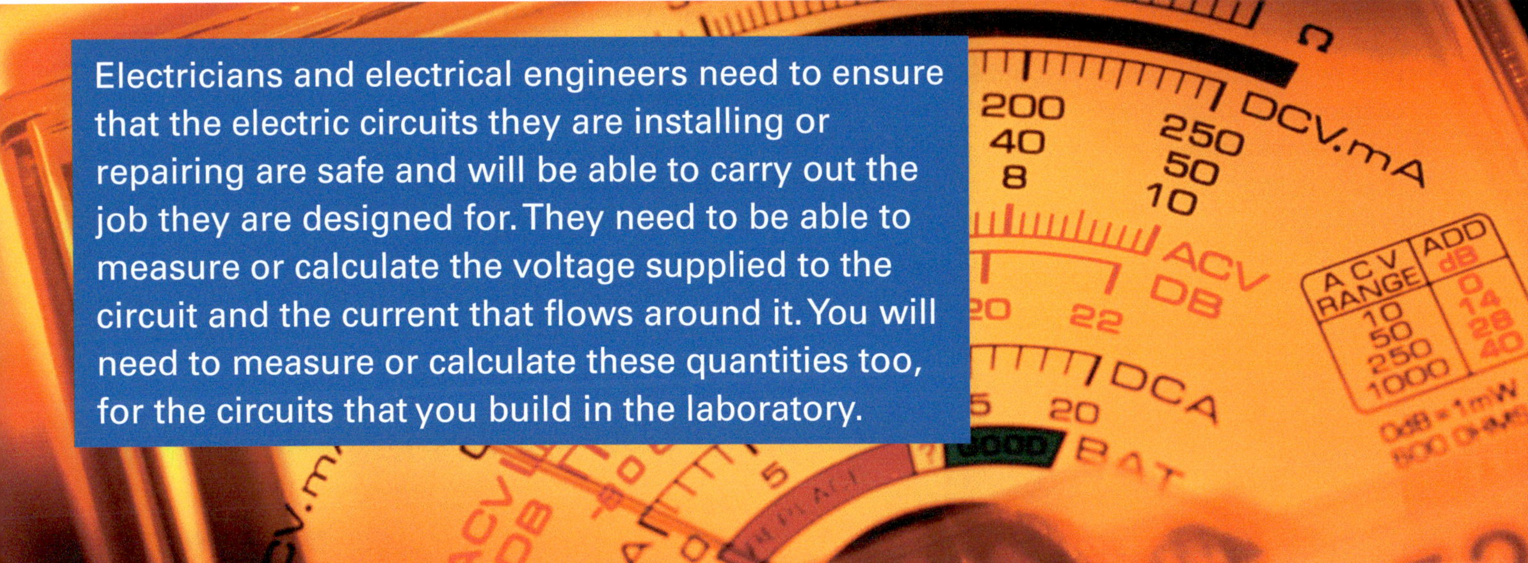

Electricians and electrical engineers need to ensure that the electric circuits they are installing or repairing are safe and will be able to carry out the job they are designed for. They need to be able to measure or calculate the voltage supplied to the circuit and the current that flows around it. You will need to measure or calculate these quantities too, for the circuits that you build in the laboratory.

Current

An electric current is formed whenever charge flows from one spot to another. In an electric circuit, this flow of charge is made up of electrons moving along the wires.

These electrons and the current they form carry energy around the circuit from the battery or power point to the different components that use it. Current can be direct (DC) or alternating (AC). In DC, all the electrons flow along the wires in the same direction. In contrast, the electrons in AC shuffle back and forth along the wire. Batteries supply DC while power points supply AC.

Electric current is measured using an **ammeter**. An ammeter measures the amount of charge that flows through it every second. The current is high if a lot of charge flows through it in one second, and low if only a small amount of charge flows through it.

The unit used to measure current is ampere (unit symbol A), which is often shortened to 'amps'.

SciFile

Nervous about electricity?

Your muscles are activated by electrical impulses sent along your nerves. The same happens in other animals too. A platypus uses sensors within its duck-like bill to detect electric currents from the muscle movements of yabbies, fish, worms and frogs.

SkillBuilder

Connecting up an ammeter

Electrons must pass through an ammeter for the charge to be detected. Therefore the ammeter needs to be in line with the rest of the circuit's components. This arrangement is known as being in **series** and is shown in Figure 6.2.1.

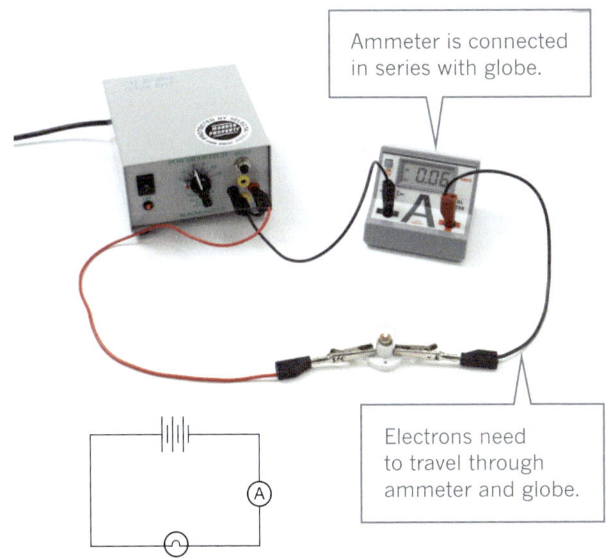

FIGURE 6.2.1 Ammeters measure the current that passes through them. An ammeter needs to be in line (in series) with the rest of the circuit.

Voltage

Voltage is a measure of the amount of energy:

- supplied to the charges by the voltage source (the supply voltage)
- used by the charges as they pass through a component such as a light globe (voltage drop). This energy is transformed into heat and light.

Voltage is measured using a **voltmeter**. The voltage is high if the electrons are supplied with a lot of energy or are losing lots of energy. The voltage is low if the electrons lack energy or lose very little. If the voltage is zero, then it means the battery is dead, the power point is turned off or the electrons are losing no energy in that part of the circuit.

The unit used to measure voltage is volts (unit symbol V).

SkillBuilder

Connecting up a voltmeter

A voltmeter compares the energy of electrons before and after they pass through a component such as a light globe. For this reason, voltmeters are connected in **parallel**. This means they are not part of the circuit itself, but instead attach across the component being measured. This is shown in Figure 6.2.2.

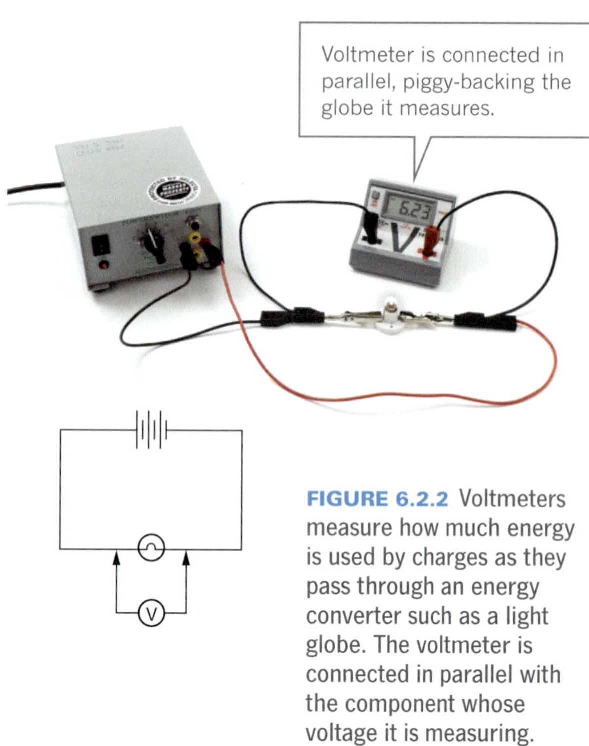

FIGURE 6.2.2 Voltmeters measure how much energy is used by charges as they pass through an energy converter such as a light globe. The voltmeter is connected in parallel with the component whose voltage it is measuring.

Supply voltage

Electrons get the energy they need to move around the circuit from the circuit's energy source. Each energy source has its own voltage. Higher supply voltages give the electrons a bigger 'push' than low supply voltages.

Mains power

In Australia, power points supply 240 V to the electrons in any circuit plugged into them. Sometimes a **transformer** is used to reduce the voltage from a power point to a more manageable voltage. For example, mobile phones typically need 5–6 V to recharge, digital cameras 6.5 V and laptops 19 V (Figure 6.2.3).

In the laboratory, power packs reduce the 240 V from a power point to the voltages required in experiments. Most power packs can be adjusted to supply a range of voltages from 1.5 V to 6 V or 12 V.

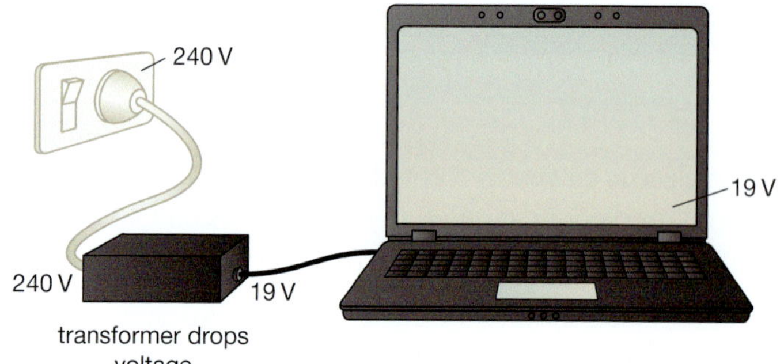

FIGURE 6.2.3 A step-down transformer reduces the 240 V provided by a power point to the 19 V that a laptop needs.

SciFile

Bright spark!

Electrons from a 240 V power point have insufficient energy to jump the gap in a switch. However, extreme voltages cause air in the gap to break down, allowing a spark to jump across it! A 1 cm gap requires about 3000 V for it to 'spark'.

Batteries

Batteries are an excellent source of portable electrical energy. Batteries are generally made of small cells or smaller 'mini-batteries'. Batteries connect to the rest of the circuit via positive and negative terminals. Cells can be classified as:

- wet cells
- dry cells
- photovoltaic or solar cells.

A **wet cell** has conducting electrodes submerged in a liquid electrolyte (a solution that conducts electricity). **Electrodes** are rods, sheets or plates made of a metal or some other conducting material like graphite. The **electrolyte** is a solution that conducts electricity. A car battery uses a set of six wet cells to supply its electrical energy. Each wet cell has two plates that act as electrodes, one made of lead and the other made of lead oxide. The electrolyte is a solution of sulfuric acid. Each cell supplies roughly 2 V and so six cells in a car battery provides 12 V. Wet cells provide relatively large voltages, last a long time and can be easily recharged. However, they are heavy and they can leak. The basic structure of a car battery is shown in Figure 6.2.4.

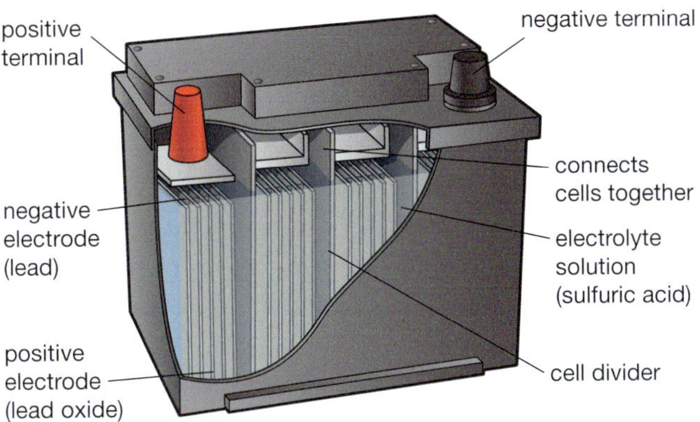

FIGURE 6.2.4 A car battery uses six wet cells to provide 12 V. The battery is used to start the car and to power lights and radio when the engine is not running. When driving, a device called an alternator recharges the battery.

The small, portable batteries used in torches, toys and remote controls are **dry cells**. Dry cells are compact because they have one electrode wrapped around another. They don't leak because they use a conducting paste as its electrolyte instead of a liquid. A typical dry cell is shown in Figure 6.2.5.

Prac 1
p. 235

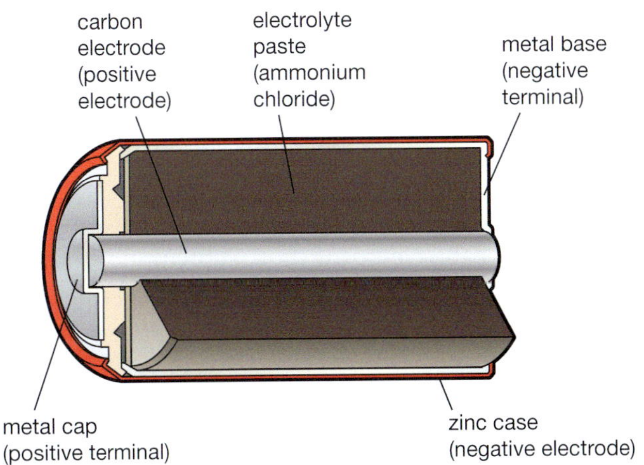

FIGURE 6.2.5 Dry cell batteries are compact and do not leak.

As Figure 6.2.6 shows, different dry cells provide different voltages. The voltage of dry cells can be further increased by placing a number of them end-to-end. For example, eight 1.5 V AA batteries arranged head to tail give the same 12 V as supplied by a car battery.

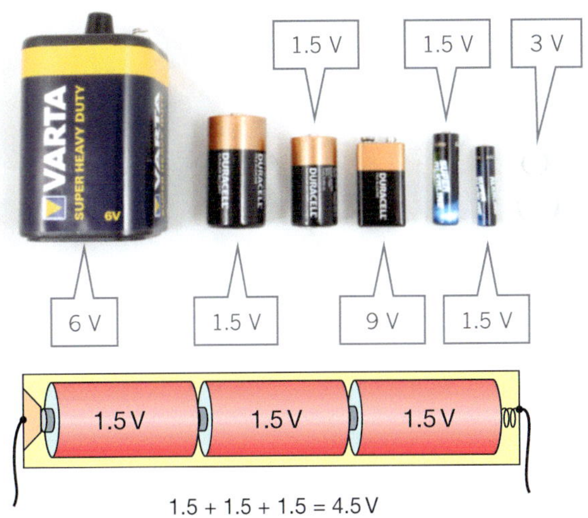

FIGURE 6.2.6 Dry cell batteries come in different shapes and voltages. Higher voltages can be obtained by arranging the batteries end to end in series.

Prac 2
p. 236

Photovoltaic cells (or solar cells) convert solar energy directly into electrical energy. Energy in the sunlight knocks electrons off silicon crystals within the cell. These electrons move away from the crystal, forming an electrical current that can be used to power appliances.

Many homes have banks of solar panels made of many photovoltaic cells to provide them with a source of renewable energy that releases no greenhouse gases. Any excess power generated by the panels is fed back into the electrical grid of the community and earns the household money. If more power is required, the house draws electricity from the grid just like any other house.

Alternatively, the current generated by photovoltaic cells can be used to recharge batteries for later use. For example, the solar garden light in Figure 6.2.7 on page 230 uses batteries that are recharged by solar cells during the day.

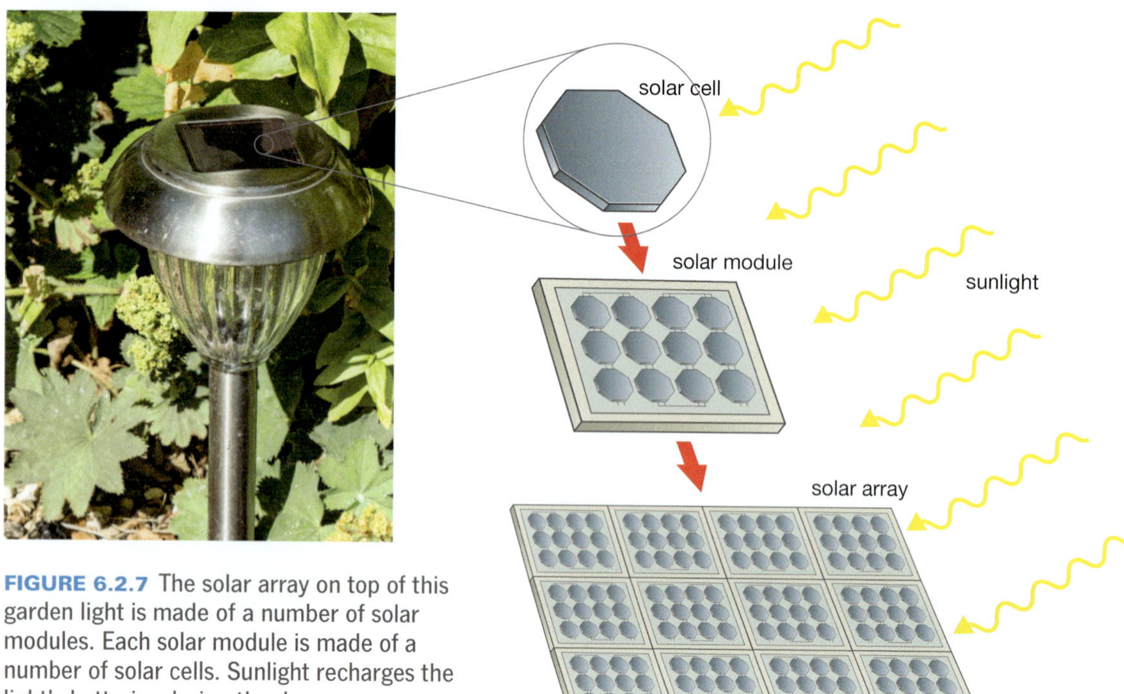

FIGURE 6.2.7 The solar array on top of this garden light is made of a number of solar modules. Each solar module is made of a number of solar cells. Sunlight recharges the light's batteries during the day.

SciFile

Photovoltaic power plant

The world's largest photovoltaic power plant is to be built near Mildura (in Victoria). The plant will use arrays of mirrors to reflect sunlight (concentrating it 500 times) onto a tower of solar cells. This is very different from existing large-scale solar power plants. They will use the reflected sunlight to boil water to turn turbines to produce electricity.

Voltage drop

Electrons lose energy as they pass through a component such as a light globe, a heating element or a motor. This results in a voltage drop across the component. This voltage drop depends on the resistance of the component.

AB 6.2

Resistance

As electrons pass along the wires of an electric circuit, their path is restricted a little by the atoms that make up the wires. This restriction is known as **resistance**. Resistance measures how difficult it is for an electric current to flow through a material or a component. A high resistance means that it is difficult for electrons to pass through the material. A low resistance means that it is easier for electrons to pass through the material.

The energy and voltage lost by electrons as they pass through a component depends on the resistance of the materials making it up. Electrons don't bump into much as they pass through low-resistance materials, and so they lose almost no energy and almost no voltage. High-resistance materials have obstacles in the way of the electrons. A little energy is lost every time the electrons are bumped off-course, and so overall a lot of electrical energy and voltage is lost as they pass through.

Resistance also affects the current flowing through a circuit. As the resistance of a component increases, fewer electrons get through it every second. This reduces the current flowing.

The resistance of a wire depends on the:

- type of material the wire is made from. For example, metals generally have low resistance, whereas rubber has an incredibly high resistance.
- length of the wire. Doubling the length of a wire doubles the number of obstacles that the electrons must pass through. This doubles its resistance.
- thickness of the wire. It is more difficult for electrons to pass along thin wires than to pass along thick wires.

Resistance is measured using the unit ohm. The unit symbol for ohms is a letter from the Greek alphabet known as omega, Ω. Resistance can be measured by a multimeter, like the one in Figure 6.2.8.

FIGURE 6.2.8 A multimeter combines an ammeter and a voltmeter, and can also measure resistance.

Resistors are components that have a known resistance (Figure 6.2.9). They are used to restrict the current flowing through them and the components in the rest of the circuit. In this way, resistors ensure that the other components receive the desired current and voltage and are not burnt out.

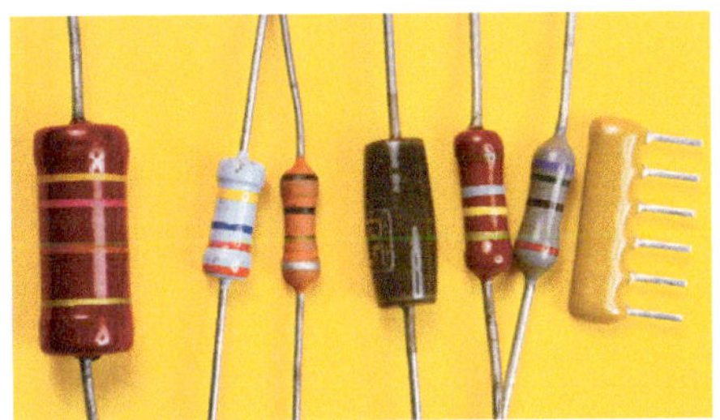

FIGURE 6.2.9 Resistors

Variable resistors allow you to change their resistance and so they can be used to control the voltage and current flowing through components in a circuit. In this way they can control the speed of motors, the brightness of globes and the volume of speakers. Hence, variable resistors are commonly used in dimmer switches and to control radio and TV volume, the speed of airconditioner fans and car windscreen wipers.

Conductors

Metals are **conductors**. This means that an electric current will pass easily through them. However, some metals are better conductors than others. It all depends on their resistance.

Copper is an excellent conductor. It has a very low resistance and almost no energy is lost from it. It is also relatively cheap. For these reasons, copper wires are used in most electric circuits around the home, in factories and in cars. Another excellent, low-resistance conductor is aluminium. Aluminium is more expensive than copper but also much lighter. This is why aluminium is used for high-voltage transmission lines strung between distant pylons (poles), like those in Figure 6.2.10.

FIGURE 6.2.10 Transmission lines need to be made of a low-resistance, light metal. Copper would be far too heavy, so aluminium is used instead.

Tungsten and nichrome alloys are metals too and so they also conduct electricity, but not as well as copper or aluminium. Tungsten and nichrome have relatively high resistances and so electrons passing through them lose much of their energy and voltage. This energy is converted into heat and sometimes light. This makes them ideal to use as heating elements in electric kettles, hair dryers, electric blankets and the filaments of old-fashioned incandescent light globes like the one in Figure 6.2.11.

FIGURE 6.2.11 The resistance of the tungsten filament of this old-fashioned light globe converts electrical energy into heat and light.

Insulators

Some materials have such a high resistance that they block electric current completely. These materials are said to be **insulators**. Examples are rubber, plastics, wood, glass and ceramics. Figure 6.2.12 shows how plastic is used to wrap electric wires and cables to insulate them from their surroundings. Glass and ceramics are used to insulate high-voltage power lines so that current doesn't pass into the poles that are holding them up.

Prac 3 p. 236 Prac 4 p. 237

FIGURE 6.2.12 Plastic coating is used to insulate each of the three wires in an electric cable. More plastic coating wraps all the three wires together, insulating them even further.

Working with Science

ELECTRONIC ENGINEER

Steve Camilleri

Electronic engineers design, develop, repair and maintain electronic systems and their parts. These engineers work in a broad range of areas, such as automotive, rail, aerospace, biomedicine, construction, defence, robotics and meteorology. Steve Camilleri uses his skills as an electronic engineer to design and develop innovative high performance motors and electronic controllers (Figure 6.2.13). Steve's interest in electronics started early, when he completed his first electronics course in Primary School. He went on to study an Engineering degree at Darwin University, where he was part of a team that developed a new kind of motor for a solar car. The motor won a Technical Innovation Award and was licensed by a company in the United States. Steve estimates that around 90% of solar cars use their motor today.

Steve and a friend from university went on to start their own electronic engineering company. They have developed motor technology for fans and electric bikes, worked in Indigenous communities installing solar technology, and have worked with NASA to develop satellite replacement technology that uses solar energy to fly. Steve's job involves desk work, hands-on jobs in the workshop and work outside, installing technology. Steve loves being able to design, build and test new electronics and motors. He hopes that his company can continue to develop environmentally friendly electronic solutions for people.

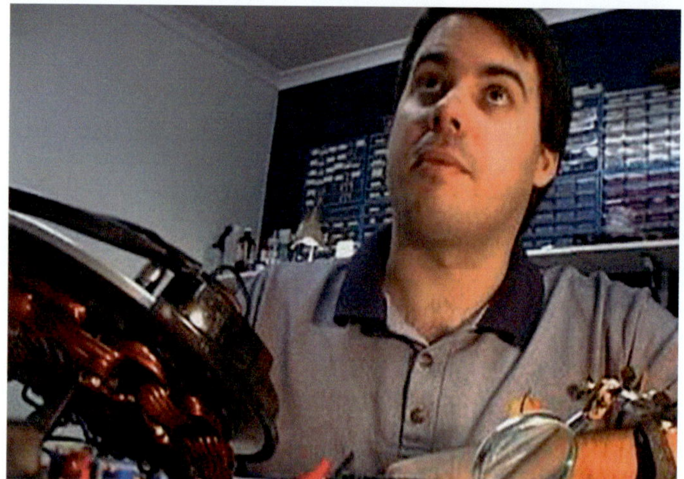

FIGURE 6.2.13 Steve Camilleri is an electronic engineer. His company designs and builds motors and electronic controllers for environmentally friendly technology.

An Engineering degree can lead you to a career in electronic engineering or the many other engineering-based careers out there (e.g. telecommunications, mechanical and materials engineers). There are also Certificate and Diploma courses in Electronic Engineering that can get you started in this career. There are a wide variety of exciting opportunities in the field of electronic engineering and job opportunities are expected to grow.

Review

1. In what other areas might an electronics engineer's skills be used?
2. Steve would like to make technology that can benefit people. What are some ways that you think electronic technology can improve people's lives?

MODULE

6.2 Review questions 

Remembering

1 Define the terms:

a current
b conductor
c resistance
d electrolyte.

2 What term best describes each of the following?

a energy provided to charges
b material that blocks an electric current
c cell that converts sunlight into electricity
d metal or graphite plates that conduct electricity in a wet cell.

3 What is the voltage of each of the following?

a voltage available from power points in Australia
b voltage usually needed by laptops
c needed to recharge mobile phones.

4 Recall units and unit symbols by copying and completing the table.

Units and symbols for current, voltage and resistance

Quantity	Unit	Unit symbol
current		
voltage		
resistance		

5 List three things that resistance depends on.

6 What is a transformer used for?

7 What are the advantages of using dry cells for a TV remote control?

8 What are two examples of:

a an electrical conductor?
b an insulator?

9 Name two metals that have:

a low resistance
b high resistance.

Understanding

10 Why does an ammeter need to be connected so that it is in line (in series) with the components of a circuit?

11 Explain why copper is used for the wiring around a house but aluminium is used for high-voltage transmission lines.

Applying

12 Identify whether a wet or dry cell would be best to power the following.

a a laptop computer
b a boat starter motor
c the lights in a caravan
d a bionic ear implant.

13 Calculate the total supply voltage of the battery arrangement shown in Figure 6.2.14.

The following key applies to questions 14 and 15.

Key
A 0 A (current is blocked)
B 1 A
C 2 A
D 4 A

14 The circuit in Figure 6.2.15 on page 234 was set up. The ammeter shows the current flowing through the globe. The supply voltage was then increased to 12 V. Use the above key to predict the new ammeter reading.

15 The globe in Figure 6.2.15 on page 234 was swapped with one of greater resistance. Use the above key to predict the new ammeter reading.

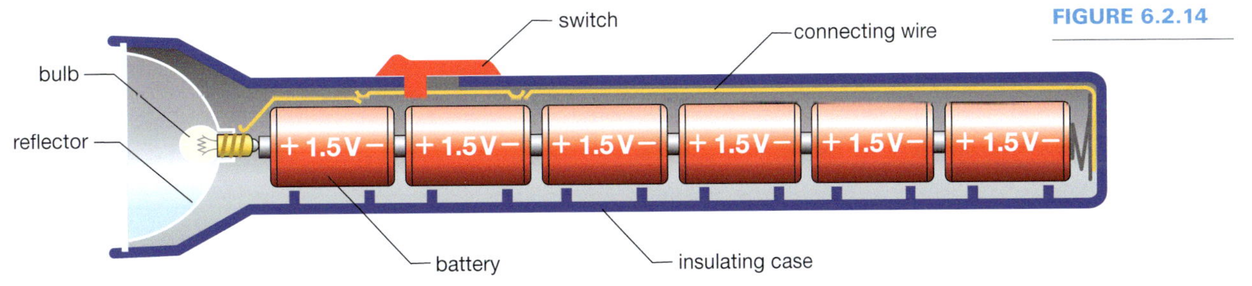

FIGURE 6.2.14

MODULE 6.2
Review questions

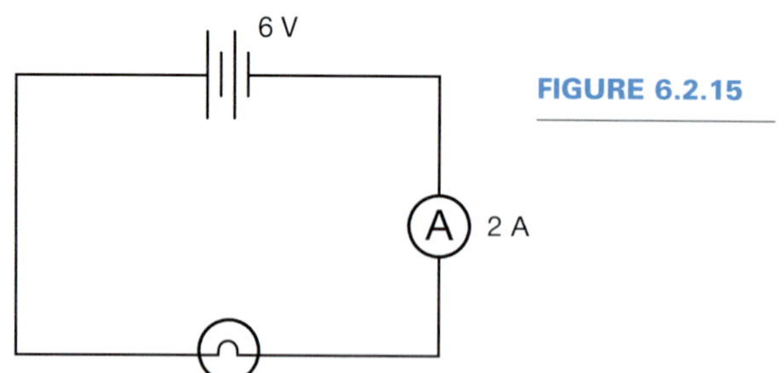

FIGURE 6.2.15

Analysing

16 Contrast a conductor with an insulator.

17 Contrast the way an ammeter and a voltmeter are connected into a circuit.

Evaluating

18 AA and AAA batteries both supply the same voltage of 1.5 V but are different sizes. Propose a reason why batteries come in different shapes and sizes, even when some of them supply the same voltage.

19 An analogy is a model that compares something that is difficult to understand with something that is easier to understand. For example, electrons moving along a wire are impossible to see and difficult to imagine. they are often compared to something easier to understand, such as cars driving along a road.

- **a** In this analogy, state what would represent:
  - **i** electric current
  - **ii** resistance.
- **b** Analyse the flow of cars along a busy single-lane road that:
  - **i** widens with extra lanes
  - **ii** is blocked by a broken-down car.
- **c** Use this analogy to predict what would happen to current if a light globe were replaced by:
  - **i** a copper wire
  - **ii** an insulator.

20 An electric current can flow when different metals touch each other.

- **a** If you place aluminium against an amalgam filling in your tooth, you will feel pain in the tooth. Propose reasons why.
- **b** Builders often work with different metals. For example, they use steel nails and screws, aluminium foil and copper wires. Propose ways in which they can keep themselves safe when working with different metals.

Creating

21 Construct a circuit diagram that has a battery, a resistor, a switch and connecting wires. Include an ammeter that measures the current that has flowed through the resistor and a voltmeter that measures the voltage drop across it.

22 Construct a circuit diagram like that shown in Figure 6.2.16 but add:

- **a** a switch that would turn all three globes on and off
- **b** a switch that will only turn globe B on and off
- **c** an ammeter that would measure the total current through the circuit
- **d** an ammeter that would measure the current that flows only through globe A
- **e** a voltmeter that would measure the voltage lost by charges as they pass through globe B.

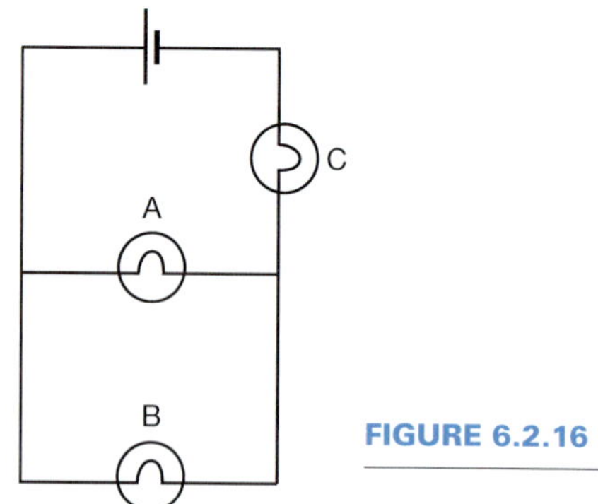

FIGURE 6.2.16

MODULE 6.2

Practical investigations

1 • Lemon cells

Purpose

To construct a cell and a battery using fruit.

Timing 45 minutes

Materials

- lemon and/or other fruits and vegetables (such as kiwi fruit, apple, tomato, melon)
- copper nail, small sheet of copper or length of stripped copper wire
- iron nail (can be galvanised) or small sheet of iron
- milliammeter, multimeter or LED (light-emitting diode)
- connecting wires with alligator clips
- paper towel

Procedure

1 Soften the inside of the lemon a little by squeezing it. Don't break the lemon's skin.

2 Insert the copper nail/sheet/wire into the lemon. Do the same with the iron nail/sheet.

3 As shown in Figure 6.2.17, connect the copper electrode to either the positive terminal of the milliammeter/multimeter or the long terminal of the LED.

4 Connect the iron electrode to the negative terminal of the milliammeter or short terminal of the LED. A current should flow immediately. Record the current flowing or describe how brightly the LED shines.

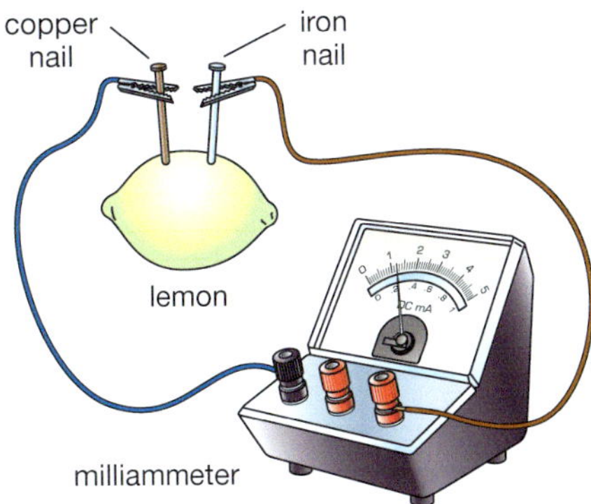

FIGURE 6.2.17

5 Remove the electrodes and pat dry with paper towelling. Repeat the experiment but with different fruits and vegetables.

6 Increase the energy supplied by connecting up a series of the same types of fruit, as shown in Figure 6.2.18. Start with two pieces of fruit, then three and so on. Once again, record the current or brightness of the LED.

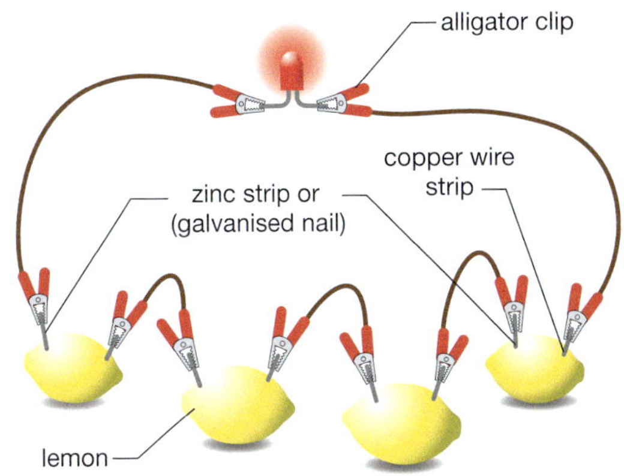

FIGURE 6.2.18

Results

In your workbook, construct a table like the one below in which to record your results.

Current generated

Type of fruit	Number of pieces of fruit	Current (mA)	Brightness of LED
lemon	1		
lemon	2		

Review

1 a Classify these fruit batteries as wet or dry cells.

b Justify your answer.

2 Identify the electrolyte (conducting liquid) in these fruit batteries.

3 Propose a reason why it was recommended that you soften the inside of the lemon a little before the experiment started.

MODULE

6.2 Practical investigations

2 • Dry cell voltages  

Planning & Conducting | Processing & Analysing

Purpose

To measure the supply voltages of different batteries.

Timing 30 minutes

Materials

- a selection of batteries with some charge left in them
- voltmeter or multimeter

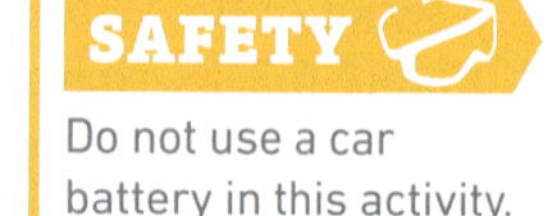

Do not use a car battery in this activity.

Procedure

1 In your workbook, construct a table like the one shown in the Results section.

2 Set the voltmeter or multimeter to its least-sensitive scale.

3 Attach or touch the voltmeter/multimeter terminals or probes to the terminals of each battery (for most batteries this will be their ends). Record your measurement in the results table.

4 Record the voltage printed on the battery.

Results

Record your results in the table.

Voltage of different batteries

Battery type	Voltage printed on battery (V)	Measured voltage (V)
AA		

Review

1 Use your results to assess the accuracy of the statement: *The supply voltage of batteries is always a little lower than the voltage printed on them.*

2 Batteries have a resistance (their internal resistance) and so they use up some of the supply voltage. Use this fact to explain your results.

3 • Graphite light globe  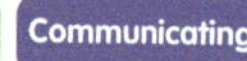

Planning & Conducting | Communicating

Purpose

To construct a light bulb using a pencil refill.

Timing 45 minutes

Materials

- refill for a mechanical pencil
- cardboard tube
- 9 V battery
- 2 wires with alligator clips
- large beaker, glass jar or drinking glass
- electrical insulation tape
- scissors
- digital camera or mobile phone (optional)

SAFETY

The pencil refill will get very hot, so do not touch it once it is connected into the circuit.

The pencil refill will break easily so take care when handling it.

Procedure

1 Set up the apparatus shown in Figure 6.2.19.

2 Connect the wires to the terminals of the battery.

Results

Record what happens.

Review

1 Pencil refills are made from graphite mixed with clay. Deduce whether the graphite–clay mix is a conductor or insulator.

2 a Do you think the graphite–clay mix has a higher or lower resistance than the copper wires?

b Justify your answer.

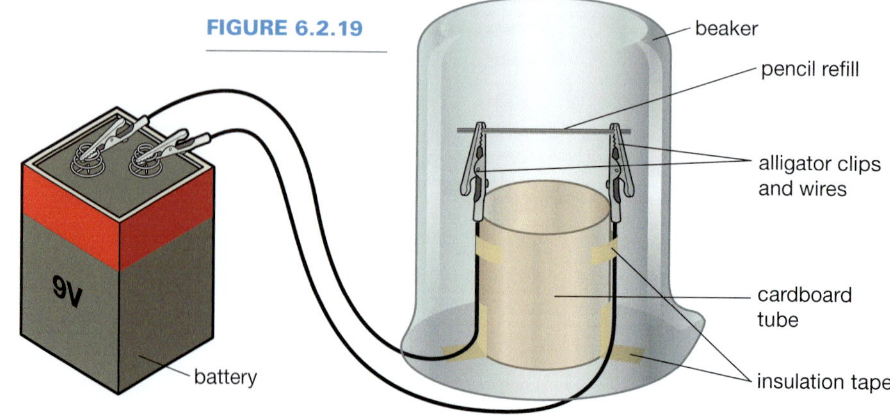

FIGURE 6.2.19

MODULE

6.2 Practical investigations

• STUDENT DESIGN •

4 • Make a dimmer

Planning & Conducting | Evaluating

Purpose

To use a variable resistor to make a dimmer switch.

Timing 45 minutes

Materials

- power pack
- 12 V light globe
- connecting wires
- variable resistor (rheostat/potentiometer) (Figure 6.2.20)

A risk assessment is required for this investigation.

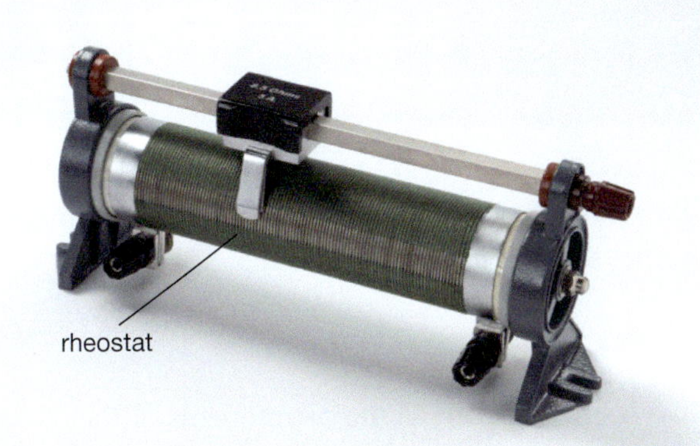

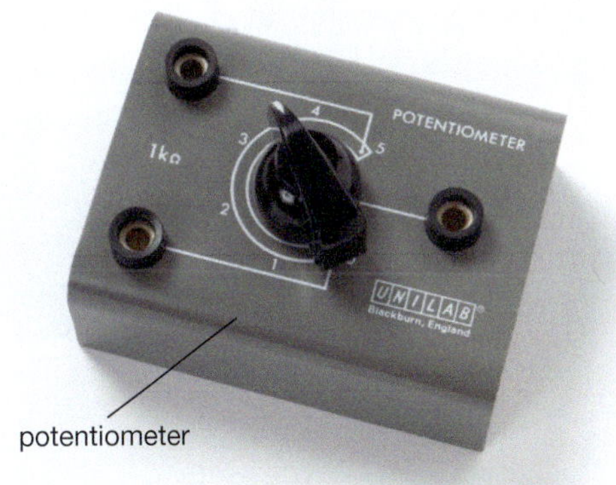

FIGURE 6.2.20 Variable resistors are also known as rheostats or potentiometers and always have a slider or dial that controls them. Sliding or turning the dial changes the resistance.

Purpose

1. Your task is to design and build a circuit that acts as a dimmer for a light globe. Your circuit needs to be able to turn the light globe from full brightness to very dim or completely off.
2. In your workbook, construct a diagram of the circuit you intend to build. Before you start building the circuit, assess any risks associated with the circuit or when you test it. Construct a Risk Assessment that outlines these risks and precautions you need to take to minimise them.

   See Activity Book Toolkit to assist with developing a risk assessment.
3. Show your teacher your planned circuit and your Risk Assessment. If they approve, then collect all the required materials and start work.

   Use the STEM and SDI template in your eBook to help you plan and carry out your investigation.

Review

1. Will the current need to be high or low for the globe to glow brightly?
2. What does a resistance do to the current flowing through it?
3. Outline how a change in the current flowing through the resistor will change the current flowing through the globe.
4. Explain how your circuit works.

MODULE 6.3

Practical circuits

The electric circuits around your home are far more complex than the simple circuits you have seen so far. Your home has many lights and power points, a TV, a washing machine, computer and dishwasher. It also has multiple switches and fuses or circuit breakers to protect the circuits—and you—if something goes wrong.

science 4 fun

Tongue circuits

YES

Collect this...

- clean, new nail
- clean copper wire

Do this...

1 Wind one end of the copper wire around near the head of the nail.

2 Place the sharp end of the nail and the free end of the copper wire on your tongue. (Be careful not to cut yourself.)

Record this....

1 Describe what you felt.

2 Explain what you think caused this feeling.

Two main types of circuit are commonly found at home: series and parallel circuits.

Series circuits

In a **series circuit**, all the components of the circuit are connected up one after another to form a single loop. Series circuits are the easiest of all the circuits to connect up. Figure 6.3.1 shows a typical series circuit, in which two identical light globes are connected in series.

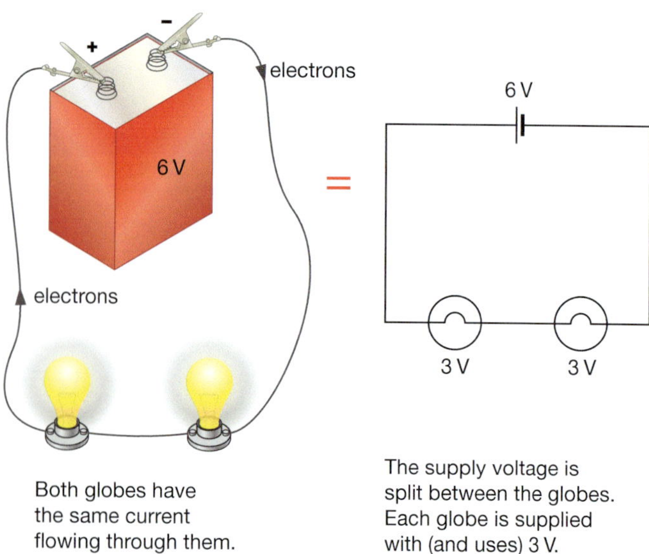

FIGURE 6.3.1 This series circuit has two light globes arranged one after the other.

When the charges leave the battery, they carry a full load of energy (in this case 6 V). Very little energy is lost in the wires because they have a very low resistance. This leaves 6 V worth of energy to be shared equally by both globes. Each globe therefore uses 3 V worth of energy. To get back to the battery, the current must pass through both globes.

In summary, components in a series circuit have the same current flowing through them but split the voltage between them. Series circuits are easy to connect up but are not very practical. This is because:

- the globes cannot be controlled individually with a switch. A switch would turn them all on or all off
- current stops flowing around them if any of the globes 'blow'. This breaks the circuit and causes all the other globes to go out too, making it difficult to locate which globe is the faulty one
- adding more globes to the circuit makes them glow duller than before. This is because the voltage is shared by more globes, so each globe receives less voltage, and so the globe is duller than before. This effect is shown in Figure 6.3.2.

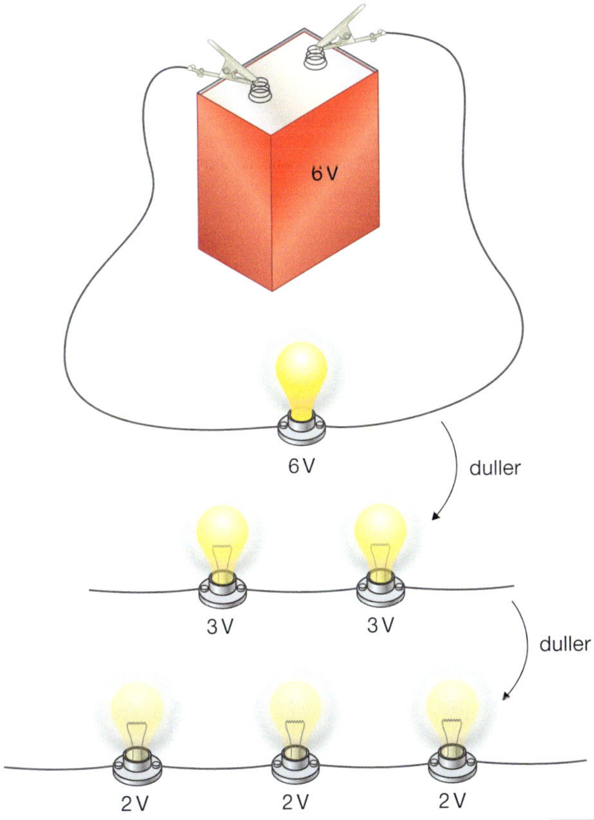

FIGURE 6.3.2 Globes get duller and duller as you add more of them to a series circuit.

AB 6.3

Parallel circuits

A **parallel circuit** has a number of branching circuits, each branch having its own components. A typical parallel circuit is shown in Figure 6.3.3.

In this parallel circuit, the current leaving the battery splits into two, with half going down each branch. An individual electron can only pass through one globe and so it uses all its energy in that one globe. Therefore, each globe receives the full 6 V supplied by the battery.

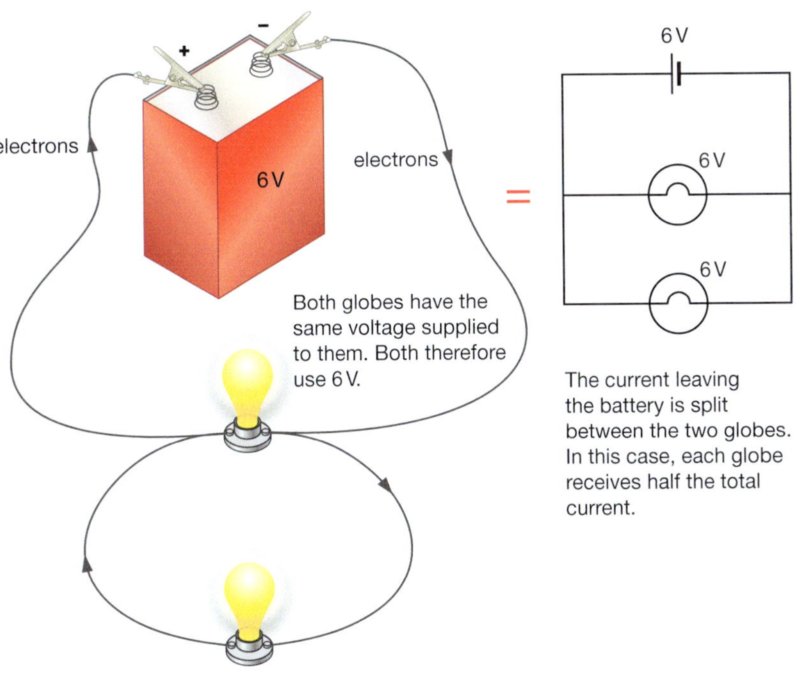

FIGURE 6.3.3 This circuit has two globes in parallel.

Parallel circuits have many advantages over series circuits. In parallel circuits:

- each branch can have its own switch. This allows each globe to be turned on or off independently of the others
- only one branch is affected if a globe 'blows'. All the others keep working. This also makes it easy to find the faulty globe
- adding extra globes does not affect their brightness. This is shown in Figure 6.3.4 on page 240. Each branch always receives the full supply voltage, regardless of how many globes there are.

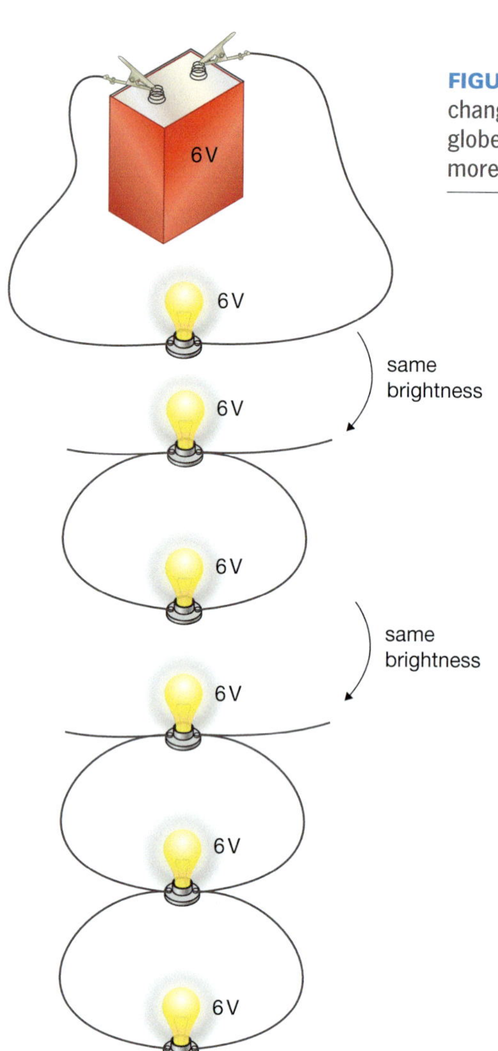

FIGURE 6.3.4 There is no change in the brightness of globes in a parallel circuit when more of them are added.

In summary:

- Components in a series circuit have the same current through them but split the voltage between them.
- Components in a parallel circuit have the same voltage across them but split the current between them.

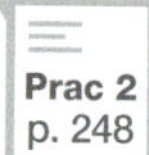

Combination circuits

Sometimes circuits have some of their components arranged in series and other components in parallel. Consider the circuit in Figure 6.3.5. In this circuit, two globes (B and C) are in series with each other. These two globes are in parallel with globe A.

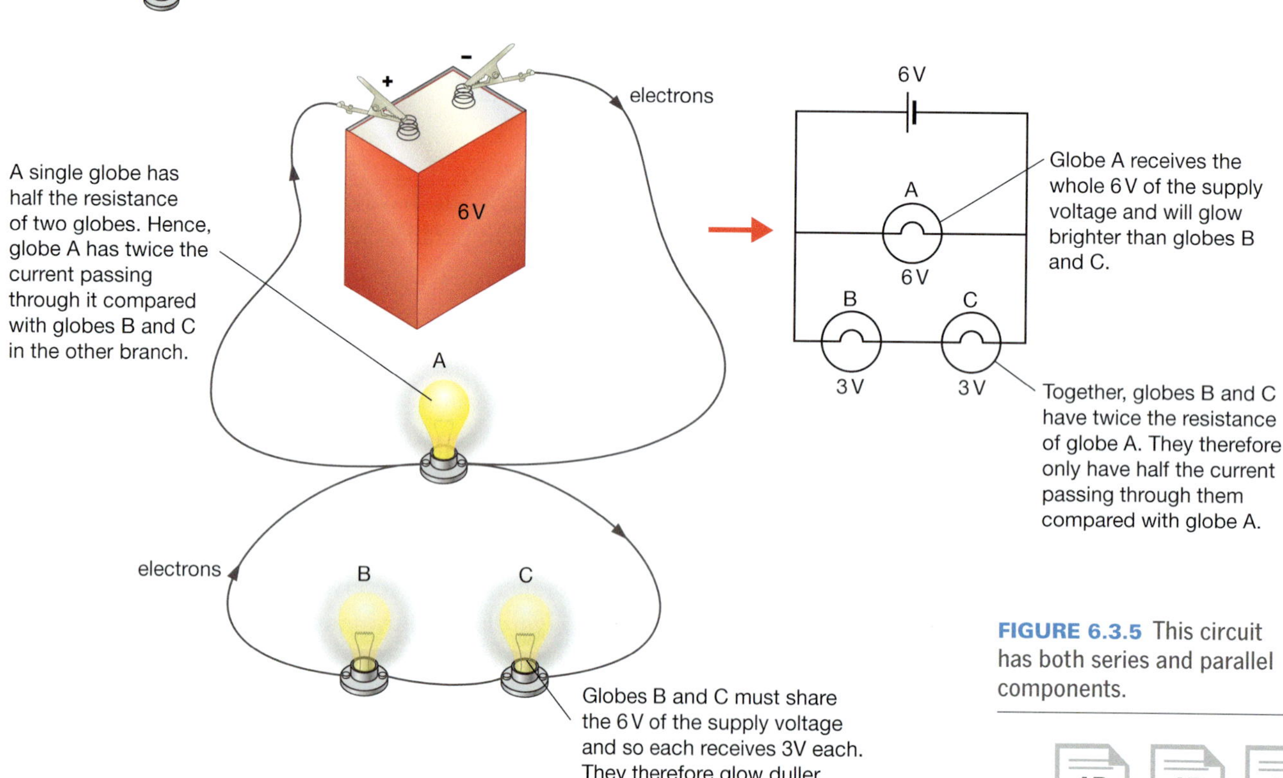

FIGURE 6.3.5 This circuit has both series and parallel components.

AB 6.4 AB 6.5 AB 6.6

Household wiring

The electrical wiring in a house or an apartment is one large parallel circuit, with each light and power point located on its own branch with its own switch. Each receives the full supply voltage of 240 V, allowing each to work at full power. A simplified version of a typical household circuit is shown in Figure 6.3.6.

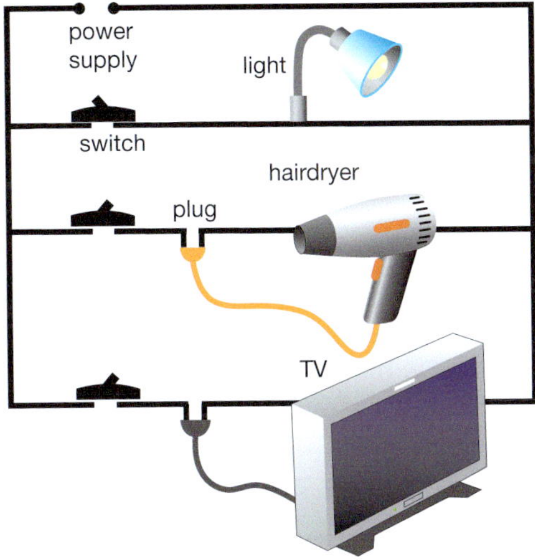

FIGURE 6.3.6 The circuits around your home are connected in parallel. This allows everything to be controlled independently and provides everything with 240 V.

FIGURE 6.3.7 Power points and cables have three wires. The active and neutral wires carry the current to and from the appliance. The earth wire is included only for emergencies. Stray currents caused by faulty appliances will pass along the earth instead of through you.

The electrical cables supplying power points are made of three separate wires. The **active wire** (coated in brown plastic) carries current to the power point, and the **neutral wire** (blue) carries current away from it. However, 240 V can be deadly if the current finds a way out of the wires and through you. This might happen if part of the circuit within an appliance breaks, allowing a wire to touch the casing or switch. You can then become part of the circuit and current will flow through you instead of down the neutral wire! The result would be an electric shock or possibly **electrocution** (death by electricity). To avoid this possibility, power points have a third wire, called the **earth wire** (coated in green and yellow plastic). This wire connects the power point (and any metal part of an appliance connected to it) to the earth beneath you. This provides a way for dangerous stray currents to flow out of the appliance without passing through you. The arrangement of active, neutral and earth wires within an extension lead is shown in Figure 6.3.7.

SciFile

Colour changes

Old wiring may have red (active), black (neutral) and green (earth) wires. Up to 8% of electricians are red/green colour blind and cannot tell the difference between a red active wire and a green earth wire. These two wires are deadly if swapped! For the safety of everyone, the colours have been changed to brown, blue and green/yellow.

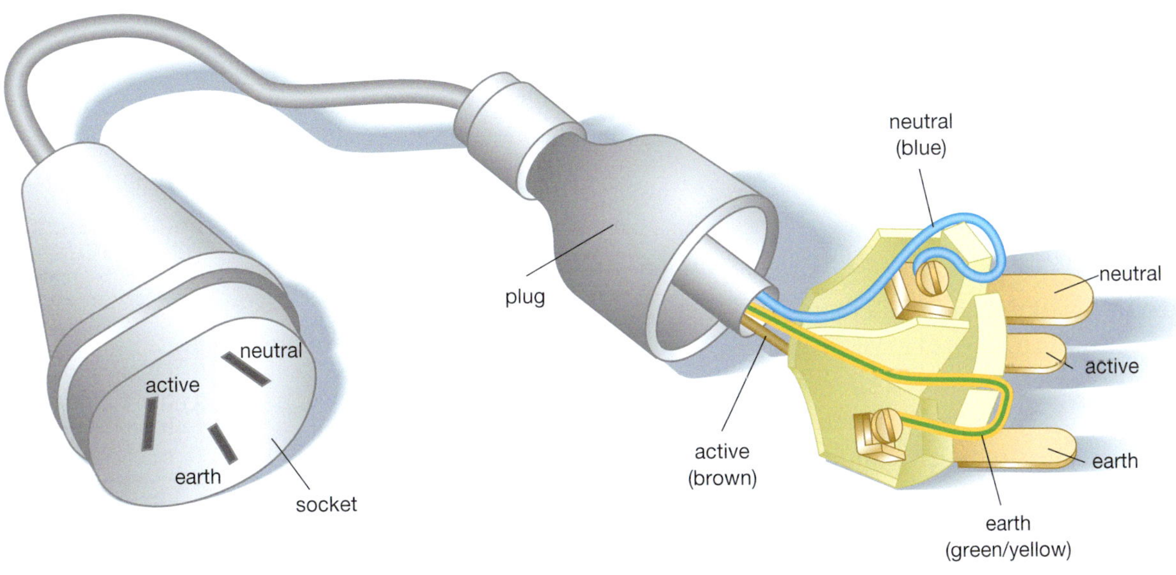

Electrical safety

Most practical circuits also have a device that deliberately breaks the circuit if a faulty appliance allows an abnormally high current to flow. This current might end up passing through you or might set the house on fire. Abnormally high currents cause wires to heat up rapidly. This might melt their plastic coatings. They can then set fire to the fluff and dust trapped with the wires in the walls and roofspace. This is what happened in Figure 6.3.8.

FIGURE 6.3.8 Old or faulty electrical wiring is the most common cause of house fires.

Fuses

A **fuse** is a wire of high resistance and low melting point. It will melt if too much current flows along it. Melting breaks the circuit and stops the current. Fuses are common in older houses, and are still used in cars and trucks and electronic devices such as music and home theatre systems. A typical car fuse box is shown in Figure 6.3.9.

Prac 3
p. 249

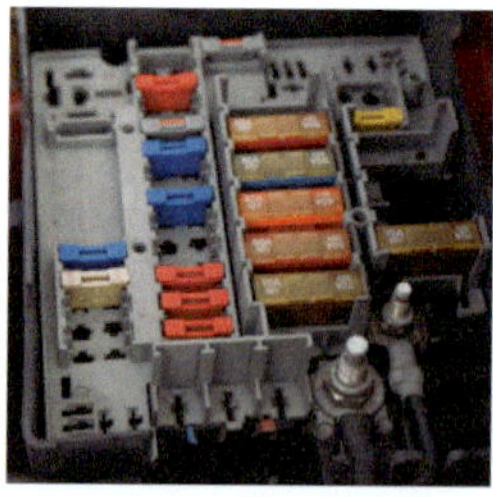

FIGURE 6.3.9 Each colour fuse in this car will melt at a different temperature and current.

Circuit breakers

New houses generally use circuit breakers instead of fuses. A **circuit breaker** is a switch that is activated by a higher-than-normal current. This most commonly happens if there is a short-circuit. A short-circuit can occur when a component breaks in a circuit which creates an easier path for the current to travel through. A short-circuit causes a massive current to flow, dangerously overheating the circuit and putting people at risk of electrocution. When this happens, the circuit breaker switches 'off', breaking the circuit. Figure 6.3.10 shows a typical household switchboard or junction box. Each circuit breaker controls a different circuit. One controls the lights while another circuit breaker controls the air conditioner. Others control the circuits of different clusters of rooms such as the family room, kitchen and bedrooms.

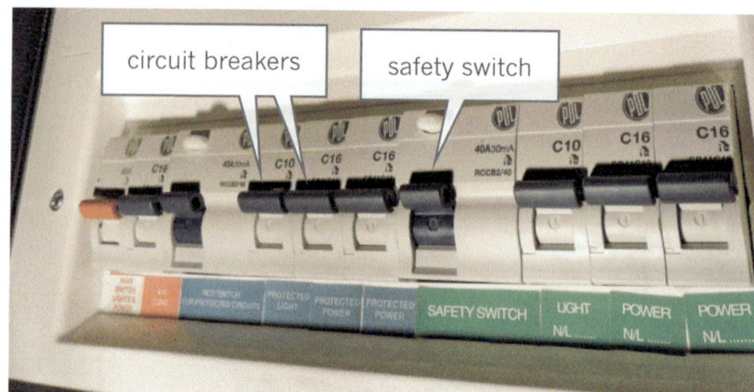

FIGURE 6.3.10 You will probably find your home's switchboard near the front door. Those in most newer homes look like this one, with a set of circuit breaker switches and safety switch.

Safety switches

Modern home switchboards also have a **safety switch** on their lighting and power circuits that monitors how much current is flowing through them. The current flowing into the house through the safety switch should be the same as the current flowing out of the house through the same safety switch. If they are different, then current is likely to be 'leaking' either into a faulty appliance or into you. When the safety switch detects a leak, it breaks the circuit within 0.03 seconds. This stops any further current from flowing. In this time, you will still receive a nasty shock but hopefully the current will be switched off fast enough to stop you being electrocuted.

Some power points around a home may also have their own safety switch, protecting you from any faulty appliances attached to it. These power points have a small blue reset/test button, as seen in Figure 6.3.11.

A safety switch is also known as a residual current device (RCD).

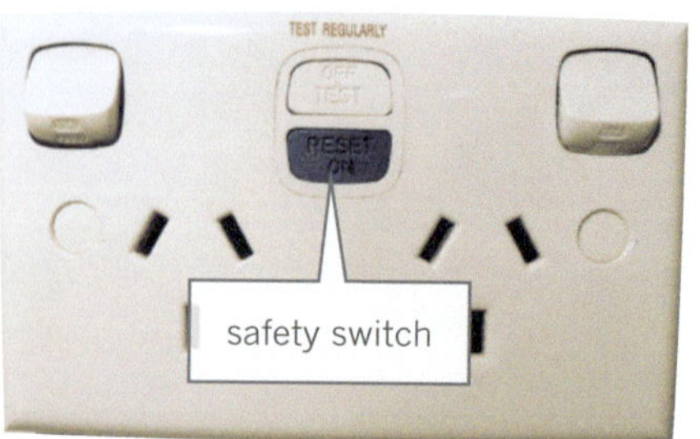

FIGURE 6.3.11 Some power points have their own safety switch.

SCIENCE AS A HUMAN ENDEAVOUR

Use and influence of science

Electronic circuits

FIGURE 6.3.12 Delicate electronic circuits control the heavier duty electric circuits that run the heater, motor and pump in a washing machine.

Most modern appliances have a combination of both electric and electronic circuits. For example, electric circuits in a washing machine will power the heavy-duty components like its water-heater, pump and motor. Electronic circuits will control these electric circuits, turning the heater and pump on and off when necessary and triggering the machine to move from wash to rinse to spin-dry cycles (Figure 6.3.12).

Electronic circuit components

Most components used in electric circuits are able to withstand relatively high voltages and currents. Around the home, electricity is usually 240 V, delivered as alternating current (AC). Electric circuits are used to power our lights, heat our water and power motors and pumps in washing machines, dishwashers and airconditioners. The components used in an electric circuit are referred to as being 'passive'. This means that they have no control over the current flowing through them.

In contrast, components in electronic circuits require much smaller voltages, typically in the range 3 to 12 V. The electronic components of smartphones, tablets and laptops would be burnt out if plugged directly into a power point—transformers need to drop the voltage from 240 V. Current through electronic circuits must also be low and most components need direct current (DC). The components in an electronic circuit are 'active'. The ability to control, block or amplify current or store its charge gives electronic circuits the ability to process and change information sent in the form of a current. The symbols for some electronic components are shown in Table 6.3.1 on page 244.

The components of an electronic circuit can be wired up just like the components in an electric circuit. However, most electronic circuits are integrated onto a board with the connecting wires replaced with conducting lines printed onto a circuit board (Figure 6.3.13).

FIGURE 6.3.13 Electronic circuits commonly use strips of metal conductor printed on an insulating material base, as shown in this magnified image.

SCIENCE AS A HUMAN
ENDEAVOUR

TABLE 6.3.1 Electronic components and their symbols

Component	Illustration	Symbol
resistor		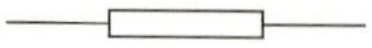
LDR		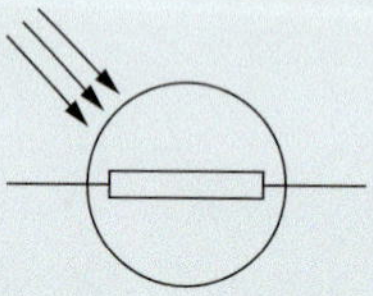
thermistor		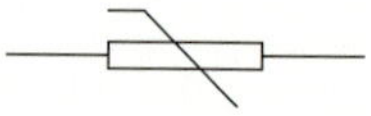
capacitor	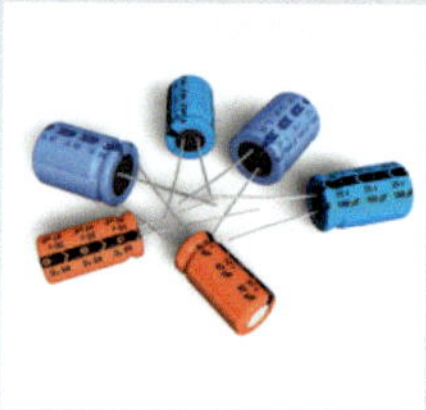	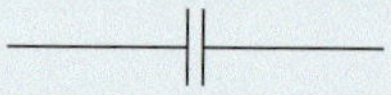
Diode		
LED	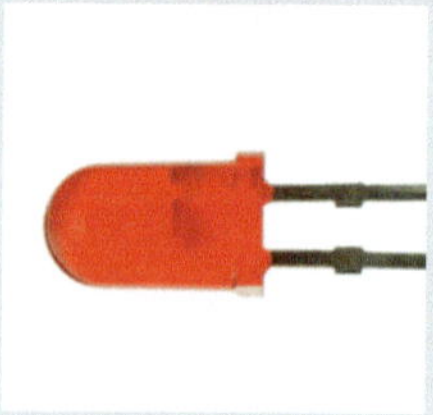	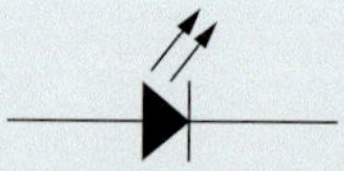
transistor		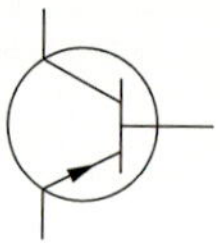

Resistors

Electronic components are fragile and large currents damage or burn them out. Resistors are used in electronic circuits to reduce the current to manageable levels.

A light-dependent resistor (LDR) has a resistance that changes as the amount of light falling on it changes—as the light brightens, its resistance decreases. LDRs are light sensors that can control when lights automatically turn on and off. They are used in street lights, solar garden lights and light-sensitive car headlights.

Thermistors are a type of resistor whose resistance changes as the temperature changes—as the temperature increases, its resistance decreases. Thermistors are a type of temperature sensor that can be used to control sprinkler systems and fire alarms.

Capacitors

Capacitors store charge. This charge is then discharged (released) a short time later. Different capacitors store different amounts of charge and discharge after different times. They are used to control timers for things like clocks and flashing lights. They are also used to block steady currents (DC) while allowing varying currents (AC) through and to 'fill in' gaps in the current if it drops below what the circuit needs.

Diodes

Diodes are components that allow current to flow through them in one direction, but block any current attempting to flow through in the other direction. The direction that current can flow is indicated by the direction of the arrow head in its symbol. In alternating current (AC), the current keeps changing direction. A diode will therefore block half of the current while letting the other half through. This is why diodes are used to convert AC into DC. Some diodes emit light. These are called light-emitting diodes or LEDs.

Prac 4
p. 249

Transistors

Transistors are used as an electronic switch or to amplify and increase currents through part of a circuit. Transistors are made from the semiconductors silicon or germanium. Semiconductors conduct electricity at normal room temperatures but act as insulators when cold.

REVIEW

1 Compare electric circuits with electronic circuits.
2 Explain why a dishwasher needs a combination of electric circuits and electronic circuits to operate.
3 What feature of thermistors and light-dependent resistors allows them to be used as control devices for fire sprinklers and automatic car headlights?

MODULE 6.3

Review questions

Remembering

1 Define the terms:
   a parallel circuit
   b fuse
   c electrocution.

2 What term best describes each of the following?
   a a circuit with all the components in a line
   b wire connecting a power point to the ground
   c switch activated by higher-than-normal current.

3 What are the advantages and disadvantages of a series circuit?

4 State the colours of the following wires used in household wiring.
   a active
   b neutral
   c earth.

5 How long does it take for a safety switch (RCD) to activate?

Understanding

6 Why is a home wired as a parallel circuit instead of as a series circuit?

7 How can electrical faults cause house fires?

8 Describe how a safety switch (RCD) detects a problem in the circuit.

Applying

9 Use the key below to fill in the following statements. Options can be used more than once.

Key
A don't change
B don't light up
C shine brighter
D shine less brightly

   a If another globe is added to a series circuit, the globes ________.
   b If another globe is added to a parallel circuit, the globes ________.
   c If one globe in a series circuit 'blows', the others _________.
   d If one globe in a parallel circuit 'blows', the others _______.

Analysing

10 Analyse the circuit in Figure 6.3.14, then use the key below to fill in the following statements. Options can be used more than once.

Key
A the same as
B twice
C half
D three times

   a The current through globe A is ______ the current through globe B.
   b The current through globe B is ______ the current through globe C.
   c The voltage across globe A is ______ the voltage across globe B.
   d The voltage across globe B is _____ the voltage across globe C.

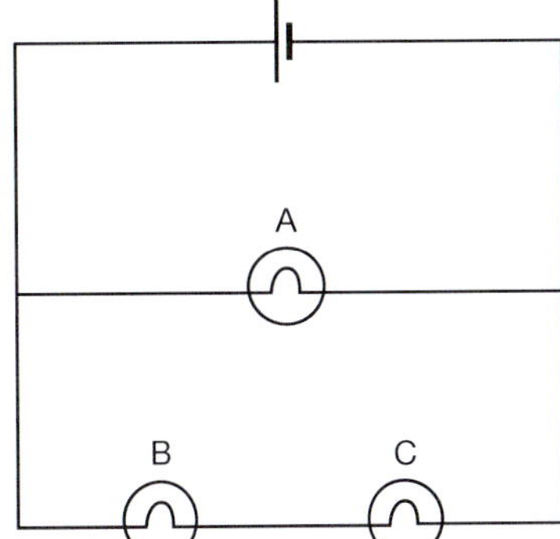

FIGURE 6.3.14

11 Compare a fuse with a circuit breaker by listing their similarities and differences.

12 A surge protector is a device that protects delicate electronic devices such as computers from surges in electric current that may occur because of lightning strikes. It detects the surge in current and then breaks the circuit, protecting any device attached to it.
Compare a safety switch with a surge protector.

MODULE

6.3 Review questions

13 Analyse the circuit in Figure 6.3.15 to complete Table 6.3.2 indicating which globes would be on or off.

Note: Closed means that current can flow through the switch. Open means that current cannot flow through the switch.

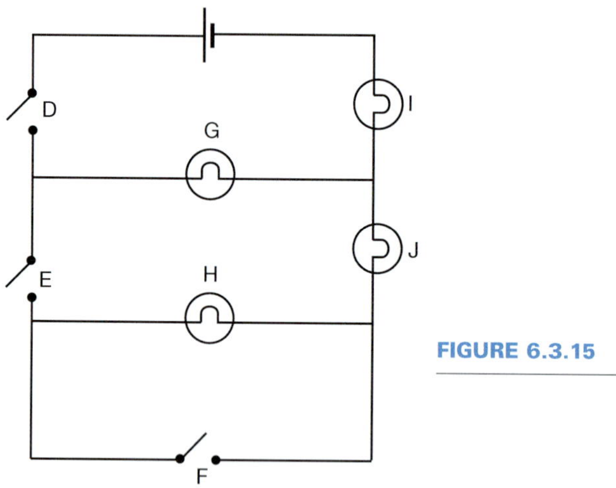

FIGURE 6.3.15

14 Predict the order of brightness (from brightest to dullest) of the globes in Figure 6.3.16.

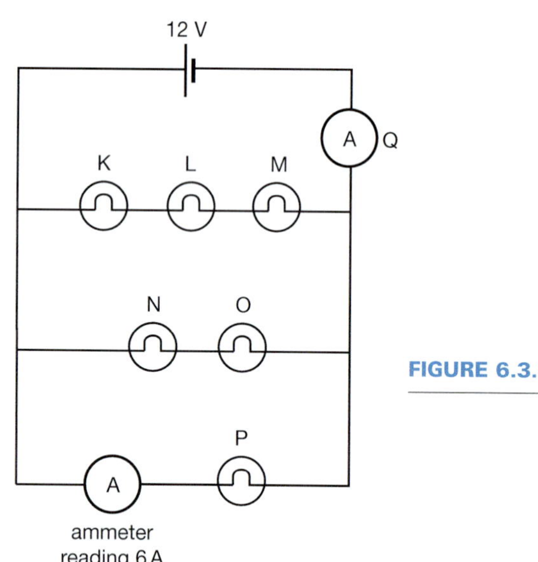

FIGURE 6.3.16

TABLE 6.3.2 Globe status for different combinations of switches D–F

	Switch D	Switch E	Switch F	Globe G	Globe H	Globe I	Globe J
a	closed	closed	open				
b	open	closed	closed				
c	closed	open	closed				
d	closed	closed	closed				

15 a Analyse the circuit shown in Figure 6.3.16 to complete the following table.

Current and voltage for globes K–P

	Globe					
	K	L	M	N	O	P
current (A)						
voltage (V)						

b Calculate the current through ammeter Q.

Evaluating

16 What do you think would happen if a car had all its electrical components (such as windscreen wipers, headlights, blinkers, internal light, radio) wired up in series, not parallel?

17 A primitive battery can form when two different metals touch each other or if they both touch something that is salty and moist. Use this information to propose a reason why your tongue tingles in the science4fun activity on page 238.

Creating

18 Construct a circuit diagram like that shown in Figure 6.3.14, but add a switch that would turn on and off:

a all globes in the circuit

b globe A only.

MODULE 6.3

Practical investigations

1 • Series and parallel circuits

Purpose

To compare series and parallel circuits.

Timing 45 minutes

Materials

- three globes (preferably 6 V)
- power pack
- connecting wires
- switch
- ammeter
- voltmeter

SAFETY

If using a power pack of variable voltage, do not go higher than 6 V.

Procedure

1 In your workbook, construct a table like the one in the Results section.

2 Connect up the basic circuit shown in Figure 6.3.17.

3 Measure the current flowing through the globe, and the voltage lost across it.

4 Add another globe to construct the series circuit shown in Figure 6.3.17.

5 Note the brightness (very bright/bright/dull/very dull) of the globes, and measure the current and voltage.

6 Remove the second globe and reconnect it so that it is in parallel as shown in Figure 6.3.17.

7 Once again, note the brightness of the globes, and measure the current and voltage.

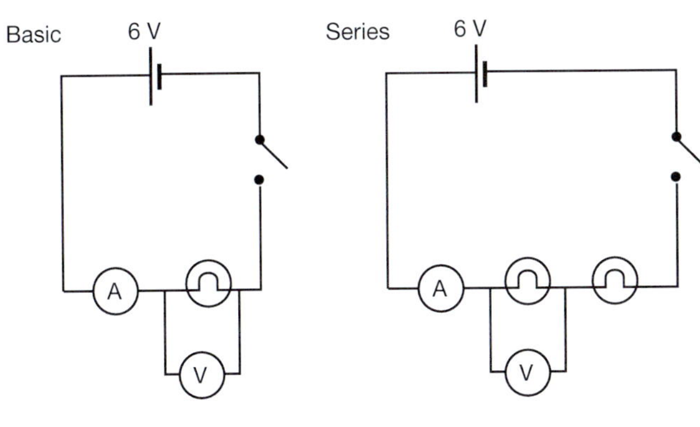

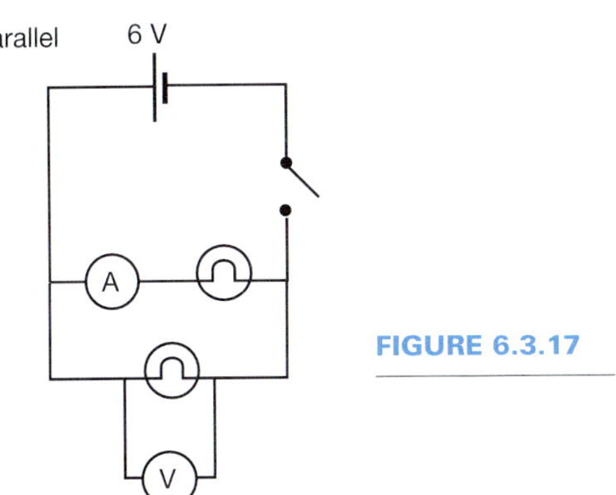

FIGURE 6.3.17

Results

Record your results in the table.

Globe brightness, current and voltage

	Single globe	Globes in series	Globes in parallel
Brightness			
Current (A)			
Voltage (V)			

Review

1 Describe what happened to the current when another globe was added in series.

2 Use your knowledge of resistance to explain why this happened.

3 Adding another globe in series makes all the globes duller. Explain why.

4 Explain why adding globes in parallel makes no difference to their brightness.

MODULE 6.3 Practical investigations

2 • Binary counting

Questioning & Predicting | Evaluating

A switch can turn a circuit off (given the symbol 0) or on (given the symbol 1). These 0 and 1 numbers are called binary numbers.

Purpose

To construct a simple circuit that can count from 0 to 3.

Timing 45 minutes

Materials

- dry cell battery, batteries or power pack
- 2 globes (of a lesser voltage than battery pack or battery)
- 2 switches
- connecting wires with alligator clips

Procedure

1 Copy the table below into your workbook.

2 Analyse the circuit in Figure 6.3.18 and predict whether each globe will be off (0) or on (1) when the switches are in the positions shown in the table. Record your predictions in columns 4 and 5 of your table.

3 Construct the circuit and test your predictions. Write your observations in column 5 and 6.

Results

1 Predict what each globe will do in each case below. Write your predictions in columns 4 and 5 of your table.

2 Write your observations in columns 6 and 7.

Review

1 Assess how accurate your predictions were.

2 Construct a circuit diagram for Figure 6.3.18.

3 Explain how this circuit counts from 0 to 3.

4 Propose how this circuit could be adapted to count higher than 3.

5 Binary is a word describing anything that has only two choices. Explain how switches are binary.

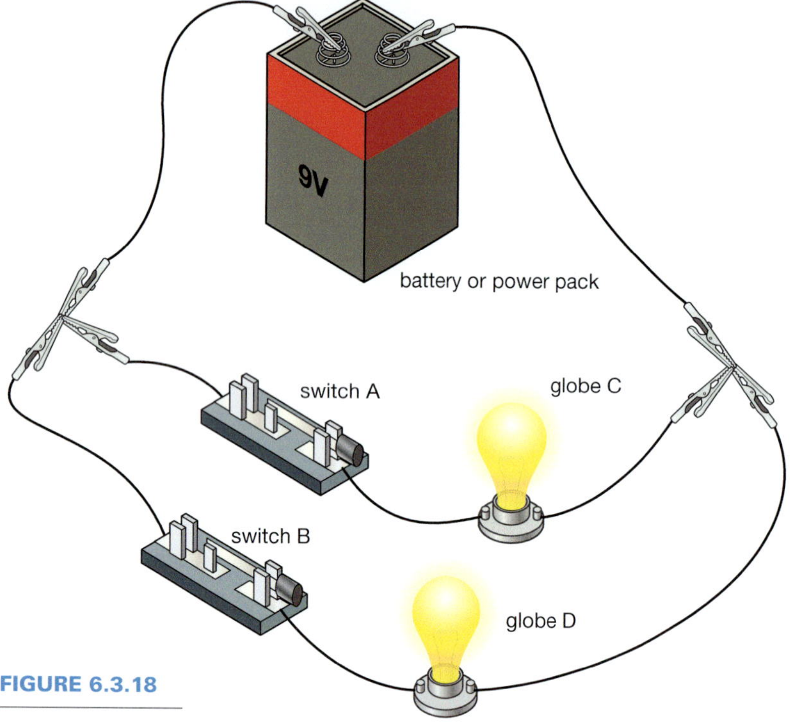

FIGURE 6.3.18

Circuit results

Switching arrangement	Switch A	Switch B	Predicted		Observed	
			Globe C	Globe D	Globe C	Globe D
0 (start)	0 (off)	0 (off)				
1	0	1 (on)				
2	1	0				
3	1	1				

MODULE 6.3

Practical investigations

3 • Fuses

Purpose

To construct and test a fuse.

Timing 45 minutes

Materials

- 2 dry cell batteries
- cork
- steel wool
- globe
- connecting wires
- sticky tape
- screwdriver with insulating handle
- access to pliers

SAFETY

Do not use a nail, screw or any uninsulated metal to cause the short circuit. See Activity Book Toolkit to assist with developing a risk assessment.

Your screwdriver must have an insulating handle.

Procedure

1. Use the pliers to strip a short section of insulation off the ends of two connecting wires.
2. Also strip off a short section of the insulation halfway down each connecting wire.
3. Construct the circuit shown in Figure 6.3.19.
4. To construct the fuse, tease out a strand of steel wool and wind it around the ends of the connecting wires.

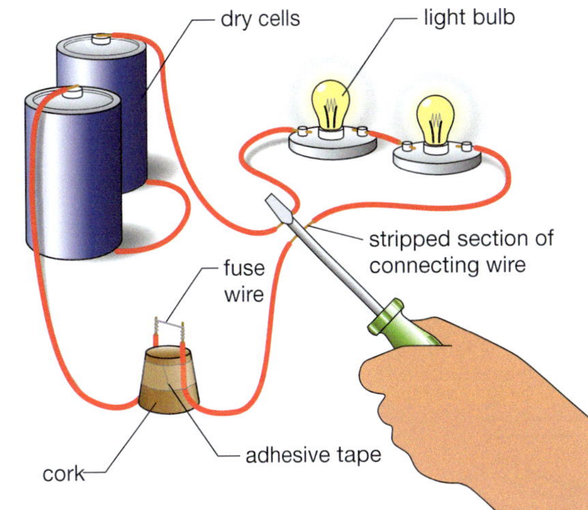

FIGURE 6.3.19

5. Deliberately short circuit this circuit by touching the screwdriver as shown across the two stripped sections of the connecting wires. As you do so, watch the fuse.
6. Thicken the wad of steel wool used as a fuse by adding some more to it. Then repeat the test.

Review

1. Explain the purpose of a fuse in a circuit.
2. Describe what the fuse did here when there was a short circuit.
3. Describe what happened when thicker 'fuses' were used.

4 • Investigating the diode

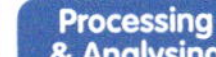

A diode (Figure 6.3.20) is an electronic component made of the semiconductor silicon with aluminium and phosphorus. A diode passes current in one direction only, blocking any current that attempts to pass in the opposite direction.

The diode only needs a small voltage to operate and it can handle only a small current. The circuit symbol for a diode is shown in Figure 6.3.21.

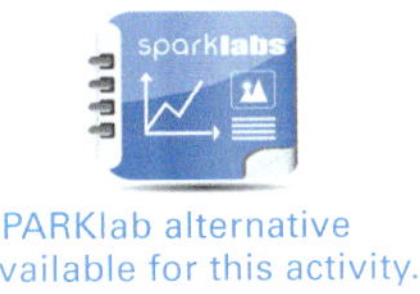

SPARKlab alternative available for this activity.

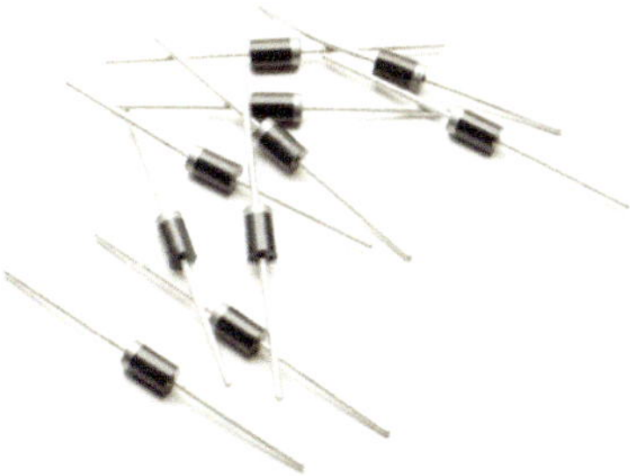

FIGURE 6.3.20 Typical silicon diodes

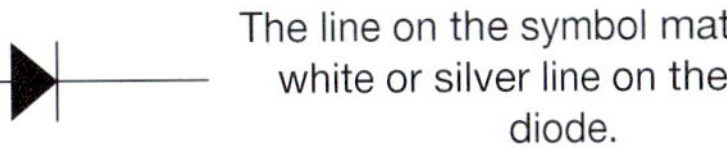

FIGURE 6.3.21 The triangle of the symbol points like an arrow in the direction the current can pass through. Current cannot pass in the opposite direction.

MODULE

6.3 Practical investigations

Other types of diodes emit light when current is passed through them. These are called light-emitting diodes or LEDs (Figure 6.3.22).

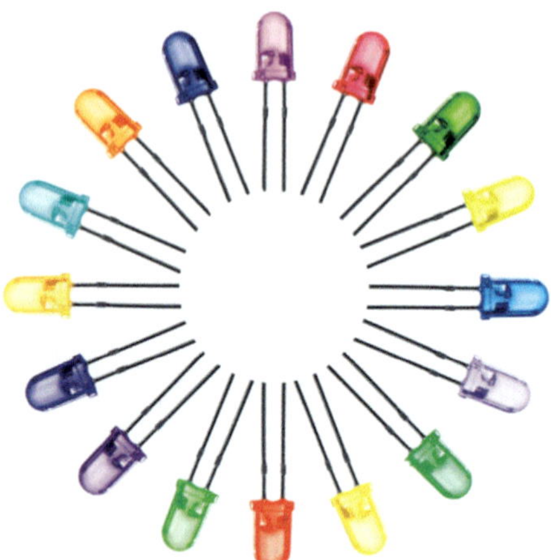

FIGURE 6.3.22 LEDs come in many different colours.

Purpose

To investigate the behaviour of a silicon diode in a simple circuit.

Timing 45 minutes

Materials

- power pack with variable supply voltage
- silicon diode
- 56 Ω resistor
- ammeter
- voltmeter (or 2 multimeters)
- connecting leads with alligator clips
- coloured LEDs (red, green, yellow etc.)

Procedure

1 Set up the circuit shown in Figure 6.3.23. Make sure the band on the diode is pointing away from the positive terminal of the power supply.

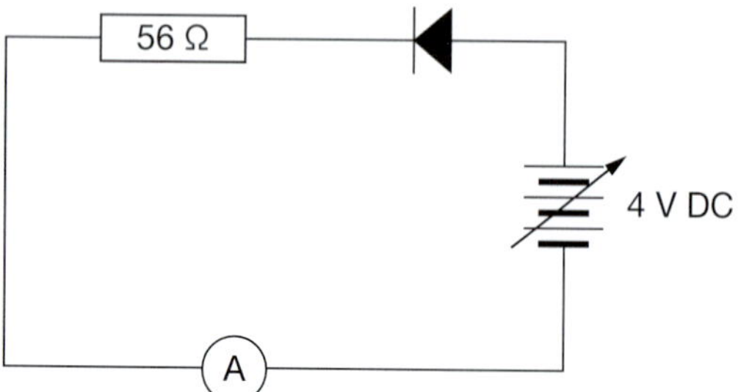

FIGURE 6.3.23

2 Set the power pack on 4 V DC. Turn it on and record the reading on the ammeter.

3 Reverse the direction of the diode. The diode should now be blocking the current so there should be no reading on the ammeter. Check the reading on the ammeter and confirm that it is zero.

4 Reverse the diode again so that it conducts. Connect the voltmeter across it so that it is parallel with the diode. Switch the power on and record the reading on the voltmeter.

5 Shift the voltmeter so that it now piggybacks the resistor instead of the diode. Once again, record the voltmeter reading.

6 Replace the diode in your circuit with an LED. The two wires on the LED are called 'legs'. The longer leg must be connected to the positive wire from the power supply.

7 Determine the voltage drop across the LED required for it to light up. The setting of the power supply can be altered from 4 V DC if needed. Check if it still works when reversed.

Review

1 Was your circuit a series or parallel circuit?

2 When a diode is connected the wrong way around, it is referred to as being 'reverse biased'.
   a Do you think this a good name for it?
   b Justify your answer.

3 Propose a reason why a diode is sometimes called a 'one way gate'.

4 Which required the largest voltage to operate, the diode or the LED?

5 Compare the behaviour of the diode and the LED, describing their similarities and differences.

MODULE

6.4 Electromagnets, motors and generators

Electricity and magnetism are related to each other. The relationship explains how electricity is produced, and how electric motors, electromagnets and transformers work.

science 4 fun

Pick-me-up!

YES

Collect this ...

- paperclips
- bolt, large nail or large screw
- C cell battery
- insulated copper wire
- access to pliers

Do this ...

1. Cut off 15 to 20 cm of electrical wire and use the pliers to strip the wire of its insulating plastic.
2. Coil the wire around a bolt, a large nail or a screw.
3. With one hand, hold the ends of the wire against the ends of the battery.
4. With the other hand, use the end of the bolt, nail or screw to try and pick up a paperclip.

Record this ...

1. Describe what happened.
2. Explain why you think this happened.

STEM 4 fun

Find a direction

PROBLEM
Can you find the direction north using a magnet?

SUPPLIES

- paper, sticky tape, cotton, bar magnet

PLAN AND DESIGN Design the solution. What information do you need to solve the problem? Draw a diagram. Make a list of materials you will need and steps you will take.

CREATE Follow your plan. Produce your solution to the problem. Take a video.

IMPROVE What works? What doesn't? How do you know it solves the problem? What could work better? Modify your design to make it better. Test it out.

REFLECTION

1. What area of STEM did you work in today?
2. In what career do these activities connect?
3. What did you do today that worked well? What didn't work well?

Magnetism

Around a permanent magnet is an invisible force field called a **magnetic field**. This field exerts forces on:

- materials containing large quantities of iron, cobalt or nickel. This explains why magnets can be used to separate iron and steel nails and screws from plastic and scrap
- other magnets nearby. Each magnet has a north pole (N) and a south pole (S). What two magnets do depends on which poles are near one another: unlike poles (N/S) attract and like poles (N/N or S/S) repel.

The direction and strength of a magnetic field is shown by its **field lines**. Field lines show the direction of the force on iron filings and compass needles. For example, the iron filings in Figure 6.4.1 have aligned (lined up) with the field lines of a bar and a horseshoe magnet.

The needle of a compass is also a magnet and it is also attracted and repelled by the magnetic field of the Earth and of nearby magnets.

FIGURE 6.4.1 Iron filings clearly show the shape of a magnetic field.

Electromagnetism

A magnetic field is also produced when an electric current flows along a wire. When compasses are placed around the wire, their needles align with the magnetic field around the wire. As Figure 6.4.2 shows, the field produced is a set of circular rings. In this case, electricity has caused magnetism. This is known as **electromagnetism**.

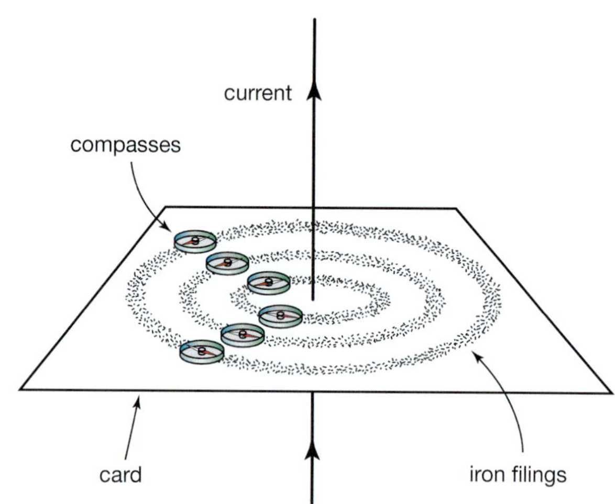

FIGURE 6.4.2 The magnetic field around a current-carrying wire forms concentric rings.

Electromagnets

If a current-carrying wire is twisted around to form a loop, the magnetic field down its centre is reinforced and made stronger. If the wire is wound many times to form many coils, then the magnetic field down its core is made stronger. This looped, current-carrying coil is known as a **solenoid**. The magnetic field it creates is shown in Figure 6.4.3. The magnetic field down the core of the solenoid is made even stronger when an iron rod is placed down it. This device is known as an **electromagnet**.

FIGURE 6.4.3 A solenoid has a magnetic field similar to that of a bar magnet. Insert an iron rod down the core and you get an electromagnet.

The magnetic field of an electromagnet depends on the current flowing through it. This means that an electromagnet can be controlled in ways that are not possible with a permanent magnet. Table 6.4.1 compares an electromagnet with a permanent magnet.

TABLE 6.4.1 An electromagnet compared with a permanent magnet

Properties of magnet	Electromagnet	Permanent magnet
magnet on or off	'on' if current flows through it 'off' if no current flows through it	always 'on'
strength of the magnet	can be altered by changing its current	cannot be adjusted
direction of the magnetic field	can be changed by changing the direction of the current passing through it	can only be changed by flipping the magnet end to end

Using electromagnets

Electromagnets are used in many ways. They are used in car starter motors, to operate automatic latches and to separate iron and steel metal in junkyards. When too much current flows through a house circuit, electromagnets in circuit breakers pull the switch open. This breaks the current running through it.

You use electromagnets every day when you listen to your smartphone or watch TV. Within every speaker, earplug or headphone is a cone that is connected to an electromagnet. As Figure 6.4.4 shows, a speaker has a permanent magnet and an electromagnet. The current fed into the electromagnet changes as the music changes its pitch and volume. The two magnets either attract or repel each other, causing the cone of the speaker to vibrate. These vibrations in turn cause the sounds you hear.

Build a loudspeaker

PROBLEM
Can you build a loudspeaker with things you find at home?

SUPPLIES
- thin wire, small magnet
- empty yoghurt cup, tape, glue, sandpaper, wire cutters or scissors
- radio or sound source with detachable speakers

PLAN AND DESIGN Produce the solution. What information do you need to solve the problem? Draw a diagram. Make a list of materials you will need and steps you will take.

CREATE Follow your plan. Draw your solution to the problem. Take a video.

IMPROVE What works? What doesn't? How do you know it solves the problem? What could work better? Modify your design to make it better. Test it out.

REFLECTION
1 What area of STEM did you work in today?
2 In what career do these activities connect?
3 What did you do today that worked well? What didn't work well?

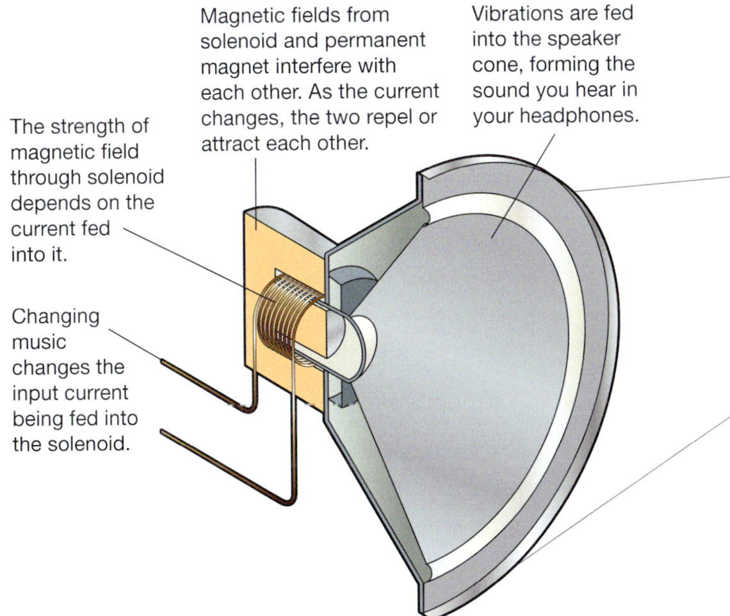

FIGURE 6.4.4 Speakers, headphones and earphones work because two magnetic fields attract or repel each other, causing the speaker to move.

Electric motors

Two magnets experience a force when placed within each other's field. Likewise, a current-carrying wire experiences a force and moves whenever it is placed in a magnetic field. In this case, electricity causes a magnetic field, which causes movement. This too is part of the phenomenon of electromagnetism.

An electric **motor** spins about its pivot because of electromagnetism. A current-carrying coil is placed within the magnetic field of a permanent magnet. The two magnetic fields interact and the coil then spins because of the forces on it. This forms the basis of a simple motor, as shown in Figure 6.4.5.

Simple motors like this are commonly used in toys such as slot-cars because these toys don't require much speed or power. Appliances like hairdryers and power drills need stronger motors and use stronger solenoids instead of permanent magnets. As Figure 6.4.6 shows, they also use multiple planes of coils.

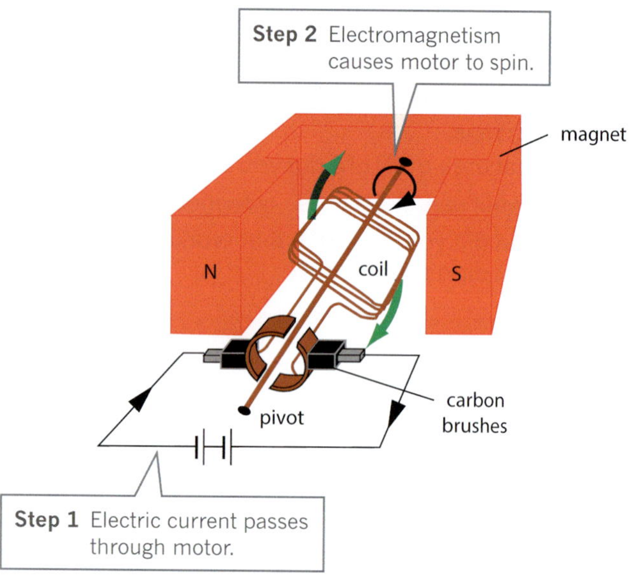

FIGURE 6.4.5 In an electric motor, current is passed through a coil. The magnetic field it produces interacts with the permanent magnetic field of the motor, causing the coil to spin.

FIGURE 6.4.6 Heavy-duty motors like the one in this electric fan use electromagnets instead of permanent magnets.

Prac 1 p. 262

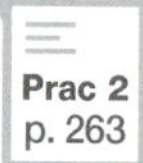
Prac 2 p. 263

Generators

The electricity available from batteries is useful for portable devices. However, the voltages and currents are far too low for most purposes around the home and in industry. This is when a generator is needed. A **generator** uses electromagnetism to generate electricity. A simple generator is shown in Figure 6.4.7. Table 6.4.2 compares an electric motor with a generator.

science 4 fun

Solenoid generators

NO

Collect this ...

- solenoid or coil of wire
- milliammeter or galvanometer
- connecting leads with alligator clips
- bar magnet

Do this ...

1. Connect the solenoid to the milliammeter/ galvanometer.
2. While watching the milliammeter/ galvanometer, quickly insert the bar magnet into the solenoid.
3. Leave the magnet there and then quickly remove it. Watch what happens.
4. Find out what happens when you insert the magnet more slowly, or when two magnets are used instead of one.

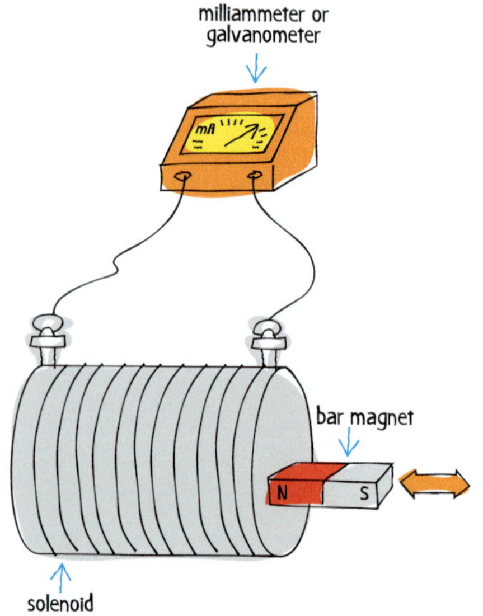

Record this ...

1. Describe what happened.
2. Explain why you think this happened.

TABLE 6.4.2 Comparing motors and generators

	Electric motor	Generator
needs	a coil	a coil
	a magnet	a magnet
	an electric current	spinning movement
produces	spinning movement	an electric current

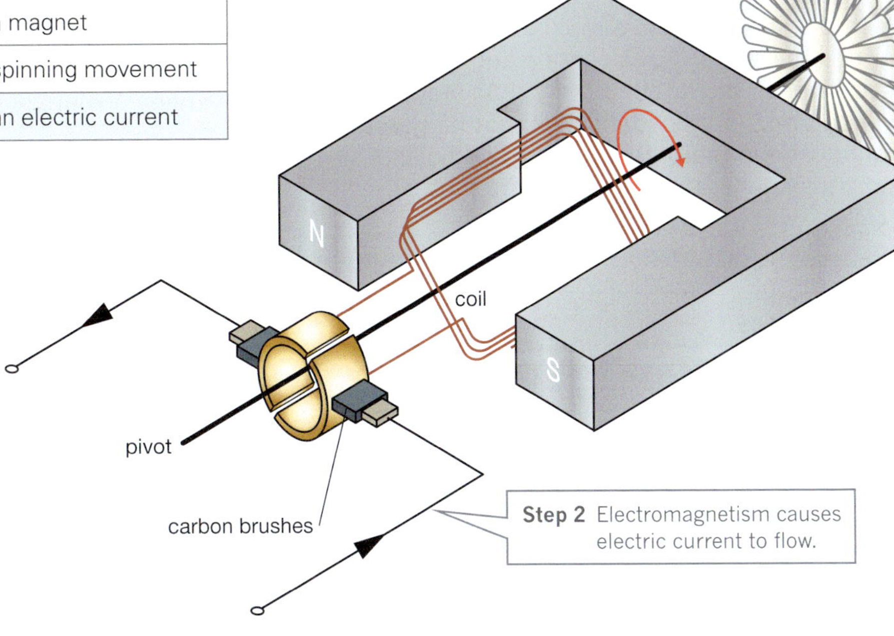

FIGURE 6.4.7 A simple generator shares many of the features of a simple motor.

Dynamos

A **dynamo** is a small generator that spins its magnet instead of its coils. Dynamos are often used to power the front lights on bicycles. As Figure 6.4.8 shows, the wheel spins a magnet alongside a coil, and this generates a current inside the coil.

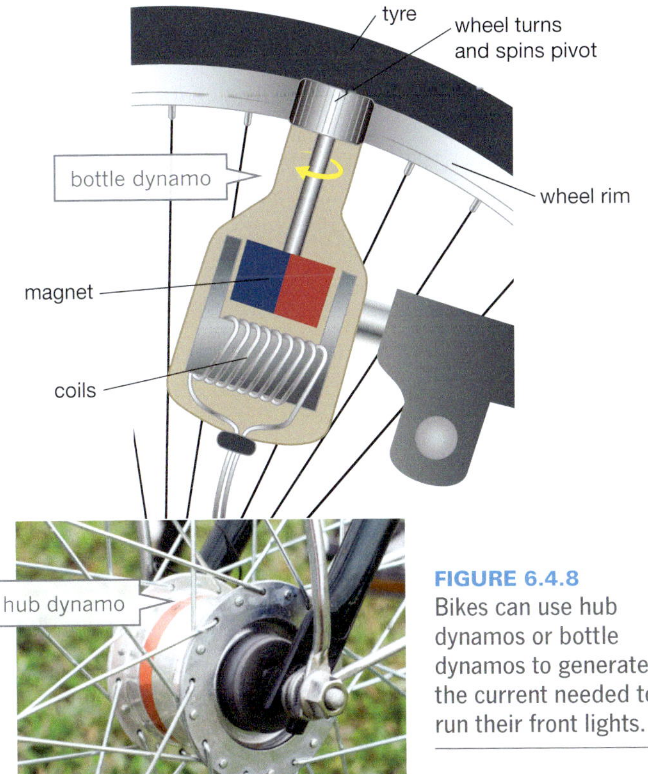

FIGURE 6.4.8 Bikes can use hub dynamos or bottle dynamos to generate the current needed to run their front lights.

Turbines

A simple generator produces electricity only if its coils spin. In contrast, a larger generator spins its magnet and keeps its coils fixed. Anything can be used to spin the coils or magnet. For example, many bicycles use a small generator known as a dynamo to generate current to power their headlights. The dynamo uses the front wheels of the bike to spin a magnet within a set of fixed coils.

A large-scale electricity generator is known as a **turbine**. Turbines need to be spun at high speeds, and different methods can be used to spin them. Figure 6.4.9 shows turbines being turned by the wind. More commonly, moving water or steam is used.

FIGURE 6.4.9 Wind turbines use a renewable resource to generate electricity.

SciFile

SOS!

Electricity is difficult to supply to the outback. For this reason, outback cattle stations once used bikes to power their emergency radios. One person would pedal while another person would use the radio. Solar panels and diesel and petrol generators now provide these stations with the electricity they need.

Moving water

There are a number of ways in which water can be made to spin a turbine.

- Hydro-electricity is generated by water falling onto the blades of the turbines, spinning them like old-fashioned but very fast water wheels.
- Wave power uses the regular swells of the ocean to rock the turbines back and forth.
- Tidal power uses the massive flows of water from the twice-daily changes of the tides to spin turbines. One design is shown in Figure 6.4.10.

FIGURE 6.4.10 Water rushing past the blades as the tide comes in and goes out spins the turbines to generate electricity. This design resembles an underwater wind turbine.

SciFile

That's fast!

In Australia, AC electricity changes direction 50 times every second. This means that the turbines that make it are spinning 50 times every second!

Steam

Most power plants around the world are basically big kettles, because they boil water and change it into high-pressure steam. This steam then spins the turbines. Afterwards, the steam is usually released via cooling towers. The heat required to run them can come from a variety of sources, such as:

- burning fossil fuels such as coal or gas
- burning biomass, which is any biological material such as wood, leftover woodchips, sugarcane waste, methane gas produced by human and animal waste
- geothermal energy, where heat energy from deep in the Earth's crust is used to heat water
- nuclear power, where a controlled nuclear reaction produces the heat necessary to generate steam
- solar power, where sunshine is reflected from mirrors onto a central furnace through which water flows.

All these methods of generating power have advantages and disadvantages. For example, burning coal is an easy and cheap way of generating steam (and therefore electricity) but it emits huge quantities of the greenhouse gas carbon dioxide (CO_2). In contrast, nuclear power emits almost no greenhouse gases but it produces wastes that are radioactive for many thousands of years.

A summary of these advantages and disadvantages is shown in Table 6.4.3.

AC/DC

Batteries and solar cells produce **direct current (DC)**. In DC, all the electrons move in one direction and with one voltage. Generators, dynamos and turbines produce **alternating current (AC)**. In AC, the electrons regularly change their direction and so just shuffle back and forth along the wire. The electricity you obtain from a power point is AC and it changes direction 50 times every second, at a frequency of 50 times every second (50 Hz). The voltage also changes. Its 'average' voltage (called RMS voltage) is 240 V, but 100 times a second it climbs to 340 V and then drops to –340 V.

AC is produced because it can generate more power than DC, is easier to transmit and its voltages can be boosted or dropped using transformers. Table 6.4.4 on page 258 compares AC with DC.

TABLE 6.4.3 Comparison of different methods of turning a turbine

	Method	Renewable?	Clean? (emitting minimal CO_2)	Other advantages	Other disadvantages
uses wind	wind	✓	✓	• energy is 'free' • can be located close to where power is needed	• output changes as wind changes • turbulence from one turbine can affect nearby turbines • noisy
uses water	hydro	✓	✓	• quick to start up • can be used to 'top up' power supply when needed	• valleys need to be dammed
	tidal	✓	✓	• energy is 'free' • reliable	• must be on the coast • needs very large tides and so can only be used in a few places
	wave	✓	✓	• energy is 'free' • minimal environmental impact • can be located near coastal cities	• largely experimental • output changes as waves change
uses steam	fossil fuels	✗	✗	• coal is abundant • coal is relatively cheap	• power plant needs to be located near coal mines • emits huge amounts of CO_2 • open-cut mining devastates the environment
	biomass	✓	✓	• removes waste and carbon dioxide from the environment • can be located near where electricity is needed • individual houses, farms and factories can power themselves	• large amounts of CO_2 emitted • biofuels use crops that could instead be used as food
	geothermal	✓	✓	• energy is 'free'	• only countries on fault lines can have large-scale plants
	nuclear	✗	✓	• clean when operating normally • uses very little nuclear fuel	• waste remains radioactive for many thousands of years • disastrous if something goes wrong
	solar power	✓	✓	• energy is 'free' • can provide power to inland cities	• needs to be in sunny areas • needs mirrors to track the Sun • takes up wide areas of flat landscape

TABLE 6.4.4 Comparing AC with DC

	AC	DC
power supplied	• suitable for industry and household appliances	• suitable for smaller appliances only
voltage changes	• AC voltages can be boosted or dropped using transformers	• DC voltages cannot be changed • they must match the voltage required by the appliance
AC/DC conversion	• AC to DC conversion is easy using an electronic rectifier	• DC to AC conversion is more difficult
motors	• AC motors are relatively simple • power output is higher than from DC motors	• DC motors are more complex than AC motors • power output is usually low
transmission losses	• low because AC voltages can be boosted for transmission	• high because DC voltages cannot be changed for transmission
storage	• AC needs to be converted to DC before it can charge/recharge batteries	• DC can charge/recharge batteries, which store the energy as chemical energy

Transmission of electricity

Most power plants produce AC voltages of around 20 000 V. This electricity must then be transmitted far away to the cities and towns that need it. All electrical wires have a resistance, and energy is wasted when current passes along them. One way to reduce wasted energy is to use a transformer that reduces the current and boosts the voltage until it is an incredible 220 000 to 500 000 V!

A **transformer** is a device that steps up AC voltage (increasing it) or steps down AC voltage (reducing it).

Transmission voltages are far too high for users at the other end, and so a series of transformers reduce it until it is the 240 V that is fed into your home. The final transformer is probably up a pole at the end of your street. Other transformers might even be needed inside your home to reduce the voltage even further to the levels needed for halogen light globes and electronic equipment such as computers or mobile phone rechargers.

Figure 6.4.11 shows where transformers are commonly used in the transmission lines.

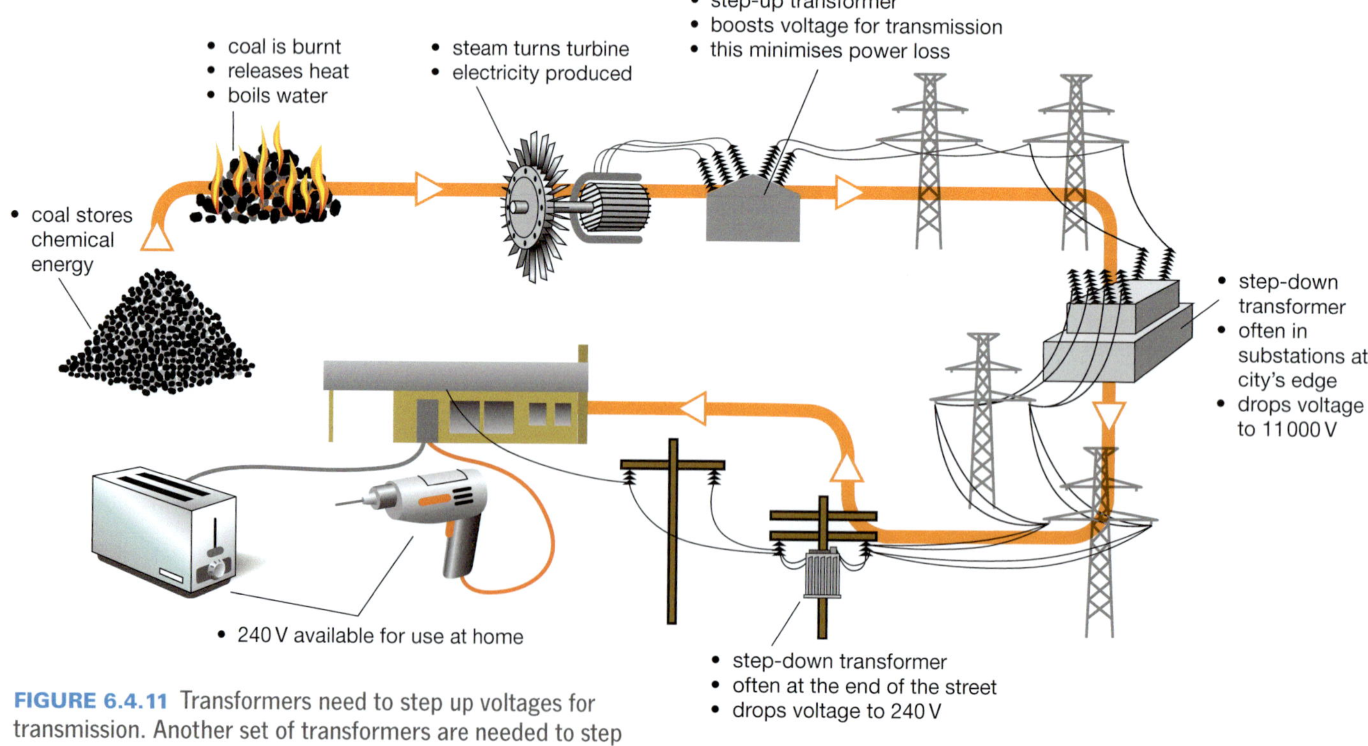

FIGURE 6.4.11 Transformers need to step up voltages for transmission. Another set of transformers are needed to step down the voltage to what is needed in the home.

SCIENCE AS A HUMAN ENDEAVOUR

Use and influence of science

Making light out of gravity

Approximately 15% of the world's population does not have access to electricity. To light their homes, many people in developing countries use kerosene lamps. These lamps are usually made from bottles or tin cans filled with kerosene and a wick that is made from a piece of rope or cloth (Figure 6.4.12).

FIGURE 6.4.12 Kerosene lamps are a fire hazard, emit dangerous fumes and the fuel is expensive.

The lamps are a major fire hazard and the burning of the kerosene causes indoor air pollution. The pollution caused by kerosene lamps causes lung and throat cancers, eye and lung infections, and low birth weights in babies whose mothers inhaled the pollution during pregnancy. It is estimated that breathing the fumes from kerosene lamps has a similar effect to smoking 40 cigarettes per day and those who are most affected are women and children. Kerosene lamps are inefficient, giving poor light for the amount of fuel used. The people have difficulty working and learning in the dim light while the ongoing costs of kerosene trap them in a cycle of poverty.

Two entrepreneurs have a mission to replace kerosene lamps. They have found a way to provide a safe source of light that has zero running costs. It uses a completely renewable source of energy, instantly provides reliable light that is five times brighter than kerosene lamps, and does not need sunlight or batteries to run. This smart innovation generates light using only gravity. GravityLight was invented in 2012 by UK designers, Martin Riddiford and Jim Reeves. It uses a weight (a bag filled with up to 12 kg of rocks or soil) attached to a cord that is threaded through the GravityLight's electricity-generating device (Figure 6.4.13 and Figure 6.4.14 on page 260). The bag hangs approximately 1.8 m above the ground and drops at a rate of around 1 mm per second. As the bag falls, it pulls the cord through the device, which rotates gears within the GravityLight, driving a generator that powers an LED. It provides a continuous source of light for around 20 minutes (until the bag reaches the ground), after which the bag can be lifted again to restart the process. The first model of the GravityLight has been trialled by over 1300 families in 26 countries and a second model is currently under development (Figure 6.4.15 on page 260). The GravityLight provides a smart solution to a global problem and has potential to improve the lives of millions of people.

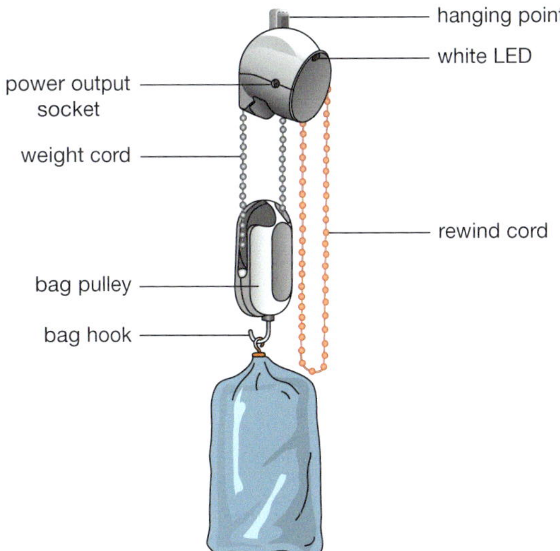

FIGURE 6.4.13 The GravityLight bag is filled with rocks to form the weight that will be used to drive the generator inside the device.

SCIENCE AS A HUMAN ENDEAVOUR

FIGURE 6.4.15 Lakshmi and her grandson, Ganeshan, trialled the first model of the GravityLight in their home in Tamil Nadu, India.

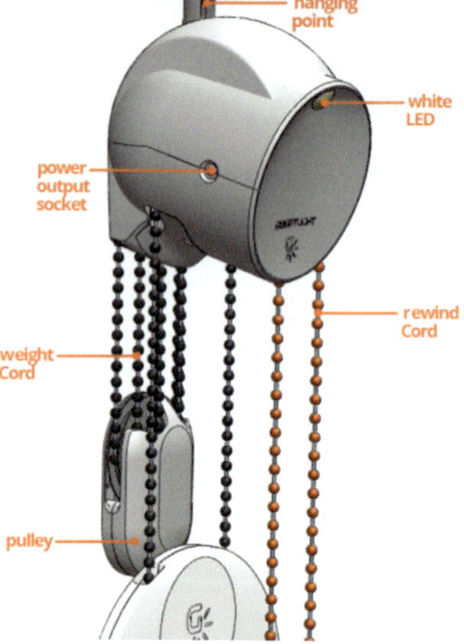

FIGURE 6.4.14 A bag filled with rocks or soil is attached to the bag hook of the GravityLight. The bag falls slowly to the ground and pulls the weight cord through the device. The movement of the cord turns gears, which drive a generator that powers an LED.

TABLE 6.4.5 Electricity usage by country

Country	Electricity usage (kWh per person in 2013)
South Sudan	39
Kenya	168
India	765
Brazil	2 529
China	3 762
Germany	7 019
Australia	10 134
United States of America	12 985

Source: The World Bank, Electric power consumption (kWh per capita, 2013)

REVIEW

1 Electricity use varies widely between countries. Table 6.4.5 shows the electricity use per person in different countries. Why do you think the electricity usage differs so much between these countries?

2 What factors do you think affect people's access to electricity?

3 Why do you think it's important that we improve our sustainable use of electricity in Australia and other parts of the world?

MODULE 6.4

Review questions

Remembering

1 Define the terms:
   a magnetic field
   b solenoid.
   c generator.

2 What term best describes each of the following?
   a solenoid with a metal rod down its core
   b large-scale electricity generator
   c electrons shuffling back and forth along a wire.

3 Do the following combinations of magnetic poles attract or repel?
   a N/N
   b S/S
   c N/S.

4 What advantages does an electromagnet have over a normal, permanent magnet?

5 List the advantages and disadvantages of electricity that is:
   a AC
   b DC.

6 List three ways in which water can be used directly to spin turbines.

7 Steam is commonly used to spin turbines. What are three commonly used ways of changing water into steam?

8 State where transformers are commonly used:
   a around the home
   b in transmitting electricity from the power plant to home.

9 Draw a simple flow chart that shows the steps in transmission of electricity from power station to a toaster in your kitchen.

10 What is the name for the space around a magnet where a magnetic force can be felt?

Understanding

11 How can you make the field lines around a magnet visible?

12 Below are six sentence fragments. Combine three of them to describe an electric motor and the other three to describe an electric generator.
   - An electric motor …
   - An electric generator …
   - … creates an electric current through a coil …
   - … passes an electric current through a coil …
   - … to cause it to spin.
   - … by spinning it.

13 Explain why toys use small permanent magnets but power drills and hairdryers use electromagnets.

Applying

14 Identify whether AC or DC current is used to power:
   a a smartphone
   b a washing machine.

Analysing

15 Compare the magnetic fields of the bar magnet and horseshoe magnet shown in Figure 6.4.1 on page 252.

16 Draw a Venn diagram like the one shown in Figure 6.4.16. Then identify which of the following terms describe a motor only, a generator only or both. Place the terms in the Venn diagram.
   - uses a coil
   - uses a magnet
   - needs movement
   - needs electric current
   - produces movement
   - produces electric current.

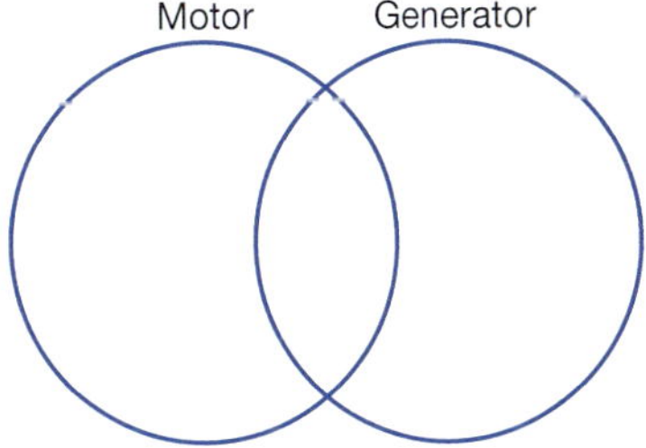

FIGURE 6.4.16

17 Compare the following.
   a a solenoid with an electromagnet
   b an electric motor with a generator.

18 Compare a bike dynamo with a simple generator.

Evaluating

19 a Assess which method of power generation is best suited to providing Australia's future energy needs.
   b Justify your answer.

20 What do you think would be the effect on the electromagnet if the current through it was increased?

MODULE 6.4 Practical investigations

1 • Force on a wire

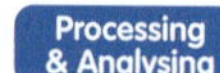

Purpose

To demonstrate electromagnetism.

Timing 45 minutes

Materials

- small sheet of cardboard
- scissors
- sticky tape
- aluminium foil
- retort stand, bosshead and clamps
- wires with alligator clips
- switch
- power pack with circuit breaker/auto cutoff
- horseshoe magnet

Procedure

1. Cut a 'picture frame' out of the cardboard and stick a single thin strip of aluminium foil across it.
2. Construct the apparatus as shown in Figure 6.4.17 and set the power pack at its lowest voltage.

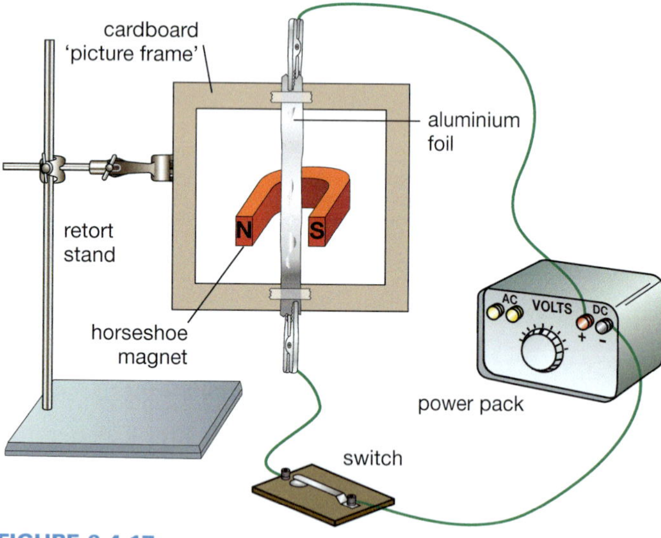

FIGURE 6.4.17

3. Hold the horseshoe magnet as shown and quickly close then open the switch. Record what happens, in a table like the one shown in the Results section.
   Note: The power pack might 'trip'. If it does, you will need to wait until it resets before attempting the rest of the prac.
4. Reverse the terminals on the power pack and repeat.
5. Reverse the orientation of the magnet (i.e. swap poles) and repeat.

Results

In your workbook, construct a table similar to the one shown here, and use it to record all your observations.

Electromagnetism experiment observations

Terminals of power pack	Magnet poles	Direction aluminium strip moves (in/out/ left/right)
as shown	as shown	
reversed	as shown	
as shown	reversed	
reversed	reversed	

Review

1. What is the purpose of the aluminium foil is in this experiment?
2. Describe what happened when:
   a. the terminals of the power pack were reversed
   b. the poles of the magnet were swapped.
3. Describe how this experiment demonstrates electromagnetism.
4. How does this experiment relates to an electric motor?

MODULE 6.4

Practical investigations

2 • Make your own motor

Purpose

To build a simple electric motor.

Timing

45 minutes

Materials

- solid copper wire (without strands and insulating plastic)
- small piece of sandpaper
- 2 large paperclips
- 2 insulated connecting wires with alligator clips on one end
- access to pliers
- bar magnet
- plastic or paper cup
- sticky tape
- 1.5 V AA battery
- rubber band

Procedure

1. Straighten one end of both paperclips, as shown in Figure 6.4.18, and tape them to the top of the cup.

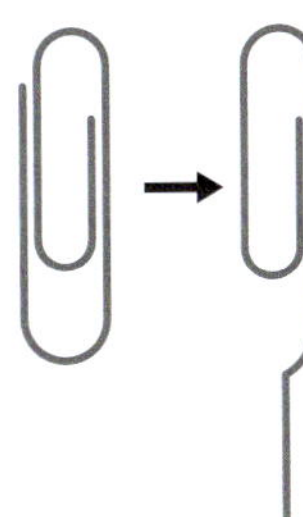

FIGURE 6.4.18

2. Lightly sand the ends of the copper wire. Then wind the copper around your little finger to form a coil. Leave two straight sections about 1 cm long on each side. It should look like Figure 6.4.19. Place your coil in the holder formed by the paperclips.
3. Strip about 1 cm of insulating plastic off the ends of both connecting wires as shown in Figure 6.4.20.
4. Use the rubber band to secure these stripped ends to each end of the battery. Your set-up should look like Figure 6.4.21.

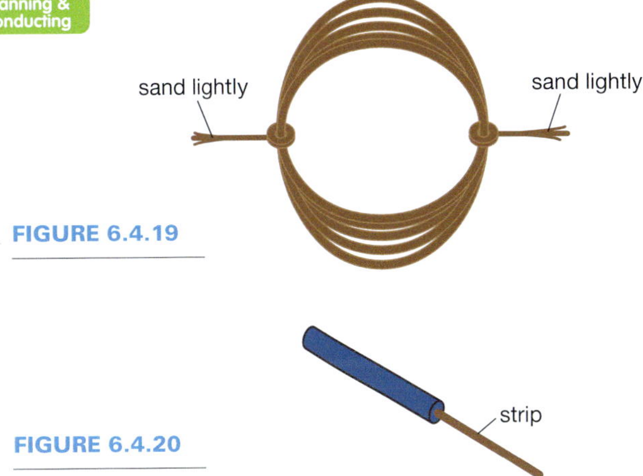

FIGURE 6.4.19

FIGURE 6.4.20

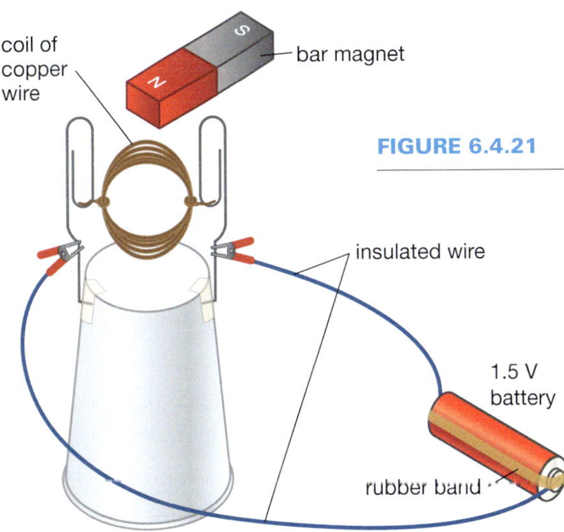

FIGURE 6.4.21

5. When ready, attach the alligator clips to the paperclips. Meanwhile, hold the bar magnet close to the coil. If your motor doesn't spin:
   - try giving the coil a small flick
   - remove the coil and twist it so that the straight parts are exactly central.

Review

1. A motor changes the form of energy. What type of energy is:
   a. provided to the motor?
   b. produced by the motor?
2. What does the passing of a current create down the core of the coil?
3. The motor won't spin without a magnet nearby. Explain why.

CHAPTER 6

Chapter review

Remembering

1 Which of the following statements are true and which are false?

- **a** Energy converters have resistance.
- **b** A current is flowing when a spark jumps from one object to another.
- **c** Radiation spins the turbine in a nuclear reactor.
- **d** Australian AC electricity changes direction 240 times a second.

2 Name the components shown in Figure 6.5.1.

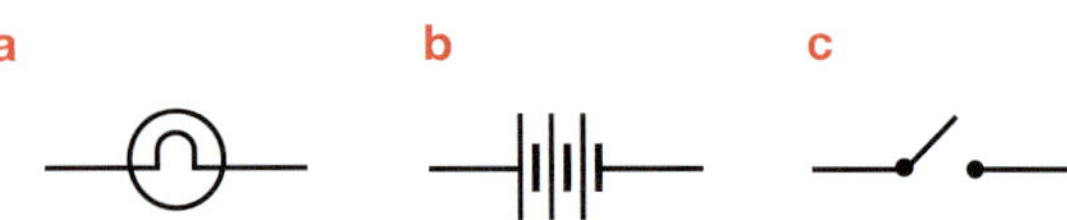

FIGURE 6.5.1

3 What happens when an abnormally high current passes through the following?

- **a** fuse
- **b** circuit breaker.

4 State the unit and unit symbol used to measure resistance.

5 Coal is Australia's main source of electrical energy. List its advantages and disadvantages.

Understanding

6 Why is a wet cell suitable for a car but not for a mobile phone?

7 Explain the advantages of a home having all its appliances connected in parallel rather than in series.

8 Voltage is boosted before electrical power is transmitted long distances. Explain why.

9 Why do laptops need a transformer when plugged in?

Applying

10 Identify three appliances around your home that use an electric motor.

11 Identify the expected current and voltage of each of the globes in Figure 6.5.2. Enter your predictions in a table like the one provided below.

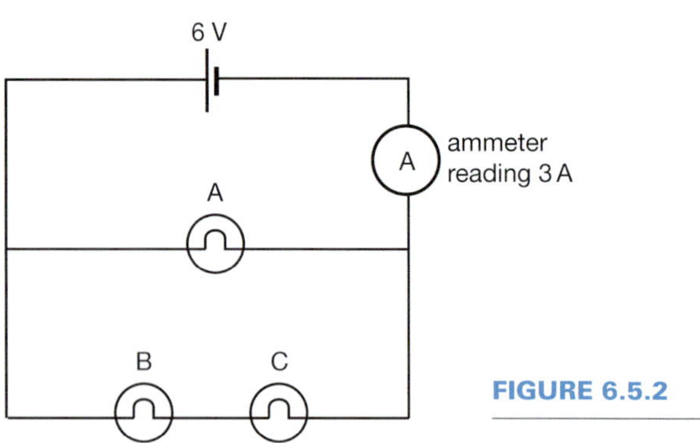

FIGURE 6.5.2

Expected current and voltage for globes A–C

	Globe A	Globe B	Globe C
expected current (A)			
expected voltage (V)			

Analysing

12 Compare an electric motor with a generator.

13 Contrast the electron flow and voltage of AC electricity and DC electricity.

Evaluating

14 Propose reasons why:

- **a** electrical wires are wrapped in insulating plastic
- **b** high-voltage power lines need to be insulated from the poles that hold them up.

15 Identify which renewable way of generating electricity would be best for your area in Australia. Justify your choice.

16 **a** Assess whether you can or cannot answer the questions on page 219 at the start of this chapter.

- **b** Use this assessment to evaluate how well you understand the material presented in this chapter.

Creating

17 Use the following ten key terms to construct a visual summary of the information presented in this chapter.

current	voltmeter	motor
voltage	magnet	generator
resistance	coil	
ammeter	motion	

AB 6.10

CHAPTER 6

Inquiry skills

Research

1 Evaluating

Three-part inquiry question

Select your entry point and complete the relevant parts of this inquiry.

In Australia, electricity is generated in different states using different sorts of power stations.

a i Investigate how electricity is generated in the various Australian states. Produce graphs which show how generation is divided among the different types: wind, coal, hydro and solar.

ii What is the major source of electrical energy in the area in which you live?

There is a general perception that electric cars are always better for the environment; however, whether or not this is true will depend on how the electricity being used to power the car is generated.

b Investigate electric cars and how much electricity they need to charge them. Discuss whether an electric car is a more 'green' option where you live.

c Discuss whether electric cars currently represent a practical option in Australia both as a city only car and as a car for use outside of the major cities.

2 Planning & Conducting

a Use the search term *electricity mania* to find interactive games in which you need to connect up household circuits.

b Some schools have a software package called 'Crocodile clips' which allows you to construct and test electric circuits. If your school has this package, test your skills in circuit construction.

3 Processing & Analysing

Not all power points and plugs around the world are the samc. Rcsearch the different types used and the voltages they work with.

Present your findings as a world map with photos or diagrams of each power point attached.

4 Processing & Analysing

Analyse the claims made in an advertisement, in a consumer report or by a salesperson about an electrical device such as widescreen TV. In particular, analyse what is said about the device's power use and efficiency.

Present your research as an email to a consumer website or magazine such as Choice.

5 Communicating

Explore how light and temperature sensors are used in robotics. Find:

- how the sensor works. For example, does it work because resistance decreases as light or temperature increases?
- the change in light or temperature that is required to trigger change. For example, what change in light or temperature will trigger a change in the robot?
- the change in action that is triggered (for example, does the robot turn off, reverse or turn 90°?)
- how this action might be useful in a practical way.

Present your research in digital form.

6 Evaluating Communicating

Research wind farms in your state. Find out:

- where they are located
- how many turbines are in each location
- how many more are planned for the future.

Present your findings as an annotated map.

7 Evaluating Communicating

Australia has 23% or the world's uranium. The main fuel used for nuclear power stations. Find:

- a list of countries to which Australia exports uranium
- how much we export to each (in tonnes or as a percentage)
- the reasons why we export uranium to some countries but not others.

Present your research as an article for a newspaper or news website. Include a table or graph in your article.

CHAPTER 6 Inquiry skills

Thinking scientifically

1 Analogue voltmeters have a needle and a dial. Many have different terminals along their base, with each terminal measuring a different maximum voltage. This gives you the best chance of getting an accurate reading.

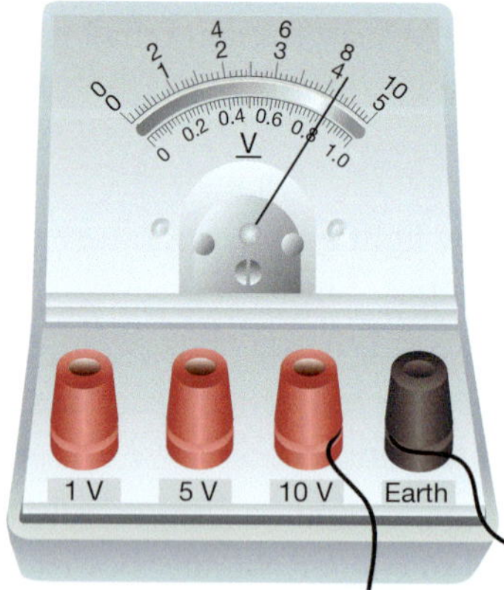

The voltmeter shown here was used to measure the voltage over a light globe. Its measurement is shown. What is the most likely reading of this voltmeter?

A 0.84 V
B 4.2 V
C 8.2 V
D 8.4 V.

2 The actual supply voltage of a 3 V battery is to be measured using the voltmeter above.

Which set of terminals would give the most accurate reading of this battery's voltage?

A 1 V and earth
B 5 V and earth
C 10 V and earth
D earth only.

3 Marge tested how long each of the electric kettles shown below took to boil water.

The kettles were almost identical. Identify the difference between them.

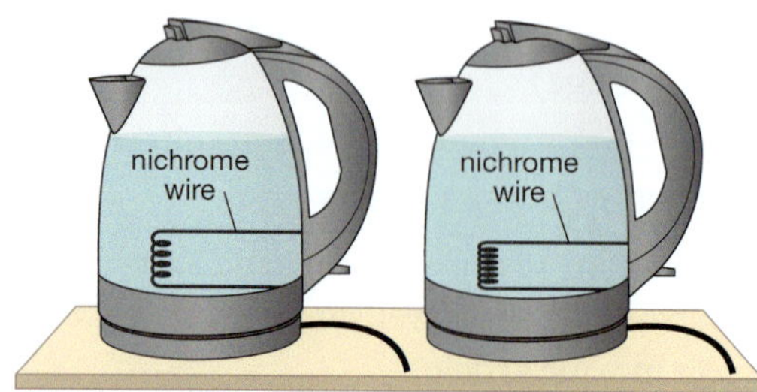

4 To run a fair test on the kettles in question 3, identify which of the following Marge would have to keep constant.

A the voltage supplied to the kettle
B the amount of water each held
C the temperature of the water at the start.
D all of the above.

5 Resistors are electronic components. Their resistance is marked on them as a series of coloured bands. What each colour band means is shown in Table 6.5.1.

TABLE 6.5.1 Colour codes for different resistor bands

Colour	Ring A	Ring B	Ring C
black	0	0	× 1
red	2	2	× 100
yellow	4	4	× 10 000
blue	6	6	× 1 000 000

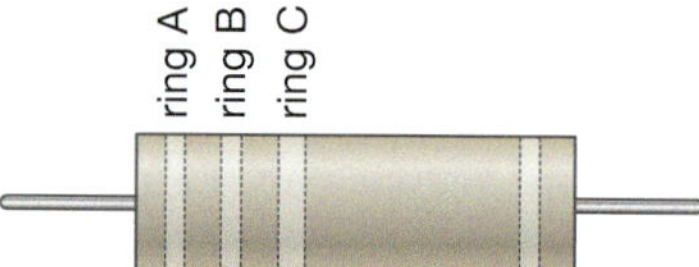

Which of the following is the most likely resistance of the resistor shown here?

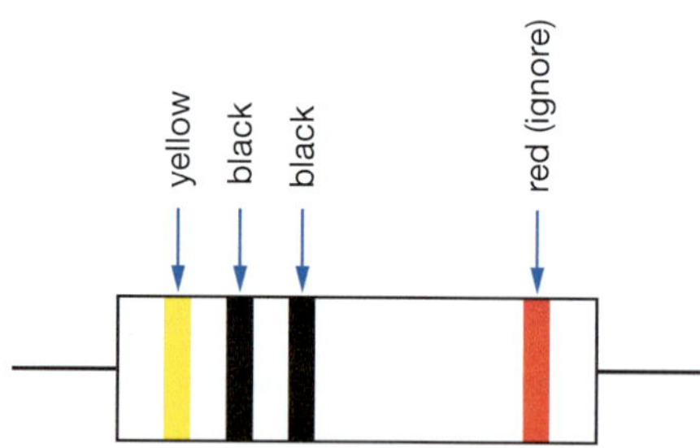

A 40 Ω
B 401 Ω
C 200 Ω
D 2004 Ω.

CHAPTER 6

Glossary

active wire: a wire that carries current to a component; it is coated in brown plastic

alternating current (AC): the current generated by electrons changing the direction in which they move

ammeter: an instrument that measures current

circuit breaker: a switch that turns off a circuit if too much current flows through it

circuit diagram: shows how all components in the circuit are connected

components: the parts of a circuit (light globes, switches, resistors, wires, globes and batteries)

components

conductor: a material that allows a current to pass

current: the flow of charge

direct current (DC): the current generated by electrons always moving in one direction

dry cell

dry cell: a compact cell that uses a paste as an electrolyte

dynamo: a small generator that spins its magnet instead of its coils

earth wire: a wire through which current only flows when there is a leak of current in an appliance; it is coated in yellow and green plastic

electric circuit: the path down which charge flows

electrocution: death by electricity

electrode: rod, sheet or plate made of conducting material in a battery

electrolyte: a solution or wet paste that conducts electricity and completes the circuit in a battery or cell

electromagnet

electromagnet: a solenoid with an iron rod in its centre

electromagnetism: the relationship between electricity and magnetism

electrons: tiny negative (−) particles spinning around the nucleus of an atom

field lines: lines that show the direction of the force on iron filings and compass needles

fuse: a wire of high resistance; it will melt if too much current flows in the circuit

generator: uses electromagnetism to produce electricity; it needs a spinning coil and a magnet

insulator: a material that blocks current

magnetic field: an invisible force field around a magnet

motor: a machine that uses electromagnetism to spin; it needs a current-carrying coil and a magnet

neutral: having no charge

neutral wire: a wire that carries current away from the component; it is coated in blue plastic

neutrons: neutral particles found in the nuclei of most atoms

nucleus: the core of an atom; it contains protons and neutrons

parallel: when components are connected in branches adjacent to one another; voltmeters are connected in parallel

parallel circuit: a circuit that has a number of branches, each with its own components

parallel circuit

photovoltaic cell: a solar cell; directly converts solar energy into electrical energy

protons: positive (+) particles found in the nuclei of all atoms

resistance: a measure of how difficult it is for current to pass; measured in ohms (Ω)

safety switch: a device that turns all household circuits off if it detects a leak in current; it is also known as a residual current device (RCD)

series: when components are connected with each other in a single line; ammeters are connected in series

series circuit: a circuit with all its components arranged in a line, forming a single loop

series circuit

solenoid: a current-carrying loop of wire

static electricity: the build-up of electric charge

transformer: a device that increases or reduces voltage

turbine: a large-scale electricity generator

voltage: a measure of the amount of energy provided to charges or used by them; measured in volts (V)

voltmeter: an instrument that measures voltage

wet cell: a cell that uses liquid electrolyte

AB 6.9

CHAPTER 7

Body coordination

Have you ever wondered ...

- how your body works without you having to think about it?
- why exercise makes your face turn red and your heart beat really fast?
- why you quickly move your hand away from something sharp or hot?
- why you get 'butterflies' in your stomach when you are nervous?
- why you shake when frightened or cold?

After completing this chapter you should be able to:

- describe how the requirements for life are provided through the coordinated function of body systems
- use models, flow charts and simulations to explain how body systems work together
- identify responses using the nervous and endocrine systems
- outline how imaging technologies have improved our understanding of the functions and interactions of body systems
- describe the impact of Australian-developed cochlear implant and bionic eye
- *identify functions of different areas of the brain*
- *model the knee-jerk reaction and explain why it is a reflex action*
- *research the causes and effects spinal cord damage.*

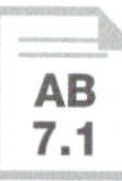

MODULE 7.1

Coordinated body systems

When you want to make something complex such as a car or a computer, all the components need to be brought together in the one place. If an essential item is missing, then it will not work, and production stops. Your body is a complex assortment of organ systems, organs and tissues. All these components work together to ensure that the cells have everything they require to build the things your body needs.

It smells!

How long does it take for a scent to travel across a room?

Collect this ...

- spray can of deodorant or air freshener
- partner
- stopwatch or watch with a second hand

Do this ...

1. Close all the windows and doors in the room to reduce air movement.
2. Stand on one side of the room, facing the wall with your eyes closed.
3. Your partner gives a short spray of deodorant into the air and records the time.
4. Call out as soon as you can smell the scent. Record the time.
5. Repeat the experiment using twice the amount of deodorant. However, you will have to wait until all the scent has cleared from the room before you do this.

Record this ...

1. Describe what happened.
2. Explain why you think this happened.

Metabolism

A large number of chemical reactions take place within an organism's cells. Collectively, these reactions are known as **metabolism**—the chemical processes that maintain life and allows an organism to grow and reproduce, maintain its structure and respond to its environment.

The chemical reactions of metabolism are divided into two groups:

- reactions that break down organic matter—examples are cellular respiration, which breaks down the glucose molecule to release energy, and the breakdown of wastes into harmless substances for excretion.
- reactions that build complex molecules from simpler substances—an example is the construction of new cells and cell components such as proteins and genetic material (Figure 7.1.1).

FIGURE 7.1.1 These plants are able to grow because their metabolism builds simple chemicals into the complex components of new cells.

Enzymes

All the reactions in your body are helped along by **enzymes**—special proteins that can speed up a reaction without being used up in the process. Without the help of enzymes, many reactions would occur too slowly to maintain life.

There are over 700 enzymes in the human body and each one is specific to one particular chemical reaction. Scientists use the 'lock-and-key' model shown in Figure 7.1.2 to explain how enzymes work.

The model shows that a larger molecule has been split into smaller ones. This is an example of a **catabolic** reaction. The opposite reaction can also occur with other enzymes, where smaller molecules are joined to form a larger one. This type of reaction is called an **anabolic** reaction.

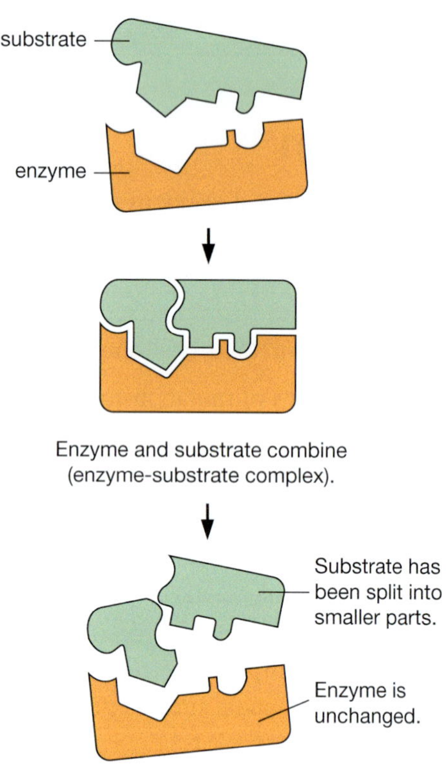

FIGURE 7.1.2 The 'lock-and-key' model explains the action of enzymes.

Each enzyme has a particular shape that allows it to attach to a specific molecule or molecules, known as the **substrate**. The substrate is what will be changed by the chemical reaction. In a reaction, the enzyme and the substrate join together and the substrate is then changed in some way.

SciFile

That's fast!

Enzymes can speed up a reaction by up to ten billion times. That's like taking a minute to do something that otherwise would take 18 000 years!

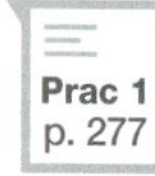
Prac 1
p. 277

Getting nutrients

The food you eat is the source of many of the raw materials your body needs for metabolism. When you chew on an apple or a slice of bread, the nutrients it contains are not always in a form that your body can use. It is the job of the digestive system to chemically change the complex molecules in the apple and the bread into simple chemical substances your body can use.

The food you eat contains complex carbohydrates, proteins and lipids (fats and oils). Through the reactions of chemical digestion, they are progressively broken down, so that when the food reaches the small intestine it is in the form of simple chemical substances.

- Carbohydrates are broken down into glucose.
- Proteins are broken down into amino acids.
- Lipids are broken down into fatty acids and glycerol.

Once broken down, into simple chemical substances, they are small enough to pass through the thin walls of the villi lining the small intestine (Figure 7.1.3). They then pass through the thin walls of capillaries and into the bloodstream.

Diffusion

Diffusion is the movement of particles in a substance from an area of high concentration to an area of low concentration. In simpler terms it means that the particles move from an area where there is a lot of that type of particle to an area where there is not much at all. When particles move this way, they are said to diffuse. Diffusion takes place in liquids and gases where the particles have enough energy to move around freely.

Prac 2
p. 278

FIGURE 7.1.3 Villi are small projections on the wall of the small intestine. These magnified images of the villi show how they greatly increase the surface area through which nutrients can be absorbed.

Figure 7.1.4 models how diffusion works. The particles are moving around, bumping into each other on side A and side B of the container. However, they are separated from one another by the barrier. Side A has a higher concentration of red particles than side B. Side B has a higher concentration of yellow particles than side A. If the barrier between the two sides is removed, then the particles can diffuse, moving freely to the other side.

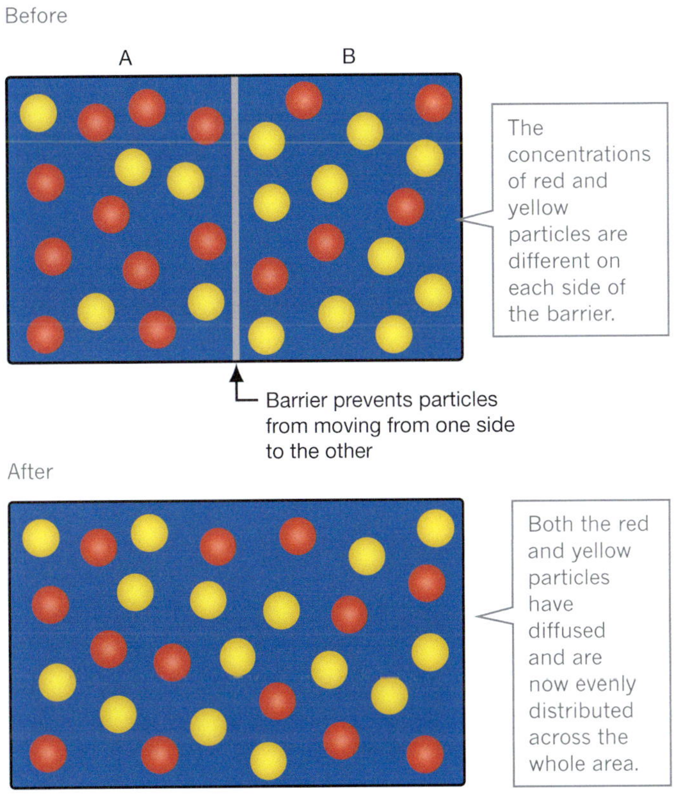

Before

A

B

The concentrations of red and yellow particles are different on each side of the barrier.

Barrier prevents particles from moving from one side to the other

After

Both the red and yellow particles have diffused and are now evenly distributed across the whole area.

FIGURE 7.1.4 In diffusion, the particles diffuse until each type is evenly spread through the container.

The red particles are now able to move into the right-hand part of the container. Eventually the red particles are evenly distributed across the whole volume. In the same way, more yellow particles are now able to move to the left. Eventually they too are evenly distributed across the whole volume.

After eating, there are high concentrations of glucose, amino acids, fatty acids and glycerol in the small intestine. In the bloodstream, there are low concentrations of these molecules. Most of the molecules move by diffusion from the small intestine, through cell membranes and into the blood capillaries in the villi.

The flow of blood in the capillaries quickly carries the digested materials away, so there is always a higher concentration in the small intestine and lower concentration in the blood. Hence, diffusion continues.

Getting oxygen

The air you breathe enters your respiratory system through your nose and mouth. It passes down the trachea, bronchi and bronchioles, ending up in the alveoli. As Figure 7.1.5 on page 272 shows, the walls of the alveoli are only one cell thick and are surrounded by blood capillaries. Oxygen dissolves at the moist surface of the alveoli and moves by diffusion across the short distance from the space inside the alveoli to the blood.

When oxygen enters the blood, it combines with haemoglobin in the red blood cells. The flow of blood carries the oxygen away, so the concentrations in the alveolus and the blood never become equal. In this way, oxygen continues to move into the blood.

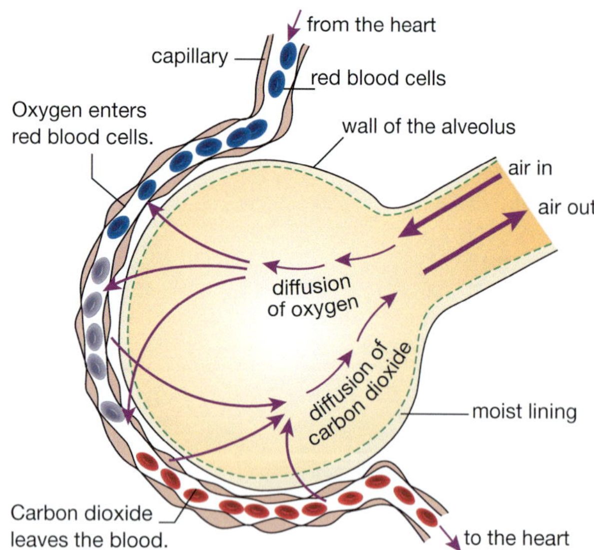

FIGURE 7.1.5 The walls of the alveoli and blood capillaries are only one cell thick. Oxygen diffuses from the alveoli into the bloodstream while carbon dioxide diffuses from the bloodstream back into the alveoli.

Circulation

Your circulatory system carries materials to and from every cell of your body via a series of blood vessels called arteries, capillaries and veins. The heart is the pump that keeps the blood moving and without it the cells would soon be starved of the materials they need to function.

Heartbeat

The heart is made up of cardiac muscle. Cardiac muscle naturally contracts and relaxes without any input from the brain. This can happen because cardiac muscle is an involuntary muscle. This means that the muscle works without you thinking about it or controlling it. The rhythm of the heartbeat is initiated by a small patch of muscle, called a pacemaker (or SA node), shown in Figure 7.1.6.

The rhythm of relaxation and contraction of the pacemaker sets the rhythm for the heartbeat. The pacemaker stimulates both atria to contract simultaneously. When the stimulus reaches the tissue between the atria and the ventricles, another small patch of specialised tissue (the AV node) stimulates both ventricles to contract. An electrocardiogram of a normal heartbeat is shown in Figure 7.1.7.

If the impulses in the ventricles become disorganised because the stimulus is not picked up correctly, then the muscles of the ventricle begin to twitch spasmodically. This is a condition known as ventricular fibrillation. Blood flow stops and unless the heart rhythm is restarted, death will follow swiftly.

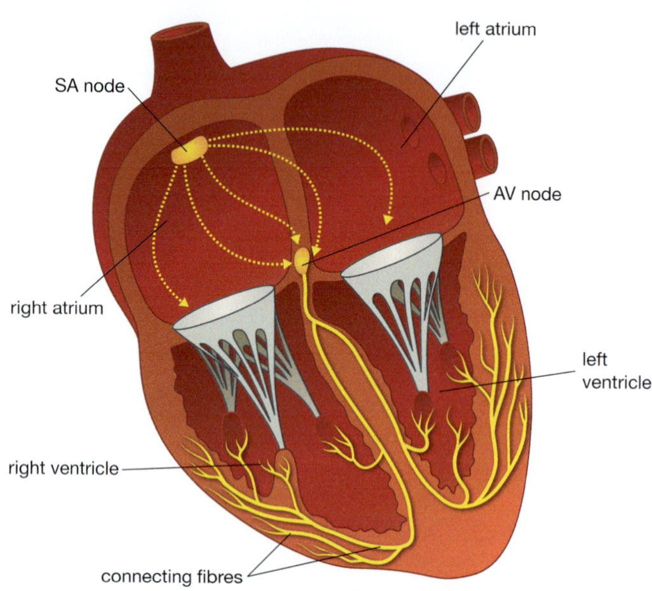

FIGURE 7.1.6 The heart has two specialised areas of muscle (the SA node and the AV node) that control and synchronise the heartbeat.

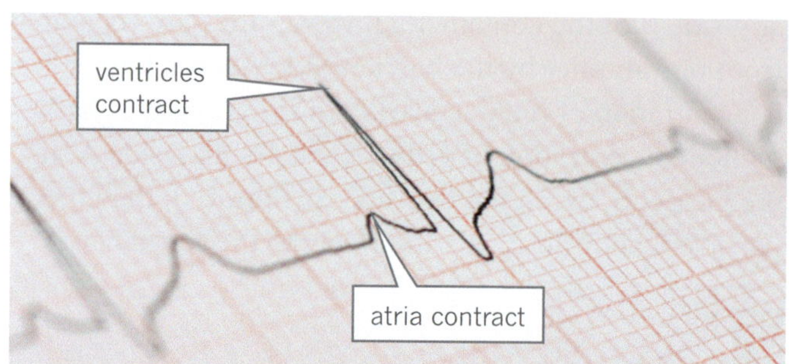

FIGURE 7.1.7 An electrocardiogram is an image of the heartbeat. The lower peak is the contraction of the atria. The high peak is the contraction of the ventricles. Then the heart rests.

A defibrillator is used to deliver electric current to the heart, in an attempt to restore the heart's natural rhythm and save the person's life. That is what is happening in Figure 7.1.8. Defibrillators are found in hospital emergency rooms, ambulances, commercial aircraft, shopping centres and surf lifesaving clubs.

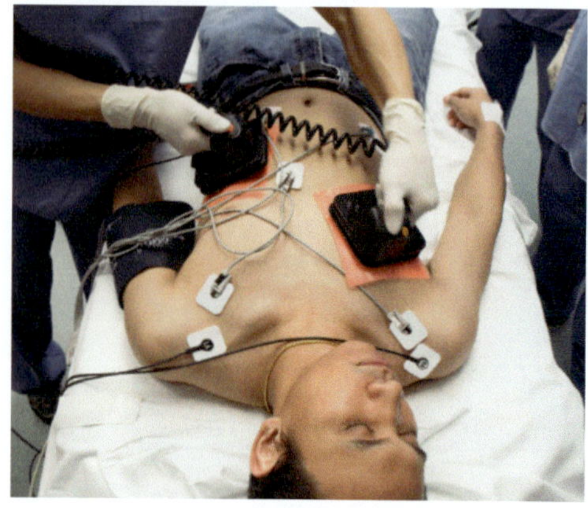

FIGURE 7.1.8 Defibrillators are used to restart the heart of a person who has suffered heart failure.

Changes to heart rate

The rate at which your heart beats changes according to the needs of your body.

Stress or fear causes the body to produce a hormone called noradrenalin. **Hormones** are chemicals that act as messengers in the body. Noradrenalin affects the heart in two ways, both of which increase blood flow. It increases the:

- rate at which the heart beats
- strength of the contractions.

Noradrenalin can stimulate the heart to pump up to five times as much blood per minute as normal.

Vigorous exercise like that shown in Figure 7.1.9 increases the heart rate in two ways:

- As cellular respiration increases, so does the carbon dioxide (CO_2) level in the blood. This change is detected by receptors. **Receptors** are specialised cells that dectect stimuli changes. In this case, receptors in the carotid artery (in your neck) and aorta (the artery leading from your heart) detect the increased CO_2 concentration and send messages to the brain. Nerves then stimulate the heart to beat faster.
- As muscular activity increases, more blood is pumped back to the right atrium. The atrium wall stretches to hold the extra blood. Stretch receptors in the wall of the atrium send nerve impulses to the brain, which stimulates the heart to beat faster.

When you stop exercising vigorously, receptors in the aorta and the carotid artery send messages to the brain to slow the heart rate.

FIGURE 7.1.9 When you exercise, receptors in your body detect changes. The nervous system then sends impulses to muscles that respond to meet the changed needs of your body.

In the cells

Your body is made up of billions of microscopic cells. Within each cell are smaller structures known as organelles, many of which can only be seen using an electron microscope. You can see some in Figure 7.1.10.

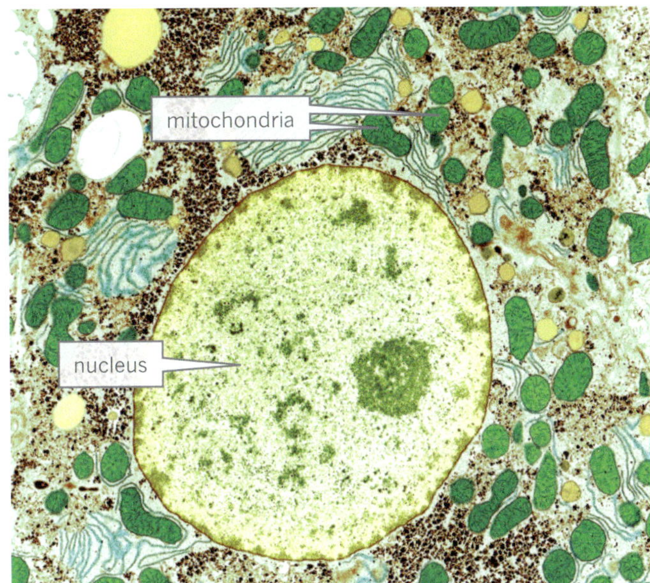

FIGURE 7.1.10 Most organelles, such as mitochrondria, are so small that they can only be seen clearly when magnifications of over x100 000 are used.

Mitochondria are the organelles in which cellular respiration takes place. Oxygen and glucose enter the cell from the blood capillaries and move through the cytoplasm to the mitochondria where they are used in cellular respiration. Cellular respiration is a series of chemical reactions assisted by enzymes. These chemical reactions release energy from glucose.

$$\text{glucose} + \text{oxygen} \rightarrow \text{carbon dioxide} + \text{water} + \text{energy}$$

$$C_6H_{12} + 6O_2 \rightarrow 6CO_2 + 6H_2O + \text{energy}$$

While the reactions of respiration are taking place in the mitochondria, chemical reactions to produce proteins are occurring on the surfaces of structures called **ribosomes**. At the ribosomes, amino acids from the proteins you have digested are reassembled into proteins your body can use. Enzymes and hormones are proteins, as are parts of cell membranes and muscle fibres.

Another type of organelle are lysosomes. Lysosomes treat wastes within cells. Cells and organelles within your body are replaced continuously. Lysosomes digest dying cells, damaged organelles, and viruses or bacteria that have invaded the cell. The products of lysosome digestion move from the cell to the bloodstream.

Removing wastes

The liver is your largest internal organ. Its position is shown in Figure 7.1.11. It carries out many different functions, some of which are related to waste treatment. The liver:

- breaks down hormones
- breaks down haemoglobin (from dead red blood cells), creating products that are added to bile and then disposed of through the digestive system
- breaks down or modifies toxic substances and most medicines—an excess of toxins such as alcohol may cause permanent damage to the liver
- converts ammonia to urea.

Most of these wastes are carried from the liver to the kidneys and are excreted in the urine.

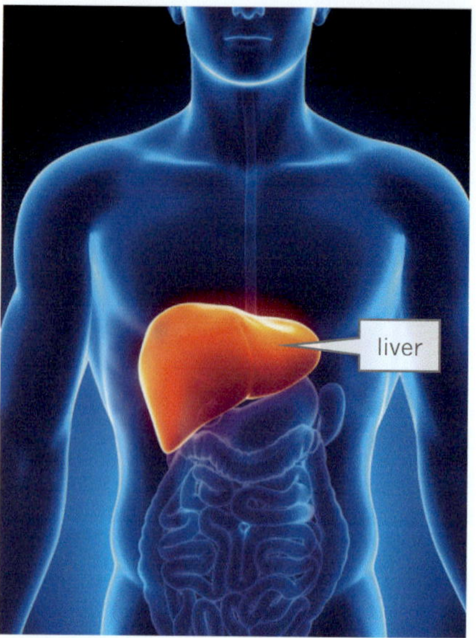

FIGURE 7.1.11 The liver has many functions, only a few of which are involved in waste disposal.

Interdependence of systems

The systems of your body are interdependent. This means that each system depends on the others and cannot function without them. Figure 7.1.12 shows the interrelationships between the excretory, digestive, circulatory and respiratory systems. The nervous and endocrine systems control the activities of these systems.

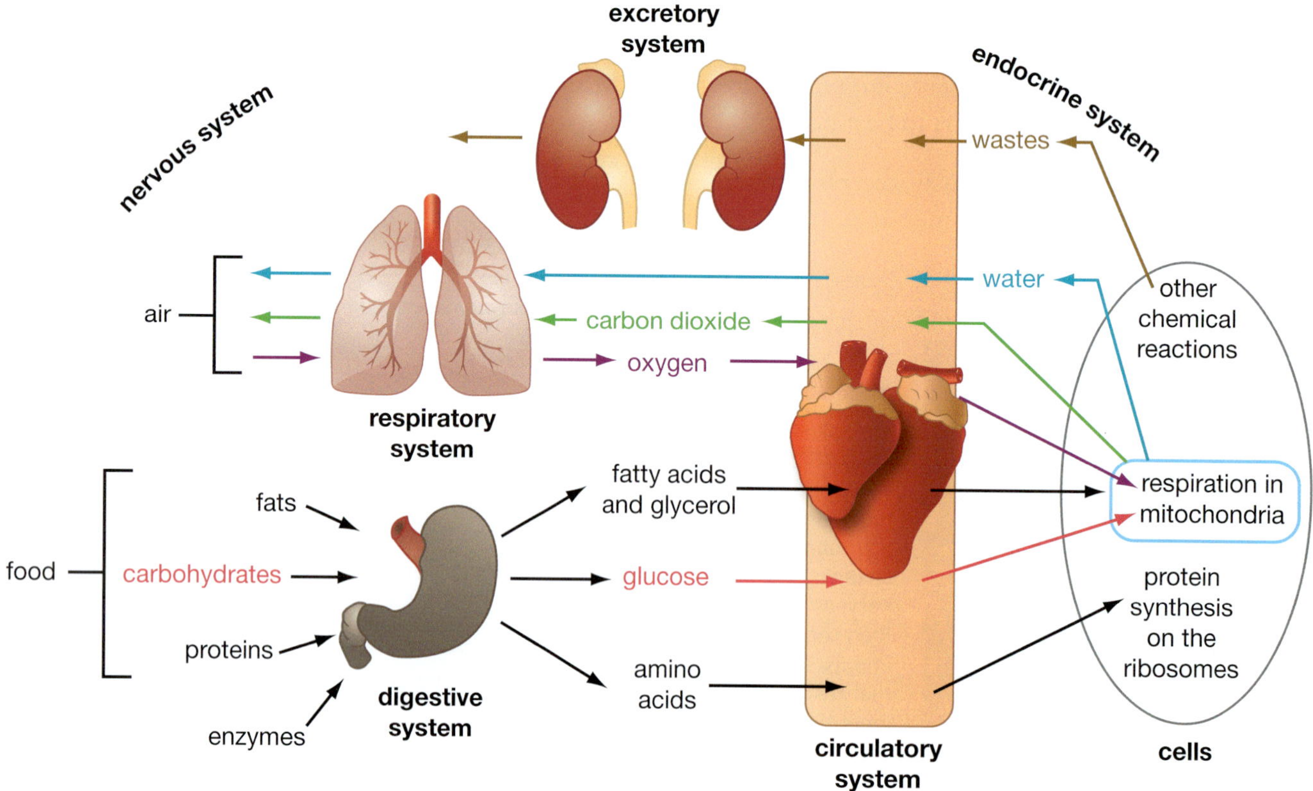

FIGURE 7.1.12 The interrelationships between the systems of your body

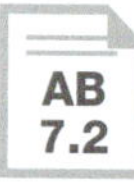

SCIENCE AS A HUMAN ENDEAVOUR

Use and influence of science

Artificial pacemakers

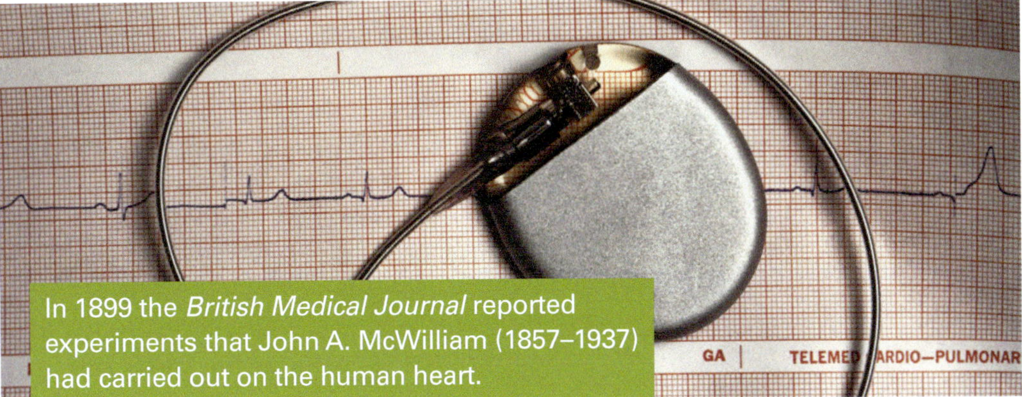

In 1899 the *British Medical Journal* reported experiments that John A. McWilliam (1857–1937) had carried out on the human heart.

FIGURE 7.1.13 Modern artificial pacemakers are inserted into the body and work only when the heartbeat becomes irregular.

McWilliam used electrical impulses to make the heart muscle contract. By stimulating the heart at the rate of 60–70 electrical impulses per minute, he was able to keep the heart muscle contracting at 60–70 beats per minute.

This research led to the development of artificial pacemakers—instruments used when the heart's natural pacemaker does not work properly. A defective natural pacemaker may cause the heart to beat too slowly, too quickly or in an irregular way. Any of these can cause health problems. Artificial pacemakers have been developed to help control abnormal heart rhythms. A typical artificial pacemaker is shown in Figure 7.1.13.

The first artificial pacemaker was invented by Australian anaesthesiologist Dr Mark C. Lidwell (1878–1969). He used his device to resuscitate a newborn baby in Sydney, in 1926. Lidwell did not patent his device and chose to remain anonymous because research of this nature was very controversial at the time. However, two years later, Lidwell worked with a physicist, Edgar H. Booth of the University of Sydney, to develop a portable artificial pacemaker that plugged into a light socket. The circuit was created by applying a pad soaked in strong salt solution to the skin and inserting a needle insulated except at its point into one of the ventricles of the heart.

Refinement of artificial pacemakers continued, but the biggest breakthrough came with the development of the silicon transistor in 1956. From that time, there was rapid development towards pacemakers that were wearable and then to ones that could be implanted in the body.

Swedish scientists pioneered the use of pacemakers in 1958. The first device failed after only three hours. A second device lasted two days. In total, the recipient, Arne Larsson, received 26 different pacemakers over a period of 43 years. He died in 2001 at the age of 86.

Modern artificial pacemakers are small devices placed in the chest or abdomen that use batteries to send electrical pulses to prompt the heart to beat at a normal rate. An electrode is placed next to the heart wall and small electrical charges travel through a wire to the heart.

Most pacemakers have a sensing device and send out signals to the heart only when necessary. The sensing device turns the signal on when the heart beats too slowly. When the heart is beating normally, the sensing device turns the signal off.

REVIEW

1 State the rate of a normal heartbeat.

2 a Name the Australian inventor of the artificial pacemaker.

 b State the year in which he first trialled his invention.

3 What was the breakthrough that enabled development of the wearable artificial pacemaker?

4 How do modern pacemakers control heart rate?

MODULE

7.1 Review questions

Remembering

1 Define the terms:
   a diffusion
   b substrate
   c receptors.

2 What term best describes each of the following?
   a the chemical processes that maintain life
   b chemicals that speed up the rate of reaction
   c the reaction that releases energy from glucose.

3 What is the name of the process by which food is broken down into molecules small enough for the body to use?

4 Copy the following table into your workbook. For each complex molecule listed, what is the simple molecule formed following digestion?

TABLE 7.1.1

Complex molecule	Simple molecule following digestion
carbohydrate	
protein	
lipid	

Understanding

5 Explain why complex molecules such as proteins in your food need to be broken down into simpler chemical substances such as amino acids.

6 a Describe the function of the SA node and the AV node in the heart.
   b Explain what is happening when the heart is fibrillating.
   c Explain why a defibrillator is used in that circumstance.

7 How does the lock-and-key model of enzyme action explains the specific nature of enzyme action?

8 a Explain how villi make the small intestine more effective.
   b Each villus has a good blood supply. How does this help the digestive system supply the cells with the nutrients they need?

Applying

9 Use diagrams similar to those in Figure 7.1.2 on page 270 to demonstrate a reaction where two small molecules are joined to create a larger molecule.

10 Use annotated diagrams to demonstrate:
   a how diffusion occurs
   b the roles of the circulatory system and diffusion in the movement of nutrients from the small intestine
   c the interdependence of the roles of the circulatory system, the respiratory system and diffusion in removing carbon dioxide from the body.

Analysing

11 The chemical reactions of metabolism are divided into two groups. Contrast these two groups.

12 a List the body systems that are involved in getting nutrients to the cells of your body.
   b Use an example to demonstrate how these body systems must work together to deliver nutrients to your cells.

Evaluating

13 In your body, there are many enzymes and each is specific to a reaction.
   a How do specific enzymes help your body to function?
   b What do you think would be the effect on your body if enzymes were not specific and instead could assist every reaction in the body?

14 It would be impossible for your body to function if oxygen, nutrients, hormones and wastes moved through your body by diffusion instead of the circulatory system. Propose reasons why.

Creating

15 Construct a flow diagram of the heartbeat, starting and ending with contraction of the atria.

16 Construct a flow diagram of the changes that take place in your body as you take part in vigorous exercise.

MODULE 7.1

Practical investigations

1 • Enzyme activity

Amylase is an enzyme that converts starch to glucose.

Purpose

To demonstrate the action of enzymes.

Hypothesis

Which do you think will react faster—a reaction with an enzyme present or the same reaction without an enzyme? Before you go any further with your investigation, write a hypothesis in your workbook.

Timing

45 minutes

Materials

- 50 mL starch solution
- amylase
- iodine
- Tes-Tape™
- 2 × 50 mL beakers
- stirring rod
- labels
- 50 mL measuring cylinder
- water bath at about 30°C

SAFETY

Iodine stains. Avoid contact with skin and clothes.

Procedure

1. Label the 50 mL beakers 'Enzyme' and 'No enzyme'.
2. Using Tes-Tape, test the starch solution for the presence of glucose.
3. Add a few drops of iodine to the solution until the starch is showing a distinct blue-black colour.
4. Add 20 mL of the coloured starch solution to the two 50 mL beakers. Your set-up should look like Figure 7.1.14.
5. Add amylase to the beaker labelled 'Enzyme'. (Use an amount equivalent to a match head.)
6. Stir the amylase and starch solution mixture.
7. Place both beakers in the water bath.
8. After about 10 minutes, note any colour change in the beakers.
9. Using the Tes-Tape, test both beakers for glucose.

Results

1. Record any change in appearance of the beakers.
2. Record whether glucose was present in either of the beakers.

Review

1. If glucose was present in either of the beakers, explain where the glucose came from.
2. Account for any colour change that occurred.
3. Why were the beakers heated to 30°C?
4. Summarise the results of this experiment, relating any changes to the action of amylase.
5. a. Construct a conclusion for your investigation.
   b. Assess whether your hypothesis was supported or not.

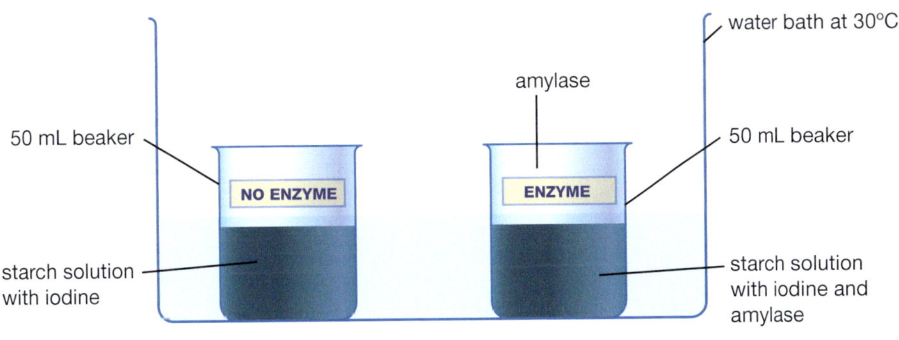

FIGURE 7.1.14

MODULE
7.1 Practical investigations

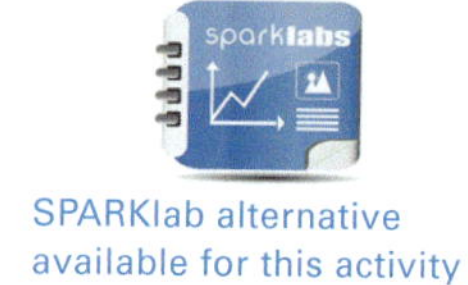

SPARKlab alternative available for this activity

2 • Diffusion

Purpose

To demonstrate the process of diffusion in cells.

Hypothesis

Which do you think will leak colour most quickly—dialysis tubing filled with strongly coloured water or tubing filled with faintly coloured water? Before you go any further with your investigation, write a hypothesis in your workbook.

Timing
45 minutes

Materials

- food colouring
- water

- 3 pieces of dialysis tubing 15 cm long
- 3 × 500 mL beakers
- 3 × 50 mL beakers
- stirring rod
- string
- 3 retort stands with clamps
- stop watch

Procedure

1. Add 300 mL of water to each of the 500 mL beakers. Set them up on the bench with a retort stand behind each one (Figure 7.1.15).
2. Let the water settle.
3. Add 25 mL of water to each of the 50 mL beakers. Label the beakers 1, 2 and 3.
4. Create solutions of different concentrations using the food colouring. Add 10 drops of food colouring to beaker 1, 20 drops to beaker 2 and 40 drops to beaker 3.

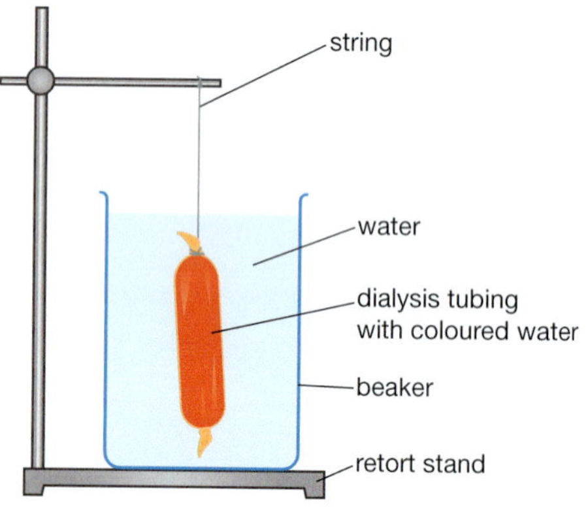

FIGURE 7.1.15

5. Run water from the tap over the three lengths of dialysis tubing.
6. Tie one end of each piece of tubing in a tight knot.
7. Rub the other end of the tubing to open it up.
8. Fill one piece of tubing with the coloured water from beaker 1.
9. Close the end of the tubing by tying it with string.
10. Rinse off the dialysis tube 'sausage' you have created so that there is no coloured water on the outside.
11. Using the other end of the string, tie the 'sausage' to the retort stand and suspend it over one of the large beakers.
12. Repeat steps 8–11, adding the other solutions to the dialysis tubing.
13. Carefully lower the dialysis tubes into the large beakers, taking care to disturb the water as little as possible.
14. Observe any changes.

Results

Record the time it takes for the:

a. colour to 'leak' out of the dialysis tubing
b. colour to move half way across the water in the beaker
c. water in the beaker to become uniformly coloured.

Review

1. Compare the rate of movement of the coloured solutions in the three beakers.
2. Explain any differences in terms of the concentration of the original solutions.
3. Identify the part of the model that represents the cell.
4. What aspect of diffusion of oxygen in the lungs, or nutrients in the digestive system, is not demonstrated by this model?
5. a. Construct a conclusion for your investigation.
   b. Assess whether your hypothesis was supported or not.

MODULE

7.2 Nervous control

Most people can walk across a room without really thinking about it. You listen to music and smell sausages cooking on barbecue without having to decide how to hear or smell. Your nervous system controls all these actions and many more.

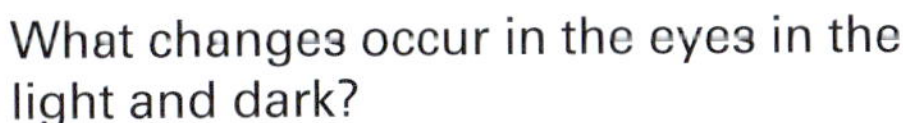

Pupils change

What changes occur in the eyes in the light and dark?

Collect this ...

- mirror
- strip of dark cloth

Do this ...

1. Stand in front of the mirror in a room with good light.
2. Look at your eyes—pay close attention to the size of the pupils.
3. Close your eyes and cover them gently with the dark cloth. Do not press on your eyes.
4. Keep your eyes closed for two minutes.
5. Remove the cloth and open your eyes. Look in the mirror immediately.
6. Observe the pupils of your eyes.

Record this ...

1. Describe what happened.
2. Explain why you think this happened.

Nervous system

The nervous system is a communication network that controls all the other systems of your body, such as the digestive and circulatory systems. The human nervous system has two main parts. These are shown in Figure 7.2.1 on page 280.

The two parts of the nervous system are the:

- **central nervous system** (**CNS**), made up of your brain and your spinal cord. The CNS receives information from all over the body, processes that information, and then sends out messages telling the body how to respond.
- **peripheral nervous system** (**PNS**), made up of the nerves that carry messages to and from the CNS and other parts of your body.

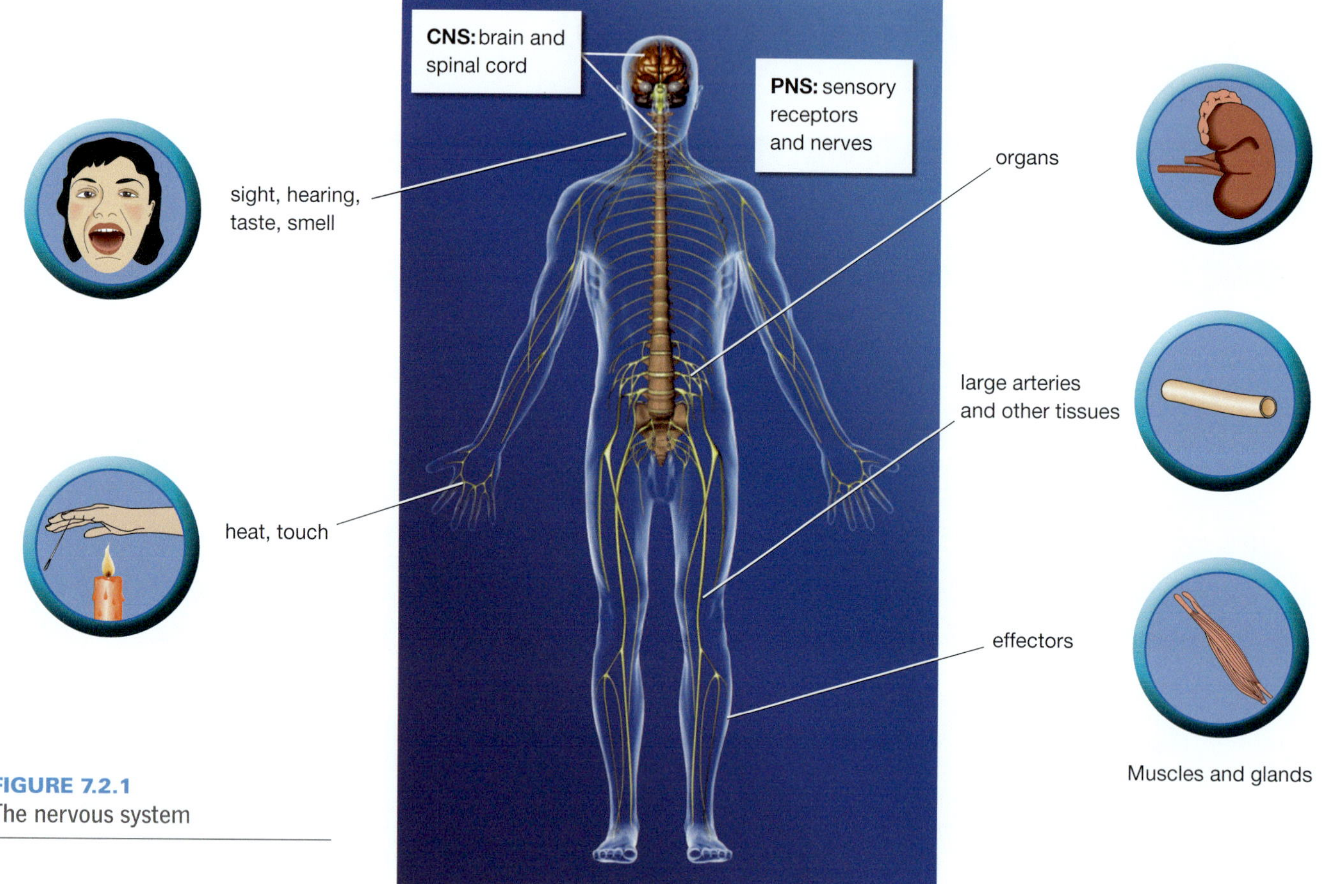

FIGURE 7.2.1
The nervous system

Nerve cells

The nervous system is made up of trillions of nerve cells called neurons. **Neurons** are specialised cells that transmit electrical messages from one part of your body to another at very high speed. These electrical messages are called **nerve impulses** and they can travel in only one direction.

A neuron has all of the usual features of any animal cell including a nucleus, cell membrane and cytoplasm. As you can see in Figure 7.2.2, a neuron also has many other specialised parts including dendrites, an axon and axon terminals.

The **cell body** contains the nucleus, which is the control centre of the cell. The **dendrites** branch out from the cell body and receive messages from other nerve cells, which are then sent on to the cell body. The **axon** or nerve fibre sends nerve impulses in only one direction—away from the cell body. The axon terminals pass the message on to the next neuron.

There are different types of neurons. Each type of neuron has a different function within the nervous system (Figure 7.2.3).

- **motor neurons**—carry messages from the CNS to effectors. **Effectors** are muscles or glands (tissues that secrete hormones) that translate the messages into actions.
- **connector neurons**—transmit messages between neurons in the CNS.
- **sensory neurons**—have specialised receptors, which are sensitive to stimuli such as heat or light. They carry messages to the brain and spinal cord from cells in the sense organs (such as your eyes, ears, tongue and skin).

The messages sent along the neuron are electrical. If all the neurons in your body touched one another, then stimulating one nerve ending would be like turning on one switch in your house and having all the lights and appliances come on. Your body needs to control which nerves 'fire' at a certain time.

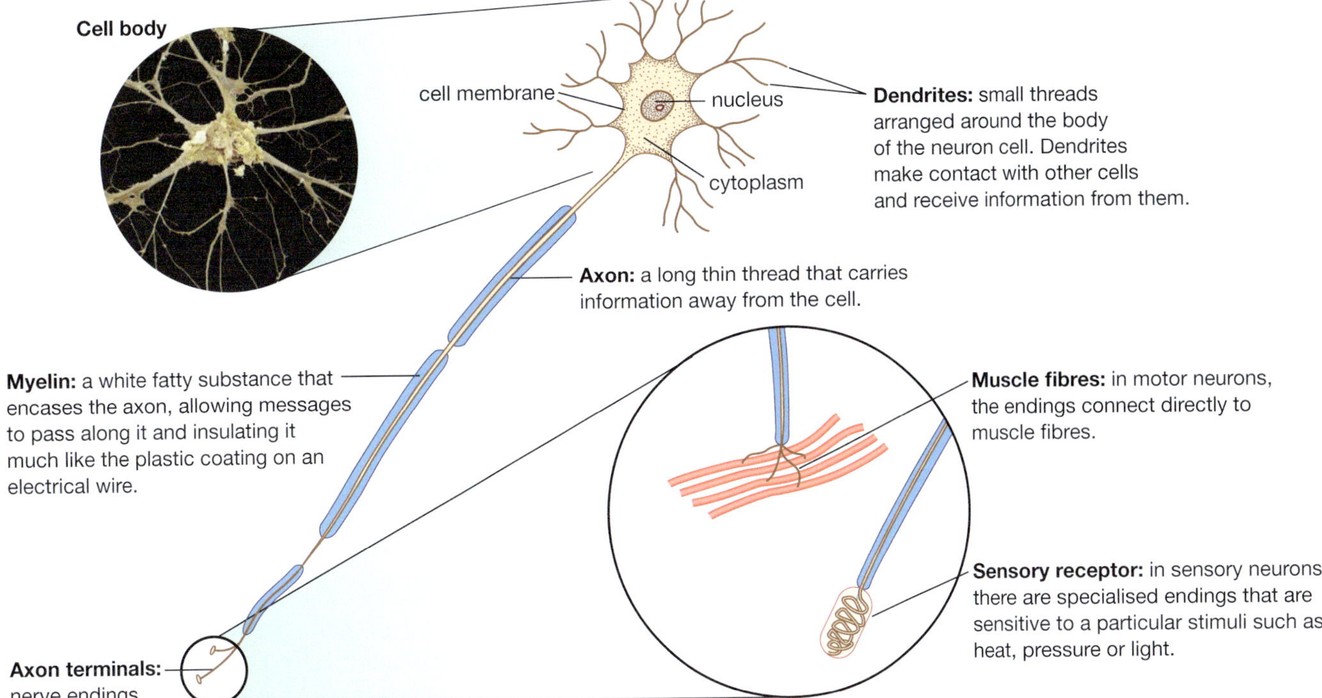

FIGURE 7.2.2 These are some of the specialised endings a typical neuron can have.

When the nerve impulse reaches the axon terminals at the end of an axon, a chemical called a **neurotransmitter** is released into the space between the neurons. This space is called a **synapse** (Figure 7.2.4 on page 282). The neurotransmitter carries the message from the axon of one neuron to the dendrite of the next neuron. The dendrite receives the chemical message and sends off an electrical signal.

About 50 different neurotransmitters carry electrical impulses across these gaps. These neurotransmitters control which nerves fire and when.

In your body, the neurons are bundled together to form nerves, as shown in Figure 7.2.5 on page 282. Neurons are covered with an insulating layer called a **myelin sheath**. The myelin sheath electrically insulates the neurons from each other and increases the speed of the nerve impulse.

The parts of the CNS that contain neurons covered in myelin are called white matter. The parts that contain mainly cell bodies are called grey matter. The outer parts of the brain are made up mainly of grey matter.

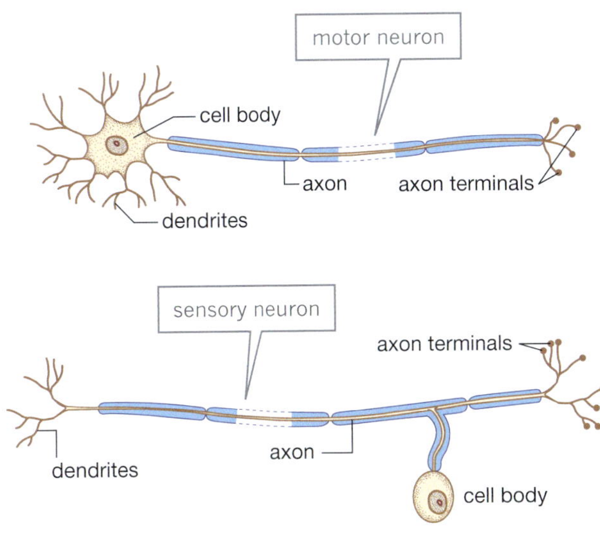

FIGURE 7.2.3 Two types of neuron

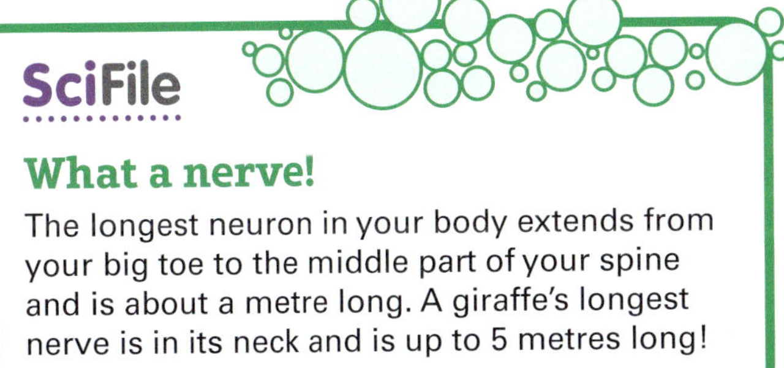

What a nerve!

The longest neuron in your body extends from your big toe to the middle part of your spine and is about a metre long. A giraffe's longest nerve is in its neck and is up to 5 metres long!

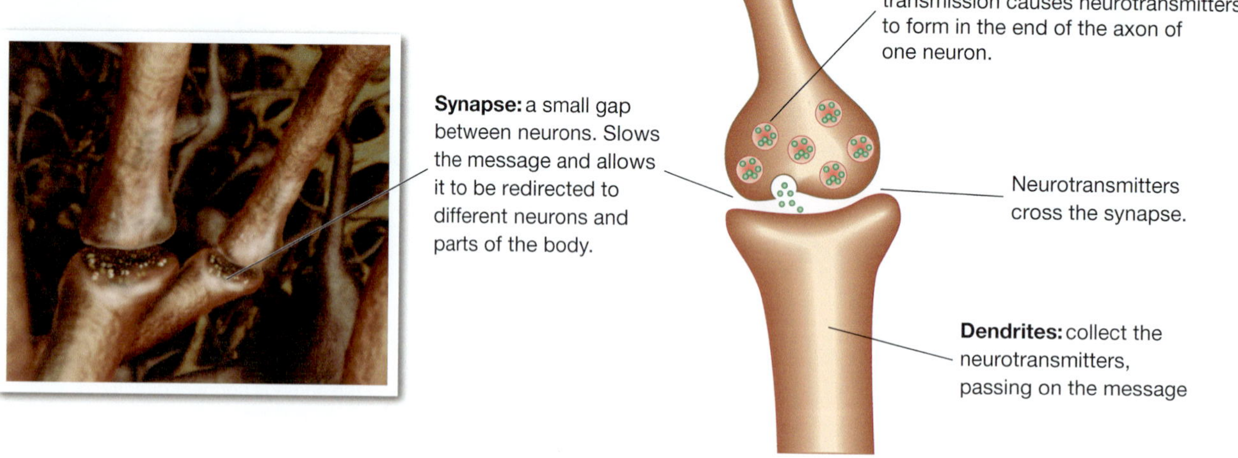

FIGURE 7.2.4 At the synapse, the electrical signal of the nerve is converted into a chemical signal called a neurotransmitter, which carries the signal across the gap. The chemical signal is then converted back to an electrical signal.

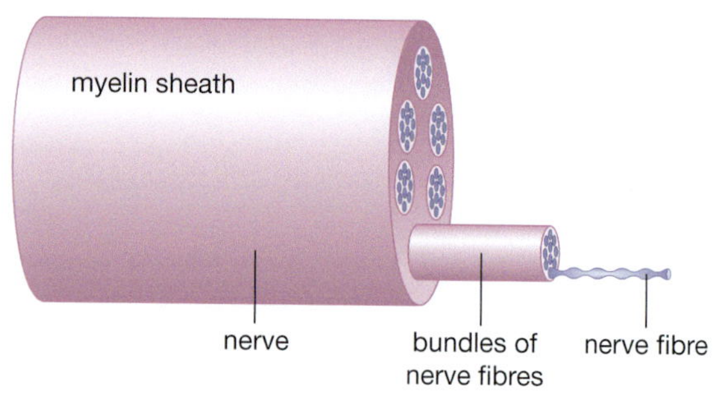

FIGURE 7.2.5 A nerve is made up of a large number or neurons, each of which is surrounded by a myelin sheath.

The brain

Compared with other animals, humans have a very large brain for their body size. The human brain contains about 100 billion neurons, and has an average volume of 1200–1400 mL. The brain controls and regulates body functions. Without it, you cannot survive.

Medical imaging techniques can look inside a living brain. For example, MRI (magnetic resonance imaging) uses strong magnetic fields to distinguish different types of body tissue. MRI is useful in diagnosing brain tumours and finding areas of brain injury.

Sometimes other parts of the brain take over the function of the damaged parts, but there are situations where brain damage is permanent. fMRI (functional magnetic resonance imaging) measures and maps brain activity though changes in blood flow. fMRI is useful to determine the effects of a stroke or disease, or to guide brain treatment.

The cerebrum

When you think of what a human brain looks like, you are probably thinking of the **cerebrum**. You can see it in Figure 7.2.6. It occupies more than 80% of the brain and contains over 10 billion neurons. Its folded surface makes surface area three times greater than if the brain's surface was smooth.

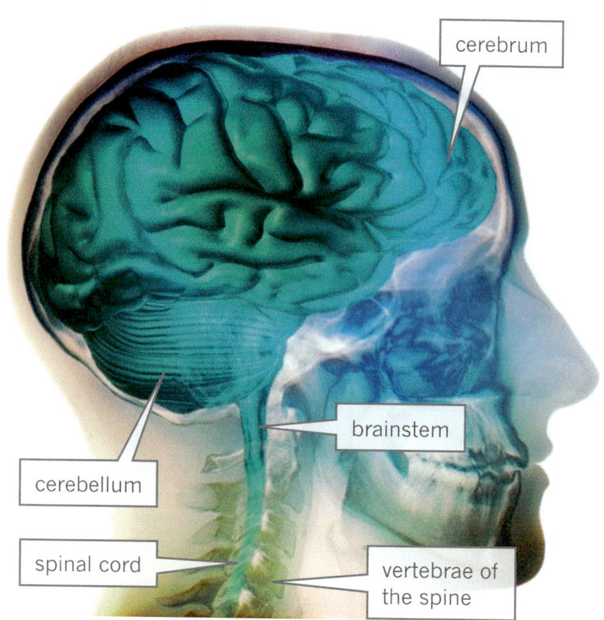

FIGURE 7.2.6 Computer-enhanced X-ray showing the external structure of a human brain.

It is in the cerebrum that the higher intellectual functions of humans take place. The cerebrum controls your conscious thoughts and the intentional (voluntary) movement of every body part. For example, if you scratch your nose then it is the cerebrum controlling your movement. The cerebrum also receives sensory messages from all body parts. For example, physical pain, the sound you hear and the light you see are all processed by the cerebrum.

The cerebrum is made up of two parts, called the right and left cerebral hemispheres. When you are performing an intended and voluntary action such as walking or hitting a ball, the right hemisphere controls the left side of your body and the left hemisphere controls the right side of your body (Figure 7.2.7). Each half of the brain can work independently, but you use both cerebral hemispheres for most activities. One side usually dominates in a particular task. For example, in most people the left side has more control over language and logical thinking, such as mathematical ability. The right side is the more creative and emotional side. Musical and artistic abilities depend on the right side of the brain.

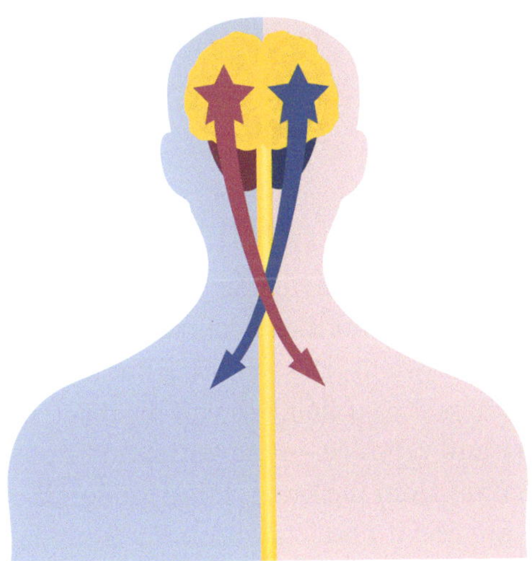

FIGURE 7.2.7 The right and left sides of the brain control the opposite sides of the body.

Figure 7.2.8 shows images of the brain created using both fMRI and PET (positron emission topography). Together these scans reveal the parts of the brain that are active during various activities. They show that the left-hand side of the brain is active during activities that involve language. They also show that different parts are active when listening, speaking, reading or just thinking about words.

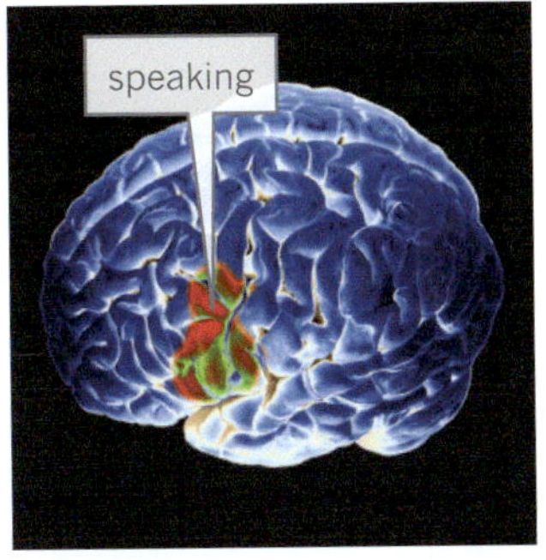

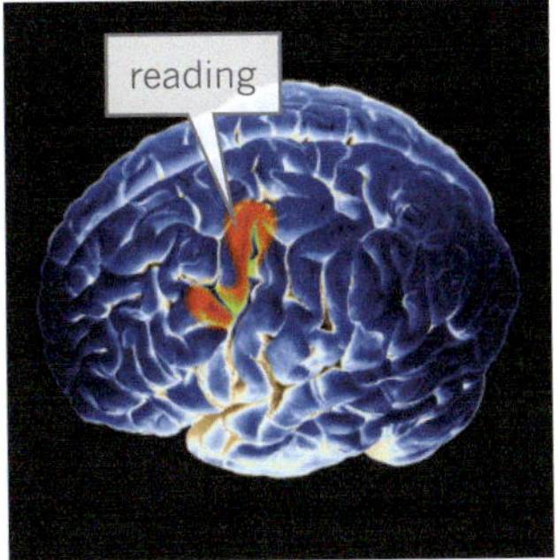

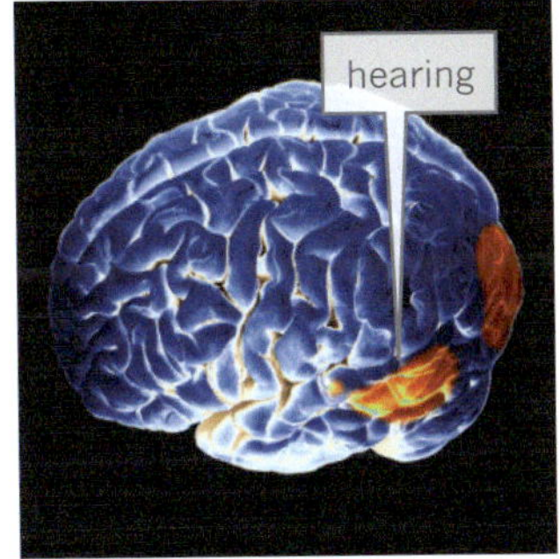

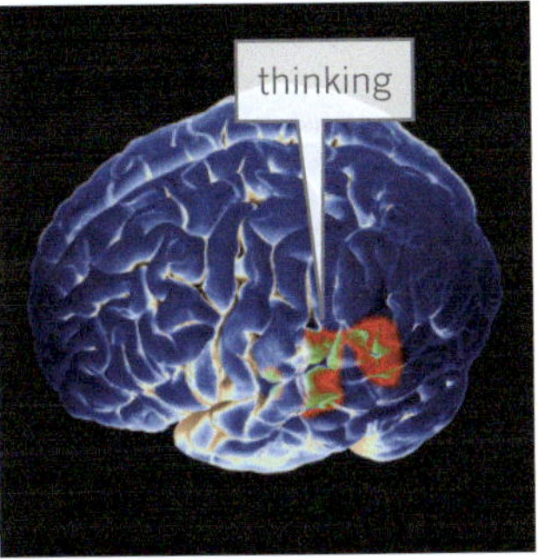

FIGURE 7.2.8 The red and green areas in these images show the areas of the brain that are active during various activities.

The cerebellum

At the base of the cerebrum is the **cerebellum**. The cerebellum is located where your skull curves inwards (Figure 7.2.9). The cerebellum is responsible for coordination and balance. Without it, walking would be impossible.

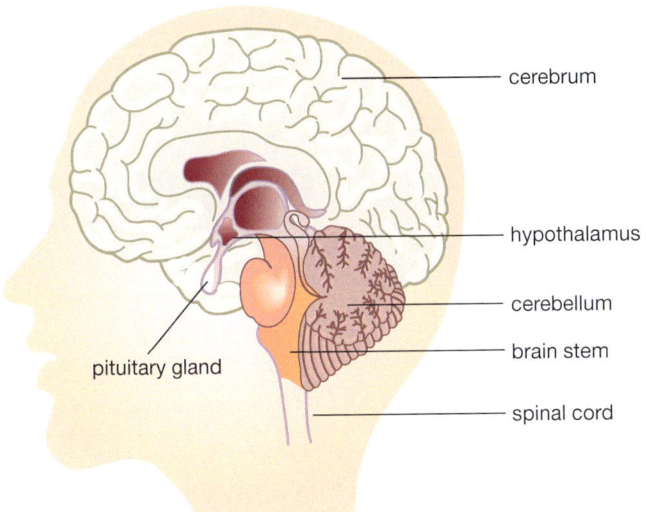

FIGURE 7.2.9 Vertical cross-section of the brain

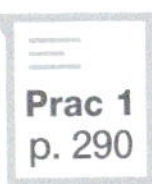
Prac 1 p. 290

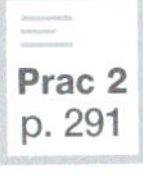
Prac 2 p. 291

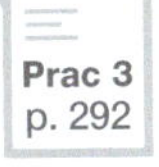
Prac 3 p. 292

The medulla and spinal cord

The lower part of the **brain stem** or **medulla** can be seen where the spinal cord widens just after it passes into the skull. It controls the body's vital functions, such as breathing, blood pressure and heart rate. Damage to this area can be fatal.

Protecting the brain and spinal cord

The spinal cord and brain are so important to body function that they have special protective structures surrounding them. A covering of bone known as the **cranium** protects the brain and the spinal cord is also surrounded in bony structures called **vertebrae**. Both the spinal cord and the brain are surrounded in a fluid called **cerebrospinal fluid (CSF)**, which provides nutrients to the neurons and acts as a shock absorber.

science 4 fun

Stroop effect

How easy is it to fool your brain?

Do this ...

As quickly as you can, name the colour the word in printed in. Do NOT read the actual words.

RED BLUE GREEN ORANGE YELLOW VIOLET

Record this ...

1 Describe what happened.
2 Explain why you think this happened.

AB 7.6 AB 7.7

Responding to stimuli

A simple model of your nervous system is a stimulus–response model. You can see it in Figure 7.2.10. Receptors stimulate the sensory neurons. This is referred to as the **stimulus**. The sensory neurons send a message to the brain. The brain works out the response that is required, then sends a message along motor neurons to the effectors.

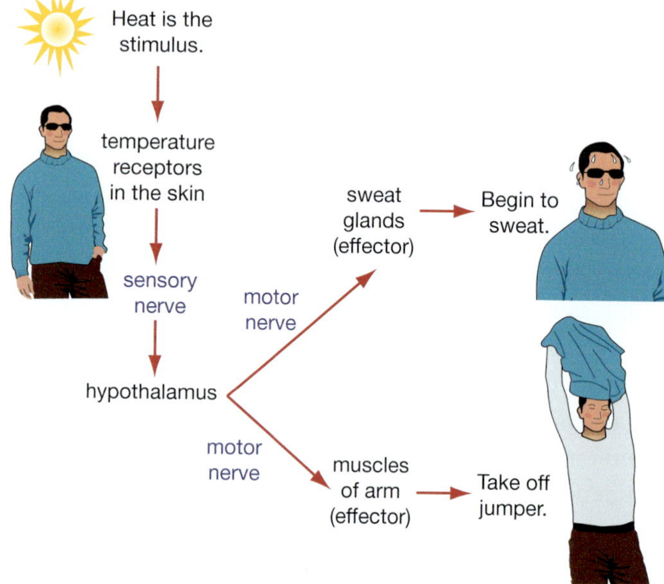

FIGURE 7.2.10 This stimulus–response outlines your body's reaction to being too hot.

Reflex actions: a rapid response

Arm muscles are normally under voluntary control. This means that you normally control what they do. However, when you touch something hot or sharp, you automatically pull your hand away. This reaction is involuntary and very fast—your arm didn't need to wait for instructions from the brain. This reaction is called a **reflex action** and it protects your body from danger.

Receptors detect the heat on your skin. This heat activates a sensory neuron, which sends nerve impulses to the spinal cord. Within the spinal cord, a relay neuron passes the message directly to a motor neuron, which sends impulses to the arm muscles, which are the effectors. The arm muscles contract, lifting your hand away from the sharp object. A message is sent to the brain shortly afterwards.

Only then can the brain register pain. The nerve pathway operating in a reflex action is called a **reflex arc**. Figure 7.2.11 shows an example of a reflex arc. Most reflex actions are rapid because they only involve a few neurons.

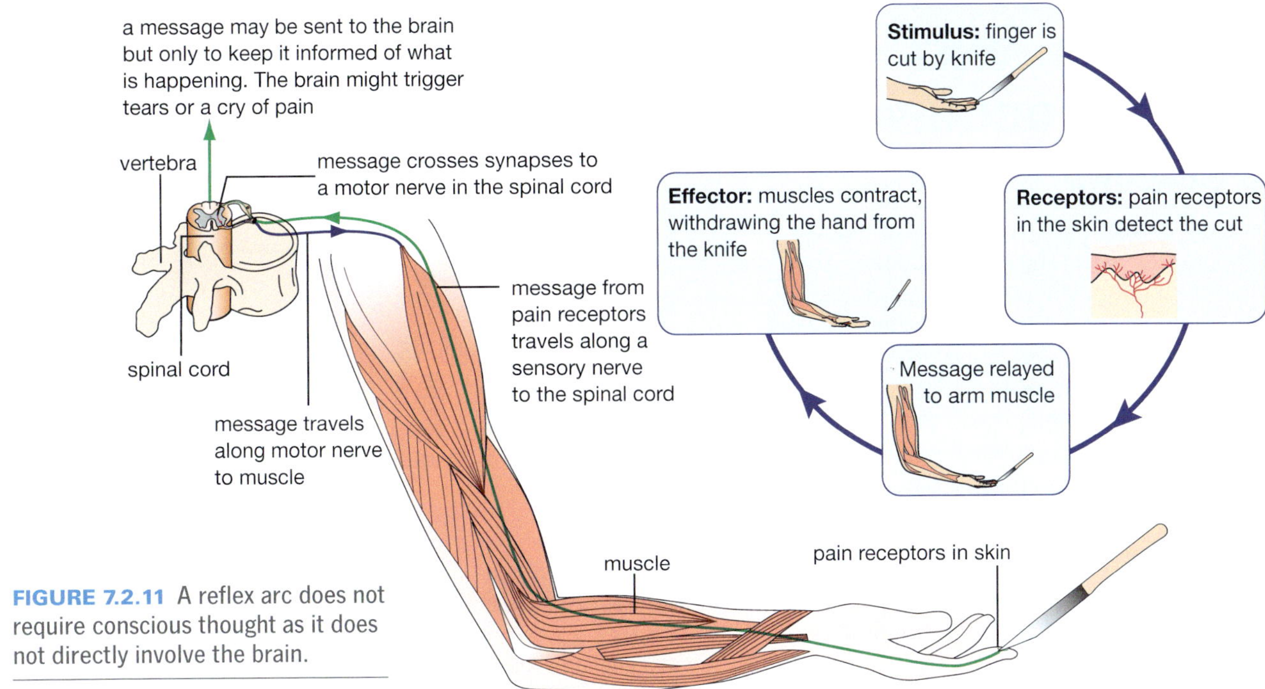

FIGURE 7.2.11 A reflex arc does not require conscious thought as it does not directly involve the brain.

AB 7.4

STEM 4 fun

Model arm

PROBLEM

Can you build a model arm?

SUPPLIES

- 2 icy-pole sticks (or cardboard strips of a similar size)
- box knife
- wire
- cotton or string
- scissors
- stationery split pin
- skewer etc.

SAFETY

Take extreme care when using the box knife.

PLAN AND DESIGN Your task is to build a model arm with icy-pole sticks for the bones and string for the muscles that act as effectors. Use it to show that muscles make the bones move.

Design the solution. What information do you need to solve the problem? Draw a diagram. Make a list of materials you will need and steps you will take.

CREATE Follow your plan. Produce your solution to the problem.

IMPROVE What works? What doesn't? How do you know it solves the problem? What could work better? Modify your design to make it better. Test it out.

REFLECTION

1. What field of science did you work in? Are there other fields where this activity applies?
2. In what career do these activities connect?
3. What did you do today that worked well? What didn't work well?

Working with Science

PAEDIATRIC OCCUPATIONAL THERAPIST

Anna Meadows

Occupational therapists (OTs) work with people of all ages with a wide range of difficulties. OTs promote independence in everyday activities, including personal care, home maintenance, work, leisure and social skills. OTs can work in hospitals, community centres, specialist clinics, government organisations, schools and private clinics. They work closely with other health professionals such as doctors, physiotherapists, speech pathologists and psychologists, social workers, teachers and architects, and with organisations such as local councils and equipment suppliers.

Anna Meadows is a paediatric (children's) OT (Figure 7.2.12). Children's occupations include self-care, going to preschool and school and, of course, play! Anna's clients may have cerebral palsy, autism spectrum disorder, global developmental delay, brain injury or sensory processing disorder, or they may simply want to improve in a particular area, such as handwriting, playing with friends or paying attention in class. Therapy is designed to be fun, motivating and meaningful. For example, a young student having trouble sitting at a desk during a colouring, cutting and pasting activity might be helped if they can develop postural and hand strength. This can happen if they hang from and swing on monkey bars.

OTs are experts in analysing activity requirements, assessing strengths and challenges around those activities and problem-solving. OTs give people the skills they need to reach their potential and improve their quality of life. OTs do this by helping people modify the way they carry out tasks or by making changes to the environments where they live or work.

FIGURE 7.2.12 Anna Meadows has a Bachelor of Occupational Therapy.

Anna is passionate about OT and loves seeing children improve and achieve their goals. She also loves that her job is so varied. She has worked in many different settings, both in Australia and overseas.

Review

1 Occupational therapists help people with many different tasks. What are five everyday tasks and skills that OTs might help their clients with? Do not include examples mentioned in the text.

2 What activities do you think paediatric OTs could do with children to help them improve the following?
   - a body coordination (such as using both hands together or moving arms and legs at the same time such as when playing sports)
   - b visual and motor skills working together (hand-eye coordination).

SCIENCE AS A HUMAN ENDEAVOUR

Use and influence of science

The bionic ear and eye

FIGURE 7.2.13 A cochlear implant

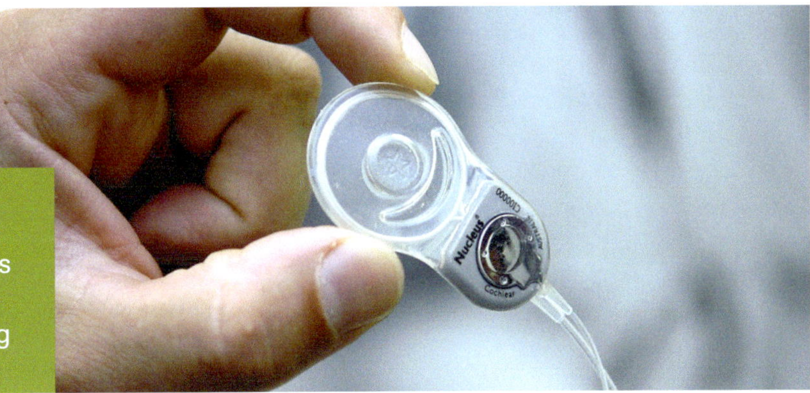

The bionic ear and the bionic eye were both developed by Australian scientists working in Australia. These technologies have made a huge impact on the lives of many who have limited or no hearing or sight.

The bionic ear

The bionic ear is more correctly known as a cochlear implant. You can see it in Figure 7.2.13. The cochlear implant was developed by an Australian scientist, Professor Graeme Clark. Clark was born in country New South Wales in 1935. His father was partially deaf, and this sparked Professor Clark's interest in the causes of deafness. He became a surgeon, specialising in otolaryngology, which is the study of diseases of the ear and throat. There was very little money available for his research, as most people believed it was impossible to restore hearing to the deaf.

The cochlear implant mimics the way that the cochlea receives sounds. A microphone and a speech processor are placed behind the ear (Figure 7.2.14). They pick up sounds and turn them into electrical signals. These signals pass into the implant, which is placed in the skull and connected to the cochlea. The cochlea then stimulates the auditory nerve to send messages to the brain.

In 1978, Clark successfully implanted the first bionic ear into a man named Rod Saunders, who had lost his hearing in a car accident. Clark's success has been recognised worldwide.

The Australian bionic ear has now provided hearing to more than 150000 people in more than 120 countries.

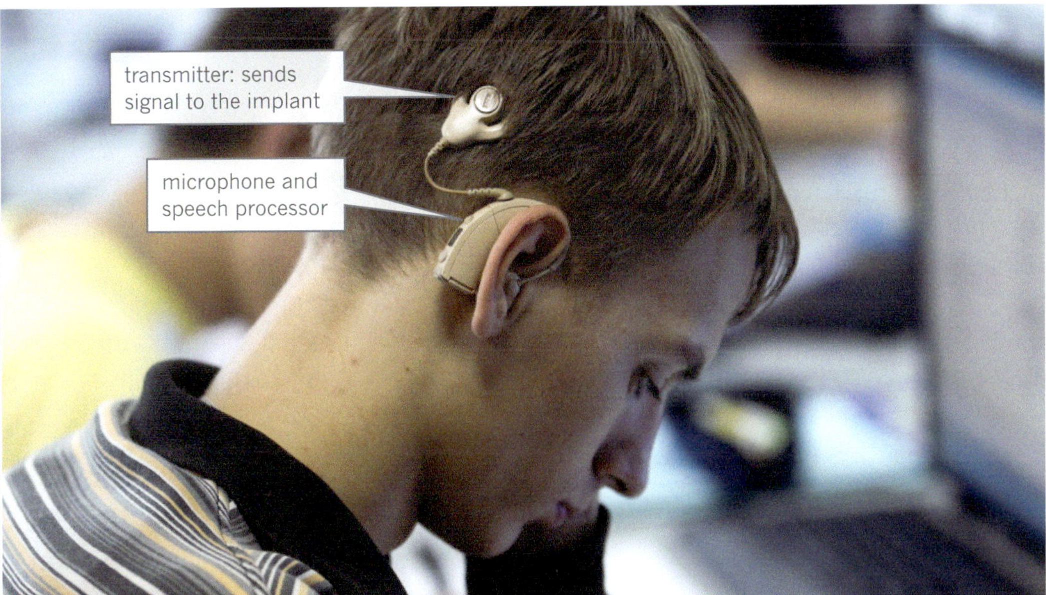

FIGURE 7.2.14 The cochlear implant or bionic ear allows some deaf people to hear again. The device is worn behind the ear.

SCIENCE AS A HUMAN ENDEAVOUR

The bionic eye

The bionic eye is being developed by Bionic Vision Australia researchers at the University of New South Wales. About 1.5 million people worldwide have a disease called retinitis pigmentosa, and about one in 10 people over the age of 55 have age-related macular degeneration. Both diseases cause cells in the retina of the eye to gradually die and the person becomes vision impaired or blind.

The bionic eye is still in the experimental stage, but it is hoped that it will be able to help people with these conditions. The bionic eye is a device that consists of a camera attached to a pair of glasses. How the device works is described in Figure 7.2.15.

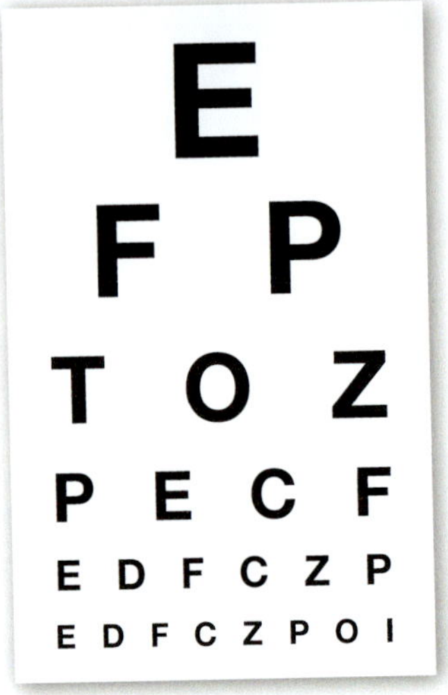

FIGURE 7.2.16 One version of the bionic eye should allow people fitted with it to read the third line of a Snellen chart.

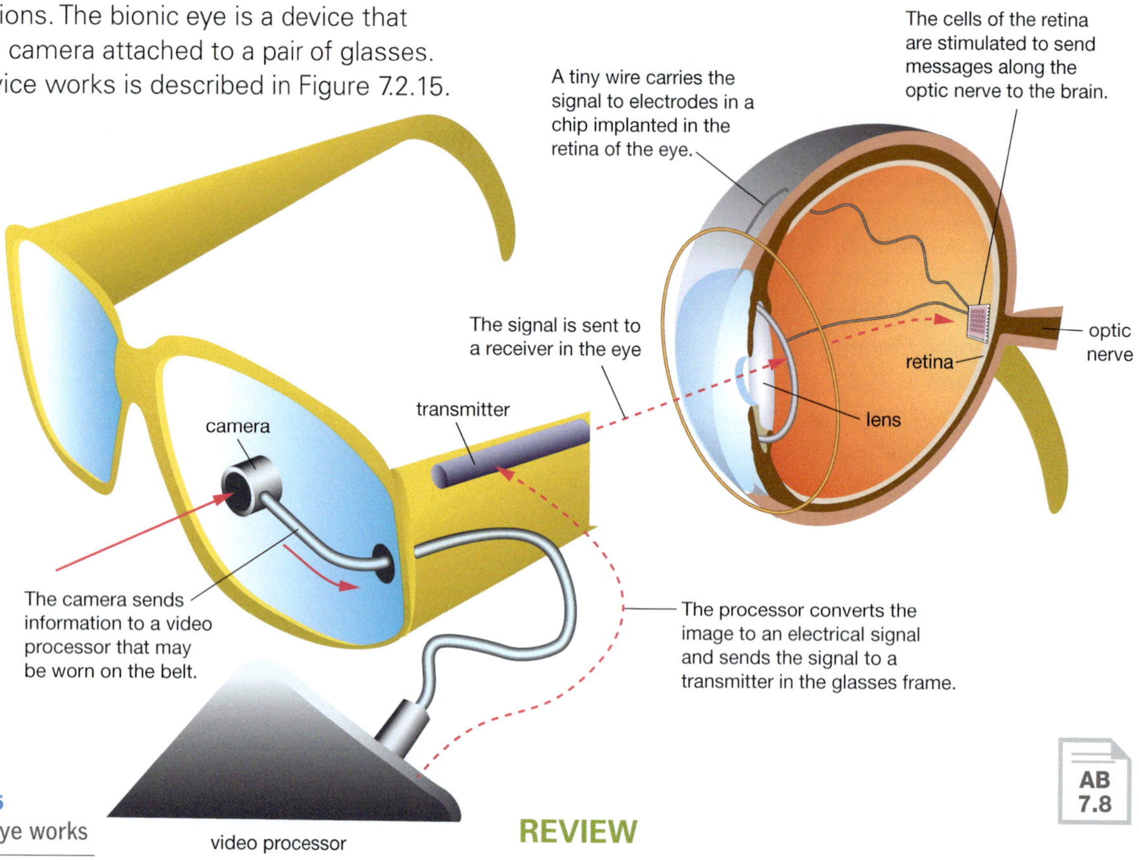

FIGURE 7.2.15 How a bionic eye works

One version of the bionic eye is designed to enable a person to distinguish light from dark. This will help the person move around large objects such as buildings, parked cars and benches or rubbish bins on footpaths.

A second version of the bionic eye will have many more electrodes. With it, the person may be able to recognise faces and read large print (Figure 7.2.16).

A different model of the bionic eye is also being developed by the Monash Vision Group at Monash University in Melbourne. This model does not act directly on the optic nerve, but offers similar results.

AB 7.8

REVIEW

1 Identify the events that led to Professor Clark's interest in the causes of deafness.
2 State the branch of medicine concerned with diseases of the ear and throat.
3 Describe what a cochlear implant does.
4 Construct a flow diagram that shows the steps from a sound being made to a person with a cochlear implant hearing it.
5 Compare the bionic ear and the bionic eye by listing their similarities and differences.
6 What do you think are the advantages for a patient of having a bionic eye? Include those mentioned in the text.

MODULE 7.2

Review questions

Remembering

1 Define the terms:
  a PNS
  b neurons
  c neurotransmitter
  d cerebellum.

2 What term best describes each of the following?
  a the layer of insulation surrounding nerves
  b the type of neuron that carries messages from the CNS to effectors
  c the space between the neurons
  d the nerve pathway operating in a reflex action.

3 What are the two parts of the central nervous system?

4 Draw a diagram of a neuron and label its main parts.

5 What types of activities does the right side of the brain have more control over?

6 What are the functions of the cerebrum?

Understanding

7 a Describe the function of dendrites.
  b Explain how their structure suits their function.

8 Describe the function of neurotransmitters.

9 Why does an injury to the left side of the brain often affect the right side of the body?

Applying

10 Use an example to describe how your brain controls an activity inside your body without your knowing about it.

11 Figure 7.2.17 shows the reflex arc for the knee jerk reflex. This is the reflex that happens when a doctor hits your knee with a small hammer. Your lower leg jerks up in response.

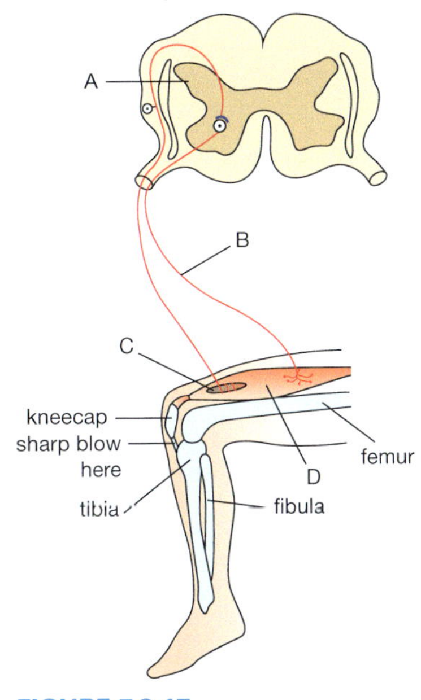

FIGURE 7.2.17

Apply your understanding of the reflex arc to identify what is happening in A, B and C on the diagram.

12 a Identify the arrow below (A or B) that represents an impulse passing along the sensory nerve.
  b Use your knowledge of stimulus and response to identify the effector in this example.

See an apple $\xrightarrow{a}$ brain $\xrightarrow{b}$ pick up the apple

Analysing

13 Compare the roles of a sensory neuron and a motor neuron.

14 Contrast a stimulus and a response.

15 Compare the roles of receptors and effectors.

Evaluating

16 When people first do the science4fun activity on page 284, most find it nearly impossible to name the colours the words are printed in. Instead, they read the actual words.
  a Which hemisphere of the brain is mainly associated with reading?
  b Deduce what side of the brain is associated with recognition of colours.
  c Propose a reason why this activity is very difficult.

17 Propose a reason why severe damage to the neck region is often fatal.

18 Many quadriplegic patients who are paralysed from the neck down can still maintain normal body functions such as breathing and digestion. Why do you think this is so?

19 During a medical test called a PET scan, a sugar solution containing a radioisotope is injected into the patient. The most active brain cells use the most sugar so they absorb more of the radioisotope. An image of the brain's activity is produced, based on where the radioisotope collects. How do you think a PET scan could be used to determine whether a person's brain is functioning normally?

Create

20 Construct a model of a neuron. Evaluate the success of your model by asking other students to identify the different parts of the neuron.

MODULE 7.2

Practical investigations

1 • Brain dissection

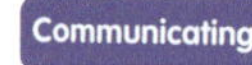

Purpose

To investigate the structure of a brain.

Timing 60 minutes

Materials

- 1 partly frozen lamb's brain

- scalpel
- dissection board
- newspaper
- disinfectant

Take extreme care with scalpels. Wear rubber gloves, safety glasses and a laboratory coat or an apron at all times.

Procedure

1 Cover your workbench in newspaper and place the dissection board on it.

2 Carefully inspect the surface of the brain. Use Figure 7.2.18 to identify the cerebrum, cerebellum, brain stem (medulla) and spinal cord.

3 The brain has a fine membrane or 'skin'. If possible, use the edge of your scalpel to lift and peel it off the brain.

4 Slice the brain in half, separating the right and left hemispheres.

5 Remove the cerebellum and brain stem (medulla).

6 If it is still sufficiently frozen, then slice the brain into thin slices. Use Figure 7.2.9 on page 283 and Figure 7.2.18 to help you identify the pituitary gland and hypothalamus.

7 Dispose of the remains and clean the dissecting board and equipment according to your teacher's instructions.

Results

Construct a table like the one below to summarise your observations about the size, colour, texture and consistency of everything you identified.

Brain structures and observations

Part	Size	Colour	Texture	Consistency
cerebrum				
cerebellum				

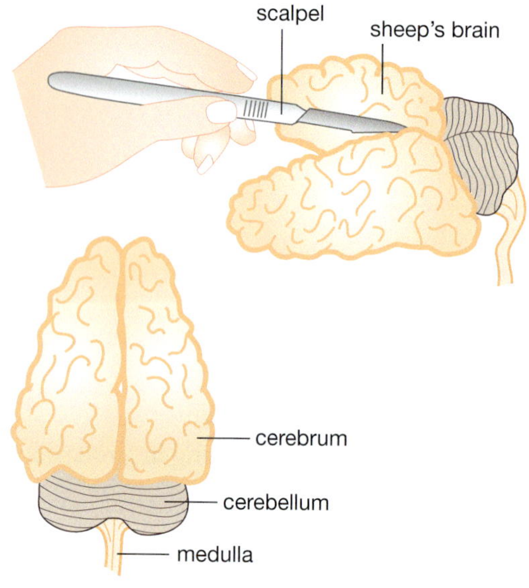

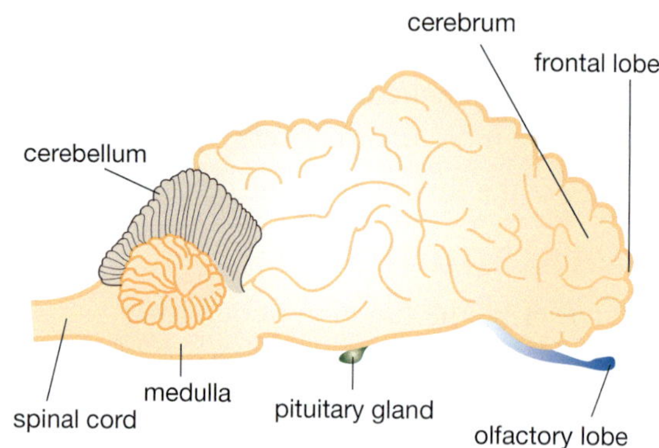

FIGURE 7.2.18

Review

1 Why do you think a frozen brain was dissected and not a fresh brain?

2 How did you identify which hemisphere was the left one and which was the right?

SkillBuilder

Using a scalpel

When using the scalpel:

- make many light cuts instead of one deep cut
- cut away from the hand holding the brain in place.

MODULE 7.2

Practical investigations

2 • A model brain

Purpose

To construct a model of a human brain.

Timing 45 minutes

Materials

- unpeeled orange
- assorted lollies (such as 1 banana, 3 jubes, 1 marshmallow, 3 snakes (different colours), 1 spearmint leaf), 2 sultanas
- toothpicks
- cotton buds
- plastic knife
- newspaper

SAFETY

Before beginning, ensure tables are wiped clean and covered.

Food should not be consumed in the laboratory.

Procedure

1 Peel the orange.

2 Identify the front of the 'brain' and attach two sultanas with toothpicks to represent the likely position of the eyes.

3 Carefully cut the orange to partly separate it into two halves/hemispheres.

4 Refer to Figure 7.2.9 on page 283 and Figure 7.2.19 and use toothpicks to attach the following lollies in their correct place in the brain as shown in this table.

Parts of the brain

Part of the brain	Lolly
cerebellum	marshmallow
medulla	banana
hippocampus	spearmint leaf
thalamus	jube
amygdala	jube (different colour
hypothalamus	jube (different colour)
motor cortex (motor area)	snake
sensory cortex (touch, smell, taste, hearing)	snake (different colour)
spinal cord	snake (different colour)

Review

1 Identify what represented the skull in this model.

2 Identify what represented the cerebrum in this model.

3 Construct a table showing which functions are carried out by each hemisphere.

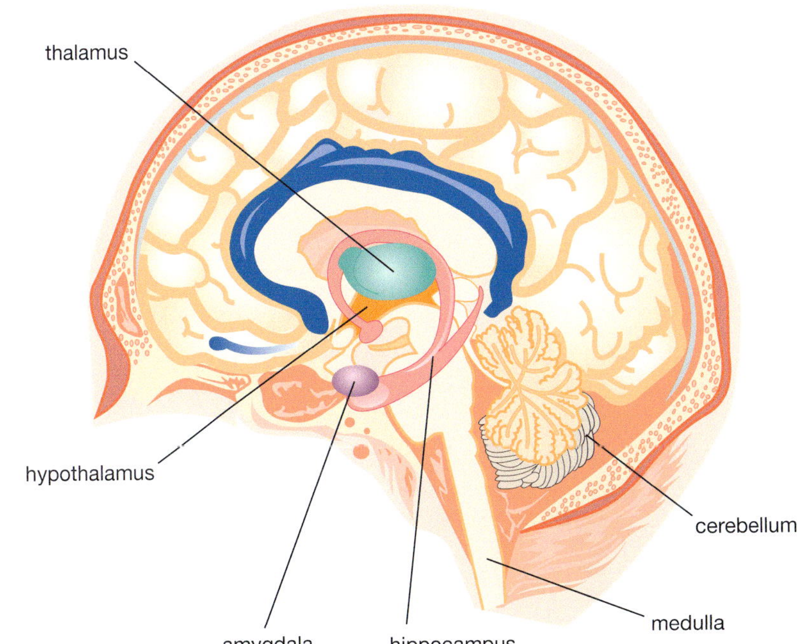

FIGURE 7.2.19

MODULE 7.2

Practical investigations

3 • How does the brain learn?

Questioning & Predicting | Evaluating

Purpose

To determine if a task will become easier if it is repeated

Timing 30 minutes

Materials

- A4 paper
- pen
- 10 cm × 10 cm cardboard
- mirror

Procedure

1. Draw a complex closed shape such as a star.
2. Draw around your shape about 1 cm distance from the original line.
3. Place the mirror behind your drawing so that you can see its reflection.
4. Punch a hole in the centre of the 10 cm × 10 cm piece of card, and fit the card over your pen.
5. Use the card to block the view of your shape so that you can only see the shape in the mirror.
6. Your task is to draw between the lines of your shape by only looking in the mirror and not at your hand.
7. Repeat the task three times (Figure 7.2.20).

FIGURE 7.2.20

Results

1. Copy the table below into your workbook.

Time taken to draw shape

Drawing	Time (s)
1	
2	
3	

2. In your notebook, record the time it takes to complete the shape.

   Add 5 seconds for every time the pen line is drawn over one of the shape lines.

Review

1. How does the time change when the experiment is repeated with an identical shape?
2. Assess whether your hypothesis was supported or not.

MODULE

7.3 Chemical control

There are small chemical messengers in your body that are responsible for big changes. Hormones control many functions of your body. Hormones are responsible for the changes of puberty as well as being responsible for controlling water balance and the volume of glucose in your blood. The glands of the endocrine system release hormones.

The endocrine system

The endocrine system is a communication system that controls the internal environment of the body. Water and glucose levels in your blood, body temperature and the changes the body goes through during puberty and pregnancy are all controlled by the endocrine system.

Hormones are chemical substances that act as messengers in the body. They are produced by the **endocrine glands**, which are located all around your body as seen in Figure 7.3.1. Together, all these glands form the **endocrine system**.

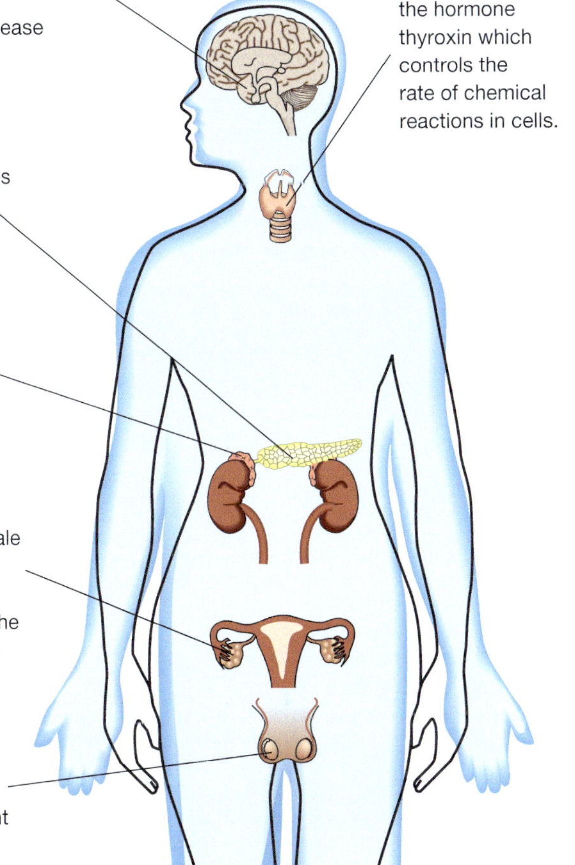

FIGURE 7.3.1 Major human endocrine glands

Chemical messengers

Hormones are secreted in very small quantities and travel through the bloodstream to all parts of the body. However, only particular cells called **target cells** will respond to particular hormones. Other cells don't have the particular receptor needed to read the message being sent. Hormones have a specific chemical structure and specific shape that fits chemically onto a receptor on the cell membrane of the target cell—just like pieces of a jigsaw puzzle. In this way they act very similar way to the lock-and-key model found in enzymes. In this way hormones target the correct type of cell without others being affected. When the hormone binds to the receptor on the target cell, this will start changes in the activities of the cell. This process is shown in Figure 7.3.2.

Hormones are only needed in very small quantities to have an effect on the body. They regulate functions like growth, water balance, regulation of glucose and sexual development.

Systems working together

The endocrine system is coordinated by the **pituitary gland**, which responds to information from the hypothalamus. The **hypothalamus** is a portion of the brain. You can see its location in Figure 7.3.3. It constantly checks the internal environment—that is, the conditions within the tissues, organs and systems of your body. If these conditions change, then the hypothalamus responds.

The most important function of the hypothalamus is to link the nervous system and the endocrine system. It secretes (releases) hormones that act on the pituitary gland.

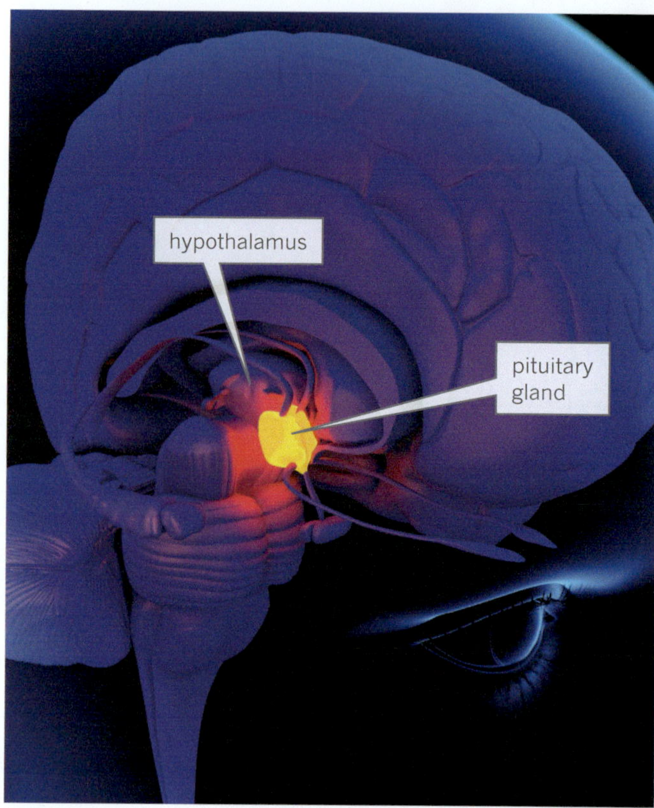

FIGURE 7.3.3 Two parts of the endocrine system: the hypothalamus, and the pituitary gland that attaches to it. These are located deep within the skull, where they are well protected.

The **pituitary gland** is often called the 'master gland' because it controls the activities of other endocrine glands such as the ovaries, the testes and the thyroid gland. The pituitary gland responds by hormones coming from the hypothalamus secreting other hormones or producing less of the hormones. Through its action on the pituitary gland, the hypothalamus controls important aspects of the body such as body temperature, rate of metabolism and water content.

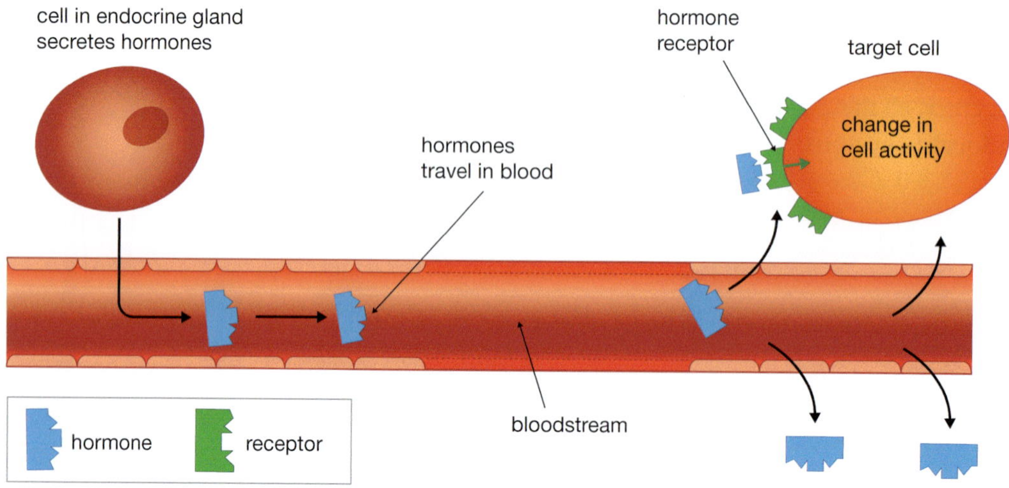

FIGURE 7.3.2 Hormones are only active in cells where the shape of the hormone and the shape of the receptors fit together like pieces of a jigsaw.

In situations of fear or stress both the nervous system and endocrine system have roles to play (Figure 7.3.4). This response is known as the 'fight or flight response'. The endocrine system releases the hormones adrenaline and cortisol, which increases heart rate, breathing rate and blood pressure. Blood glucose levels also rise to provide muscles with required energy for this response. The nervous system acts swiftly through the network of neurons to work with the endocrine system to increase the rate of breathing and increase the heart rate. Pupils will dilate (widen) to improve vision, sweat glands will produce more sweat and digestion will slow down or cease.

FIGURE 7.3.4 The actions of the nervous system and the endocrine system cause the sensations associated with fear or stress.

Controlling the internal environment

Your body works most efficiently when its internal environment is kept constant. This means that factors such as temperature, water content, available energy, available oxygen and concentration of wastes in the blood are all controlled. The process of maintaining a constant internal environment is known as **homeostasis** (*homeo* means 'same' and *stasis* means 'state'). Homeostasis involves receptors that are sensitive to a particular stimulus, and effectors, muscles or glands that have an effect on the same stimulus.

This type of control is known as a **feedback system**.

To understand this, consider a reverse-cycle air conditioner like the one shown in Figure 7.3.5 as an example of a machine that maintains a constant environment. A sensor called a thermostat is set at a particular temperature range, such as 21–23°C.

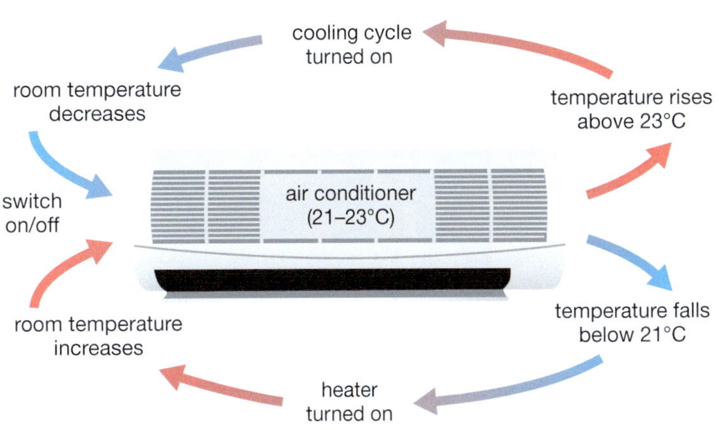

FIGURE 7.3.5 A reverse-cycle air conditioner maintains a preset temperature in a room by heating or cooling the air in response to temperature detectors.

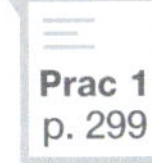
Prac 1
p. 299

If the temperature of the room goes above 23°C, the air conditioner switches on and cools the room until the required temperature is reached. The sensor detects this and then the air conditioner switches off until it once more detects a rise in temperature. If the sensor detects a lower temperature than 21°C, the heating system turns on. The temperature in the room rises and the heater turns off once the set temperature range of 21–23°C is reached.

Your body responds to the environment in a similar way. When temperatures rise, the body reacts by sweating and dilating blood vessels in an attempt to cool a person down. When the temperature decreases, the body responds by contracting blood vessels, creating goosebumps and making hairs stand up on end in an attempt to insulate the body and prevent heat loss. In this way, the body maintains its optimal temperature.

Controlling body temperature

Digestion, growth and repair, respiration and manufacture of hormones are some of the chemical reactions taking place inside your body. The heat they produce as a by-product maintains your body temperature regardless of the temperature of your surroundings. This is illustrated in Figure 7.3.6 on page 296. Because you can maintain a constant body temperature, you are said to be **endothermic**. However, if the temperature inside your body was to increase by more than a few degrees above 37°C, your metabolism would stop, and you would die. If your body temperature fell below 37°C, your metabolism would slow down.

FIGURE 7.3.6 Your body temperature stays the same regardless of the temperature of the environment. There are things you can do to help your body, such as adding clothing when it is cold, and removing clothing and using shade when it is very hot.

One of the most important examples of homeostasis is the regulation of body temperature. Not all animals can do this as a normal body function. Animals that can maintain a constant internal body temperature such as birds and mammals are called endothermic. Endothermic animals are sometimes called warm-blooded animals because they maintain their body temperature within narrow limits. These narrow limits are controlled by the nervous and endocrine system; they maintain an internal body temperature between 36 and 40°C. Animals that cannot maintain a constant internal body temperature such as reptiles are called **ectothermic**. Figure 7.3.7 demonstrates the relationship between body temperature and environmental temperature for ectothermic and endothermic animals.

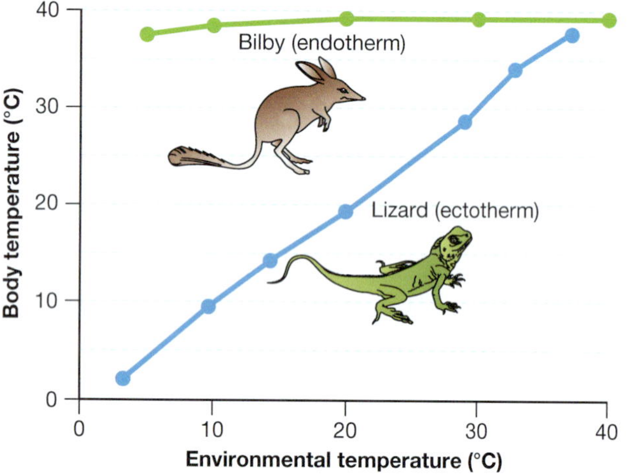

FIGURE 7.3.7 Endothermic animals can maintain their body temperature in a changing environment. Ectothermic animals have a body temperature similar to that of their surroundings.

Ectothermic animals can also have very warm blood; however, they regulate their body temperature through behavioural mechanisms such as by lying in the sun when cold and moving into the shade when hot, as seen in Figure 7.3.8. Some other ectotherms such as bees and tuna use extended muscular activity to increase body temperature. These behavioural mechanisms of controlling body temperature can be very effective especially when combined with internal mechanisms that ensure that the temperature of the blood going to vital organs such as the brain and heart is kept constant.

FIGURE 7.3.8 Like all lizards, the Australian water dragon is ectothermic. This dragon has flattened itself out to maximise the amount of sunlight falling on it.

Hormonal control of temperature

The hypothalamus acts on the pituitary gland to control body temperature through the action on another endocrine gland—the thyroid gland.

The hypothalamus receives information from temperature receptors in the skin as shown in Figure 7.3.9 and from internal receptors, including the hypothalamus itself.

If the hypothalamus detects a fall in body temperature, it produces a hormone that causes the pituitary gland to secrete more thyroid-stimulating hormone (TSH). TSH stimulates the thyroid to release more of the hormone thyroxine. Thyroxine travels in the blood to all cells and causes the rate of metabolism in the cells to increase.

Increased metabolism generates more heat and warms the body. Producing the hormones that cause these changes takes time, and therefore the endocrine system does not have immediate control over body temperature.

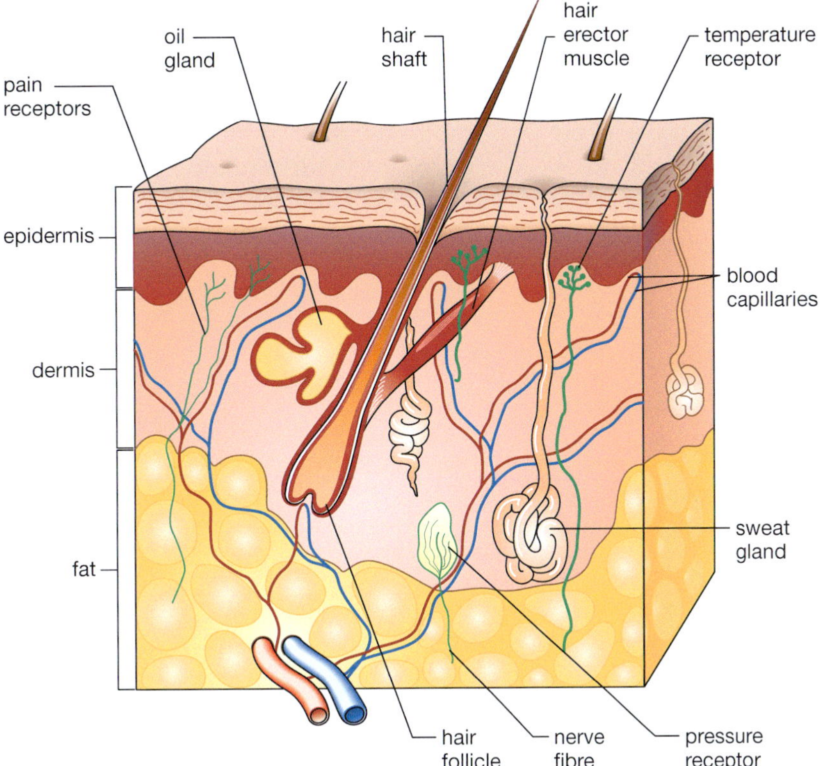

FIGURE 7.3.9 The skin has many receptors and provides you with a lot of information about your surroundings.

Nervous control of temperature

Body temperature is also controlled by the nervous system and this is a more immediate response. When the hypothalamus detects a drop in temperature, it sends nerve impulses to muscle groups around vital organs such as the heart and lungs. Small shaking movements begin in these muscles. Eventually, the shaking movements extend to the large muscles of the arms and legs, and you begin to shiver. Shivering increases the activity of muscle cells, producing heat and raising body temperature. This is the body's way of creating warmth by using energy.

Another aspect of nervous control is the process that reduces blood flow to your skin when you are cold. The sympathetic nervous system causes a narrowing of the blood vessels near the surface of the skin. This reduces blood flow and therefore heat flow to the skin. If the external temperature is very low, the blood flow to the fingers, toes, nose and ears is reduced further and you can lose feeling in them. This causes your toes and fingers to go numb.

When the hypothalamus detects a rise in body temperature, nerve messages are sent to the sweat glands and blood vessels. Blood vessels close to the skin dilate (increase in diameter). This change allows more blood and the heat it carries to reach the skin surface. The extra blood near the surface makes your skin red.

The message from the hypothalamus causes the sweat glands to produce more sweat. Heat from your body causes the sweat to evaporate. The rate at which heat is lost by evaporation depends on the difference in temperature between the body and the surrounding air, and the relative humidity of the air.

MODULE 7.3

Review questions

Remembering

1 Define the terms:

- **a** hormone
- **b** endothermic
- **c** effector
- **d** receptor.

2 What term best describes each of the following?

- **a** unable to maintain a constant internal body temperature
- **b** the type of cells that respond to hormones
- **c** the process of maintaining a relatively constant internal environment in the human body
- **d** the gland in the body referred to as the 'master gland'.

3 What three functions in your body are regulated by hormones?

4 Refer to Figure 7.3.9 on page 297 and list three types of receptors found in the skin.

Understanding

5 Describe the relationship between hormones and the endocrine system.

6 **a** What is the function of the hypothalamus?

- **b** Why can the hypothalamus be thought of as part of both the nervous system and the endocrine system?

7 Explain why sweating is an efficient way for the body to lose heat.

8 **a** Why is it important that your body temperature remains constant?

- **b** Describe two involuntary reactions that keep your body temperature from rising.
- **c** Describe one involuntary reaction that prevents your body temperature from falling.

9 How are hormones transported in the body?

10 The body makes more than 20 hormones, each with a specific function. Outline the process by which these hormones recognise which tissues and organs to communicate with.

Applying

11 Identify the endocrine glands labelled A–G in Figure 7.3.10.

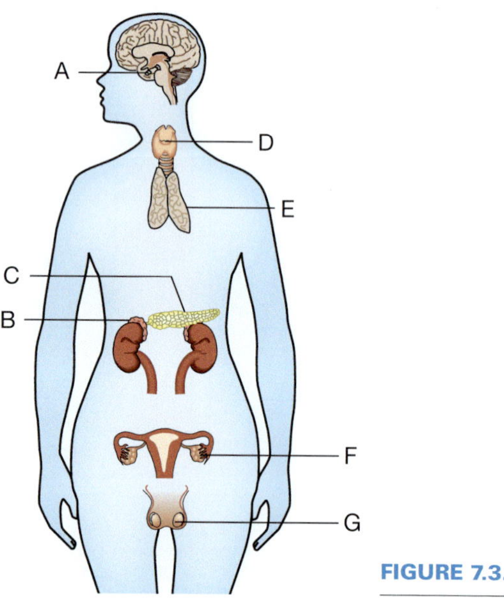

FIGURE 7.3.10

12 Use a diagram to help you explain how hormones recognise their target cells.

Analysing

13 Use examples to compare the way that endothermic and ectothermic animals control body temperature.

14 Analyse why:

- **a** ectothermic animals generally live in warmer climates
- **b** endothermic animals can live in a wide range of environments, from hot to cold.

Evaluating

15 Explain why the pituitary gland is often referred to as the 'master gland'.

Create

16 Create a model or digital animation to demonstrate the action of hormones. Begin with the hormones being produced in the gland and end with the target cells responding.

17 Create a diagram to represent the response your body would have to an increase in environmental temperature.

MODULE 7.3

Practical investigations

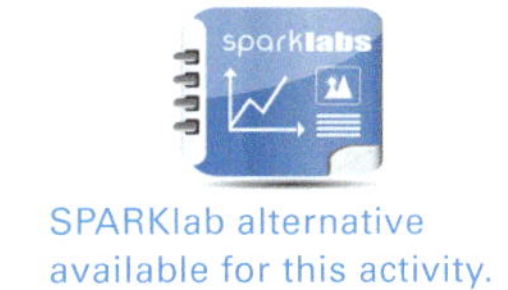

SPARKlab alternative available for this activity.

1 • Changing temperature

Purpose

To investigate the effect of exercise on body temperature.

Hypothesis

What do you think happens to your skin and internal body temperatures as you exercise? Before you go any further with this investigation, write a hypothesis in your workbook.

Timing 60 minutes

Materials

- stopwatch
- electronic clinical thermometer (if available, data-logging equipment could be used for this experiment)
- area of the school where you can run around

SAFETY

Students with health problems may not be able to take part in this activity.

Procedure

1 Work in pairs. One person is the subject and the other is the recorder.

2 Record the skin temperature of the subject at the start of the experiment. Measure the skin temperature by holding the thermometer inside a bent elbow. Measure the internal temperature at the ear using an electronic clinical thermometer.

3 The subject undertakes 10 minutes of vigorous exercise, enough for them to feel hot and possibly turn red in the face.

4 The recorder records the subject's skin temperature and internal temperature.

5 Observe and record any other changes resulting from the exercise in an appropriate table.

Results

Record all measurements and observations.

Review

1 Describe the changes to the skin temperature during the experiment.

2 Describe the changes to the internal body temperature during the experiment.

3 Compare the changes in skin temperature with the changes in internal body temperature.

4 Describe any other changes that were observed.

5 How could these other changes contribute to the changes you saw in temperature?

6 a Construct a conclusion for your investigation.

b Assess whether your hypothesis was supported or not.

CHAPTER 7

Chapter review

LS

Remembering

1 Define the terms:
   a metabolism
   b homeostasis.

2 Name the two parts of the human nervous system.

3 What is the 'master gland' of the endocrine system?

4 Name the body system that:
   a delivers materials to cells
   b gets oxygen into the body
   c removes wastes from the body.

Understanding

5 Explain the relationship between proteins and amino acids.

6 Explain why the human body needs the following systems.
   a digestive system
   b respiratory system
   c excretory system
   d nervous system
   e endocrine system
   f circulatory system.

7 a What is the role of the 'pacemaker' in the heart?
   b Predict what could happen to the heart rate if the pacemaker stopped working.

8 Describe two situations in your body where substances move from one place to another by the process of diffusion.

9 Describe a situation in which the response of the body is controlled by both the nervous and endocrine systems.

Applying

10 Use the concept of diffusion to explain the process of oxygen and carbon dioxide exchange in the lungs.

11 a Describe the changes that occur in your body as you start to do some vigorous exercise.
   b Identify the systems that are causing the changes.
   c Explain how the changes are brought about.
   d Explain how the changes help your body maintain that level of activity.

12 Use an example to demonstrate that multicellular organisms rely on coordinated and interdependent internal systems.

Analysing

13 a Contrast the roles of the central and peripheral nervous systems of the body.
   b Why are both systems required?

14 Compare the nervous and endocrine systems, listing their similarities and differences.

Evaluating

15 a Assess each of the statements below and decide whether they are true or false.
      i Motor neurons carry messages from muscles and glands to the central nervous system.
      ii The peripheral nervous system has two main parts: the somatic nervous system and the autonomic nervous system.
      iii The digestive system can work independently of the other systems of the body.
      iv The kidneys are the only organs involved in removing wastes from your body.
   b For each statement you decided was false, justify your decision.

16 It takes time for your eyes to react to the change from dark to light.
   a Do you think this reaction is caused by the nervous system or the endocrine system?
   b Justify your response.

17 a Assess whether you can or cannot answer the questions on page 268 at the start of this chapter.
   b Use this assessment to evaluate how well you understand the material presented in this chapter.

Creating

18 Use the following ten key terms to construct a visual summary of the information presented in this chapter.

AB 7.11

central nervous system	axon
peripheral nervous system	receptors
nervous system	metabolism
neurotransmitter	homeostasis
diffusion	reflex actions

CHAPTER 7

Inquiry skills

Research

1 Planning & Conducting | Evaluating

Three-part inquiry question

Select your entry point and complete the relevant parts of this inquiry.

a i Describe the roles of the sympathetic and parasympathetic nervous systems in the control of heartbeat.

ii Investigate the pacemaker or SA node and atrioventricular (AV) nodes and then discuss how they control the passage of electrical messages through the heart.

iii How might damage to the AV node affect heartbeat?

iv Discuss the likely effects of AV node damage on blood pressure and oxygen supply to the tissues.

b i What is an artificial pacemaker?

ii Although many patients who have had heart transplants develop good natural heartbeat control a significant number require artificial pacemakers explain why. Remember to include the relevance of the SA and AV nodes in your answer.

c Electrocardiography (ECG) is used to monitor heartbeat. A normal electrocardiogram trace (ECG) is shown in Figure 7.4.1.

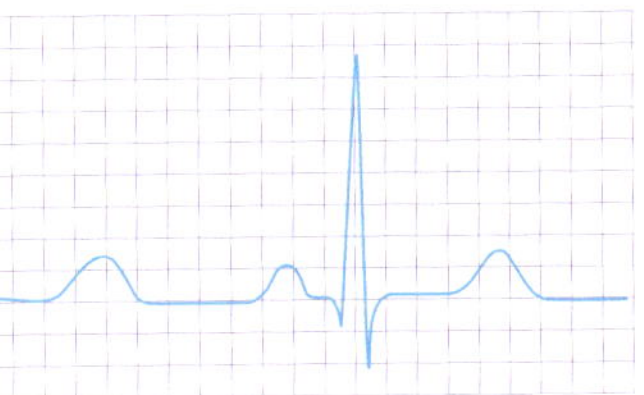

FIGURE 7.4.1 A normal ECG trace

i On the ECG, identify the contraction of the atria and ventricles.

ii Damage to the SA or AV nodes stops messages being received by the ventricles from the brain. This does not stop the ventricles contracting. Another group of cells in the heart take control. These are called Purkinje fibres. However, these produce contractions at a very slow rate of less than 40 beats per minute. Also contractions of the ventricles and atria occur completely independently of each other. How might this affect blood flow both within the heart chambers and to the body? Explain in detail.

2 Processing & Analysing | Evaluating

Search the internet to find games that measure your reaction times. Use a variety of different games to determine your reaction times in each. Use these times to calculate your average reaction time.

3 Processing & Analysing | Evaluating

Research the work done in the body by the liver.

- List the functions of the liver.
- Explain how each function helps your body to work.
- Describe ways in which lifestyle can affect the functioning of the liver.

Present your findings as an information pamphlet of the type available in doctors' waiting rooms or in digital form.

4 Processing & Analysing | Evaluating

Research the life and work of the Australian heart surgeon Victor Chang.

- List his major achievements.
- Explain how his work helped improve the survival rate of heart transplant patients.
- Describe Dr Chang's contribution to the Australian medical profession.

Present your research as a feature article for a magazine or as a web page.

5 Processing & Analysing | Communicating

Research the Stroop effect. As part of your research:

- find which hemisphere of the brain is associated with reading and speaking
- find which hemisphere of the brain is associated with recognising colours (not words)
- explain how the different hemispheres are involved in the science4fun activity on page 284
- propose reasons why there is confusion while you try to complete this task
- repeat the task test but with the words for these colours written in another language. Compare this task with your earlier task. Did you have the same problem?

Present your research as a set of answers to the questions above.

CHAPTER 7

Inquiry skills

Thinking scientifically

Questions 1 and 2 refer to Figure 7.4.2.

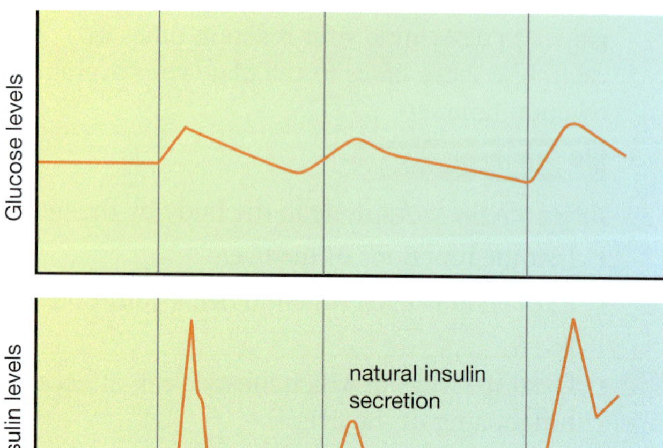

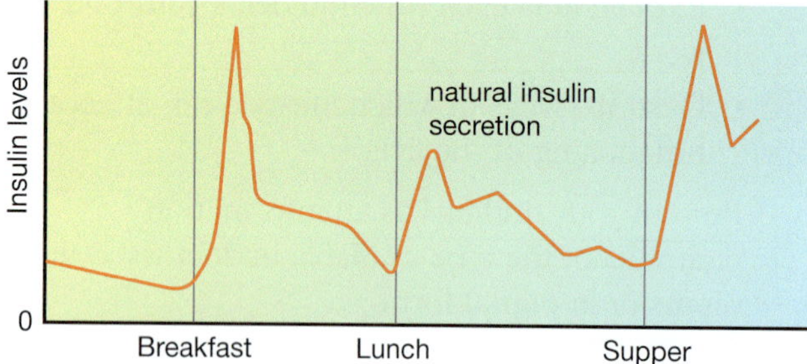

FIGURE 7.4.2

1 There was a large spike in insulin levels in the morning because:
   - A food had just been consumed
   - B there had been no food consumed overnight
   - C the body was preparing for a meal
   - D none of the above.

2 Just before meal times, blood glucose levels:
   - A decreased
   - B increased
   - C did not change
   - D decreased and then rapidly increased again.

3 Given the data in Table 7.4.1, which of the following hypotheses is supported?
   - A Group 1 who exercised produced more urine than Group 2 who sat in the sun.
   - B The volume of water lost through exercise is less than the volume of water lost when sitting in the sun.
   - C Group 2 who sat in the sun produced more urine than Group 1 who exercised.
   - D Both groups produced the same volume of urine.

TABLE 7.4.1 Volume of urine produced

Time (min)	Volume of urine produced (mL)	
	Group 1	Group 2
0	50	50
30	53	50
	vigorous exercise for 30 minutes	sat in the sun for 30 minutes
60	60	30
90	10	20
120	8	20
150	35	25
180	40	23

4 The graphs in Figure 7.4.3 show the level of activity of a human enzyme at different temperatures and pH.

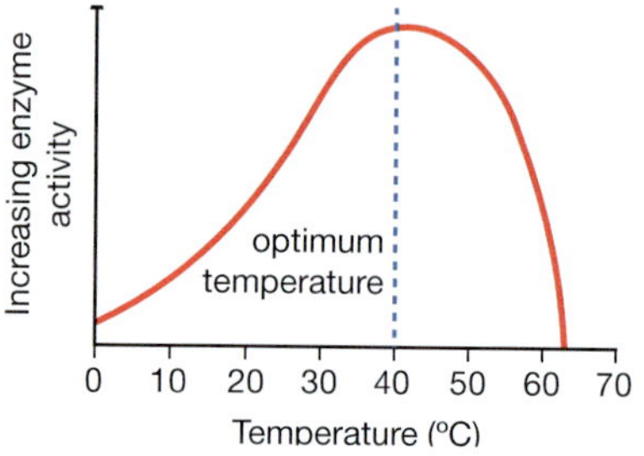

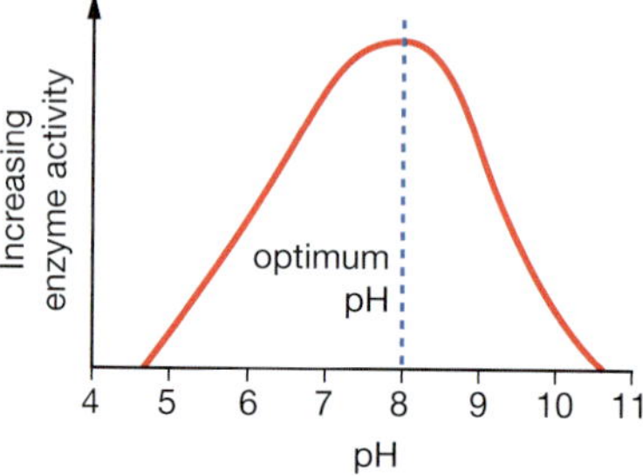

FIGURE 7.4.3

The normal internal temperature of the human body is 37°C. The pH of the mouth is about 7.5, in the stomach it is about 3, and in the small intestine it is about 8.

Select the statement that best fits all the data presented.
   - A The enzyme works best at low pH and high temperature.
   - B Temperature and pH have no effect on activity of the enzyme.
   - C The enzyme would be able to function best in the small intestine.
   - D The enzyme is most likely a digestive enzyme from the stomach.

CHAPTER 7 Glossary

anabolic: a process involving chemical reactions that produce complex molecules from simpler substances

axon: a nerve fibre that sends nerve impulses away from the cell

brain stem: the part of the brain where the spinal cord enters the skull; it controls the body's vital functions such as breathing, blood pressure and heart rate

catabolic: a process involving chemical reactions that breaks down complex molecules into smaller ones

cell body: the part of the neuron that contains the nucleus

central nervous system (CNS): the brain and spinal cord

cerebellum: the part of the brain that is responsible for coordination and balance

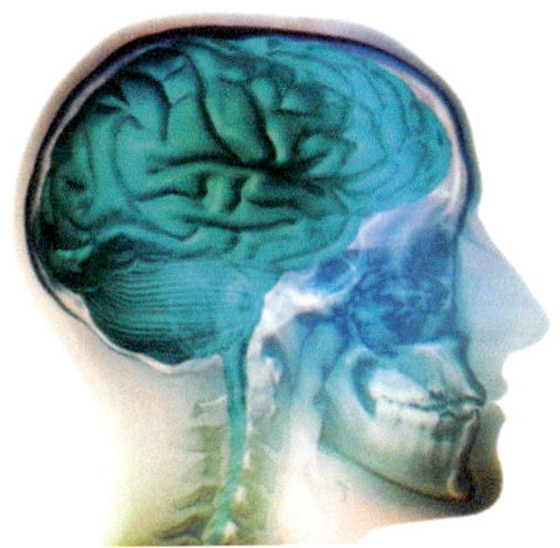

cerebellum

cerebrospinal fluid (CSF): a fluid surrounding the brain and spinal cord, which provides nutrients to the neurons and acts as a shock absorber

cerebrum: the part of the brain that controls conscious thoughts, and the movement of every body part, and receives sensory messages from each body part

connector neurons: these neurons transmit messages between neurons in the CNS

cranium: a bony structure that surrounds the brain

dendrites: branches from the cell body that receive messages from other neurons

diffusion: the movement of particles of a substance from an area of high concentration to an area of low concentration

ectothermic: animals that rely on environmental temperature to regulate body temperature. Ectothermic animals have a body temperature similar to that of their surroundings

effectors: muscles or glands that put the messages into effect

endocrine glands: glands that produce hormones

endocrine system: all the endocrine glands of the body

endothermic: able to maintain a constant body temperature

endothermic

enzyme: a chemical that speeds up a rate of reaction

feedback system: body systems regulate themselves by monitoring and self-correction adjusting output depending on stimulus

homeostasis: the process of maintaining a constant internal environment

hormones: chemical substances that act as messengers in the body

hypothalamus: a portion of the brain that constantly checks the internal environment of the body

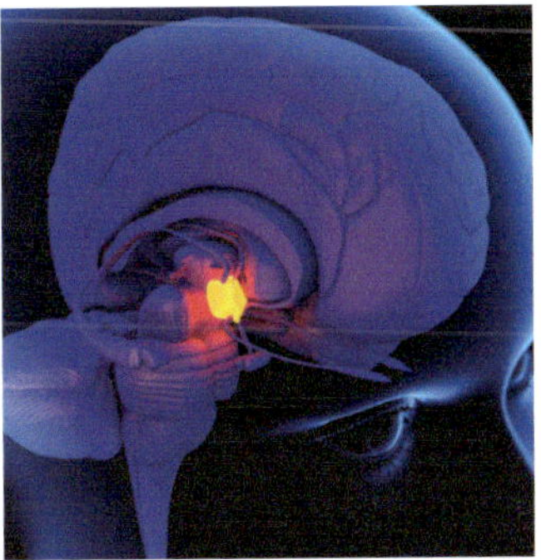

hypothalamus

medulla: the lower half of the brain stem

metabolism: all the chemical reactions occurring in the cells

mitochondria: organelles where cellular respiration occurs

motor neurons: nerve cells that carry messages from the CNS to effectors

CHAPTER 7

Glossary

myelin sheath: the insulating layer that covers a neuron

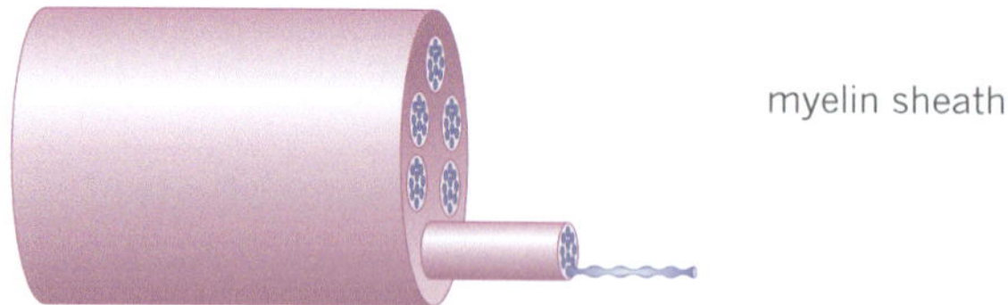
myelin sheath

nerve impulse: the electrical message carried by a nerve cell

neuron: a nerve cell

neurotransmitter: a chemical message released at the end of an axon to be received by the next neuron's dendrites

peripheral nervous system (PNS): the nerves that carry messages to and from the central nervous system and other parts of the body

pituitary gland: the endocrine gland that controls the activities of other endocrine glands; it is often called the 'master gland'

receptor: a specialised cell that detects stimuli (changes)

reflex actions: quick, automatic actions that protect the body from danger; also known as reflexes

reflex arc: the nerve pathway operating in a reflex action

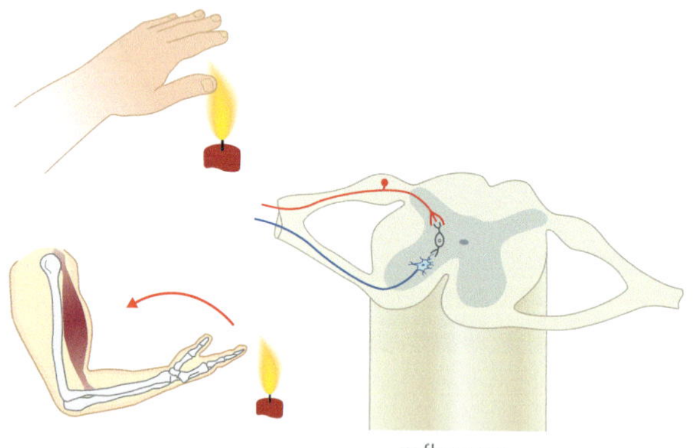
reflex arc

ribosome: the structure where proteins are manufactured

sensory neurons: nerve cells that carry messages from cells in the sense organs to the CNS

stimulus: any factor that stimulates a receptor and brings about a response

substrate: the molecule that is going to be changed by a chemical reaction involving an enzyme

synapse: the space between two neurons

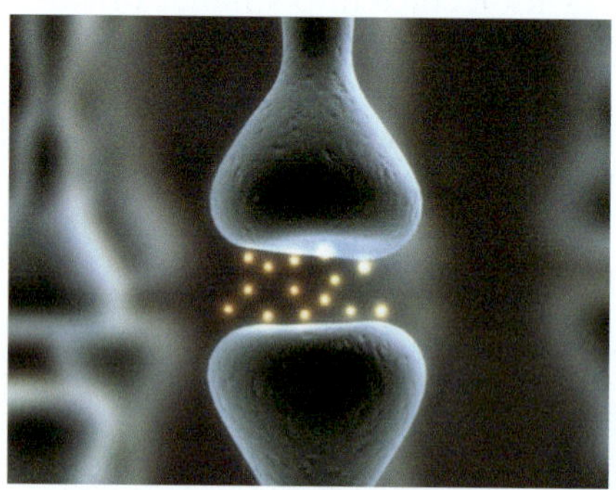
synapse

target cells: the cells on which a hormone acts

vertebrae: bones that surround the spinal cord and provide attachment for muscles

AB 7.10

CHAPTER 8

Disease

Have you ever wondered ...

- what causes you to get sick?
- how vaccines work?
- why you should wash your hands before eating?
- what is meant by a healthy diet?

LS

After completing this chapter you should be able to:

- investigate the response of your body to changes as a result of the presence of microorganisms
- explore how ideas about disease transmission have changed from medieval time to the present as knowledge has developed
- investigate the work of Australian scientists in the area of disease prevention and treatment
- investigate the use nanotechnology in medicine, such as delivery of pharmaceuticals
- use knowledge of science to test claims made in advertising or expressed in the media.

AB 8.1

MODULE
8.1 Infectious disease

Everyone gets sick at some time. Most of us get better quickly. However, sometimes it takes longer and help is needed to get better. Sometimes the reason for being sick is not clear and further tests and treatments are needed.

science 4 fun

What makes you sick?

If you have been sick this year, what do you think caused it?

Collect this ...

- pen and paper

Do this ...

1 Draw up a three column table with the headings Month, Symptoms, Suggested cause.

2 Complete the table for the month of illness and what you had or the symptoms. You may need to gather information from other people such as your parents.

Record this ...

1 Describe the symptoms that showed that you were sick in each case.

2 Explain what you think caused you to become sick in each case by completing the final column of your table.

SciFile

On me?

There are more living organisms on the skin of a single human being (even a clean one) than there are human beings on the surface of the Earth! In 2017, Earth's population is estimated to be approximately 7.5 billion.

Disease

Disease is defined as any condition in which the body or parts of the body do not function properly. There are many different factors that can cause disease. These include infection by microscopic organisms, more commonly known as microorganisms, and environmental and lifestyle factors.

Bacteria

One group of microorganisms that can cause disease is bacteria.

Bacteria are microscopic, unicellular (single-celled) organisms. Thousands of different species of bacteria have been discovered, but scientists are still finding many, many more. This has led them to believe that most have not yet been discovered.

Bacteria are an important part of the natural environment. They are decomposers, which means they convert dead plant and animal matter and wastes into nutrients that plants use to grow (Figure 8.1.1). Bacteria living in the intestines of herbivores such as cows and kangaroos help with digestion. Humans use bacteria to make medicines and to break down pollutants such as oil and plastics. However, a small percentage of bacteria are harmful and cause disease. These bacteria are known as **pathogenic bacteria** or **pathogens**.

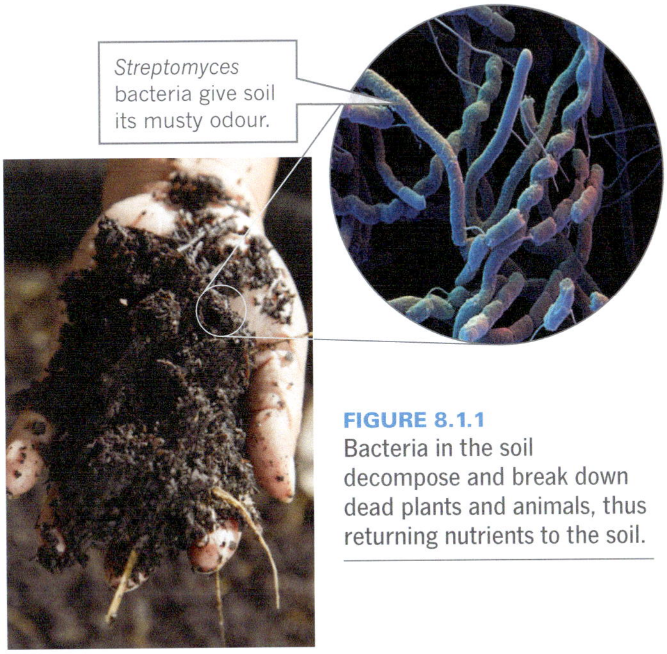

Streptomyces bacteria give soil its musty odour.

FIGURE 8.1.1 Bacteria in the soil decompose and break down dead plants and animals, thus returning nutrients to the soil.

Bacterial diseases

Pathogenic bacteria cause hundreds of diseases, such as whooping cough, tetanus, diphtheria, impetigo, pneumococcal and meningococcal disease, and typhoid fever. These are all **infectious diseases**, because they are caused by infection with pathogenic microorganisms.

Some infectious diseases are more easily spread than others. Those that are able to spread readily by close contact with an infected person are described as **contagious**. Figure 8.1.2 shows impetigo, a contagious disease that is more common in children than in adults. For this reason, it is commonly known as school sores. Touching someone with impetigo could result in you becoming infected. For this reason infected children are put into **quarantine**—they are isolated from healthy people to prevent the spread of the disease. This means they are not allowed to attend school or day care centres until they have started treatment for the disease. Even then they can only return to school if the sores are covered with watertight dressings. Quarantine is used to prevent the spread of disease within a community, between communities and between countries.

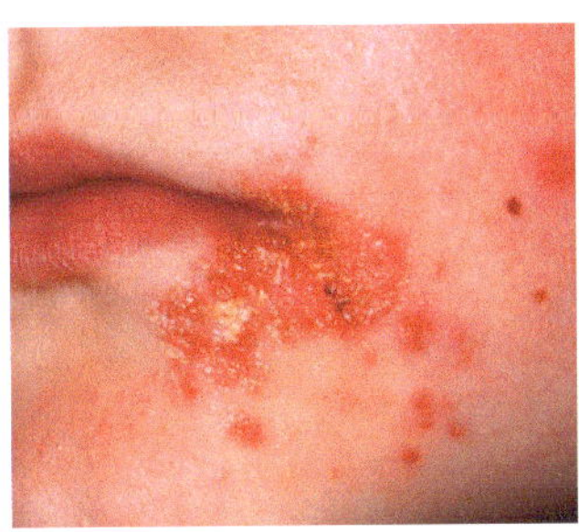

FIGURE 8.1.2 Impetigo remains contagious until the sores have healed. The wounds should heal within five days of antibiotic treatment.

Bacterial infections are treated with **antibiotics**—substances that kill or prevent the growth of bacteria. The first successful antibiotic was **penicillin**. Penicillin was made from *Penicillium* mould (Figure 8.1.3). Penicillin was not commonly used until the late 1940s. Before the development of antibiotics, you would have had to depend on your own body's **immune system** to fight off infections. Even today, antibiotics don't always work. It is important to take a full course of antibiotics as more resistant bacteria survive and reproduce when a course is not finished.

FIGURE 8.1.3 *Penicillium* is a common mould that forms on bread. It was from this mould that the first antibiotics were made.

Some people are allergic to penicillin, and not all types of bacterial infections can be treated with it. For these reasons, other antibiotics have been developed. Penicillin and these other antibiotics have saved the lives of millions of people since they were discovered.

Prac 1 p. 314 | AB 8.2 | AB 8.3

SciFile

Accidents that work

Penicillin was discovered in 1928 by accident. The Scottish scientist Alexander Fleming left some culture plates on which he was growing bacteria on a bench while he went on holiday. Mould grew on the plates. Where one particular mould grew, the bacteria didn't grow. In 1940, the Australian scientist Howard Florey and Ernst Chain developed penicillin into a useful medicine.

The immune system

Pathogens can enter your body in a number of ways, as shown in Table 8.1.1.

TABLE 8.1.1 Methods of entry of pathogens

Method of entry	Examples of disease
food and water	food poisoning, cholera
breathing in	flu, pneumonia, tuberculosis
cuts and wounds	tetanus, blood poisoning
sexual contact	gonorrhoea, syphilis, HIV
other contact	anthrax, leprosy

The first line of defence is to prevent the pathogens from entering the body.

- Skin is an effective barrier against pathogens and harmful chemicals.
- Fluids such as tears and saliva have mild antiseptic properties and help to wash away pathogens, dust and harmful substances.
- Air entering through the nose is filtered by hairs in the nostrils. Other unwanted particles in the air are then trapped in the mucous lining of the trachea (windpipe). Coughing and sneezing help to get rid of these foreign particles.
- Pathogens entering the digestive system are usually killed by the acid in the stomach. Vomiting is a quick way of getting rid of something undesirable in the stomach. Diarrhoea is a rapid way of getting rid of the body of pathogens that have got past the stomach.

Once a pathogen enters the body tissues, the affected area becomes red, hot and swollen—it is inflamed. Inflammation is a response of the body to infection. Certain immune cells release **histamine**, a chemical that causes more blood to flow to the infected area. Within the blood are a type of white blood cell called **neutrophils** that consume and destroy bacteria (Figure 8.1.4). Another type of white blood cell found in tissues is called a **macrophage** (Figure 8.1.5).

Consuming large numbers of bacteria causes many neutrophils and macrophages to die, which forms the yellow pus that collects around infected wounds. This has happened in the wound shown in Figure 8.1.6.

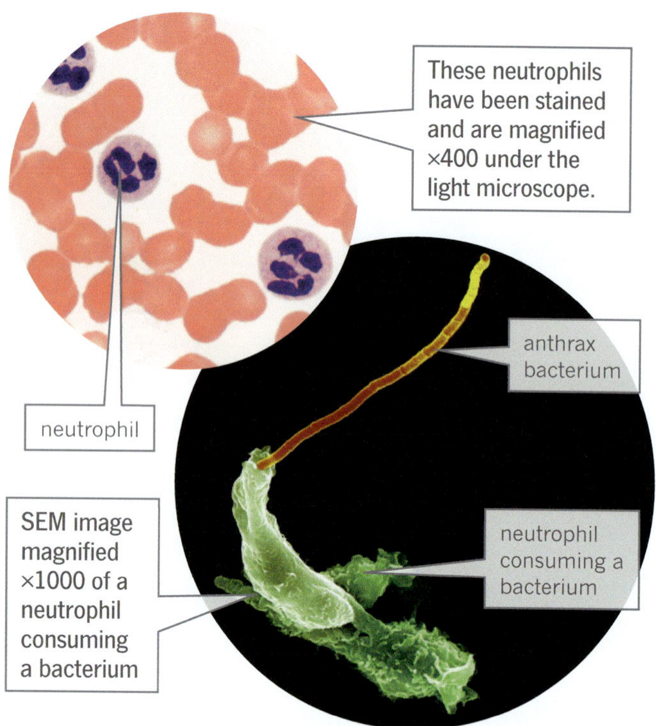

FIGURE 8.1.4 Neutrophils are white blood that can change shape and engulf bacteria.

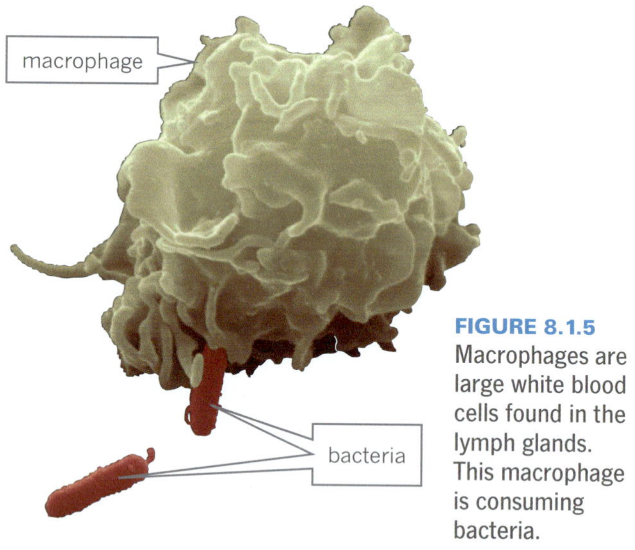

FIGURE 8.1.5 Macrophages are large white blood cells found in the lymph glands. This macrophage is consuming bacteria.

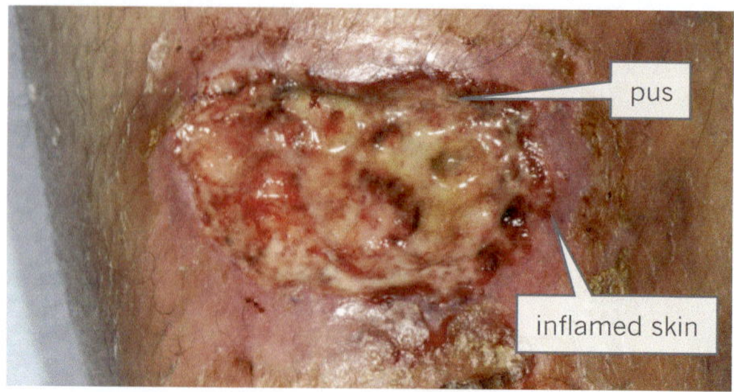

FIGURE 8.1.6 This skin is inflamed and the yellow pus indicates that neutrophils have been at work.

The lymphatic system

The lymphatic system is a series of vessels and capillaries that carry fluid from around your cells back to your heart. In areas of the lymphatic system there are nodules called lymph nodes.

Lymph nodes contain a large number of different types of white blood cells, including lymphocytes. The function of lymphocytes is to help destroy pathogens and to protect the body in the future.

Some **lymphocytes** respond by making a protein called an **antibody**. Antibodies cause pathogens to clump together, allowing the macrophages to destroy more of the pathogens at any one time.

The antibodies made are specific to that particular pathogen. Because the antibodies are made to be specific, they take time to produce, and meanwhile you may get sick. After the infection is cleared, memory lymphocytes remain. The next time your immune system meets the same pathogen, your immune system 'remembers' the pathogen and the memory lymphocytes are able to make the same antibodies quickly, meaning the pathogen is destroyed before it can make you unwell. You are now **immune** to that pathogen, meaning you are more likely to stay healthy if you meet it again. Figure 8.1.7 shows how immunity to a particular disease is developed.

During an infection, the lymph nodes closest to the site of infection become enlarged and tender. You may be able to feel them in your neck, armpits and groin.

To help you fight an infection, your body temperature is set higher than normal and you develop a fever. Pathogens that enter your body function best at normal body temperature. When your body temperature is higher, the pathogens are not able to function as well and your immune system can fight them more easily.

The body's immune defences are summarised in Figure 8.1.8 on page 310.

Vaccination

Some diseases are so serious that you cannot rely on your body developing immunity by itself. To help it out, vaccines are used. **Vaccines** are substances that cause your body to react as if it had met a pathogen.

Some ways vaccines can be made include taking a small amount of the toxin produced by a bacterium and making it inactive, or by using weakened or killed bacteria. The inactive toxin and the weakened or killed bacteria are harmless, but your immune system responds to the vaccine and you become immune to the pathogen.

In Australia, young children routinely receive vaccines against tetanus, diphtheria and whooping cough (pertussis). This process is known as being vaccinated or immunised. Vaccinations are also available for some bacterial diseases and are recommended for at-risk groups. These vaccines include those against meningococcal and pneumococcal disease and typhoid fever.

AB 8.4

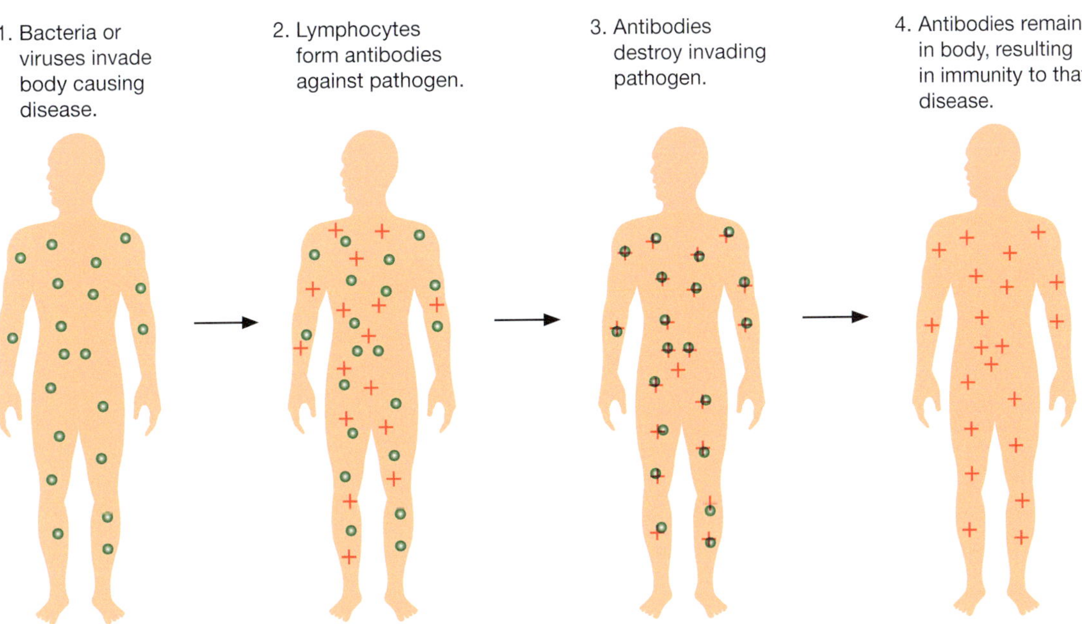

FIGURE 8.1.7 Antibodies are formed against invading microorganisms. Memory lymphocytes provide immunity against future infections.

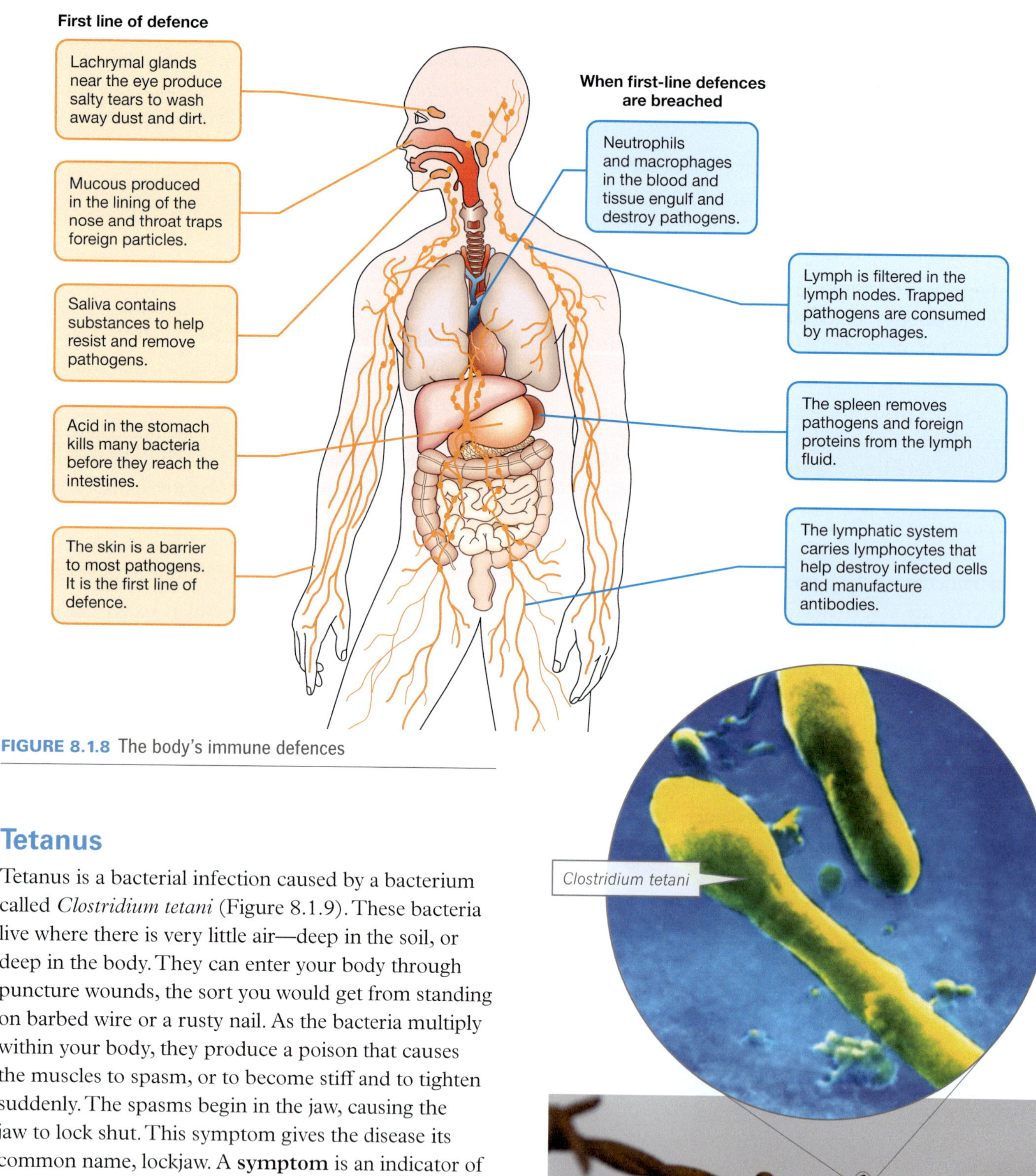

FIGURE 8.1.8 The body's immune defences

Tetanus

Tetanus is a bacterial infection caused by a bacterium called *Clostridium tetani* (Figure 8.1.9). These bacteria live where there is very little air—deep in the soil, or deep in the body. They can enter your body through puncture wounds, the sort you would get from standing on barbed wire or a rusty nail. As the bacteria multiply within your body, they produce a poison that causes the muscles to spasm, or to become stiff and to tighten suddenly. The spasms begin in the jaw, causing the jaw to lock shut. This symptom gives the disease its common name, lockjaw. A **symptom** is an indicator of a particular disease.

Immunity weakens over time, so booster vaccines are used to trigger a strong immune response that results in long-lasting immunity. The vaccination against tetanus can give you immunity for up to ten years. After that you need a booster.

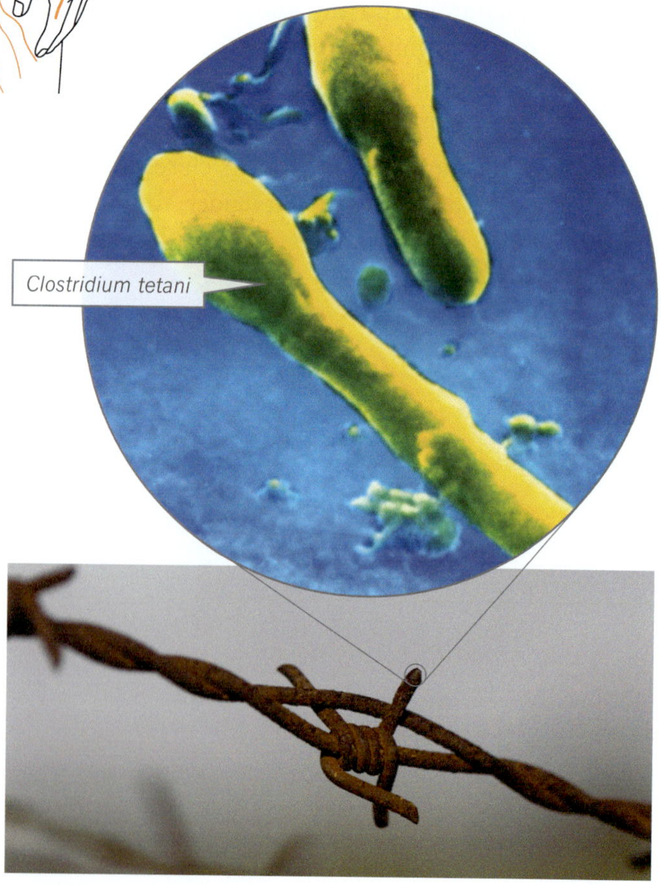

FIGURE 8.1.9 Barbed wire can cause deep puncture wounds that allow *Clostridium tetani*, tetanus bacteria, to enter the body.

Hygiene

Some diseases can be prevented by practising good hygiene. If you wake up one morning feeling sick and start vomiting, then you could be infected with *Salmonella enteritidis*—a bacterium that causes gastroenteritis. This bacterium is shown in Figure 8.1.10.

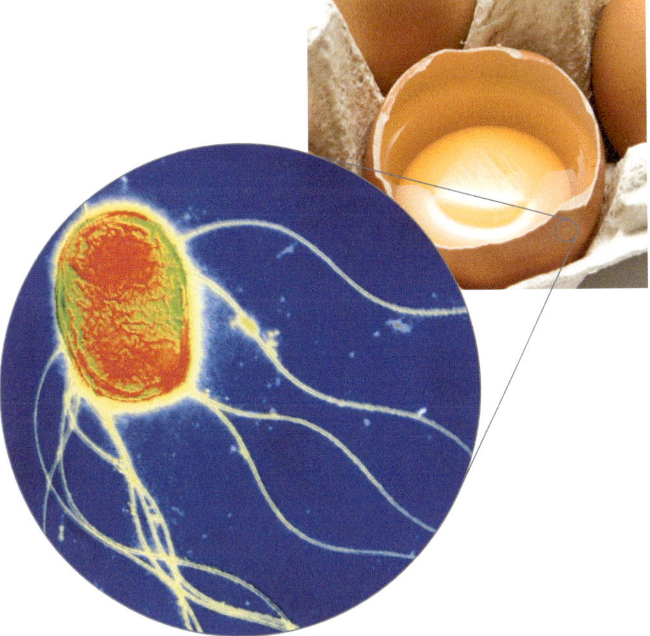

FIGURE 8.1.10 *Salmonella enteritidis* bacteria are found naturally in faeces and on dirty surfaces such as egg shells.

Salmonella enteritidis bacteria live naturally in the intestines of humans and other animals, especially birds. For this reason, salmonella is found in faeces and on dirty egg shells. If the bacteria get into food you eat, then they will multiply in your stomach, producing poisonous wastes called toxins. The toxins cause fever, headache and stomach pains. Your body tries to get rid of the toxins as quickly as possible through vomiting and diarrhoea. Vomiting and diarrhoea cause you to lose a lot of water, so dehydration (a lack of water in the body) may become an additional problem. So no matter how ill you feel, you must drink plenty of water.

Salmonella bacteria are responsible for over 9000 cases of gastroenteritis per year in Australia.

Gastroenteritis is a disease that can be avoided if you:

- wash your hands thoroughly and frequently, after handling animals, and going to the toilet, and before handling food
- thoroughly wash all surfaces on which food is prepared (Figure 8.1.11)
- keep foods such as meat, fish and dairy products refrigerated and separate from one another
- use different chopping boards and plates for foods such as meat and fish that are to be cooked and foods such as salads that are to be eaten raw.

Prac 2
p. 315

FIGURE 8.1.11 Washing and cleaning food preparation surfaces helps prevent gastroenteritis.

SCIENCE AS A HUMAN ENDEAVOUR

Use and influence of science

Stomach pains

Stomach ulcers can occur if the lining of the stomach is damaged. The stomach produces strong acid, and when this comes in contact with the damaged area, the result is pain and further damage.

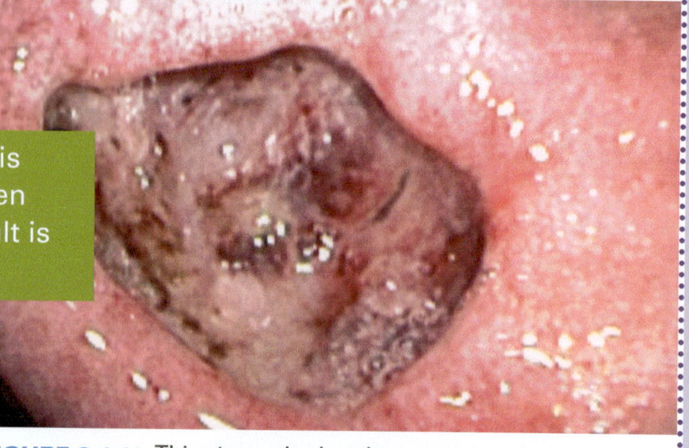

FIGURE 8.1.11 This stomach ulcer has eaten away the lining of the stomach, leaving the tissue underneath exposed to the strong acid of the stomach. This will cause intense pain and can lead to life-threatening bleeding.

Figure 8.1.11 shows a stomach ulcer. Stomach ulcers have been a major medical problem throughout the world. Until recent times there was no known cure. As well as causing severe pain and discomfort, stomach ulcers were also known to increase the risk of stomach cancer. For some time, doctors believed that stress, poor diet, alcohol, smoking or too much caffeine could all be part of the cause.

In 1979, Dr Robin Warren was working as a pathologist at the Royal Perth Hospital. Pathologists study the causes and effects of disease. Dr Warren found that an unusual bacterium was common in the stomachs of patients suffering from ulcers. He suggested the possibility of a link between bacterial infection and ulcers. Most people in the medical profession dismissed this suggestion as they did not believe that bacteria could survive in the acidic environment of the stomach.

Dr Barry Marshall is a gastroenterologist—a doctor who studies diseases of the stomach and intestine. In 1981, he and Dr Warren isolated a strange bacterium from the stomach and cultured (grew) it in the laboratory. It was a new species of bacterium, which they called *Helicobacter pylori* (*H. pylori*), shown in Figure 8.1.12. The two doctors were convinced that *H. pylori* was causing ulcers, but the rest of the medical profession was still not convinced.

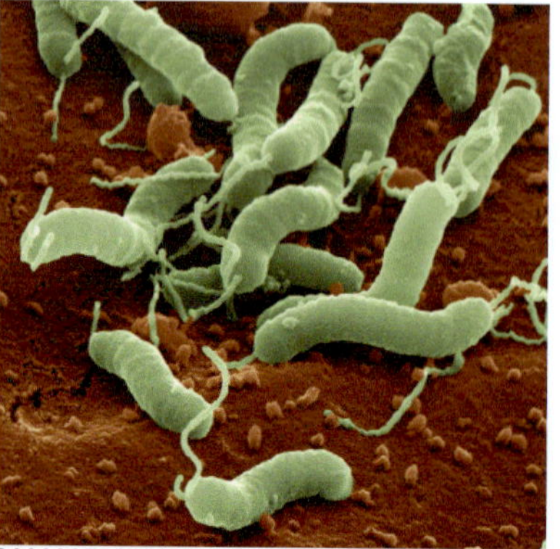

FIGURE 8.1.12 *Helicobacter pylori* was identified as a new species of bacterium and the cause of stomach ulcers.

Dr Marshall tested the hypothesis that *H. pylori* caused ulcers by infecting himself with the bacteria. He swallowed a culture of the bacteria, and a week later he began to suffer the symptoms of gastritis. Gastritis is the infection that comes before an ulcer develops. He then treated himself with antibiotics to destroy the bacteria. He soon recovered.

The discovery of the link between *H. pylori* and ulcers has been described as possibly the most significant event in medicine in Australia in the past 20–30 years.

It was now possible to cure a disease that doctors had previously considered incurable.

In 2005, Dr Robin Warren and Dr Barry Marshall were awarded a Nobel Prize for their contribution to medicine.

REVIEW

1 Explain how a stomach ulcer develops.

2 Why were many doctors unwilling to accept the idea that bacteria were the cause of stomach ulcers?

3 Summarise the evidence that Dr Warren and Dr Marshall used to suggest a link between the bacterium and ulcers.

4 The discovery of the link between *H. pylori* and ulcers has been described as *'possibly the most significant event in medicine in Australia in the past 20–30 years'*. Why do you think this discovery is considered to be so significant?

5 Construct a flow diagram showing the major events between Dr Warren first observing the bacterium and the awarding of the Nobel Prize.

MODULE 8.1 # Review questions LS

Remembering

1 Define the terms:
   a disease
   b infectious disease
   c antibody
   d pathogen.

2 What term best describes each of the following?
   a bacteria that cause disease
   b indicator of a disease
   c a chemical that causes your body to react as if it had encountered a pathogen.

3 Name two common diseases caused by bacteria.

4 What is the function of antibiotics?

5 List diseases that health authorities advise all Australian children be immunised against.

Understanding

6 a What are bacteria?
   b What does it mean to describe a type of bacteria as pathogenic?
   c Are most bacteria pathogenic? Explain your answer.

7 Why can severe cases of gastroenteritis cause dehydration?

8 Outline the process used to make a vaccine

9 Explain how vaccines work. Use a diagram or flow diagram if this helps to explain your answer.

10 What does it mean to be immune to a disease such as tetanus?

11 Why is it important that you wash your hands after playing with pets or going to the toilet?

Analysing

12 Compare:
   a a contagious disease and an infectious disease
   b antibiotics and vaccines.

13 Compare the way you become immune through natural reactions of your body and through administering a vaccine.

Evaluating

14 Some parents think that immunisation is not necessary because the diseases it is used against are so rare in Australia.
   a Evaluate whether this attitude is reasonable or not.
   b What would you recommend if someone asked you whether or not they should be immunised against a disease?

15 Three friends went for a meal at a restaurant. Next day, all were feeling very unwell and were diagnosed as having gastroenteritis.
   a Suggest the possible causes of the illness.
   b How would you determine which of your suggestions was the most likely cause?
   c What treatment would you recommend?

16 a What do you think could happen if kitchen benches are not cleaned regularly and thoroughly?
   b Justify your answer.

Creating

17 Construct a series of diagrams to demonstrate how a tetanus vaccine protects you against the disease.

18 Construct a poster to inform people about tetanus and how it can be prevented.

MODULE

8.1 Practical investigations

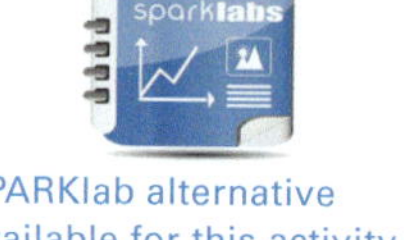

SPARKlab alternative available for this activity

1 • Growing bacteria

Purpose

To show that bacteria are common in our environment.

Hypothesis

Which area of the school do you think would have the largest number of bacteria—your classroom, the library, outside or somewhere else? Before you go any further with this investigation, write a hypothesis in your workbook.

SAFETY

Treat all bacteria as potentially dangerous.

All plates must be completely sealed with tape.

Do not open the lids of the agar plates once you have grown the bacteria. Your teacher will dispose of the plates correctly. As an added precaution, the agar plates could be sealed within a zip-lock plastic bag, as shown in Figure 8.1.13.

Timing

30 minutes + 30 minutes

Materials

- prepared agar plates
- marker pen
- cotton buds (1 per agar plate)
- zip-lock bag
- access to an incubator

Procedure

1 Take one agar plate for each group of three or four students.

2 As a group, decide which surfaces around the school you are going to take samples from. Each group should test a different surface in a different part of the school. Do not collect samples where pathogenic organisms may exist (for example, toilets).

3 Wipe a clean and dry cotton bud across the surface to be tested.

4 Open the agar plate, but do not touch the agar jelly inside. Lightly wipe the cotton bud across the agar jelly, making a series of stripes across it. Turn the agar plate 90°, and lightly wipe the cotton bud across the jelly again to form another series of stripes. As bacteria grow, the two sets of stripes will form a pattern of up-down and left-right stripes.

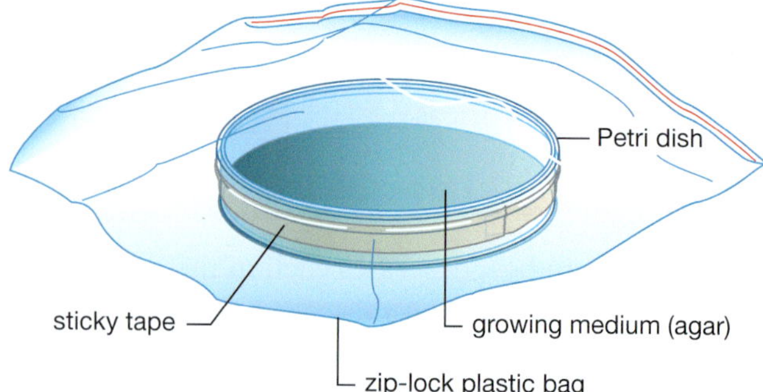

FIGURE 8.1.13

5 Replace the lid on the plate and seal it with tape. Mark the bottom of the plate with the name of your group and the place where the plate was exposed.

6 Place the plate top down in an incubator at 20–25°C and leave it overnight. If no bacterial colonies are visible, leave the plate for another 3–4 days.

Results

1 Count the number of different colonies on the plate, using the following characteristics as a guide.

- Bacterial colonies tend to be smooth and round.
- Different bacteria may have different colours.
- Fluffy areas like cotton wool are colonies of fungi.

2 Gather results for other groups in the class and construct a table that compares the number and variety of colonies from different parts of the school. Graph these results using a column graph.

Review

1 Describe the appearance of the colonies on your group's plate.

2 a Compare the number and types of colonies found in the different parts of the school.

b Account for the differences observed.

3 a Construct a conclusion for your investigation.

b Assess whether your hypothesis was supported or not.

4 Discuss the idea that one area of the school is more of a health hazard than another area.

MODULE 8.1

Practical investigations

2 • The milk's off!

Questioning & Predicting | Evaluating

Bottled milk in the supermarket has been pasteurised. This means that most of the bacteria present in the milk have been killed. Some remain, however, and in the right conditions for growth they will multiply rapidly and cause the milk to spoil and clot. Acid produced by bacteria causes the clotting.

Purpose

To test how quickly milk spoils under different conditions.

Timing

45 minutes + 30 minutes observation

Hypothesis

How will heating affect how fast milk goes off—will heating make milk spoil quicker or slower? Before you go any further with this investigation, write a hypothesis in your workbook.

> **SAFETY**
>
> When conducting experiments with microorganisms, treat them all as if they could cause disease.
>
> Do not touch the milk. If you do, wash your hands thoroughly.
>
> Do not taste, drink or sniff any of the milk.

Materials

- full-cream milk
- universal indicator paper

- access to a refrigerator
- 3 tall, heat-resistant beakers
- plastic wrap
- 3 rubber bands
- saucepan
- spoon
- electric hotplate or stove
- measuring cup
- masking tape
- access to a marking pen

Procedure

1. Use the masking tape to label each beaker 'Cold', 'Room temp/Control' or 'Boiled'.
2. Test the milk with universal indicator paper and record its colour. Note the results—acid, base or neutral—by comparing the colour of the wet strip with the chart provided with the indicator strips.
3. Pour one cup of milk into the beaker labelled 'Cold'. Cover it in plastic wrap secured with a rubber band. Place this in the refrigerator (Figure 8.1.14a).
4. Pour one cup of milk into the beaker labelled 'Room temp/Control'. Cover it in plastic wrap secured with a rubber band. Place this in an area at room temperature where it will not be disturbed (Figure 8.1.14b).
5. Pour one cup of milk into the saucepan and bring the milk to a simmer.
6. Stir continuously while letting the milk simmer for one minute.
7. Test the boiled milk with universal indicator paper and record the colour of the paper.
8. Pour the hot milk into the beaker labelled 'Boiled'. Cover it in plastic wrap secured with a rubber band. Place this in an area at room temperature where it will not be disturbed (Figure 8.1.14c).

a

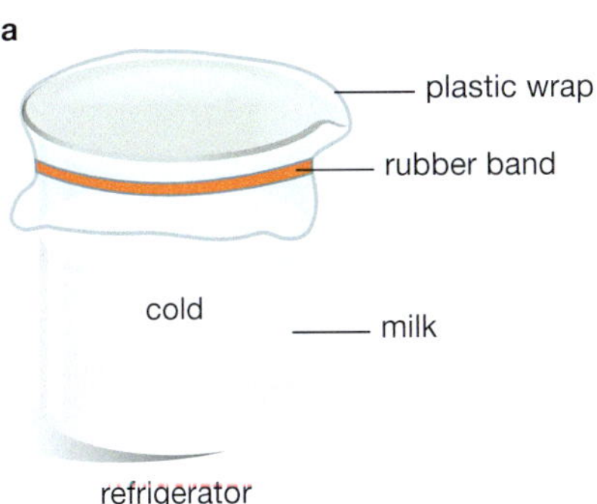

b

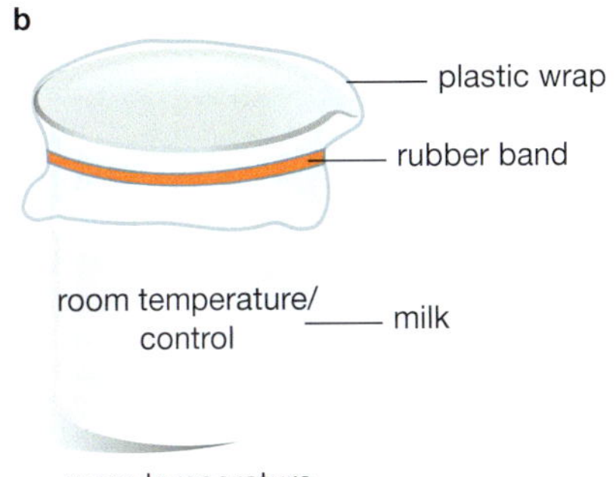

c

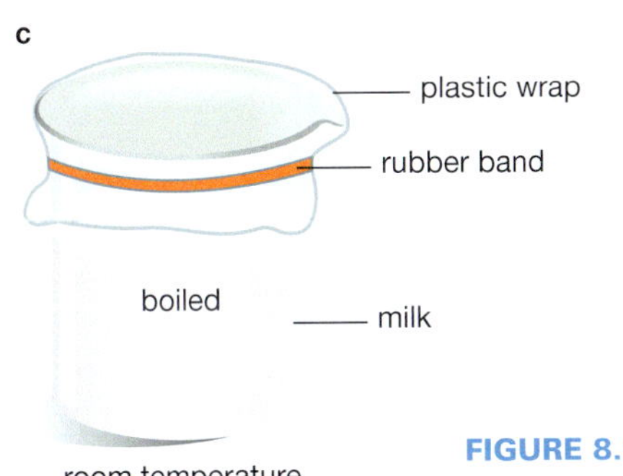

FIGURE 8.1.14

MODULE

8.1 Practical investigations

9 Record the appearance of each milk sample every day for 4–5 days. Do not remove the plastic wrap or shake the beaker.

10 At the end of the experiment, when at least one of the samples of milk has separated, open the samples with care. There may be large quantities of bacteria in the milk. The samples should not be shaken and you should not sniff the milk.

11 Place an indicator strip in each beaker and note the colour of each strip.

12 Dispose of the milk solutions as directed by your teacher.

Results

1 Construct a table similar to the one below in which your observations can be recorded.

2 Compare the colour of each strip of indicator paper with the chart provided with the indicator strips. Classify each as acidic, basic or neutral.

Review

1 Summarise your observations for each beaker of milk.

2 Suggest reasons for these observations. What do you think has happened in each case?

3 Use your results to explain why milk needs to be kept refrigerated.

4 Use your results to explain why foods spoil faster in warm weather than in cool weather.

5 From these results, propose recommendations on how other food products such as meat and cream should be stored.

6 a Construct a conclusion for your investigation.

b Assess whether your hypothesis was supported or not.

Obervations

	Cold milk	Room temperature milk (control)	Boiled milk
universal indicator colour at start			
acidic, basic or neutral at start			
appearance after 1 day			
appearance after 2 days			
appearance after 3 days			
appearance after 4 days			
appearance after 5 days			
universal indicator colour after 5 days			
acidic, basic or neutral after 5 days			

MODULE

8.2 Other sources of infection

Bacteria were the first type of microorganism to be shown to cause disease. But other microorganisms cause disease too. Some of these are hard to observe and the diseases they cause can be very hard to treat.

Viruses

After it was discovered that bacteria were the cause of many infectious diseases, scientists began to isolate and identify the bacteria responsible.

However, in some in cases no bacteria could be found. It was suggested that these diseases were caused by a pathogenic microorganism that was too small to be seen with microscopes available at that time. The name *virus* (Latin for 'poison') was given to these unseen microorganisms. This idea turned out to be true, but scientists had to wait for the invention of the electron microscope in 1931 to actually see a virus. Since then, more than 5000 types of viruses have been described in detail.

Viruses are pathogens and are about one-hundredth the size of bacteria. They do not need nutrients, produce wastes or exchange gases with the environment. The only characteristic of life a virus shows is when it invades a host cell and uses it to make thousands more identical viruses. The **host cell** is damaged or destroyed when it releases new viruses that spread throughout the body, infecting other cells. This is shown in Figure 8.2.1.

Viruses cause many common diseases, including colds and flu (influenza). They also cause measles, mumps, rubella, polio, chickenpox and cold sores (herpes).

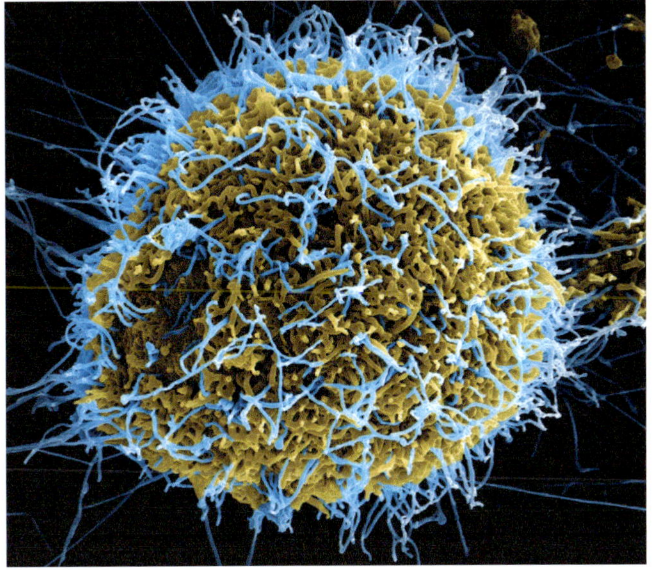

FIGURE 8.2.1 Coloured scanning electron micrograph of Ebola virus particles (blue) leaving from an infected cell (yellow). Ebola is a severe and often fatal disease with symptoms that include fever, vomiting, diarrhoea and internal haemorrhaging (bleeding).

Colds and flu

Over 200 different viruses cause colds. There can be thousands of microscopic virus particles in a droplet that is sneezed or coughed out by an infected person. The droplets in a sneeze travel out of your mouth at over 160 km/h! You can see them in Figure 8.2.2 on page 318. If you breathe in an infected droplet, then the virus has entered your body. Catching a cold is as easy as that.

FIGURE 8.2.2 Coughing and sneezing pass viruses onto others nearby, quickly spreading the cold.

Although colds are difficult to avoid, you can reduce your chances of catching one or spreading it to others by:

- covering your mouth when you cough or sneeze
- washing your hands frequently
- not sharing personal items if you or the other person are ill
- avoiding close contact with people who are coughing or sneezing or have a runny nose
- staying at home if you are sick so that you do not infect other people.

The cold virus attacks the lining of the nose and throat. Extra mucous is produced, so your nose keeps running or becomes blocked, making it difficult to breathe. Your throat gets sore and red, and you feel unwell.

You can catch flu in the same way as catching a cold—by breathing in air containing virus particles, such as those in Figure 8.2.3. Or you may have put your hands near your mouth after touching contaminated surfaces. Flu is not the same as a cold. Both are caused by viruses, but flu develops more quickly and can be more severe. Infection with the flu virus causes a high temperature, and your whole body aches.

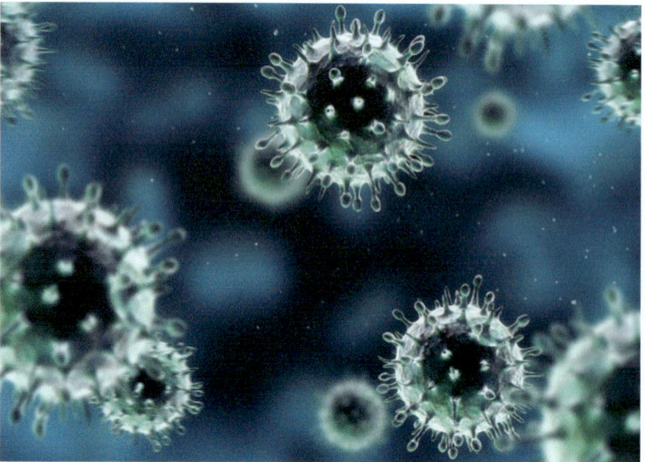

FIGURE 8.2.3 An artist's impression of the virus that causes flu. The symptoms of the disease are a high temperature and aching muscles.

STEM 4 fun

Spreading disease

NO

PROBLEM

What happens to spread disease?

SUPPLIES

many choices including: plastic cups, water, food dye, two types of tokens or other equipment of the student's design

PLAN AND DESIGN Your task is to model the spread of disease in your classroom. Modelling the spread of disease in your class could involve simulating the presence of a disease carried by one person as the presence of food dye in a cup of water or as a coloured token that may be exchanged on contact with other people or anything you can think of.

Design the solution. What information do you need to solve the problem? Draw a diagram. Make a list of materials you will need and steps you will take.

CREATE Follow your plan. Produce your solution to the problem. Take a video of it.

IMPROVE What works? What doesn't? How do you know it solves the problem? What could work better? Modify your design to make it better. Test it out.

REFLECTION

1. What area of STEM did you work in today?
2. In what career do these activities connect?
3. What did you do today that worked well? What didn't work well?

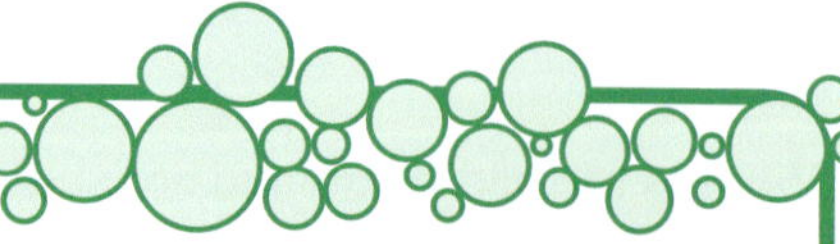

SciFile

Warts

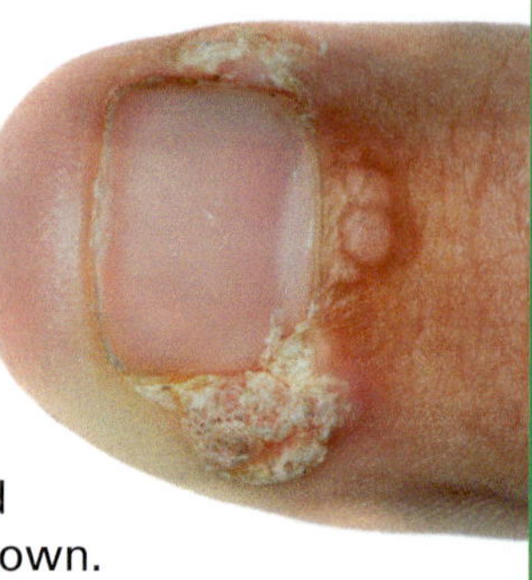

Common warts are abnormal growths caused by a virus that infects skin cells. The virus is transmitted by contact and enters the skin through cuts. It is relatively harmless and mostly disappears on its own.

Fighting viruses

Antibiotics do not work against viruses. However, sometimes when you have a viral infection, bacteria invade the body and cause secondary infections. These can usually be treated with antibiotics. Bronchitis and pneumonia are common secondary bacterial infections of a body weakened from fighting a virus.

Your body can develop immunity to viruses in the same way as it does for bacteria. Despite this you can get flu more than once. This is because viruses are able to mutate, changing into new strains (varieties). For each new strain of flu virus, the body has to make new antibodies. Also, medical researchers have to make new vaccines for the new strains.

Flu infections can make you feel very unwell and they can be fatal for people with low immunity, the elderly, the very young or those with other illnesses. Flu causes more than 4000 deaths in Australia each year. It is recommended that people at risk should be vaccinated each year.

Antiviral drugs and nanomedicine

Viruses are difficult to treat as they do not show any signs of life outside a living cell. Antiviral drugs do not target viruses directly. Instead, they work by stopping the reproduction of viruses. This is achieved by stopping the viruses from entering into a host cell or interfering with the virus once inside the host cell. Great care must be taken not damage the host cells and other cells of an infected person. Also many antiviral drugs are only partially effective and for some viral diseases, such as Ebola, there are no antiviral drugs available.

Nanomedicine

One promising area of treating viral disease is **nanomedicine**.

Nanotechnology involves the study and use of extremely small things in the range 1–100 nanometres (nm). One nanometre is very small—it is one-billionth of a metre or one-millionth of a millimetre. Nanomedicine involves the applications of nanoparticles.

In one approach, scientists are working to develop 'nanotraps'. Nanotraps are molecules that imitate the surface of a cell membrane that viruses use to enter the cell. The viruses bind to the nanotraps rather than entering and infecting the cell and are then cleared away by the body's own defences.

The molecule that is used to create the nanotraps are found naturally in the body so they are unlikely to cause side effects. Nanotraps can be easily administered and they are inexpensive to manufacture.

In another area of research, scientists are developing nanoparticles that deliver an enzyme that stops the reproduction of viruses.

AB 8.5 AB 8.6

Childhood diseases

When you are born, your body has not yet had the chance to build up the immunity you need to keep you healthy. This is why children tend to come down with lots of diseases that rarely affect adults.

Measles

Measles is a viral disease spread by infected people coughing and sneezing. It starts with a runny nose and sore eyes. A couple of days later a rash like the one in Figure 8.2.4 appears. Severe cases of measles can result in permanent hearing problems or brain damage.

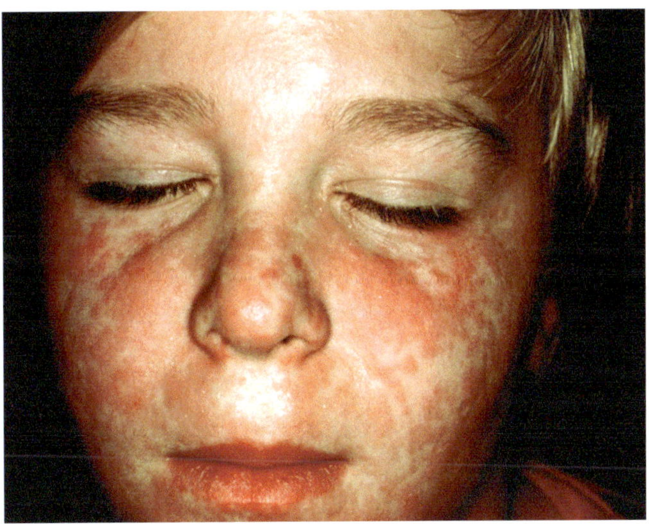

FIGURE 8.2.4 Measles rash

When European settlers first came to Australia, they brought measles with them. The Aboriginal population at that time had no natural immunity, and so measles killed many, especially children.

Measles is one of the most contagious diseases. Since 1966, children have been routinely immunised, resulting in the disease becoming very rare in Australia.

Chickenpox

The chickenpox virus causes a runny nose and a slight fever, followed by a rash of small, very itchy blisters. Scratching the blisters can lead to permanent scarring or to secondary bacterial infections of the blister.

After a person has been infected, the chickenpox virus can remain inactive in the nerve cells of their body for many years. Twenty, thirty or more years later, the virus can become active again, causing shingles—a very painful rash that can last for weeks. Figure 8.2.5 shows what the blisters of chickenpox and the rash from shingles look like.

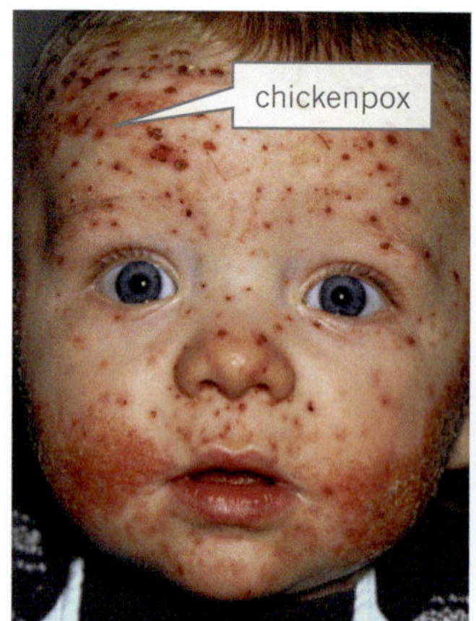

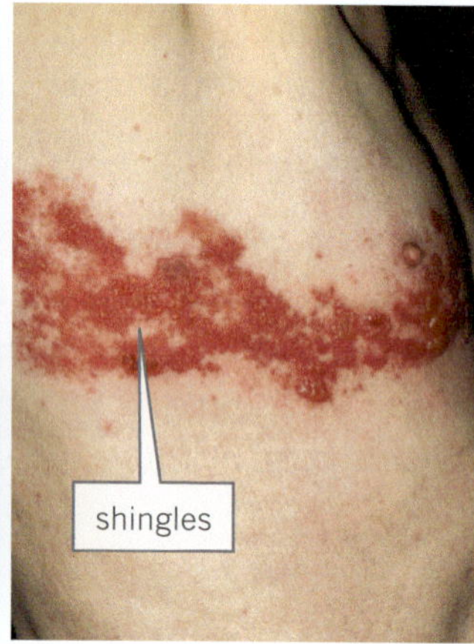

FIGURE 8.2.5 Immunisation against chickenpox protects you from getting chickenpox. It also protects you from developing shingles later in life.

Routine immunisation against chickenpox did not start until 2006, so infections still occur. However, the number of infections should decline as more and more children are immunised.

Parasitic diseases

A parasite is an organism that lives on or in the body of another organism, called the **host**, and takes nutrients from it. The host gets nothing beneficial in return and may be harmed. Some parasites can cause serious disease in humans.

Malaria

Malaria is an infection caused by a single-celled organism called *Plasmodium*. *Plasmodium* is a member of the Protist kingdom. Mosquitoes carry the *Plasmodium* from one host to another (Figure 8.2.6). As the mosquito pierces the skin, it injects a chemical into the host's body to prevent the blood from clotting. The chemical is known as an anticoagulant. The *Plasmodium* is injected along with the anticoagulant.

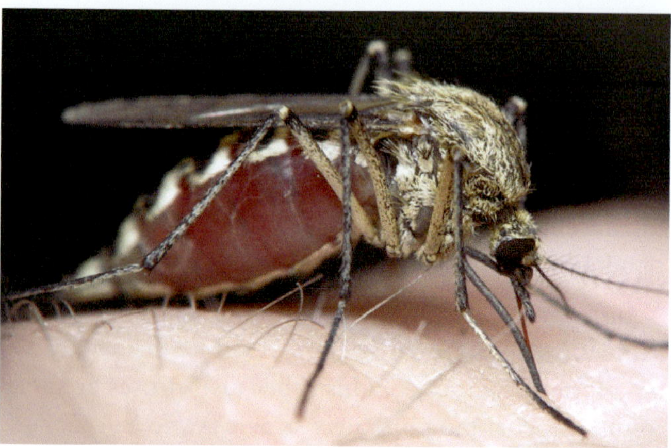

FIGURE 8.2.6 Mosquitoes carry the *Plasmodium* from one host to another, infecting a new host as it bites them and sucks their blood.

Malaria is one of the most widespread human diseases caused by a parasite. In the 1950s, there were 250 million cases of malaria each year, with 2.5 million deaths.

In 1955 the World Health Organization (WHO) embarked on a global program to put an end to malaria. The program focussed on using chemicals to kill the mosquitoes that transmit malaria. Malaria was eradicated (removed) from many areas, including Europe, North America and Australia. In 1988 'only' 110 million new cases were reported worldwide, but this trend has changed and now there are up to 220 million new cases of malaria each year.

In Australia, complete eradication was declared in 1981. This means that you cannot currently become infected with malaria in Australia. Mosquitoes capable of transmitting the *Plasmodium* live in northern Australia above the latitude of 19°S (Figure 8.2.7). However, currently the *Plasmodium* itself is absent from this mosquito population. About 700–800 people are hospitalised with malaria in Australia each year. These people were all infected elsewhere, mostly in Papua New Guinea.

FIGURE 8.2.7 Travellers to Papua New Guinea sometimes bring malaria back into Australia.

Preventing infection means preventing mosquito bites in areas where the disease could return. If visiting northern Australia or Southeast Asia:

- wear protective clothing
- use insect repellent
- if your windows are not screened, use a mosquito net when sleeping
- empty any standing water where mosquitoes could breed.

Insecticides can be used to control mosquito numbers. However, insecticides may also kill useful insects such as bees, so they need to be used with caution.

SciFile

They will find you!

Mosquitoes are attracted by carbon dioxide and body heat. So it is no wonder you often hear that annoying buzz around your head when you are trying to sleep.

Mosquitoes are also attracted to the colour blue more than twice as much as any other colour!

Amoebic dysentery

Most of Australia is supplied with fresh, clean water, and sewage is treated effectively. This protects us from many diseases found in other parts of the world. You might come into contact with these diseases if you travel overseas, or if some disaster at home damages the sewage system or contaminates the drinking water (Figure 8.2.8). One disease spread through contaminated water is amoebic dysentery. This disease causes 50 000–100 000 deaths per year worldwide.

Amoebic dysentery is caused by a unicellular (single-celled) organism that is most common in tropical areas. People become infected by swallowing a cyst containing the parasite in contaminated food or water. The cyst is one stage in the life cycle of the parasite.

People who contract the disease can remain infectious for years, so it is best to prevent infection when travelling in areas where amoebic dysentery occurs— only drink boiled water or sealed, bottled water, don't have ice in your drinks and don't eat fruit or vegetables that may have been rinsed in tap water and not cooked.

FIGURE 8.2.8 There is a greater chance of contracting amoebic dysentery when there is poor (or no) sewage treatment and the water supply is contaminated.

Fungi

Some fungi, like the mushrooms you buy at the supermarket, are useful as a source of food. Others, like mould, are decomposer organisms in the environment. Some fungi cause disease. Very few of the diseases they cause are life-threatening but can be difficult to treat.

Fungi are dispersed (spread) by **spores**, which are made of a single cell with a tough skin. Fungal spores are everywhere. A cloud of spores can be seen leaving the puffball fungus in Figure 8.2.9.

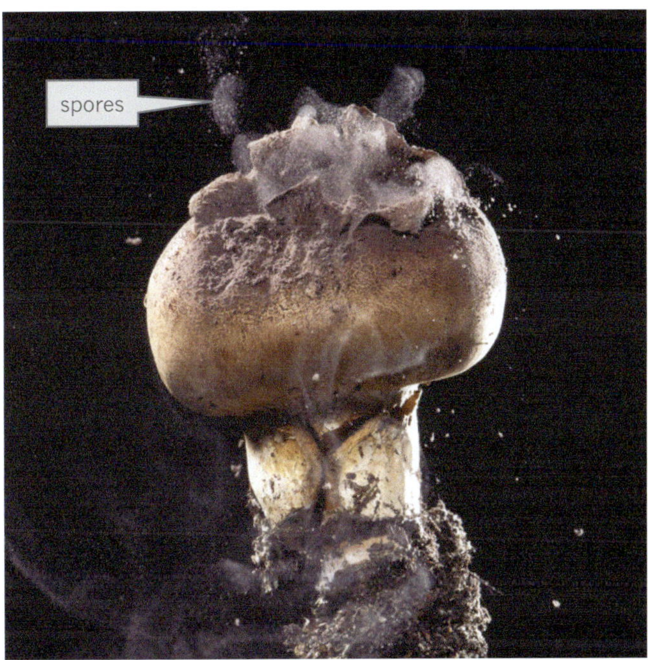

FIGURE 8.2.9 This large puffball fungus does not cause disease. However, like all fungi, it releases spores that are light and easily blown about in the wind.

The spores just need to find a warm, moist environment and they will start to grow. The warmest, most sweat-prone parts of the human body are the feet and the groin. It is there that pathogenic fungi such as tinea and thrush are most likely to grow.

Fungal infections are contagious. They can be passed from one person to another through skin-to-skin contact, the sharing of towels, or walking on floors that an infected person has walked on.

Tinea

Tinea is a fungus that can grow on the skin, hair or nails. It grows out from a centre, producing a red, inflamed ring of skin, as shown in Figure 8.2.10. Tinea infection is often called ringworm, but no worms are involved. The tinea fungus feeds on dead skin cells. When it runs out of dead cells it will attack the living cells, causing the skin to become red and itchy. If not treated, the skin will crack and bleed. Figure 8.2.11 shows an example. The infection can be treated with a **fungicide**—a chemical that kills fungi.

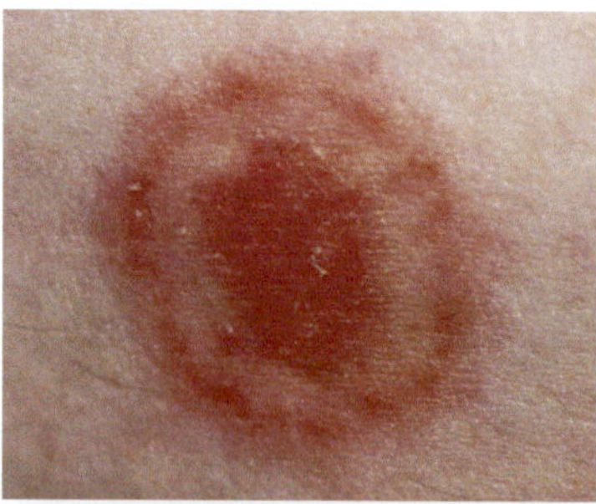

FIGURE 8.2.10 Tinea infection is often called ringworm because of the shape of the inflamed area.

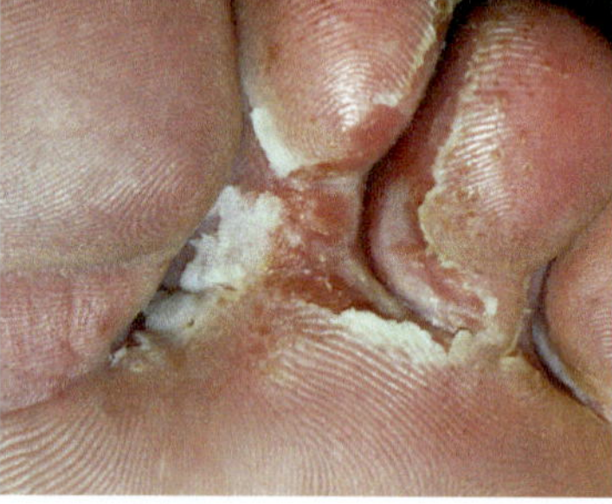

FIGURE 8.2.11 Tinea infection between the toes.

Thrush

Thrush infection is caused by fungus that is normally found in your body. Sometimes it grows out of control, causing problems. Thrush is not serious, but is very itchy and uncomfortable. It can be found as white patches on the tongue or inside the cheek, causes nappy rash in babies, and infects the vagina.

Thrush sometimes develops when you are taking antibiotics because the bacteria that naturally control fungi have been destroyed by the antibiotics.

science 4 fun

Where do spores come from?

Collect this ...

- mushroom that is open, with dark brown gills under the cap
- sheet of white paper

SAFETY

Dispose of the mushroom and the paper in the bin when you are finished.

Wash your hands well after handling the mushroom.

Do this ...

1. Place the mushroom on the sheet of paper with the gills facing down.
2. Leave it in an area where it will not be disturbed for two days.
3. Without moving the paper, lift the mushroom carefully off the sheet of paper. There should be a deposit of black spores on the paper.

Record this ...

1. Describe the pattern on the paper. You could draw a picture or take a photograph.
2. Explain where the spores came from and what would normally happen to them.

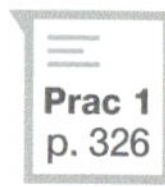
Prac 1 p. 326

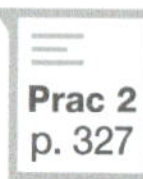
Prac 2 p. 327

SCIENCE AS A HUMAN ENDEAVOUR

Use and influence of science

Medieval medicine

Humans have always feared disease and have looked to find both causes and cures. Many of the diseases which are routinely treated and cured today were fatal in ancient times. Long before the invention of the microscope and the discovery of microorganisms, ancient scientists worked to discover the causes of disease. Hippocrates' theory of the four humours of the human body, was one of the most popular medical theories until the advent of modern medicine in the 1800s.

Hippocrates (460–377 BCE) and Galen (129–199 or 217 CE) were two influential scientists from ancient times. Galen defined disease as 'impairment of bodily activities'. Hippocrates believed that characteristics of the environment, such as weather and drinking-water, caused disease. Although these ideas reflect some of what is known today, ideas about disease have changed many times between then and now.

Medieval times were the years from 500 to 1350 CE. From 500 to 1000 CE the Roman Catholic Church was extremely influential and promoted the idea that illness was the result of sinful behaviour. The plague (black death) killed between one-third and half of the population of Europe in the 1300s. At the time, the disease was commonly believed to be a punishment from God.

FIGURE 8.2.12 Each humour was associated with a particular temperament. Blood was associated with fun, yellow bile with ambition, phlegm with a calm disposition, and black bile with melancholia.

As Europe's population grew during this time, the ideas of Galen, Hippocrates and others were revisited. Few people had access to education. The main centres of learning tended to be monasteries or were restricted to the rich and highly educated people. Hippocrates had put forward the idea that four humours controlled the health of individuals. These were phlegm, blood, yellow bile and black bile. The idea of humours was reinstated in the 1200s. It was thought that the humours were balanced in a healthy person and unbalanced where there was disease. Diagrams such as Figure 8.2.12 depicted the humours and their associations with the seasons, universal elements and certain qualities. These associations are listed in Table 8.2.1.

TABLE 8.2.1 The humours

Humour	Season	Universal element	Qualities
yellow bile (coleric)	summer	fire	hot and dry
black bile (melanc)	autumn	earth	cold and dry
phlegm (flegmat)	winter	water	cold and moist
blood (sangvin)	spring	air	hot and moist

SCIENCE AS A HUMAN ENDEAVOUR

Fever was described as a hot, dry disease caused by too much yellow bile. More of the opposite humour (phlegm) was required, to cure the problem. The patient was ordered to take cold baths.

Herbal drugs were used if the treatment failed. The vomiting and diarrhoea the herbs often caused were a sign that the imbalance of the humours had passed out of the body.

Medieval physicians believed that bad smells caused disease. Getting rid of the smell would reduce the threat of disease (Figure 8.2.13). Some town authorities tried to clear the streets of rubbish and sewage even though the link between waste and disease wasn't fully understood. However, industries that produced foul smells, such as butchery, dyeing and tanning, were located side by side with homes.

FIGURE 8.2.13 During the plague, doctors wore a beaked mask. The beak was filled with aromatic herbs and spices to overpower foul smells that were thought to cause disease.

Aristotle (384–322 BCE) believed that living things could generate spontaneously from non-living things, such as maggots (pictured here) from rotting meat. Snakes and crocodiles were thought to form from the mud of the river Nile in Egypt. Rats and mice 'appeared' from old rags, and maggots underwent spontaneous generation from rotting meat. This means that they grew out of non-living material. (Figure 8.2.14). This idea was not disproved until Louis Pasteur (1822–1895) completed his experiments in 1859. These experiments showed that all organisms come from existing organisms and disproved the idea that they were generated spontaneously.

FIGURE 8.2.14 Until the nineteenth century, scientists believed that animals like these maggots arose spontaneously. They even wrote recipe books for making animals! The theory was finally proven wrong in 1859, by Louis Pasteur.

Building on the work of earlier scientists such as Francesco Redi (1626–1697), Louis Pasteur provided enough evidence to convince scientists in Europe that there was a link between bacteria and disease.

REVIEW

1 Propose reasons why many diseases that are easily treated today were fatal during medieval times.

2 During medieval times, people lived in villages and rarely travelled far from their village. The arrival of a stranger was often regarded with suspicion and fear. Propose reasons for this.

3 Describe the medieval beliefs about the four humours and their relationship to good health.

4 Medieval physicians believe that foul smells could carry and cause disease.

   a Do you think they were correct, partly correct or incorrect?

   b Justify your answer.

MODULE

8.2 Review questions

Remembering

1 Define the terms:
   a fungicide
   b host
   c spore.

2 What term best describes each of the following?
   a a cell invaded by viruses
   b a pathogen that shows no life sign outside another cell
   c using nanotechnology in medicine.

Understanding

3 Outline some simple precautions you can take to avoid contracting an infectious disease.

4 a Explain how viruses reproduce. In your answer include the role of the host cell and how viruses spread in a viral infection.
   b Although they can reproduce and cause disease many scientists do not consider viruses to be living things. Explain why scientists might believe this.

5 Fungi reproduce and spread via spores.
   a Describe the structure of a fungal spore.
   b Suggest why it is difficult to destroy spores.
   c What type of conditions do spores need to grow?
   d Are fungal infections contagious? Explain.

6 Outline how malaria is spread.

7 Explain how the virus that causes colds is spread.

8 Explain how someone can get the flu even though they were vaccinated against it.

9 Explain why measles caused so many deaths when it was first introduced to the Australian Aboriginal population.

Applying

10 Use labelled diagrams to show how walking barefoot can spread tinea.

Analysing

11 Describe the similarities and differences between bacteria and viruses.

12 Use a series of dot points to contrast colds and flu.

Evaluating

13 Jan said that mosquitoes cause malaria. Kai said that was not a true statement.
   a Evaluate the statements and decide who is more accurate.
   b Justify your answer.

14 A family in northern Queensland knows that dengue fever is carried by mosquitoes and that there are cases of it their area. What do you think they can they do to protect themselves from the disease?

15 Why do new vaccines need to be produced for influenza each year?

Creating

16 Construct the scenario (story) for a 30-second television advertisement that makes people aware of how viral or fungal infections are spread, and ways they can prevent the spread.

17 Cholera is a disease that causes severe diarrhoea and dehydration. It is caused by the ingestion of food or water contaminated with the bacterium *Vibrio cholerae* (Figure 8.2.15). If left untreated, death can occur in hours.

   Cholera can be treated successfully with oral rehydration salts given with clean water and antibiotics if necessary. Despite this it is estimated that there are over 100 000 deaths per year from cholera. Cholera is most common in poorer countries in Africa and South and Southeast Asia particularly after heavy rain and floods. Cholera is virtually unknown in Australia.
   a Propose reasons why cholera would be common in some countries and not others.
   b Create a travel brochure that gives advice to people going to countries where cholera is present.

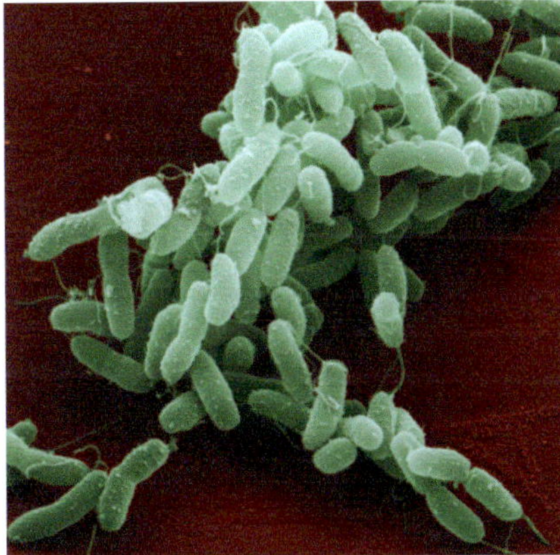

FIGURE 8.2.15 Cholera is caused by the bacterium *Vibrio cholerae*.

MODULE

8.2 Practical investigations

1 • Growing fungi

Questioning & Predicting | Evaluating

Purpose

To investigate what causes fruit to rot and to find out if all fruits rot in the same way.

Hypothesis

Which do you think will rot faster—cut fruit or whole fruit? Before you go any further with this investigation, write a hypothesis in your workbook.

SAFETY

Dispose of all the fruit used in this investigation according to your teacher's instructions.

Wash your hands thoroughly after handling the fruit.

Timing 30 minutes + 30 minutes

Materials

- a selection of fruit including:
  - 2 apples
  - 2 pieces of soft fruit such as a strawberry
  - 2 oranges

- paper towel
- marker pen
- 2 plastic boxes such as take-away containers
- hand lens or magnifying glass

Procedure

1. Read through all the steps of this procedure before you do anything.
2. Use the marker pen to label each container 'Cut fruit' or 'Whole fruit'.
3. Wash all the fruit carefully and pat it dry with the paper towel. Do not bruise the fruit.
4. Place one apple, one of the soft fruits and one orange into the plastic container labelled 'Whole fruit'.
5. Using a clean, sharp knife, make a cut in the skin of the other apple. Wash the knife and then make a cut in the orange. Be sure to cut through the skin into the segments. Wash the knife and then cut the soft fruit.
6. Place the cut fruit into the container labelled 'Cut fruit' cut side up.
7. Leave the fruit in a well-ventilated area for 3–5 days.

Results

1. Write what you predict will happen to the fruit in the two different treatments.
2. Each day, observe the fruit and record any changes in its appearance. You could use photographs to support your notes.

Review

1. Compare the changes in the whole fruit with the changes in the cut fruit.
2. What could have caused the changes?
3. Compare the changes in the apple and the orange.
4. Compare the changes in the apple and the soft fruit.
5. Explain what caused the changes in the fruit.
6. Explain why the changes were not the same for all the fruit.
7. a Construct a conclusion for your investigation.
   b Assess whether your hypothesis was supported or not.

FIGURE 8.2.16 Whole fruit and cut fruit

MODULE

8.2 Practical investigations

• STUDENT DESIGN •

2 • Testing hand sanitisers

Questioning & Predicting | Evaluating

Personal hand sanitisers are becoming increasingly popular and are presented as an alternative to soap and water. Many make claims such as 'kills 99.9% of germs on your hands'.

Purpose

To test the effectiveness of alcohol-based hand sanitisers as means of controlling disease.

Hypothesis

Do you think alcohol-based hand sanitisers are as effective as washing your hands with hot water and soap? Before you go any further with the investigation write a hypothesis in your workbook.

Timing 30 minutes plus 30 mins to follow up

SAFETY

A risk assessment is required for this investigation.

Materials

- alcohol hand sanitiser
- soap
- prepared agar plates if required

Procedure

1. Design an experiment that will test whether alcohol-based hand sanitisers are effective in controlling the spread of disease.
2. Brainstorm in your group and come up with several different ways to investigate the problem. Select the best procedure and write it in your workbook.
3. Before you start any practical work, assess all risks associated with your procedure. Construct a risk assessment that outlines these risks and any precautions you need to take to minimise them. Show your teacher your procedure and your risk assessment. If they approve, then collect all the required materials and start work.

See Activity Book Toolkit to assist with developing a risk assessment.

Use the STEM and SDI template in your eBook to help you plan and carry out your investigation.

Results

Construct a table to record the number of bacterial and fungal colonies in different samples.

Review

1. List the variables that you controlled (kept constant) during this experiment.
2. a Which variable did you change?
   b Describe the way in which this variable changed.
3. Construct a conclusion for your experiment.
4. Assess whether your hypothesis was supported or not.
5. Evaluate your procedure. Pick two other practical groups and evaluate their procedures too, identifying their strengths and weaknesses.
6. Based on your finding, make a recommendation regarding the use of alcohol-based hand sanitisers. In your recommendation include:
   - advantages of alcohol-based hand sanitisers
   - disadvantages of alcohol-based hand sanitisers
   - when and where alcohol-based hand sanitisers should be used.

MODULE
8.3 Environmental diseases

Many diseases are not caused by pathogens and so are not infectious. Instead, they are caused by factors in your environment. These factors include what you eat and drink, chemicals around you and your lifestyle.

Nutrition

When people talk about a diet, they are usually talking about what they would eat (and not eat) to lose weight. However, to scientists and nutritionists, a **diet** describes whatever people eat, regardless of whether it is healthy or not. What everyone needs from their diet is **nutrition**—the food necessary for health and growth. **Nutrients** are substances essential for healthy growth and maintenance of your body. The nutrients you need from your food are proteins, carbohydrates, fats, minerals and vitamins. Water is also an essential part of any diet.

- **Protein** is used for growth and repair of tissues and comes from meat, dairy products and fish (Figure 8.3.1).
- **Carbohydrates** are your main source of energy. Starches and sugars found in grains (such as wheat, rice and oats) as well as in fruit and vegetables are all carbohydrates.
- **Fats** provide twice the amount of energy as carbohydrates. Your body stores energy as fat. Oils and the fat found in meat and dairy products such as cheese are sources of fat.
- **Minerals** such as iron, calcium, sodium, potassium and phosphorus do not provide energy but they are important for your health. Iron is part of haemoglobin in red blood cells. Calcium and phosphorus build strong bones and teeth. Phosphorus is essential for effective metabolism. Sodium and potassium balance the water content of your body. Nerves and muscles, including the muscles of your heart, need sodium to work effectively.
- **Vitamins** control many of the chemical reactions in the body. They are required in only tiny amounts.
- Water provides no energy or other form of nutrition but it is essential for health. Your body is more than 60% water. All the chemical reactions of metabolism take place in the watery environment of your cells. Water in the blood carries nutrients around the body to where they are used.

Prac 1
p. 339

FIGURE 8.3.1 This meal of salmon and rice will provide protein and carbohydrates.

SciFile

Water of life

If you lose 10% of the water in your body, then you will become ill. Losing 20% usually means death. You can survive up to 40 days without food but no more than 3 days without water.

science 4 fun

Look at the labels

Collect this ...

labels of four of your favourite snack foods or drinks

Do this ...

1 Read the nutrition information on each label.
2 Identify which essential nutrients each snack or drink provides.

Record this ...

1 Describe the snacks in terms of the nutrients they provide.
2 Explain which snack you would choose to eat and why.

Healthy diet

Disease can occur when people do not get enough to eat. It can also be the result of eating too much or eating the wrong things. The Australian Government has produced *The Australian Guide to Healthy Eating* to provide guidance on what to eat and how much to eat in a healthy diet. This guide classifies foods into five groups:

- bread, cereals, rice, pasta, noodles
- vegetables, legumes
- fruit
- milk, yoghurt, cheese
- meat, fish, poultry, eggs, nuts, legumes.

Food should be eaten from each of the five groups every day. However, you do not need equal quantities from each group. The proportions of each section on the plate in Figure 8.3.2 indicate the relative amounts of each food group that should be consumed.

Sweets, cakes and hot chips can be part of a healthy diet as long as they are eaten only occasionally.

FIGURE 8.3.2 Enjoy food from each sector of the plate every day. The size of the sector indicates the relative amount of each group to include in your diet.

SciFile

Eat your veggies!

One in ten adults in Australia does not eat the recommended five serves of vegetables per day, and half the adult population does not eat the recommended two serves of fruit.

Under-nutrition

Malnutrition occurs when the nutrition provided by diet does not meet the needs of the body. Malnutrition can mean not having enough of the nutrients the body requires. This is called **under-nutrition**. Many children in the world are not healthy because they do not get enough food and so lack the nutrients needed for normal growth and development and the energy required to play normally.

Kwashiorkor

When food is in short supply, it is often protein that is missing from the diet. Without protein the body cannot build up muscle and brain development is slowed. Children lacking protein may develop a disease called **kwashiorkor**. The child in Figure 8.3.3 has kwashiorkor as shown by the pot-belly, which is caused by weak stomach muscles. Other symptoms include hair loss, swollen legs and no energy. Kwashiorkor kills 60% of the children who suffer from it.

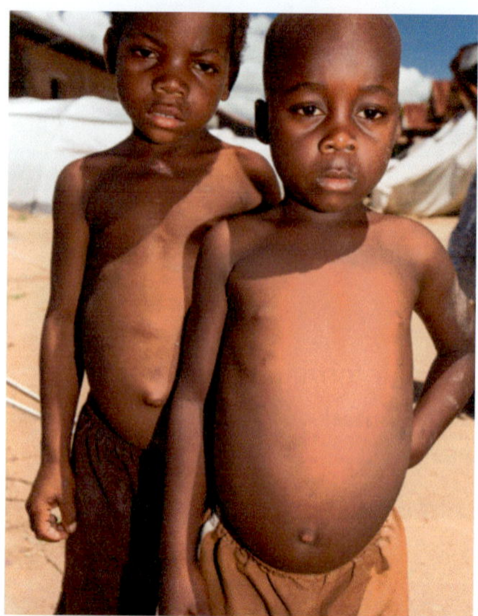

FIGURE 8.3.3 A child with kwashiorkor looks as if they are fat because of their swollen belly. In reality they are starving.

Scurvy

Sailors are often at sea for very long periods of time. In the past, sailors did not have access to fresh fruit and vegetables, so they developed a disease called **scurvy**. Their gums bled, their teeth fell out, their joints became sore and swollen, and wounds were slow to heal and became infected. You can see the effect of scurvy in Figure 8.3.4. Scurvy is caused by a lack of vitamin C, a vitamin found in many fruits and some vegetables. Particularly high quantities of vitamin C are found in blackcurrants and in citrus fruits.

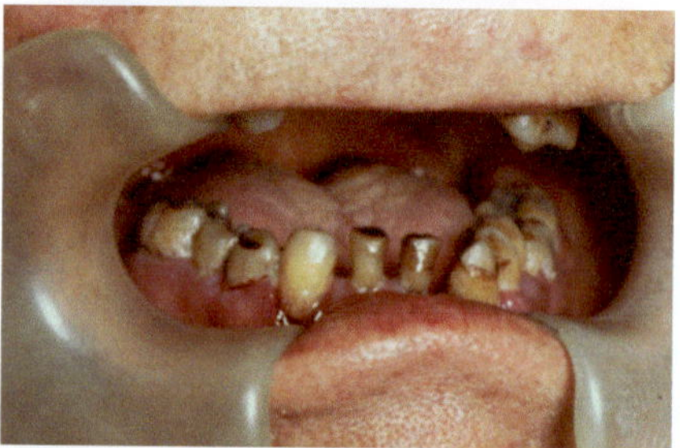

FIGURE 8.3.4 Scurvy is caused by a lack of vitamin C. Teeth fall out as a result.

Rickets

Children lacking vitamin D may develop **rickets**, a disease that causes the bones to remain relatively soft. Rickets is made even worse if the diet is lacking in calcium and phosphate. When babies learn to walk, their legs are carrying weight for the first time. If a baby has rickets, then the bones bend outwards and the legs take on the shape seen in Figure 8.3.5. This is typical of a person suffering from rickets.

Calcium hardens the bones and vitamin D helps the body to use the calcium. Calcium is found in milk and green vegetables. Vitamin D is found in fish oils. Your body can also produce vitamin D in the lower layers of the skin by using the ultraviolet rays of sunlight.

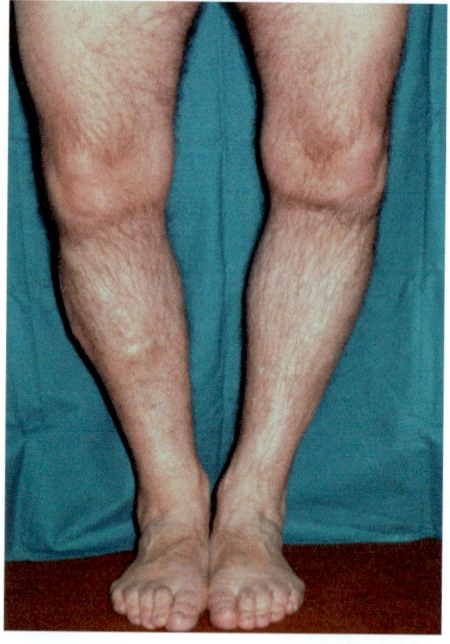

FIGURE 8.3.5 The femur (thigh bone) and the tibia (shin bone) are bent outwards in this person with rickets.

Vitamin C and curing colds

Scientific research has shown that taking vitamin C does not prevent or shorten the duration of a cold. However, adequate amounts of vitamin C in the diet are important for maintaining healthy muscle, bone and blood vessels.

Over-nutrition

Over-nutrition is another form of malnutrition. It is when the body gets so much nutrition that it does not work properly.

Over-nutrition results from eating too much, eating too many of the wrong things, not exercising enough, or taking too many vitamins or other dietary replacements. Over-nutrition can lead to many chronic diseases such as diabetes and high blood pressure. **Chronic diseases** last for a long time. Some cannot be cured. They can only be managed.

Overweight and obese Australians

A person is considered to be **overweight** when they have more body fat than is considered healthy. This depends on a lot of factors but a rough estimate can be made by measuring a person's BMI. BMI stands for body mass index and compares a person's height to their weight. It is calculated by dividing a person's weight in kilograms by their height in metres squared or by using a chart such as the one shown in Figure 8.3.6. Generally a person is potentially overweight when their BMI is between 25 and 30. A person is considered to be **obese** when their BMI is over 30.

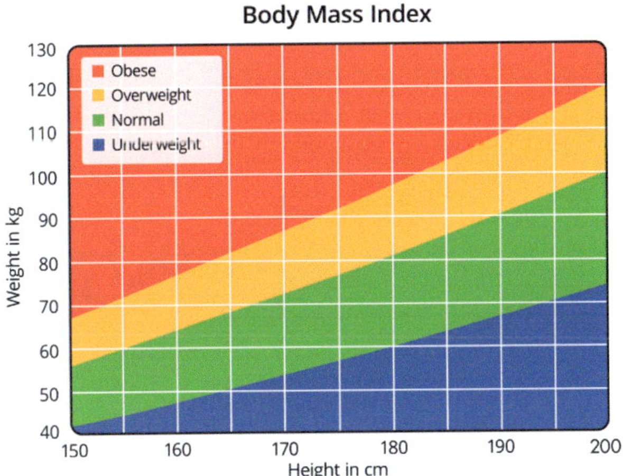

FIGURE 8.3.6 BMI can be estimated using the charts like this. Note that height is given in centimetres in this chart.

BMI has a number of limitations. It does not directly measure body fat and a person's weight will also be affected by their amount of muscle mass and bone density (Figure 8.3.7). It is only relevant for people over 18 years old and does not take into account factors such as age, gender, ethnicity and body type.

FIGURE 8.3.7 This sprinter is extremely fit and has little body fat. He has a BMI of 27. Taken on its own this would indicate he is overweight.

Worked example

Calculating BMI

Problem

A 25-year-old man is 180 cm tall and has a mass of 80 kg. Calculate his BMI and determine if he is overweight.

Solution

Thinking: Determine the man's mass.

Working: His mass is 80 kg.

Thinking: Determine the man's height in metres.

Working: His height is 180 cm = 1.8 m.

Thinking: Calculate BMI by dividing his mass (kg) by his height (m) squared.

Working: $\frac{mass}{(height)^2}$

$$\text{Mass} = 80\,\text{kg}$$
$$\text{Height} = 180\,\text{cm} = 1.8\,\text{m}$$
$$\frac{80}{(1.8)^2} = \frac{80}{3.24} = \text{BMI} = 24.7$$

Thinking: If his his BMI is less than 25, he is of normal weight.

Working: His BMI is less than 25 so he is of normal weight.

Try yourself

1 Calculate the BMI of a person 160 cm tall and weight 70 kg.
2 Calculate the BMI of a person 170 cm tall and weight 100 kg.
3 Use the BMI chart in Figure 8.3.6 to estimate the BMI of a person 175 cm tall and weight 50 kg.

Doctors may use a person's BMI to determine whether other tests including waist circumference, body fat and blood tests are needed.

In 2016 a global study by the medical journal, *The Lancet*, suggested that overweight and obese people now outnumber underweight people worldwide. Australians are amongst the most overweight people in the world—according to the Australian Government 63% of Australians are overweight or obese. This is an increase of 10% compared to 20 years ago.

Australia, Canada, Ireland, New Zealand, UK and the US are home to 20% of the world's obese people.

Over the last 30 years Australians have increased their energy consumption. At the same time, they have reduced their energy needs because they are not so physically active.

For many, watching television, surfing the internet, social media and playing computer games take up more time than activities such as sport and active recreation. If the Australian population is to remain healthy, then this trend will need to change.

Too much salt

For many people the first thing they do when they sit down to a meal is add salt to the food. A small amount of salt is important in your diet. Your kidneys need salt to balance the water and salt in your body. This balance ensures that there is the correct amount of water in your cells and tissues and the correct volume of blood circulating around your body.

Fruit and vegetables have salts in their cells and there is salt in meat and whole grains. By eating these foods, you gain all the salt your body requires.

Many processed foods have a high salt content and you can taste the salt. A small bowl of breakfast cereal like the one in Figure 8.3.8 has about the same amount of salt as a small packet of plain chips, and some sweet biscuits have as much salt as savoury biscuits.

Most Australians consume eight to nine times the amount of salt that they need each day and 75% of this comes from processed foods. Lots of salt makes the kidneys work too hard and can lead to kidney failure. It is also linked to high blood pressure.

FIGURE 8.3.8 Simple processed foods such some breakfast cereals contain considerable amounts of salt.

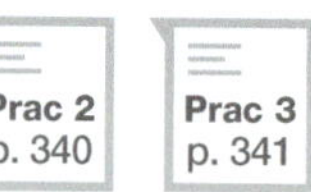

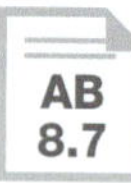

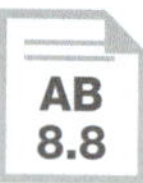

High blood pressure

Your heart is the pump that causes your blood to circulate within your body. When your heart beats, it forces blood out of the heart and into blood vessels known as arteries. Your blood is under pressure when it enters your arteries but between heartbeats the pressure in your arteries falls. During exercise your heart works very hard and your blood pressure increases and then falls again when you rest. Therefore, it is normal for your blood pressure to vary throughout the day.

SciFile

Vitamin supplements—good or bad?

Many people take vitamin supplements in the belief that it is good for them. Additional supplements are only required if your diet is inadequate. A healthy diet should contain all the vitamins and mineral the body needs. Excess vitamins are excreted by the body and in fact some vitamins are toxic when consumed in excess.

When the pressure in the arteries stays high between heartbeats and during rest, the person is suffering from **high blood pressure**. The constant high pressure strains the artery walls. It also causes fats from the blood to stick to the artery walls. The arteries become clogged, causing the heart to pump even harder to get the blood through. This increases the blood pressure further.

Doctors regularly check patients' blood pressure because high blood pressure does not always produce symptoms (Figure 8.3.9). Without diagnosis, people may not know they have high blood pressure until something major happens such as:

- **heart attack**—when part of the heart muscle is damaged or dies because the blood supply is blocked or severely reduced
- **stroke**—when part of the brain is damaged or dies because the blood supply is blocked or severely reduced.

FIGURE 8.3.9 Regular measurements are needed to ensure that blood pressure is within healthy limits.

SciFile

Long way

The human heart creates enough pressure to squirt blood 9 metres. This could happen to you if you cut a major artery.

Diabetes

Diabetes is a complex disease caused by a lack of insulin or an insensitivity to insulin. **Insulin** is the hormone that lowers the level of glucose in the blood. A person who suffers from diabetes is a **diabetic**.

The energy your body uses comes from glucose produced by the digestion of the food you eat. Glucose moves from the digestive system into the blood. From the blood, glucose moves into the cells, where it is broken down in the process of cellular respiration to release energy. The process is shown in Figure 8.3.10. The hormone insulin moves the glucose from the blood to the cells. Without insulin, glucose remains in the blood and the energy needed for growth, repair, and other essential functions is not released.

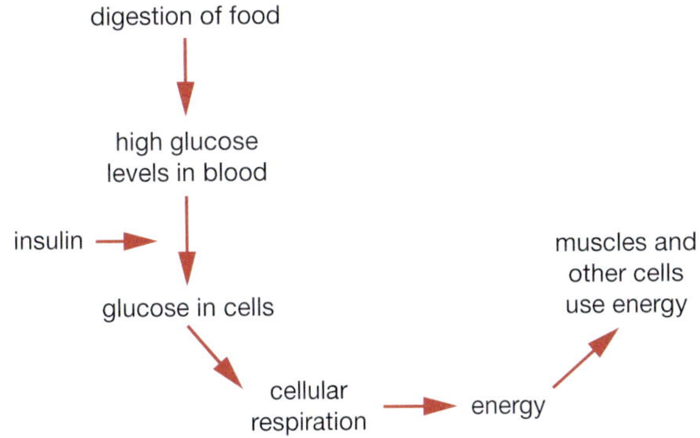

FIGURE 8.3.10 The hormone insulin controls the movement of glucose into the cells where it can be used.

There are two main types of diabetes—type 1 and type 2.

Type 1 diabetes

In **type 1 diabetes**, the body stops making insulin. The exact reasons for this are not known, but it is known that the immune system destroys the insulin-producing cells in the pancreas. A person with type 1 diabetes may eat well but their body is undernourished because they cannot release any energy from their food. Part of the treatment for people with type 1 diabetes is to regularly test the levels of glucose in their blood (Figure 8.3.11 on page 334). They must then inject themselves with insulin to replace the insulin that their body fails to make. The diabetic can then get the nutrition they need from the food they eat.

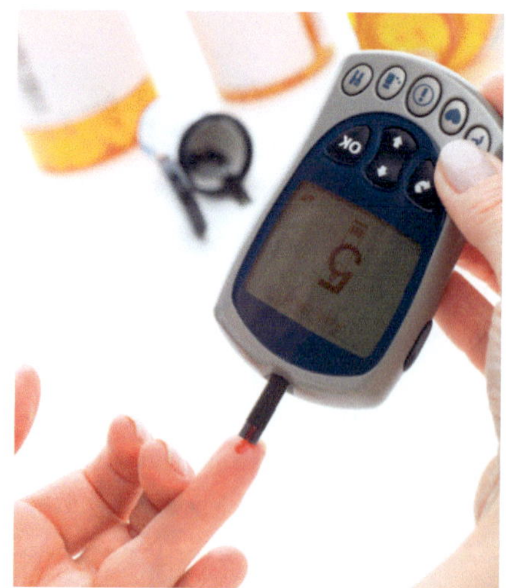

FIGURE 8.3.11 People with diabetes must check the glucose levels in their blood regularly so they know how much insulin to take.

Type 1 diabetes accounts for 10–15% of cases of diabetes in Australia. It is usually diagnosed in people under the age of 30. There could be a genetic link to type 1 diabetes, meaning that the disease is caused by inherited characteristics. Type 1 diabetes is not linked to lifestyle, but eating well and exercising regularly is essential for managing the disease.

Type 2 diabetes

In **type 2 diabetes**, the body has become resistant to insulin. The levels of insulin in the blood are normal or may be higher than normal but the body cannot use the insulin. This is the form of diabetes found in 85–90% of people with diabetes.

Glucose levels in the blood of type 2 diabetics are permanently high. When the blood glucose level gets to a certain point, the body converts the glucose to fat.

FIGURE 8.3.12 Regular exercise is a way of preventing type 2 diabetes.

People with type 2 diabetes can therefore gain weight very easily. When the disease is first diagnosed, people with type 2 diabetes can often control the problem with diet and exercise (Figure 8.3.12). Later they may also have to inject themselves with insulin just as people with type 1 diabetes do.

Type 2 diabetes may have a genetic link. However, it is also closely connected to lifestyle. A diet high in fat, sugar and refined foods increases the risk of type 2 diabetes.

Asbestos-related diseases

Our environment includes substances that can affect our health. These include naturally occurring minerals and human-made chemicals. For example, asbestos is a naturally occurring mineral that can cause severe health problems.

Asbestos miners, people who worked with asbestos products and people who lived in mining areas developed illnesses with very similar symptoms. They had a persistent cough, shortness of breath and blue colour to their lips.

These people were diagnosed with **asbestosis**, which is a lung disease caused by breathing in asbestos fibres. Asbestos fibres are shown in Figure 8.3.13. Each asbestos fibre is 50–200 times thinner than a human hair and when the fibres are floating in the air they cannot be seen. When the fibres are breathed in, they remain in the lungs and cause inflammation and scarring. The hard, inflexible scar tissue makes it difficult to get enough oxygen into the body, causing the lips to turn blue.

FIGURE 8.3.13 Fibres of asbestos look like cotton wool, but they are not soft. Asbestos fibres are fire-resistant and very strong. They were used to make corrugated roofing, water pipes, fabrics, and brake linings.

Asbestos can also cause mesothelioma—a cancer that develops 30–40 years after exposure to asbestos.

Before it was recognised as being dangerous, asbestos was mined and used in the manufacture of corrugated roofing, water pipes, fabrics and car brake linings. However, from 2003 the use of all forms of asbestos was banned in Australia. Although asbestos cannot be used in new buildings, it is still found in old buildings and is gradually being replaced by safer materials. Anyone working with asbestos needs to wear safety clothing like that shown in Figure 8.3.14.

FIGURE 8.3.14 Workers removing asbestos from buildings must wear safety clothing and a mask.

Working with Science

DIETITIAN

Dietitians are experts in human nutrition and help people modify their diets to improve their health. The relationship between diet and disease can be very complicated. Dietitians play important roles in the healthcare industry as experts in nutrition and its effects on the human body. Nutrient imbalances (under-nutrition or over-nutrition) can be the cause of many diseases (e.g. anaemia caused by iron-deficiency or obesity caused by excess-nutrition). Nutrient imbalances may also result from diseases that limit the digestion and absorption of nutrients (e.g. inflammatory bowel disease). Dietitians are trained to assess, diagnose and treat nutrient imbalances by helping people make adjustments to their diet and recommend nutritional supplements.

Dietitians have a diverse range of roles. They might provide advice and feeding support to people recovering from surgery, deliver educational programs on healthy eating habits to community groups or schools, develop meal plans for individuals to help them manage or prevent disease (Figure 8.3.15), or conduct research on nutritional biochemistry and disease. Dietitians also work in a broad range of settings, from hospitals, nursing homes, schools, community health centres and childcare services, to universities, as researchers.

To become a dietitian, you will need to complete a Bachelor of Nutrition and Dietetics.

FIGURE 8.3.15 Dietitians help people manage or prevent disease by assessing their health and providing nutritional support and advice.

You can also enter the field as a dietetic or nutrition assistant by completing a Certificate or Diploma in Nutrition and Dietetics. Good communication skills, critical thinking and the ability to support and relate to patients are important qualities in this career. The work of a dietician is diverse and rewarding and employment opportunities are growing across Australia and overseas.

Review

1. What are some widespread health problems in our society that you think dietitians can help with?
2. What are three other professionals that dietitians might work with?

SCIENCE AS A HUMAN **ENDEAVOUR**

Use and influence of science

LifeStraw

Access to clean drinking water is a basic need for health and sanitation, yet over 1 billion people around the world do not have reliable access to safe sources of water.

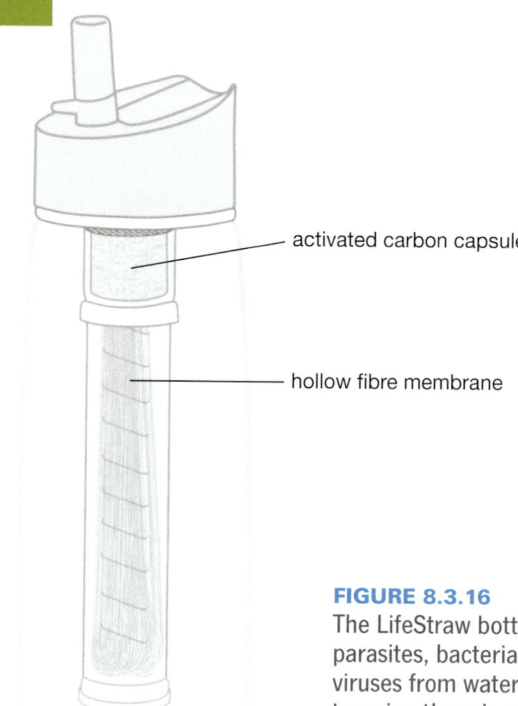

FIGURE 8.3.16 The LifeStraw bottle filters parasites, bacteria and viruses from water by trapping them in a hollow fibre membrane.

Access to clean drinking water is a major problem for people living in developing countries without the resources to build proper water treatment facilities. People often have to walk long distances to the nearest water sources, which are usually polluted with agricultural runoff and human waste because of the lack of sewage systems. Disease-causing parasites and microorganism contaminate the drinking water, leading to illness and millions of deaths every year.

The LifeStraw was developed to make safe drinking water accessible to people in areas without water treatment facilities or where water sources have been contaminated due to natural disasters. It is a lightweight, personal water filter device that can remove 99.99% of parasites, bacteria and viruses from water.

The LifeStraw uses a membrane made of hollow fibres to trap disease-causing microorganisms and parasites living in the water. Because the pore size of the fibres is too small for parasites, bacteria and viruses to pass through, they are filtered out of the water, making it safe to drink. Figure 8.3.16 shows the LifeStraw bottle and its key parts. The LifeStraw can filter up to 1000 litres of water, lasting around a year with regular use. Its simple design means that it does not need batteries or external parts to function, which is important in communities with limited access to resources. Its small size and light weight also make it ideal for travelling and camping.

SCIENCE AS A HUMAN ENDEAVOUR

FIGURE 8.3.17 The LifeStraw is a personal water filter that makes water safe for drinking by removing 99.99% of parasites, bacteria and viruses.

FIGURE 8.3.18 The 50-litre LifeStraw Community water purifier

Figure 8.3.17 shows the portable LifeStraw designed for personal use in action. It is simple and easy to use. LifeStraw also comes in a 5-litre portable gravity-fed device, an 18-litre bench-top water purifier and a 50-litre Community water purifier (Figure 8.3.18). LifeStraw also makes a refillable water bottle, a 5-litre portable gravity-fed device and an 18-litre home bench-top filter. The company behind LifeStraw uses funds from the sale of LifeStraw products to provide water purifiers to schools in Africa. So far, over 3500 LifeStraw Community water purifiers have been distributed to more than 361 000 students in 631 schools in Africa. LifeStraw is improving the lives of people around the world by giving them access to safe, clean water and helping to eradicate water-borne diseases.

REVIEW

1 Why is it important that our water sources are treated or filtered to remove parasites, bacteria and viruses?

2 Water-borne diseases are common in developing countries. What are some factors that affect people's access to clean water?

3 Water is essential for our personal health and many of our daily activities. It is also important for the health and functioning of our society, environment and economy. Fill in the table below with uses for water that are important for health, society, environment and the economy. Note that some uses may be important in more than one area.

Health	Society	Environment	Economy

4 Lack of access to clean water can have severe impacts on the productivity and health of a population, trapping people in a cycle of poverty. How do you think access to clean water impacts people's lives, health, income and education?

MODULE 8.3

Review questions

Remembering

1 Define the terms:
   a asbestosis
   b obese
   c rickets.

2 What term best describes each of the following?
   a what a person eats, regardless of whether it is healthy or not
   b not having enough nutrients in a diet
   c diseases that last a long time.

3 State the function in the body of:
   a vitamins
   b calcium
   c iron
   d protein.

4 a Name two diseases caused by a lack of vitamins.
   b Name the vitamin that is lacking in both diseases.

Understanding

5 You cannot survive for more than a few days without water. Explain why water is an important part of any diet.

6 a What is rickets?
   b Describe the symptoms of rickets.
   c Discuss the possible causes of rickets in Australia?

Applying

7 a What do the letters BMI stand for?
   b Explain how you would calculate a person's BMI.
   c Calculate the BMI for each of the following people:
      i Person 1: height 190 cm, mass 85 kg
      ii Person 2: height 165 cm, mass 90 kg
      iii Person 3: height 185 cm, mass 60 kg.
   d Use the Body Mass Index chart on page 331 to comment on each person's BMI.
   e Why must you be careful when interpreting a person's BMI?

8 Identify reasons why an adult could develop mesothelioma even though they had not been near asbestos since they were a child.

Analysing

9 Contrast over-nutrition and under-nutrition.

10 Compare a non-infectious disease and an infectious disease.

Evaluating

11 Figure 8.3.19 graphs the effect of diabetes on the health of the population against age.
   a Suggest why the line for type 1 diabetes is fairly flat.
   b Propose a reason for type 1 diabetes having less effect on the health of the population.
   c Use the graph to determine the age group that is most seriously affected by type 2 diabetes.
   d Use your knowledge of type 2 diabetes to describe what this graph might look like in 20 years time if Australians do not change their eating habits.

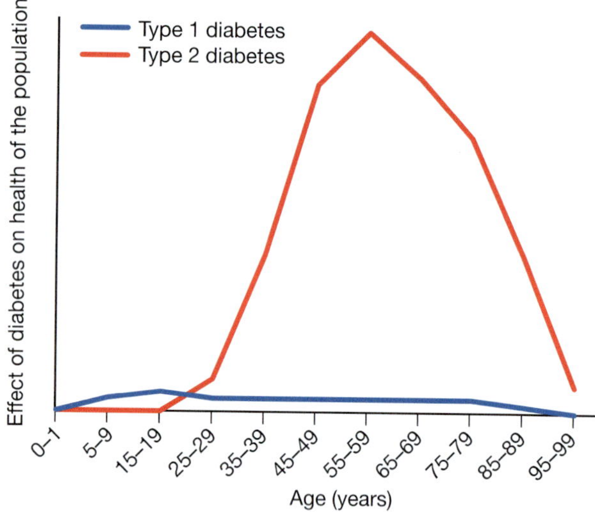

FIGURE 8.3.19 The more diabetes there is in a population, then the greater the effect diabetes has on the functioning of that population.

Creating

12 Construct a flow diagram to show how eating too much leads to weight gain.

MODULE
8.3 Practical investigations

1 • Testing for nutrients

Purpose

To carry out standard tests for fats, carbohydrates (starch and glucose) and protein.

Timing 45 minutes

Materials

- 10% glucose solution
- 10% gelatine solution (protein)
- 10% starch solution
- vegetable oil
- iodine solution
- water

SAFETY

Iodine is toxic, do not ingest. Avoid contact with skin and eyes. Wear disposable gloves and safety glasses.

- white tile
- Albustix™ paper
- Tes-Tape
- brown paper
- eyedroppers—use a new eyedropper for each test

Procedure

WATER TEST (CONTROL)

1 Use an eyedropper to place one drop of water in three places on the white tile and one drop onto a piece of brown paper.

2 Test the water using the Albustix, Tes-Tape and iodine solution.

GLUCOSE TEST

3 Use an eyedropper to place one drop of glucose solution on the white tile.

4 Dip the end of the Tes-Tape in the solution. Note the colour change.

5 Wash the eyedropper thoroughly in water.

STARCH TEST

6 Use an eyedropper to place one drop of starch solution on the white tile.

7 Add a drop of the iodine to the starch solution. Note the colour change.

8 Wash the eyedropper thoroughly in water.

PROTEIN TEST

9 Use an eyedropper to place one drop of gelatine solution on the white tile.

10 Dip the end of the Albustix in the solution. Note the colour change.

11 Wash the eyedropper thoroughly in water.

FAT TEST

12 Use an eyedropper to place one drop of vegetable oil on the brown paper.

13 Leave the paper for a few minutes and then note any change in the paper.

Results

Record your observations from each test.

Review

1 Summarise your results by completing the following sentences.

a Starch solution turned from ________ to ________ when iodine was added.

b Tes-Tape changed colour from ________ to ________ in the presence of glucose.

c Albustix turned from ________ to ________ when protein was present.

d A translucent stain was left on brown paper when ________ was present.

2 a Summarise the results for each test with water.

b Why was it important to know how water reacted to the tests?

MODULE

8.3 Practical investigations

2 • A balanced diet

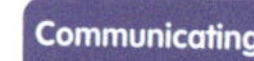

Purpose

To compare your diet with a recommended diet.

Timing 30 minutes plus 15 minutes per day for 4 days

Materials

- a list of the food eaten each day for four days

Procedure

1 Construct a blank version of the 'Healthy food plate' to fit an A4 page. Use Figure 8.3.20 as a guide.

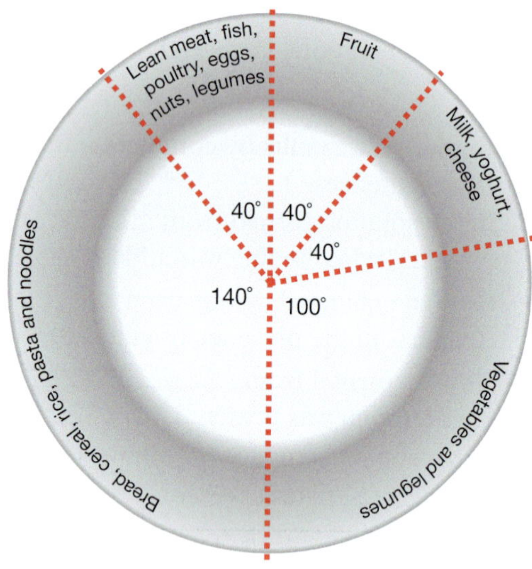

FIGURE 8.3.20

2 Over the next four days, record everything you eat and drink (including all snacks). Place a cross in the appropriate segment of the plate to represent each serve of food in that category. Use Table 8.3.1 as a guide.

TABLE 8.3.1 Size of one serving of different foods

Food	Serving size
bread	1 slice
roll	1
English muffin or hamburger bun	½
rice, pasta, cooked cereal	½ cup
ready-to-eat cereal	¾ cup
chopped fresh fruit	½ cup
canned fruit	½ cup
100% fruit juice	¾ cup
raisins or dried fruit	¼ cup
cooked vegetables	½ cup
chopped, raw vegetables	½ cup
raw, leafy vegetables	1 cup
vegetable juice	¾ cup
cooked lean meat, fish or poultry	60–90 g, size of palm of hand
egg	1
peanut butter	2 tbsp, 30 g
cooked dried peas or beans	½ cup
nuts, seeds	⅓ cup
milk	1 cup
yoghurt	1 cup
cheese	1 slice, 30 g
cottage cheese	½ cup
ice-cream	½ cup
vegetable oil	1 tsp
butter	1 tsp
peanut butter	1 tsp
ripe olives	8
salad dressing	1 tbsp

Results

1 Calculate the percentage of the 'Healthy food plate' occupied by each food type. Measure the angle of the segment—for example, the angle of the fruit sector is 40°. The complete circle is 360°; therefore, the fruit segment occupies $\frac{40}{360}$. Convert this into a percentage:

$$\frac{40}{360} \times \frac{100}{1} = 11\%$$

2 Add this information to the first two columns of the Table 8.3.2.

3 a Complete column three of the Table 8.3.2 by adding the number of crosses for each food type.

b Calculate the percentage of your diet represented by each food type. Use the following equation to work this out for each food type:

$$\frac{\text{number of crosses for food type}}{\text{Total number of crosses}} \times \frac{100}{1}$$

Practical investigations

TABLE 8.3.2 Results

	Angle of segment (°)	% of plate	Number of crosses	% of diet
fruit	40	11		
vegetables				
lean meat, fish, poultry, eggs, nuts and legumes				
milk, yoghurt and cheese				
fat				
bread, cereal, rice, pasta and noodles				

Review

1. Explain why the information for this activity was collected over four days rather than only one day.
2. Compare the diet you have recorded with the recommended diet.
3. Assess the quality of your diet.
4. Discuss the differences you have identified. What changes would be required to bring your diet closer to the recommended diet?

3 • Nutrients in food

• STUDENT DESIGN •

Purpose

To test a range of foods for fats, carbohydrates (starch and glucose) and protein.

SAFETY

A risk assessment is required for this investigation. Do not use foods that contain nuts.

Timing 45 minutes

Materials

- samples of various foods (such as apple, cheese, milk, egg white, butter, meat, orange juice, potato, bread, boiled lolly and sweet biscuit)
- equipment as chosen by students

Procedure

1. Design an experiment that will test all the selected foods for fat, sugar, starch and protein.
2. Brainstorm in your group and come up with several different ways to investigate the problem. Select the best procedure and write it in your workbook.
3. Before you start any practical work, assess all risks associated with your procedure. Refer to the SDS of all chemicals used. Construct a risk assessment that outlines these risks and any precautions you need to take to minimise them. Show your teacher your procedure and your risk assessment. If they approve, then collect all the required materials and start work.

Hint

- Practical investigation 1 may provide you with some ideas for your experiment.
- For dry foods such as bread, biscuit, meat and cheese, you will need to grind the food in a mortar and pestle with a few drops of water before testing.
- Make sure that you use clean equipment for each test.
- Use the STEM and SDI template in your eBook to help you plan and carry out your investigation.

Results

Construct a table to show the nutrients present in each food sample.

Review

1. Which food contained the most nutrients?
2. Which food contained the fewest nutrients?
3. Assess the contribution each food makes to a balanced diet.
4. Discuss the idea that it is better to eat foods containing more than one nutrient than foods containing only one nutrient.
5. A negative test for a particular nutrient does not necessarily mean that the nutrient is not present. Why might this be true?
6. Evaluate the effectiveness of the procedure you used and suggest any improvements that could be made.

CHAPTER 8

Chapter review

Remembering

1 a What is a pathogen?
  b Name two different types of pathogens.

2 Recall the names of two diseases caused by single-celled organisms other than bacteria.

3 What type of pathogen causes tinea and thrush?

4 What is the cause of the following diseases?
  a scurvy
  b kwashiorkor.

Understanding

5 Explain how washing your hands can protect you from disease.

6 Describe ways a viral disease may be transmitted from an infected person to a healthy one.

7 Explain the term *contagious*, using one contagious and one non-contagious disease in your explanation.

8 How does vaccination control the spread of disease?

9 The skin, gastric juices and mucous membranes are sometimes called the body's first line of defence.
  a What does first line of defence mean?
  b How does the first line of defence help prevent infection from pathogens?
  c What happens if pathogens get past the first line of defence?

10 a List some of the different types of white blood cells found in the immune system.
  b For each, describe their role.

11 a Define the terms:
    i malnutrition
    ii overweight
    iii obese.
  b Explain how a person who eats a good diet can be suffering from malnutrition.

12 a What does the term *blood pressure* mean?
  b What is meant by high blood pressure?
  c What are the dangers of high blood pressure?

13 Explain why few people in Australia should have rickets.

14 Water is not a nutrient. Explain why water is such an important part of a healthy diet.

15 Explain the link between asbestos mining and the disease mesothelioma.

Applying

16 Demonstrate how the following behaviours would protect you from disease.
  a Protect yourself from amoebic dysentery by eating fruit only if it can be peeled before eating.
  b Avoid tinea infection by not sharing towels.
  c Use insect repellent to avoid malaria.

17 During a regular check-up a man was told by his doctor that he was overweight. The man, who was 192 cm tall and weighed 95 kg, was surprised by this.
  a Was the doctor correct? Give reasons for your answer.
  b What advice would you give to the man?

Analysing

18 Classify the following as things that cause disease, or things that are part of the immune system.

pathogens, macrophages, viruses, neutrophils, skin, nose hairs, gastric juices, Plasmodium

Evaluating

19 Propose reasons why some vaccinations against viral diseases such as polio can give you lifelong immunity, whereas it is recommended that you get a flu vaccination each year.

20 a Assess whether you can or cannot answer the questions on page 305 at the start of this chapter.
  b Use this assessment to evaluate how well you understand the material presented in this chapter.

Creating

21 Use the following ten key terms to construct a visual summary of the information presented in this chapter.

pathogen	over-nutrition
immunity	under-nutrition
virus	lymphocytes
bacteria	disease
vaccination	chronic disease

AB 8.10

CHAPTER 8 Inquiry skills

Research

1 Processing & Analysing | Evaluating

Three-part inquiry question

Select your entry point and complete the relevant parts of this inquiry.

Present your research as a digital slide show.

a Mosquitoes not only spread malaria but also many other viruses that cause serious disease. One mosquito that is a major cause of disease is *Aedes albopictus*. This mosquito has been labelled the 'Asian tiger mosquito' because of its striped appearance. It has a vicious bite and can survive in a large variety of climates. Investigate *Aedes albopictus*.

- **i** Explain why authorities are so concerned about its possible introduction into Australia.
- **ii** Zika virus is one of the world's many mosquito-carried diseases. It has been implicated in a number of serious foetal malformations. Most notably, there is strong evidence from Brazil that it causes the serious condition called microcephaly. Babies with microcephaly are born with very small heads and brains and as a result are severely intellectually impaired. *Aedes albopictus* has been implicated in the spread of Zika virus in Africa but currently could not do so in Australia. Explain why.

b Spread of disease is influenced by many different factors. The incidence of malaria across the Pacific is a good example of how environment influences disease. Malaria is caused by a protozoan of the genus *Plasmodium*. It relies on mosquitoes of the genus *Anopheles* for its transfer from host to host.

Consider the graph in Figure 8.4.1, which shows the incidence of malarial infections across the Pacific between 1990 and 2009.

- **i** What does the graph show happened to the number of cases of people infected with malaria in the year 2000 across the Asia–Pacific region? Investigate the weather that occurred in Australia and the Pacific at that time, how might this link to a rise in the number of cases of malaria?

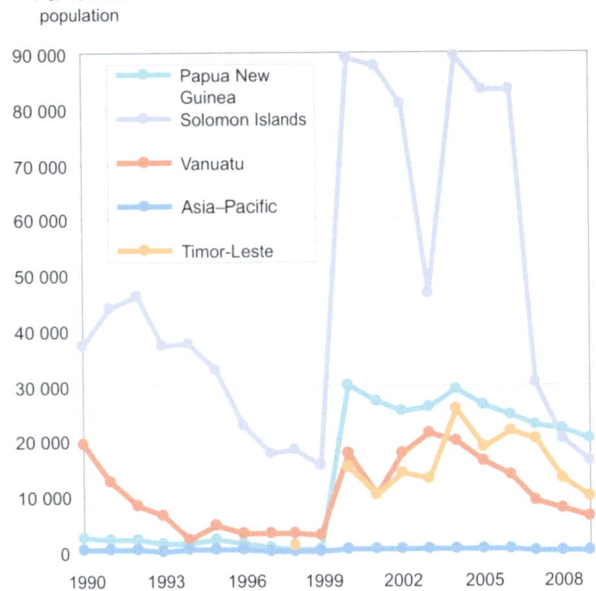

FIGURE 8.4.1

- **ii** What is different about the incidence of malaria in Solomon Islands compared with the rest of Asia–Pacific in 2000? Investigate the events occurring in Solomon Islands in the period 1999 to 2004 and then suggest how these events may have resulted in the difference in malarial cases.

One way to combat infections is to vaccinate people. Currently there are no vaccinations available for either malaria or Zika virus but there are vaccinations for many diseases. Some diseases for which vaccinations are routinely given in Australia are measles, mumps, rubella, pertussis, diphtheria and tetanus. Unfortunately, the rates of vaccination in some areas have fallen significantly in recent years.

c
- **i** Investigate the reasons given by parents for not vaccinating their children.
- **ii** Summarise the arguments for and against vaccination.
- **iii** Explain whether you would vaccinate your child, if you had one, giving reasons for your opinion.

CHAPTER 8

Inquiry skills

2 Processing & Analysing | Communicating

Investigate the work of Australian scientists Fiona Wood and Marie Stoner on artificial skin.

- Summarise the significant events in their research.
- Describe the types of injuries they have been able to treat.
- Describe how this treatment has affected the lives of their patients.
- Describe the benefits to society of this research.

Present your research in digital form.

3 Processing & Analysing | Communicating

Investigate the use nanotechnology in medicine. Include:

- what nanotechnology, nanoscience and nanomedicine are
- what disease or condition is treated
- how the treatment works
- the benefits of using nanotechnology in the treatment
- any disadvantages in using nanotechnology in the treatment.

Present your research as an annotated poster.

4 Processing & Analysing | Communicating

Gather information about viral diseases such as dengue fever, Ebola, Murray Valley encephalitis, Ross River fever, rubella, poliomyelitis and HIV/AIDS. As part of your research:

- find where in the world these diseases occur and are a problem
- find how these diseases are transmitted
- find what is being done on a local or worldwide scale to control these diseases
- identify the disease you think should have most resources allocated for future research. Justify your decision.

Present your findings as a poster.

5 Processing & Analysing | Communicating

Professor Ian Frazer was named Australian of the Year in 2006. Research:

- why he received this award
- the benefit to society of his medical research.

Present your findings as a media release.

6 Evaluating | Communicating

Research vitamin supplements. Include:

- what vitamin supplements are
- who manufactures vitamin supplements
- whether there are any benefits to taking vitamin supplements
- whether there are any concerns or dangers involved with taking vitamin supplements.

Present your findings as an information booklet that includes recommendations on taking vitamin supplements.

CHAPTER 8

Inquiry skills

Thinking scientifically

Questions 1 and 2 refer to the following information.

During an outbreak of flu the local doctor was recording the number of people in the area diagnosed each day. Figure 8.4.2 is the graph the doctor created.

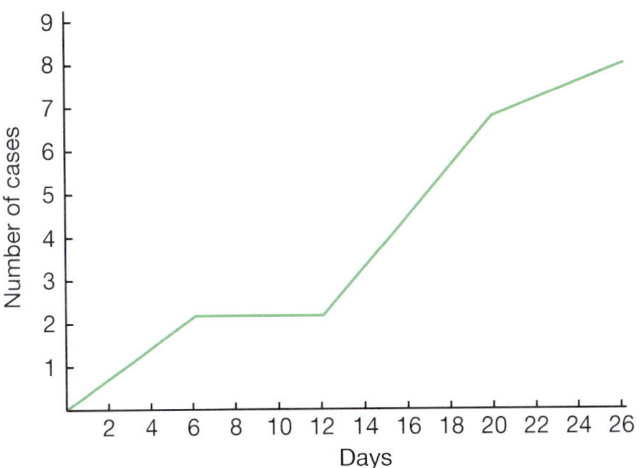

FIGURE 8.4.2

Key
A Days 0–6
B Days 6–12
C Days 12–20
D Days 20–26

1 Use the key to identify the period of time when there were no more new cases of flu diagnosed.

2 Use the key to identify the period of time when the number of cases increased most rapidly.

3 A class of students was arguing about the number of colonies of bacteria they had on the agar plate they had exposed on the windowsill of their classroom. Figure 8.4.3 is a drawing of the plate.

How many bacterial colonies can be seen on the plate?

A 3
B 6
C 7
D 10

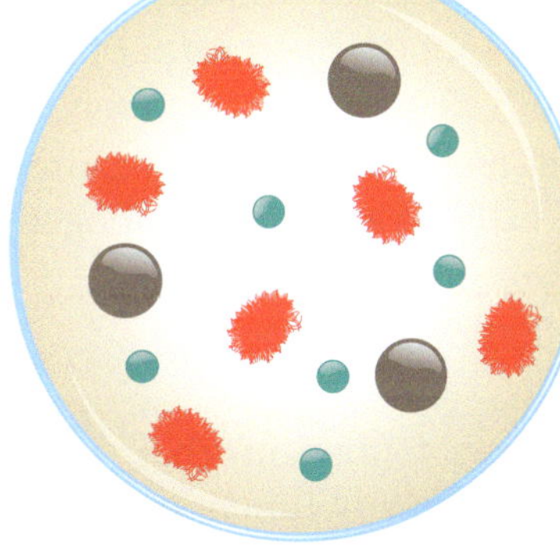

FIGURE 8.4.3

4 Figure 8.4.4 is a graph of the temperature of a person suffering from smallpox.

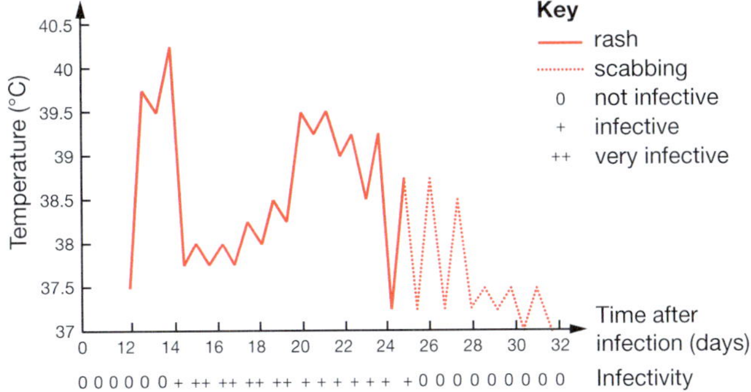

FIGURE 8.4.4

Identify the statement that is consistent with the data presented.

A The person was most infective when their temperature was highest.

B Once the scabs started to form, the temperature was consistently lower than when the rash was present.

C The total rise in temperature from day 12 to day 14 was almost 3°C.

D Thirty days after infection, the temperature was consistently back to normal.

5 Bacteria in the mouth can cause bad breath. The bacteria in a mouth were counted before and after using four mouthwashes. The results are given in Table 8.4.1.

TABLE 8.4.1 Bacteria in mouth before and after mouthwash

Mouthwash	Bacterial count	
	Before	After
1	25	10
2	80	25
3	45	6
4	60	15

Calculate which mouthwash killed the greatest percentage of bacteria.

A 1
B 2
C 3
D 4

CHAPTER 8

Glossary

antibiotic: a substance that kills bacteria or prevents the growth of bacteria

antibody: a chemical made by the immune system that makes it easier for white blood cells to destroy pathogens

asbestosis: a lung disease caused by breathing in asbestos fibres

bacteria: microscopic, single-celled organisms

carbohydrates: nutrients used as the main source of energy for the body

chronic disease: a disease that lasts for a long time

contagious: an infectious disease that is readily communicable (spread by close contact)

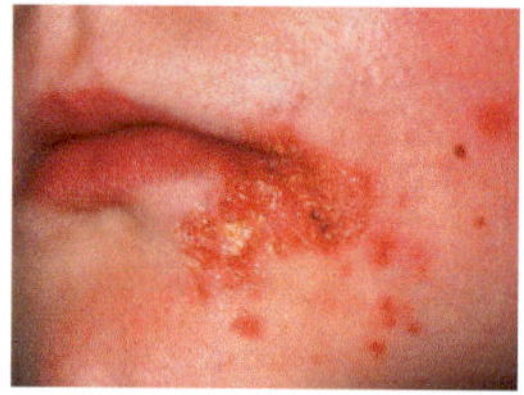

contagious disease

diabetes: a complex disease caused by a lack of or insensitivity to insulin

diabetic: a person who has the disease diabetes

diet: what a person eats

disease: anything that causes your body to stop working properly

fats: nutrients used as a source of energy and an energy store in the body

fungicide: a chemical that kills fungi

heart attack: when part of the heart muscle is damaged or dies because the blood supply is blocked or severely reduced

high blood pressure: when the blood pressure in the arteries remains high between heartbeats and during rest

histamine: a chemical that is made by cells in response to injury

host: the organism a parasite lives in

host cell: a cell invaded by viruses

immune: able to make the antibodies to a pathogen before it can make you unwell

immune system: the system in your body that fights infections

infectious disease: a disease caused by a microorganism that may or may not be communicable

insulin: the hormone that lowers the level of glucose in the blood

kwashiorkor: a disease caused by a lack of protein

lymphocyte: a white blood cell that makes antibodies

macrophage: a white blood cell that consumes pathogens

macrophage

malnutrition: when the nutrition provided by the diet does not meet the needs of the body

minerals: nutrients required for various functions in the body

nanomedicine: using nanotechnology in medicine

neutrophil: a type of white blood cell that consumes pathogens

nutrients: protein, carbohydrate, fats, minerals and vitamins

nutrition: the food necessary for health and growth

obese: excessively overweight, BMI above 30

over-nutrition: a form of malnutrition in which the body is getting more nutrients than it requires

overweight: more body fat than is considered healthy, BMI between 25 and 30

pathogen: an organism that causes disease

pathogenic bacteria: bacteria that cause disease

penicillin: the first antibiotic

proteins: nutrients used for growth and repair of the body

quarantine: isolation to prevent the spread of a disease

rickets: a disease caused by a lack of vitamin D

scurvy: a disease caused by a lack of vitamin C

spore: a single cell used by fungi to spread

stroke: when part of the brain is damaged or dies because the blood supply is blocked or severely reduced

symptom: indicator of a particular disease

type 1 diabetes: a form of diabetes in which the body stops producing insulin

type 2 diabetes: a form of diabetes in which the body has become resistant to insulin

under-nutrition: a form of malnutrition in which the body is not getting enough of the essential nutrients

vaccine: a chemical that causes your body to react as if it had encountered a pathogen

virus: a pathogen about 100 times smaller than a bacterium

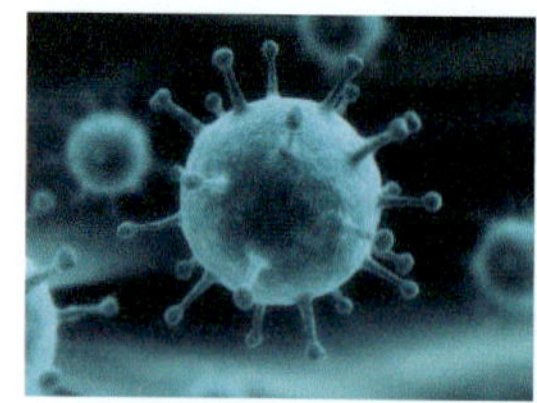

virus

vitamins: nutrients that control many functions in the body

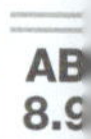

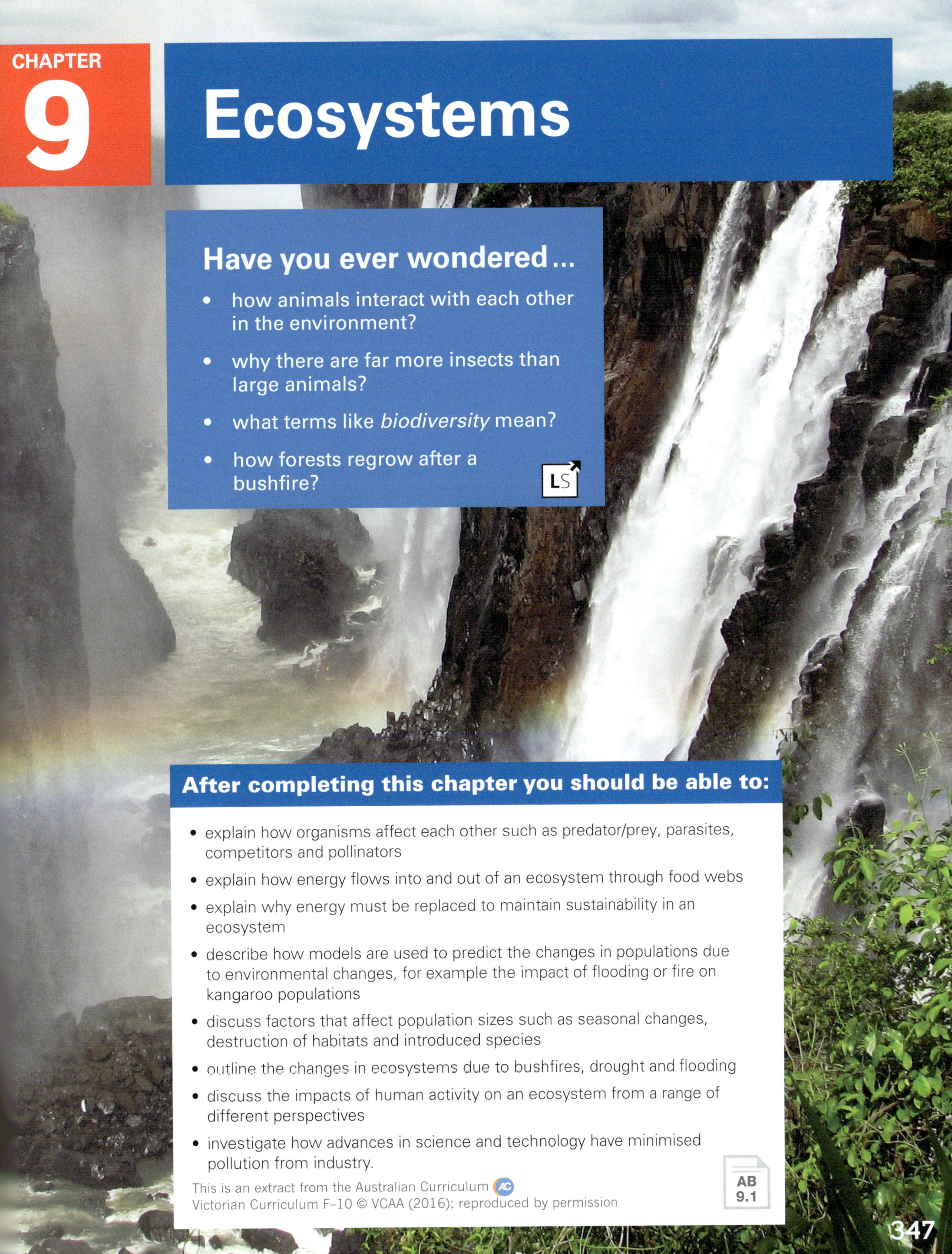

CHAPTER 9

Ecosystems

Have you ever wondered ...

- how animals interact with each other in the environment?
- why there are far more insects than large animals?
- what terms like *biodiversity* mean?
- how forests regrow after a bushfire?

LS

After completing this chapter you should be able to:

- explain how organisms affect each other such as predator/prey, parasites, competitors and pollinators
- explain how energy flows into and out of an ecosystem through food webs
- explain why energy must be replaced to maintain sustainability in an ecosystem
- describe how models are used to predict the changes in populations due to environmental changes, for example the impact of flooding or fire on kangaroo populations
- discuss factors that affect population sizes such as seasonal changes, destruction of habitats and introduced species
- outline the changes in ecosystems due to bushfires, drought and flooding
- discuss the impacts of human activity on an ecosystem from a range of different perspectives
- investigate how advances in science and technology have minimised pollution from industry.

AB 9.1

MODULE
9.1 Components of an ecosystem

Organisms live surrounded by other organisms and by non-living things such as rocks, water and soil. Animals may be chased by predators, attacked by diseases and battered by storms. Plants can be eaten, suffer drought or be destroyed by fire. Whatever their situation, organisms are affected by their surroundings and each other but somehow manage to live together.

Ecology and the environment

When biologists talk about the **environment**, they are talking about all the factors that affect an organism's chances of survival over its lifetime. Some of these factors are visible, such as the landscape and the types of rock, soil, plants and animals found there. The environment also includes less visible factors such as the number of days of sunshine, the amount of rainfall and the number of predators in the area. Every organism has its own unique environment.

An environment is not an object. A rock is an object, but it is only considered to be part of an organism's environment if the rock affects the organism's survival. For example, the rock may provide a hiding place for lizards or a surface on which they can warm up.

Ecology is the study of how organisms interact with each other and with their non-living surroundings. To interact means to affect each other, in either harmful or helpful ways. These interactions form the environment in which an organism lives. Scientists who specialise in ecology are known as ecologists.

Ecosystems

An **ecosystem** is a place where the organisms and their physical surroundings form an environment that is different from other environments nearby.

When ecologists study ecosystems they are looking at environments and the factors that affect the survival of organisms. The components of an ecosystem are these survival factors and the living and non-living things like animals, plants, rocks and water.

Natural ecosystems can exist on their own. They are balanced, meaning that they keep working without any outside help from humans. The lake in Figure 9.1.1 is an example. Humans can create artificial ecosystems such as the aquarium in Figure 9.1.2, but these need to be managed to keep them balanced. This may involve adding food materials and removing wastes.

FIGURE 9.1.1 A freshwater lake is a natural ecosystem.

FIGURE 9.1.2 An aquarium is an artificial ecosystem. Without human help, it quickly becomes unbalanced.

Factors influencing organisms

Organisms in an ecosystem are affected by two main sets of environmental factors:

- **abiotic factors**: These are non-living, physical factors such as air quality and humidity, the amount of sunlight, rainfall, wind, tides, waves, lightning and fires.
- **biotic factors**: These are living factors such as predators, parasites, fungi, infectious bacteria and viruses, competitors for food and the availability of breeding partners. Biotic factors also include dead organisms and their wastes. This means that leaves, rotting logs, faeces and urine are biotic factors too.

Some abiotic and biotic factors in the human environment are shown in Figure 9.1.3.

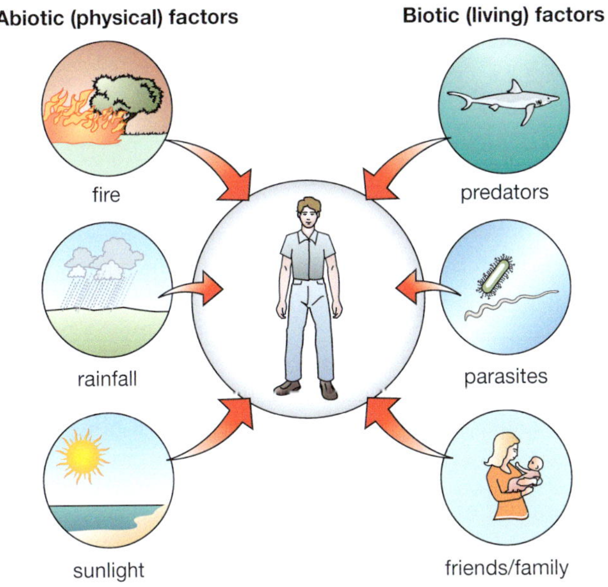

FIGURE 9.1.3 Abiotic and biotic factors affect all organisms, including humans.

Abiotic factors

Water, temperature, fire, light, soil type and oxygen levels are important abiotic factors for most organisms.

Water

Water is essential for all living organisms. It is the solvent for all materials in cells and allows numerous chemical reactions to occur. Chemicals must move around to participate in chemical reactions—water allows this movement to occur. When water is limited, chemical reactions within the cells cannot occur and cells die. As a consequence, the organism also dies.

Water affects any organisms that live in it, such as fish, sponges, seaweed and algae. Water does not evaporate from their body surfaces as it does from land organisms. Instead, aquatic animals and plants lose water through diffusion. Water diffuses out of their body cells into the ocean, river or lake.

Animals that live on land in dry areas such as deserts often avoid water loss by being nocturnal. Many like the desert hopping mouse live in burrows during the day and come out at night to feed (Figure 9.1.4). Less water evaporates from their body surfaces while in the cooler humid burrow than if they were in the sunlight. Cooler air temperatures at night also results in less evaporation.

FIGURE 9.1.4 The desert hopping mouse of Australia is nocturnal. It stays in a burrow during the day to avoid water loss and only comes out at night when it is cooler.

Water provides buoyancy (uplift), so organisms that live in water need less support for their bodies than creatures that live on land. However, water is very difficult to move through and so marine animals are usually streamlined to minimise water resistance.

Prac 1
p. 358

Temperature

Temperature affects the speed of chemical reactions in the cells. As temperature increases, the rate of a reaction usually increases too.

Fish, amphibians and reptiles like the lizard in Figure 9.1.5 on page 350 are referred to as being ectothermic. **Ectothermic** means these organisms must obtain heat from their environment rather than by generating it internally through body chemistry. Ectothermic animals are often referred to as being cold blooded but many of these animals are not 'cold'. Instead their body temperatures vary as the environment warms and cools.

These animals can also regulate their temperatures by lying on warm rocks or in sunlight to heat up, or by hiding in burrows if they need to cool down.

Birds and mammals such as humans and kangaroos are **endothermic**. Endothermic means that the organisms are warm blooded—they have the ability to generate heat internally and control heat loss to keep their body temperature constant.

FIGURE 9.1.5 A reptile is ectothermic. Its body temperature will be very low overnight and on cold mornings, and high after it has been lying in the sunlight.

SciFile

The great white hunter

Most shark species are ectotherms, allowing them to be active in warm waters. However, the great white shark is active in cold water too, allowing it to hunt for seals and penguins. It can do this because it can generate extra heat in its muscles and reduce its heat loss. This makes the great white shark one of only a few species of fish that is an endotherm.

Fire

Some bushfires start because of lightning or because of human activity. Australian Aborigines have used fire for many thousands of years to control the growth of plants and trees and to improve the growth of plants. They knew that many Australian plants regrow quickly after fire.

A bushfire like the one in Figure 9.1.6 can kill some plants, but it may help others. Some plants flower better after a fire and some drop their seeds. Many Australian plants will germinate, growing and sprouting shoots after a fire in response to the chemicals released in the smoke.

FIGURE 9.1.6 Fire can kill some organisms but help others.

Prac 2 p. 359

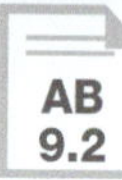

Light

Light is necessary for photosynthesis. **Photosynthesis** is the process by which plants manufacture their food materials using water, carbon dioxide and light.

Any change in the amount of light will therefore affect the growth rate of plants. If the amount of matter and energy available decreases, then the growth rate of the plant will slow down.

The amount of light does vary over the seasons. In winter, the hours of daylight received is much less. Most plants slow in their growth. Some plants, such as deciduous species, stop growing because their leaves die and drop off the plant (Figure 9.1.7).

FIGURE 9.1.7 In autumn, the hours of daylight decrease. The leaves of deciduous plants change colour as a result.

As Figure 9.1.8 shows, the northern sides of hills in Australia receive more light than the southern sides of hills. Hence plants tend to grow taller on the northern slopes than on the southern slopes. This happens in parts of Australia which lie below the tropic of Capricorn, including NSW, Victoria, South Australia, Tasmania and southern WA. In these places, the Sun never passes directly overhead. Instead, it moves across the sky in a path that lies towards the north, exposing the northern slopes of hills to more light than the southern slopes.

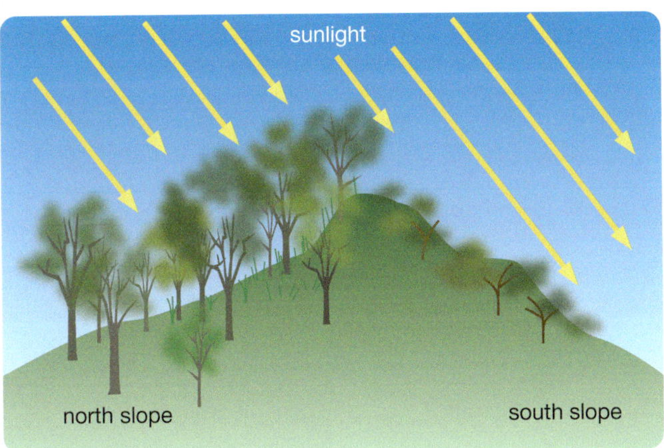

FIGURE 9.1.8 In the southern hemisphere, plants on a north slope usually grow larger because they receive more sunlight over a year.

Likewise, the Sun's rays do not penetrate into valleys early in the morning or in the late afternoon so plants growing in valleys receive less light than plants growing on hilltops.

The changing amount of light also influences the flowering of most plants. The days get longer as winter progresses into spring and many plants respond by flowering.

Not all species flower in spring or summer, some species flower when the daylength is actually decreasing in autumn or winter.

Light also affects animal behaviour. For example, rock lobsters (crayfish) avoid bright moonlight. Many other animals stay out of direct sunlight during the hottest part of the day to avoid overheating.

Soil type

Plants usually grow in soil that provides them with the water and minerals they need to help make their food. Not all soils are the same—they differ in mineral content, water-holding ability and acidity.

Two nutrients needed by plants are nitrate and phosphate. Where there are only tiny amounts of these in the soil, only specially adapted species such as banksias can survive (Figure 9.1.9).

FIGURE 9.1.9 Banksias are an Australian native plant that can grow in very poor soils low in nitrogen and phosphorus

Loam soil usually contains more plant nutrients than sandy soil because the clay particles in loam have more plant nutrients in them than sand grains do. While fertilisers stick to the clay particles in loam, they are often washed away and lost from sandy soils.

Some soils hold more water than others. Loam soils hold onto water more strongly than sandy soils. Therefore, it is more difficult for a plant to extract the water it needs from loam soils than it is from sandy soils.

The chemical composition of the soil is important too. For example, azaleas, camellias and most Australian native plants need acidic soils with a pH less than 7. High alkalinity will kill them or prevent them growing well. Other plants, such as some acacias (wattles) and eucalypts (gum trees), grow better in alkaline soils with a pH of more than 7.

Oxygen levels

The amount of oxygen in the environment affects most organisms because they require oxygen to carry out respiration, the reaction that provides their cells with energy. There is usually enough oxygen in the air for land-based organisms, but the amount dissolved in water can change greatly and affect aquatic organisms living there.

The amount of oxygen in water depends on several factors.

- **Temperature**: there is more dissolved oxygen in cold water than warm water.
- **Movement**: moving water dissolves a lot of gases from the air. Hence, a fast-flowing stream has more dissolved oxygen than a swamp.
- **Depth**: as water gets deeper, there is less mixing of water with the air, which reduces the amount of dissolved oxygen available to fish, and marine animals like prawns. Animals living in the depths of the ocean tend to move slowly because the energy available from respiration is limited by a lack of oxygen (Figure 9.1.10).

FIGURE 9.1.10 Anglerfish are slow-moving fish that live in the deep ocean where there is limited oxygen available.

Biotic factors

Living things affect each other's survival. For example, termites depend on microscopic organisms called flagellates. These can be seen in Figure 9.1.11. Flagellates live in the termite's intestines and digest the wood for the termite. The termite needs the flagellates to do this because it cannot digest wood even though wood is all the termite eats. The termite provides a moist, stable place for the flagellates to live and so the flagellates also depend on the termite.

FIGURE 9.1.11 Termites and the flagellates that live in their intestines are interdependent. They rely on each other for survival.

Unlike the termite and flagellate, organisms do not always help each other to survive. A hawk that kills and eats a mouse is dependent on the mouse for its survival. The mouse is dependent on plants for its food. Therefore, the hawk also depends on the plants because they keep its food source alive. The hawk, mouse and plant are therefore interdependent, because a change affecting one will affect all three. **Interdependence** means that all organisms in a food web are interconnected, affecting each other's survival in helpful and harmful ways.

Organisms rarely live alone—they are surrounded by other plants, animals and microorganisms. All the living things in an ecosystem are known as a **community**. Different relationships exist between the organisms in a community and these relationships are classified by how the organisms interact.

There are many different interactions between living organisms. These interactions involve biotic factors, and they play an important role in the survival of all species. Sometimes organisms assist each other, and sometimes they harm each other.

Competition

Organisms are said to be in **competition** when they both try to obtain the same resource, which may only exist in limited amounts. Competition occurs between different species and members of the same species. For example, the chicks in Figure 9.1.12 are in competition with each other for food. There is only a limited supply of food and resources, and so some individuals will not survive. In natural communities, competition is often fierce. There is a constant struggle for survival, and many die, especially the young, the old and the weak.

FIGURE 9.1.12 Baby birds compete with each other for food by trying to attract their mother's attention.

Predation

When one organism kills and eats another, the attacker is called the predator and the one being eaten is called the prey. This feeding relationship is known as **predation**. An example is shown in Figure 9.1.13.

FIGURE 9.1.13 A praying mantis preying on a cricket. The praying mantis is the predator and the cricket is its prey.

Mutualism

Mutualism is a relationship where two organisms live closely together and both benefit. The flagellates in a termite's guts are a good example. Without the flagellates, the termite would not have any food. The flagellates receive food and the correct temperature and moisture levels for survival. Both organisms depend upon each other for their survival.

The cleaner shrimp shown in Figure 9.1.14 eats parasites on the skin of the fish. Both the cleaner shrimp and the fish it cleans benefit, so this is an example of mutualism.

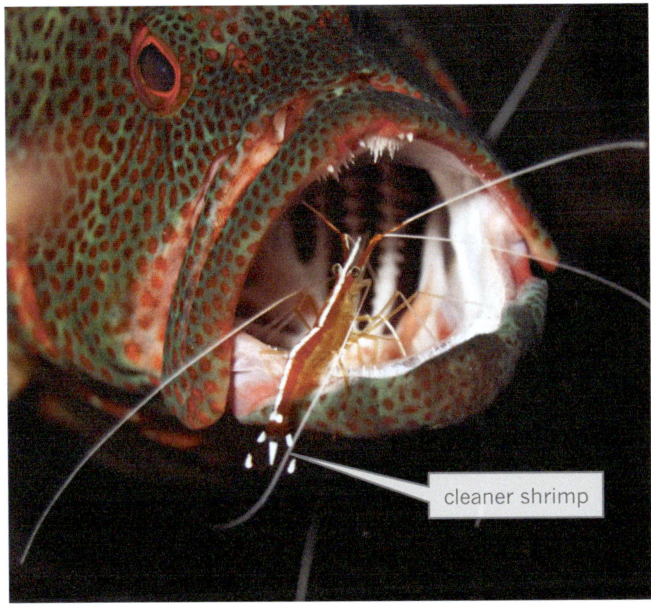

FIGURE 9.1.14 Cleaner shrimp and the fish they clean are an example of mutualism.

Pollination is another example of mutualism. Many flowering plants depend on animals to pollinate them. Pollination is the transport of pollen (containing the male sex cells) to the female parts of the flower. Pollination results in seeds. Animals such as honeyeater birds benefit from the relationship by obtaining food in the form of a sugary liquid (nectar).

Parasitism

Parasitism is a relationship where one organism benefits from the relationship, and the other is harmed. The parasite lives on or in another organism, otherwise known as the host, and feeds off it. The parasite cannot survive without the organism in which it lives. The parasite usually harms the host, but rarely kills it. The caterpillar in Figure 9.1.15 on page 354, which has been attacked by a wasp parasite is an example of parasitism killing the host.

FIGURE 9.1.15 A parasitic wasp has laid its eggs on this caterpillar. They will hatch and eat the caterpillar.

SciFile

River blindness

River blindness is caused by a parasitic worm that lives in human eyeballs. The worm is transmitted by a bite from a type of fly called a blackfly. The worm is estimated to have made about half a million people blind.

Commensalism

Commensalism is a relationship in which one organism benefits and the other is unaffected. An example of commensalism is the relationship between a whale and the barnacles it carries, shown in Figure 9.1.16.

FIGURE 9.1.16 Barnacles attached to the tail of this whale collect food but do not harm the whale. This relationship is an example of commensalism.

The barnacles attach themselves to the outside of the whale, but do not harm the whale or gain food from it. The barnacles collect their food by filtering microscopic animals out of the water as the whale swims. The whale does not eat the same animals as the barnacles, and the barnacles' presence does not harm the whale and so the whale is unaffected by them.

Adaptations

Organisms are able to cope with the biotic and abiotic factors in their environment because they have special features known as adaptations that assist them in their survival. An **adaptation** is any feature that assists an organism to survive and reproduce in its environment. Adaptations are classified as structural, behavioural or functional features of the organism.

A structural adaptation is a body part that helps an organism to survive. An example is the wing of a bat. The long fingers and the thin skin stretched over them form wings, helping the bat hunt flying insects.

A behavioural adaptation is a helpful habit, action or feature organism displays.

For example, the bat in Figure 9.1.17 emits clicking sounds when it hunts for flying insects. These sound waves reflect back off the insect to the bats ears, helping it locate the insect.

FIGURE 9.1.17 The greater mouse-eared bat hunts using echolocation.

A functional adaptation is a feature of the way an organism's body works. For example, when the bat flies, its heart automatically beats faster so that more blood is supplied to its muscles. This is a functional adaptation because it is controlled automatically, meaning the bat cannot consciously change it.

Environmental changes and populations

When the environment changes, the population sizes of all species can be affected. These changes involve both biotic and abiotic factors. The most common environmental changes are those due to seasonal changes in abiotic factors such as temperature, light and water.

In southern Australia, native plants have a growth spurt in spring when there is adequate water and light and temperatures are around 20°C. Many plants also flower at this time, providing food for animals that drink nectar or eat flowers. Therefore, the changes in abiotic factors affect the biotic factors. Increased quantities of plant matter result in increased numbers of herbivores (plant-eaters) and the carnivores (meat-eaters) that feed on the herbivores and on each other.

The size of a population changes because environmental factors affect the rates of birth, death, immigration and emigration. Some examples are shown in Table 9.1.1.

TABLE 9.1.1 Factors that affect population size in an area

Factor	Definition of factor	Example of how the environment changes the factor
birth rate	number of individuals born per thousand of the population	The number of animal births often increases when food is plentiful. Adults have more energy and breed. More breeding adults can survive because more food is available.
death rate	number of individuals who die per thousand of the population	When food is plentiful, more animals survive and less die, which keeps the population high. In contrast, exposure to unusually low temperatures can increase the death rate (Figure 9.1.18).
immigration	number of individuals moving into an area per thousand of the population in the area	More animals move into an area when there are more resources such as food and water.
emigration	number of individuals moving out an area per thousand of the population in the area	Animals usually move out of an area when they do not have enough food and water.

FIGURE 9.1.18 Emperor penguin adults and young huddle together to keep warm in an Antarctic blizzard. Very low temperatures increase the death rate, decreasing the size of the population.

MODULE

9.1 Review questions 

Remembering

1 Define the terms:
- a environment
- b interdependence
- c parasitism
- d commensalism.

2 What term best describes each of the following?
- a a non-living factor of the environment, also known as a physical factor
- b a relationship between organisms that are trying to use the same limited resource
- c a place where organisms and their physical surroundings form an environment that is different from others nearby
- d animals that can generate body heat internally.

3 What are the two main components of an ecosystem?

4 List five different abiotic (physical) factors that affect organisms.

Understanding

5 Describe three abiotic factors of soil that affect land plants.

6 Farmers find that adding fertilisers to loam soils increases crop yield, while adding them to sandy soils does little. Explain why.

7 Explain how seasonal changes in abiotic factors can alter the rate of birth, death, immigration and emigration.

8 Why does cold weather affect reptiles more than birds and mammals?

9 Explain how fire can be beneficial in the Australian bushland.

Applying

10 Use a eucalypt (gum tree) to show how an organism is affected by abiotic and biotic factors.

11 Identify the interaction between the penguins in Figure 9.1.18 on page 355.

Analysing

12 Compare:
- a a community with an ecosystem
- b abiotic and biotic factors
- c predator and prey
- d parasite and host.

13 a Discuss five biotic factors and five abiotic factors in your environment.
- b Compare these factors with those for a kangaroo.

14 Most of the land animals found in the Arctic and Antarctica are birds or mammals. Few are reptiles or frogs. Compare these animals and explain this observation.

15 Compare the effects of the following five abiotic factors on plants in beach sand dunes and plants in a rainforest: temperature, humidity, soil moisture, sunlight and wind.

Evaluating

16 Which type of animal would require more food—an endotherm or an ectotherm? Justify your answer.

17 a For each of the following pairs of animals, what do you think the type of interaction is between the two animals?
- i hawk and budgerigar
- ii tick and bobtail lizard
- iii human and pet budgerigar
- iv tinea and human
- v dingo and wedge-tailed eagle
- vi sheep and bacteria in its gut
- vii soldier ant and worker ant in a colony.

- b Justify each answer.

18 a What do you think are the relationships between the following species: grass, rabbits, foxes and wedge-tailed eagles?
- b Explain how these four species are interdependent.

19 Why are mosses more likely to grow on the southern side of trees in Australia, whereas they usually grow on the northern side of trees in Europe?

MODULE 9.1

Review questions

20 Lots of rock lobsters (crayfish) are able to be caught on nights that are overcast or when there is no moon. If the moon is full and the sky is clear, then very few lobsters are caught. What do you think is a likely reason for this difference?

21 Plants that are being attacked by aphids release special chemicals into the air. These chemicals can be detected by ladybirds that prey on the aphids. The ladybirds then fly to the plant and kill the aphids. However, some ants drink juice given off by aphids and defend the aphids by attacking the ladybirds.

a Use the above information to classify the relationship between the following pairs of organisms:

i aphid and plant

ii aphid and ladybird

iii aphid and ant

iv ant and ladybird

v plant and ladybird.

b Justify each decision.

Creating

22 Imagine the environmental conditions of Earth changed to those listed below. Design an animal predator that would have adaptations that would allow it to survive in this environment.

Assume your predator is a mammal or reptile and that it preys on land animals. Draw your predator and label ten adaptations that would help the animal to survive.

- The atmosphere has only 15% oxygen.
- The temperature increases by 10°C.
- The evaporation rate doubles.
- There are very few animals around to eat.
- Their prey can run twice as fast.

MODULE

9.1 Practical investigations

1 • Fish shapes

Purpose

To discover how the shape of plasticine 'fish' affects their movement through water.

Hypothesis

You are going to make some plasticine models of fish and see how fast they move through the water. What shape do you think will be best for fish to move faster through the water? Before you go any further with this investigation, write a hypothesis in your workbook.

Timing

45 minutes

SAFETY

It is best to conduct this experiment outside.

Materials

- 5 sticks plasticine
- test container shown in Figure 9.1.19
- bucket (at least 5 L capacity)
- timer
- knife

Procedure

1. Cut five cylinders of plasticine about 2 cm wide by 2 cm long.
2. Use the plasticine to make different fish shapes (with a tail). You can make shapes such as an oval, a sphere, a cube and a pyramid. Draw the shapes that you made.
3. Take the materials outside. Fill the test container with water. Drop one fish shape into the top of the container and time how long it takes each to reach the bottom. Record your result.
4. Drop each of the other fish shapes and time them. Record your results.
5. Carefully tip the water out of the top end of the test container into the bucket and collect the fish.
6. Refill the test container and repeat the measurements in steps 4 and 5.

Results

1. Construct a table of your results showing fish shape and time taken.
2. Record all the class results on butcher's paper or on the whiteboard.

Review

1. What fish shape was the fastest?
2. What fish shape was slowest?
3. a. Construct a conclusion for your investigation.
   b. Assess whether or not your hypothesis was supported.
4. Explain your results.
5. Use the results of this experiment to propose a reason why many fish have a thin, flattened body that is 'streamlined'.

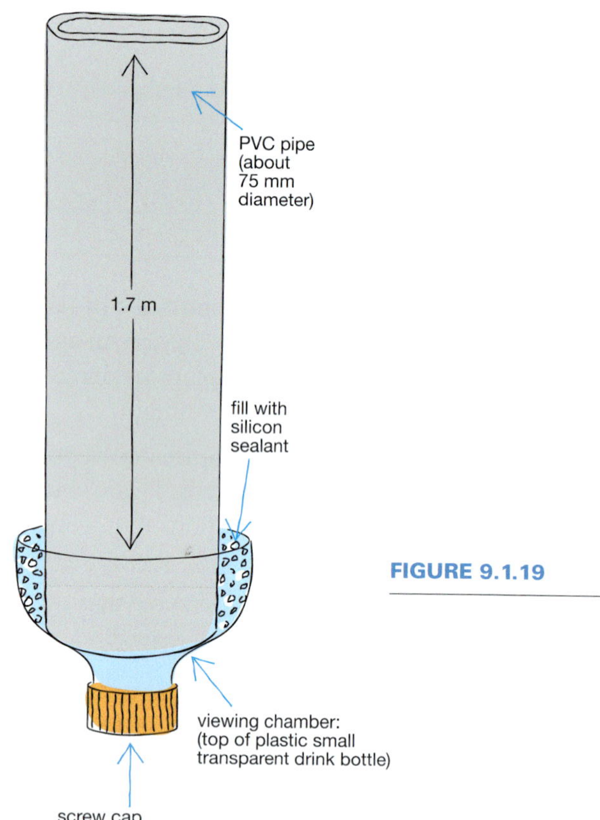

FIGURE 9.1.19

MODULE
9.1 Practical investigations

• STUDENT DESIGN •

2 • Plants and fire

Some Australian native plants germinate after a fire because of a chemical in the smoke and not because of the heat from the fire. The chemical is released into the air in smoke and can also be found in ash produced after a fire.

Purpose

To test whether ash makes seeds germinate rather than the heat of the fire.

Timing 60 minutes

Materials

- as selected by students (probably wattle seeds, featherflower (*Verticordia*) or fringe myrtle (*Calytrix*) seeds, containers, matches, newspaper or leaves to burn, can, beakers, water)

SAFETY

A risk assessment is required for this investigation.

Fires must only be lit outside the classroom in an appropriate container. Teachers must supervise at all times.

Procedure

1. Design an experiment that will test whether it is the ash or the heat from a fire that causes seeds to germinate.
2. Brainstorm in your group and come up with several different ways to investigate the problem. Select the best procedure and write it in your workbook. Draw a diagram of the equipment you need.
3. Before you start any practical work, assess all risks associated with your procedure. Construct a risk assessment that outlines these risks and any precautions you need to take to minimise them. Show your teacher your procedure and your risk assessment. If they approve, then collect all the required materials and start work.

   See Activity Book Toolkit to assist with developing a risk assessment.

Hints

- Your team will need to decide aspects of your design such as:
  - how you will test the effect of heat on the seeds (putting them directly in a fire will kill them)
  - whether you will subject the seeds to a fire (which would need to be done in a tin can outside the classroom) or some other source of heat such as hot or boiling water
  - how you will capture chemicals from ash so that you can then use on the seeds (capturing smoke is a bit difficult in a school setting without special equipment).
- Use the STEM and SDI template in your eBook to help you plan and carry out your investigation.

Results

Construct a table showing your results and the results of other prac teams.

Review

1. Construct a report on the experiment.
2. How may your results help with repairing damaged ecosystems?
3. Evaluate your procedure. Pick two other prac groups and evaluate their procedures too, identifying their strengths and weaknesses.

MODULE 9.2 Sustainability

An untouched rainforest is a natural ecosystem. It is also sustainable. This means that the rainforest maintains itself and helps all the living things within it by providing everything they need to survive, grow and reproduce. The seas once were sustainable too but overfishing has severely disrupted essential food chains in them. Natural ecosystems are amazingly complex and scientists are still discovering how they function and sustain themselves.

science 4 fun

Ant food

YES

Do ants eat sugar?

Collect this ...

- 5 mL of sugar water (about 10% sugar)
- food colouring in dropper bottles (several different colours)
- waxed paper 10 cm square (taped to cardboard to resist wind)
- digital camera or mobile phone with camera (with zoom function)

SAFETY

Some ants give a painful sting. Do not touch them. Try not to step on a nest.

Do this ...

1. Add about 10 drops of the food colouring to the sugar water. Different groups can use different colours.
2. Go outside with your equipment and find an ant nest or ant trail.
3. Place the paper next to the nest or in the trail.
4. Add about 5 drops of different-coloured sugar water to the paper (you can swap colours with other groups). Space the drops about 2 cm apart.
5. Observe what the ants do for several minutes. Observe their abdomens carefully for evidence that they drank the liquid. Take a close-up digital photo if you can.
6. Look carefully at the photos you took. Zoom in to see their abdomens more clearly.

Record this ...

1. Describe what happened.
2. Explain why this happened.

What is sustainability?

Natural ecosystems are usually sustainable. **Sustainability** means that an ecosystem has the ability to maintain suitable living conditions for the community. For an ecosystem to sustain itself, it needs:

- a supply of substances necessary for survival and growth such as water and nitrogen
- an input of energy, usually from sunlight
- a wide range of species living in the ecosystem.

Food in ecosystems

In any ecosystem, food is vital. Food contains the matter and energy required by living organisms. To understand how ecosystems function, it is important to understand how the matter and energy in food enter and flow through the community.

Producers and consumers

Every community must have a source of food. This source is made by organisms known as **producers**. Most producers are green plants such as flowering plants, conifers, moss and algae. Some producers are bacteria. Producers are essential for the community. Without them there would be no life. This is because other organisms cannot make their own food whereas producers can.

Animals and fungi require a ready-made source of food. For this reason, animals and fungi are known as **consumers**. Humans are consumers because we eat plants directly as vegetables or fruit and we eat animals that depend on plants. For example, we eat sheep (lamb), cows (beef) and pigs (pork). All these animals eat grass or plant seeds. Humans also eat tuna, salmon, shark (flake) and prawns. These animals eat other smaller animals that depend on plants such as seaweed for their food. Even the food we manufacture depends on plants. For example, bread and pasta are made from wheat seeds. Regardless of its type, all our food originally comes from a producer.

Apart from some bacteria, all other producers make food by photosynthesis (Figure 9.2.1). This is a chemical process that takes place inside cells containing a green pigment called chlorophyll.

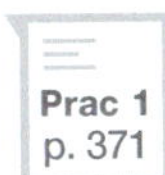
Prac 1
p. 371

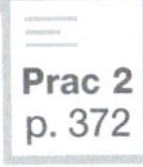
Prac 2
p. 372

FIGURE 9.2.1 Trees are producers. They make their food by photosynthesis.

Photosynthesis is a complicated process of many chemical reactions, but it can be summarised in a word equation or balanced formula equation:

$$\text{carbon dioxide + water} \xrightarrow[\text{chlorophyll}]{\text{sunlight}} \text{glucose + oxygen}$$

$$6CO_2 + 6H_2O \longrightarrow C_6H_{12}O_6 + 6O_2$$

For land plants, carbon dioxide (CO_2) is absorbed from the air, and water (H_2O) is absorbed through the roots. When sunlight falls on the cells containing chlorophyll, the cells start making a sugar called glucose ($C_6H_{12}O_6$).

The glucose made by photosynthesis is vital to the plant. It is used to make all the other materials that a plant needs, such as proteins, fats and vitamins. Often the sugar is turned into starch and stored in the leaves until it is needed. Many plants also store starch in seeds. This is why wheat is used to make bread and other foods.

Food chains and webs

A **food chain** is a diagram showing the sequence of organisms feeding on each other. The arrows in the diagram show the direction in which the food materials move from one organism to another. In this way, the matter and energy needed by organisms pass along food chains.

In this food chain shown in Figure 9.2.2, the food materials made by the grass (such as proteins, fats and vitamins) are passed to the grasshopper when it eats the grass. The grasshopper uses these food materials to build and maintain its own body. When the grasshopper is eaten by the frog, the food materials in the grasshopper's body are used by the frog to build and maintain its body. Similarly, when the frog is eaten by the snake, food materials pass to the snake. So the grasshopper, the frog and the snake all depend on the grass to make the food materials they need. Without grass, the snake has no food.

FIGURE 9.2.2 The arrows in a food chain show the direction in which matter and energy pass from each organism to the next as one consumes the other.

> **SciFile**
>
> **Living in rock**
>
> Scientists have recently found bacteria that live nearly 2 km underground in solid rock. These bacteria live in extremely hot water in cracks in the rocks. The bacteria are producers, making food using energy in chemicals rather than the energy of sunlight. These bacteria are known as chemosynthetic bacteria.

There are many food chains in communities and they are all interconnected. All the connected food chains are known as a **food web**. An example is shown in Figure 9.2.3. Food webs are usually complex, with each organism appearing in many food chains. For example, the mouse and snake are in four food chains.

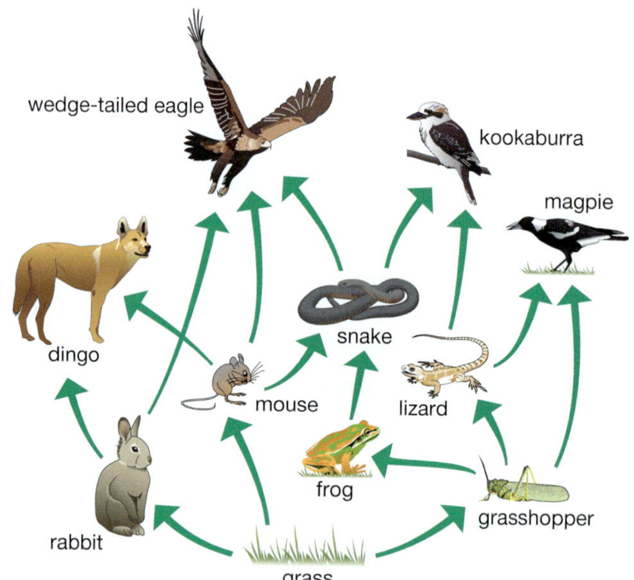

FIGURE 9.2.3 Part of a food web in a eucalypt woodland community

> **SciFile**
>
> **Big eater**
>
> The blue whale consumes tiny shrimp-like creatures called krill. It eats about 3500 kg of krill per day, as much as the mass of an Asian elephant or the equivalent mass of about fifty humans every day!

Recycling in nature

A vital group of organisms in any ecosystem is the **decomposers**. These organisms are bacteria and fungi like the ones in Figure 9.2.4. Decomposers break down dead bodies and wastes, and recycle matter for the producers to re-use.

You can see the process in Figure 9.2.5. Beginning with producers, the arrows show that all matter in food materials is eventually passed through decomposers and back to the producers. Without decomposers ecosystems would run out of resources.

FIGURE 9.2.4 Decomposers like these fungi are vital for the ecosystem to keep functioning.

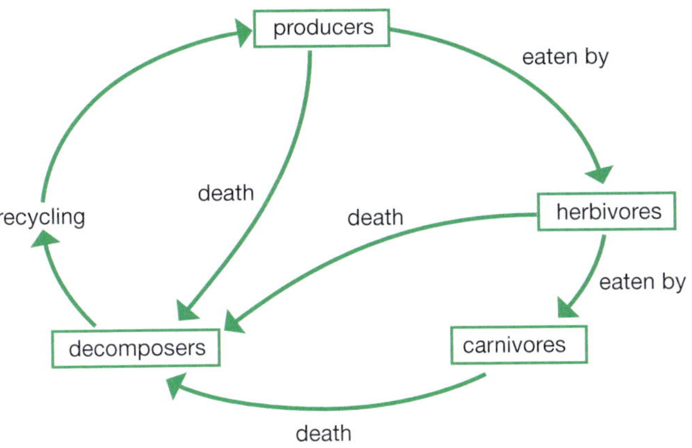

FIGURE 9.2.5 Decomposers recycle materials back to the ecosystem for re-use.

Energy flow in ecosystems

Energy does not cycle through ecosystems like matter does. Instead, energy flows continuously through the ecosystem, being lost as fast as it is gained.

To understand how energy flows through ecosystems, think about what you eat. You eat food for two reasons:

- to build new cells needed for growth and repair
- to provide energy for movement and processes such as generating heat to keep you warm.

In ecosystems, food flows from one organism to whatever organism eats it. This food carries energy so energy flows too from one organism to another. However, energy is lost from the community every time an organism eats another organism. For example, when a mouse eats grass seeds, it uses some of the energy in the seeds to build and repair its body cells. If the mouse is eaten by an eagle, then the energy contained in the body cells of the mouse flow to the eagle. The eagle will use that energy to build new cells, to move about and to keep warm. However, the mouse also used some of the energy it got from the seeds to move around and to keep itself warm. This energy is lost from the mouse and is not transferred to the eagle when the eagle eats it.

STEM 4 fun

Food chain cups

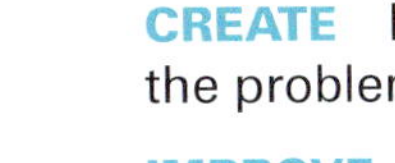

PROBLEM

Can you create a food chain display?

SUPPLIES

- polystyrene cups, Textas, pictures of organisms in a food chain, glue

PLAN AND DESIGN Your task is to design a way of demonstrating and displaying a food chain using a set of polystyrene cups.

Design the solution. What information do you need to solve the problem? Draw a diagram. Make a list of materials you will need and steps you will take.

CREATE Follow your plan. Produce a solution to the problem.

IMPROVE What works? What doesn't? How do you know it solves the problem? What could work better? Modify your design to make it better. Test it out.

REFLECTION

1 What field of science did you work in? Are there other fields where this activity applies?
2 In what career do these activities connect?
3 What did you do today that worked well? What didn't work well?

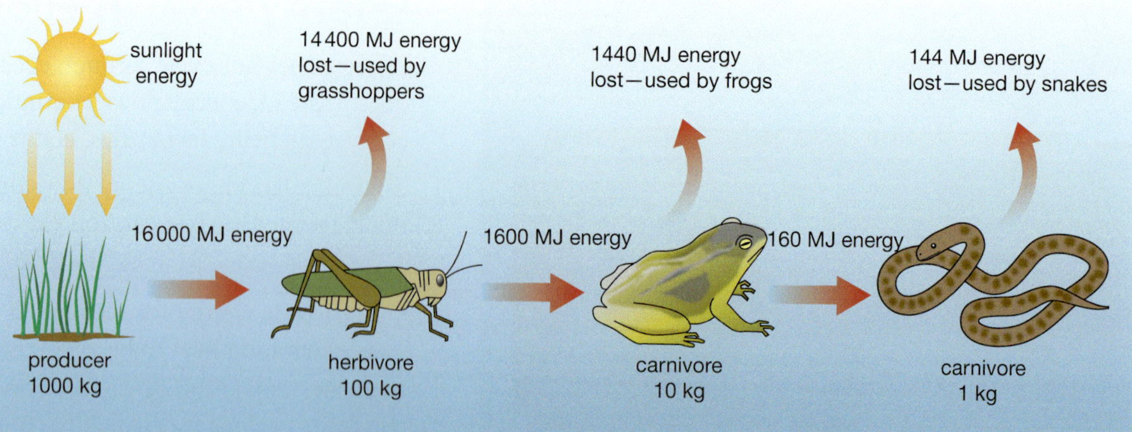

FIGURE 9.2.6 Energy flow through food chains results in energy losses. In this example, 90% of the energy is lost at each level.

Figure 9.2.6 demonstrates how energy is lost whenever one organism eats another in a food chain. The units of energy shown here are megajoules (MJ). A megajoule is 1000 kilojoules. Figure 9.2.6 shows the energy in 1000 kg of grass can only sustain about 100 kg of grasshoppers feeding on that grass. This is because each grasshopper uses up about 90% of its food for energy, rather than for making chemicals that form part of its body cells.

The 100 kg of grasshoppers can only feed about 10 kg of frogs. The frogs use 90% of its food materials to maintain themselves and, again, only about 10% of the energy ends up in the cells of the frogs. The same losses of energy occur when the frogs are eaten by the snakes.

So, the 16 000 MJ of energy originally in the grass ends up as only 16 MJ of energy in the snakes. This food chain demonstrates that there is a great loss of energy along it.

FIGURE 9.2.7 This Kookaburra will only gain a small proportion of the energy that was originally in the snake.

The large loss of energy along food chains explains why the chains are short—there is little energy available to the organisms at the end of the chain. It also explains why there are fewer higher-order consumers such as snakes than lower-order consumers such as insects—there is not enough energy available to keep high numbers of large organisms alive. Figure 9.2.7 shows that kookaburras eat snakes. Snakes have used up a lot of energy before the kookaburras eat them. That is why kookaburras need to eat a lot of snakes and this is why there are fewer kookaburras and more snakes in an ecosystem.

AB 9.5

SciFile

Humming along

Hummingbirds use energy faster than any other animal. Hummingbirds have to drink their weight in nectar every day. They are about as small as an endothermic animal can be. If they were any smaller, then they could not consume enough food to generate the heat to maintain their body temperature.

Pyramids of energy

The energy losses along a food chain can be represented in a diagram called a **pyramid of energy**. The pyramid in Figure 9.2.8 represents the energy in the food materials at each level in the food chain. The area of each rectangle represents the amount of energy.

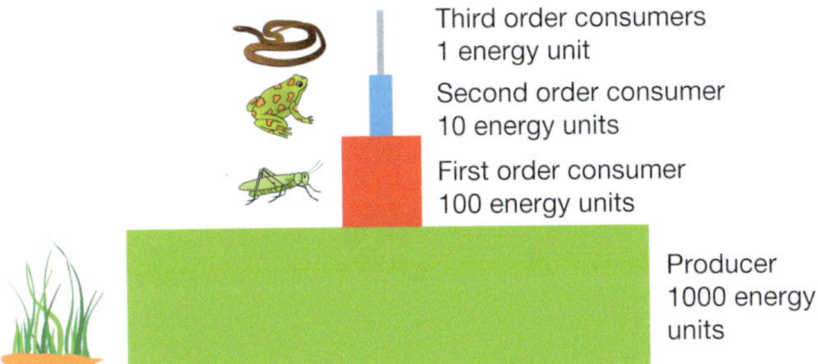

FIGURE 9.2.8 An energy pyramid showing the relative amount of energy available at each level of a single food chain.

Ecologists can prepare similar diagrams for whole ecosystems. These diagrams give an idea of the **productivity** of an area, meaning how well an area supports life. The diagrams are called productivity pyramids. A productivity pyramid shows how much energy is used per square metre by each feeding level in a year. An example is shown in Figure 9.2.9.

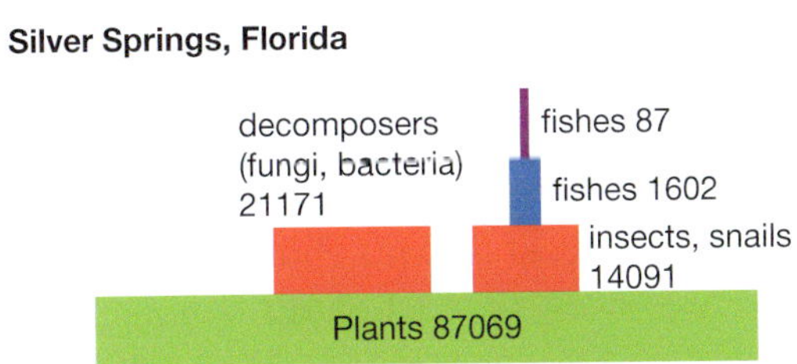

FIGURE 9.2.9 In a productivity pyramid, the area of each rectangle shows the amount of energy used per square metre per year in that particular feeding level.

Biodiversity

Biodiversity refers to the number of different species in an ecosystem and the population of each species. If an ecosystem is biodiverse then it has large numbers of many different species with a range of different characteristics. Figure 9.2.10 is an ecosystem that has high biodiversity.

FIGURE 9.2.10 Coral reefs are one of the most biodiverse ecosystems on Earth. A large number of different species are found in a small area.

A biodiverse ecosystem will be likely to continue over time. The populations of each species may change dramatically, but the ecosystem continues. The ecosystem is stable, will not collapse and most species survive and continue to be part of the food web. A biodiverse ecosystem is also less likely to be disrupted by environmental changes. For example, Paterson's curse is an introduced weed that competes with native plants for resources. It is the purple flowering plant in Figure 9.2.11. This weed can easily invade bushland containing only a few native plant species because it can out-compete the native plants.

FIGURE 9.2.11 When an introduced species such as this Paterson's curse invades an ecosystem, the ecosystem is more likely to resist the invasion if there are many different species in the ecosystem.

However, Paterson's curse has less chance of invading a biodiverse ecosystem because the weed is unlikely to be superior to all the different species present.

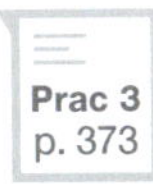

Prac 3
p. 373

Interactions in a food web also contribute to stability in the ecosystem. Feeding interactions tend to smooth out variations in numbers of a species over a long period of time. The numbers of a particular species might increase, but then return to near their previous levels. As an example, look at Figure 9.2.12. A change in the weather might lead to an unusually heavy flowering of eucalypts. This is likely to lead to an increase in the number of bees. This is because more food is available and so more bees can survive. If the number of bees increases, then there is more food for honeyeaters, Jacky winters and rainbow bee eaters like the one in Figure 9.2.13. As the bee numbers start dropping again, it becomes harder for the birds to find food. So the bird numbers decrease as the birds die or migrate away.

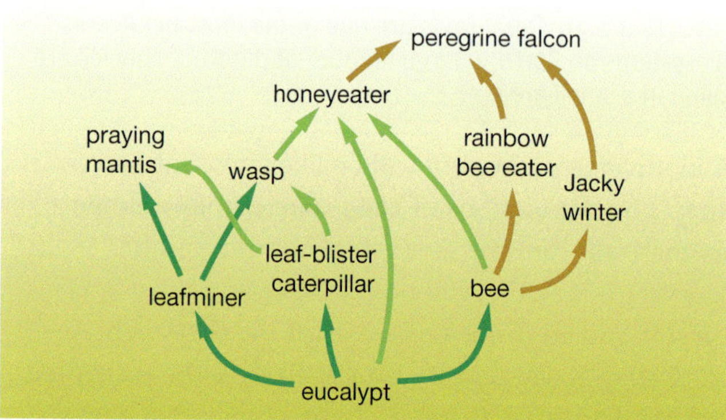

FIGURE 9.2.12 In a food web, each organism can appear in many food chains.

FIGURE 9.2.13 Rainbow bee eaters are predators that eat bees and so affect the population size of bees in an ecosystem.

So the changing numbers of a predator and its prey directly impact each other. You can see this in Figure 9.2.14. A rise in prey numbers is followed by a rise in predator numbers as there is more food available to support more predators. As the predator numbers increase, the prey numbers decrease because more prey is being eaten. The predator numbers then decrease (as they have little to no food), which triggers another rise in the prey.

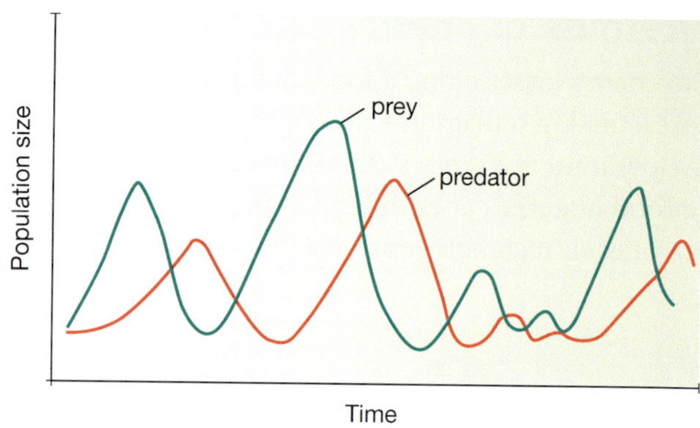

FIGURE 9.2.14 Predator and prey numbers affect each other.

If a food web only has a few species, then a change to one species could have a big impact on others. Any one species may be the only food source for others. If that species dies out, then there are no others available and the community dies out. In a biodiverse ecosystem, the loss of one food source is not a disaster because there are other food sources available.

Protecting ecosystems

AB 9.6

Although Australians now take it for granted that humans need to protect natural ecosystems, the idea is relatively new in Western civilisations. For this reason, the science of ecology has only been taken seriously over the last 50 years. Four reasons for protecting ecosystems are:

- Cultural value: some species have a value as part of the way of life of a country or region. For example, a kangaroo, emu and golden wattle are included in the Australian coat of arms shown in Figure 9.2.15. These organisms are part of our national view of ourselves and Australians would not allow these animals to be hunted to extinction or the plant killed off. Likewise, New Zealanders would be furious if the kiwi or the silver fern became extinct.

FIGURE 9.2.15 The Australian coat of arms shows some of the organisms we value as part of our culture.

- Economics: Some plants and animals have an economic or financial value. People can make money from ecosystems in many ways. For example, many tourists want to see untouched ecosystems and so these ecosystems provide income and jobs. Harvesting from the wild or farming species such as eucalyptus trees, prawns or kangaroos also creates wealth. New pharmaceutical drugs are still being discovered using plants that had not been previously tested. Some animals could be possible biological control agents for controlling new pests.
- Survival: We get our oxygen, water and our food from the environment around us. Pollution and overfishing reduces the food available to us and clearing trees for agriculture can result in eroded soils. As a result, less land is available so less crops can be grown. Air pollution makes the air unfit to breathe and produces acid rain that can kill crops.
- Compassion is humans feeling sympathy for other organisms (Figure 9.2.16). The idea of compassion is that all organisms have a right to live, and that humans have no right to exterminate any species. If humans are the only organism on Earth capable of understanding the possible fate of all life, then surely we have a responsibility to make wise decisions.

FIGURE 9.2.16 This whale became stranded on a beach and was rescued by compassionate humans.

Working with Science

ENVIRONMENTAL OFFICER

Environmental officers monitor and assess environmental quality and make recommendations to protect and preserve the environment. Their work often involves balancing urban development, such as land clearance for building roads and houses, with environmental and Indigenous cultural values. Their work can be very diverse and bring new challenges every day. An environmental health officer's job can involve pest plant and animal management, threatened species protection, site restoration, air, water, soil and noise pollution testing, cultural heritage management, and working with Indigenous communities for the best environmental outcomes.

A Diploma of Conservation and Land Management or a Bachelor of Science, majoring in environmental management, conservation or botany, will give you the skills needed to become an environmental officer. These qualifications can also lead to careers as an environmental scientist, environmental consultant or park ranger. Qualities that are important in these positions are excellent communication skills, the ability to work in a team, and a love of being outdoors, working to protect the environment.

FIGURE 9.2.17 An environmental officer monitors and assesses environmental quality to ensure that environmental and cultural values are protected and preserved.

Review

1 Have you ever noticed road works around your home or school? What are some of the environmental and cultural values of that land that you think are important to protect? (Hints: Are there any threatened species in that area? Were trees removed? Who are the Indigenous communities connected to that land and what are their values? Are there concerns about noise, air, water or soil pollution?)

2 There are many different people involved in major projects like road works and housing development. Can you think of three different groups of people that might be affected by urban development projects? Considering this, why do you think communication skills are an important part of an environmental officer's job?

SCIENCE AS A HUMAN ENDEAVOUR

Use and influence of science

Models and population changes

Ecologists study how environmental factors, such as fires and drought, may affect populations of animals such as kangaroos in natural ecosystems. They also want to know if hunting is having serious effects on the kangaroo populations.

FIGURE 9.2.18 An aircraft used for counting kangaroos. The red weights and cord attached to the wing struts are used in the count.

The model

To manage kangaroos in a sustainable way, scientists need to estimate how many there are in a particular area. Scientists also consider how much food and water are available and how weather affects the ecosystem. This information is used to construct a model that can help ecologists predict kangaroo numbers in the future when particular environmental factors occur. The model is a mathematical one using data such as: a sample of kangaroos from particular places, the temperature range that kangaroos can tolerate, the water availability, the quantity of food available and the known home range of particular kangaroos.

To obtain samples of population numbers in particular places, kangaroos can be counted from the air using aircraft and helicopters.

The survey

There is no way that all the kangaroos in a wide area could be directly counted. So an estimate needs to be made instead.

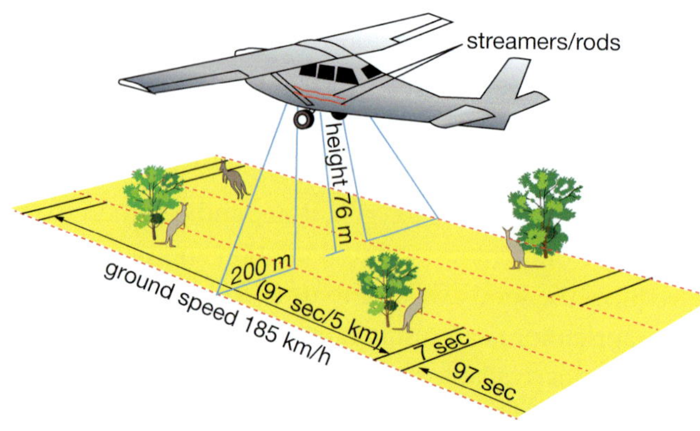

FIGURE 9.2.19 Using a strip transect to count kangaroos from fixed-wing aircraft.

This estimate is made using a 'strip transect', in which an aircraft flies in a straight line at 185 km/h, 76 m above the ground. Two observers are used, one viewing each side of the plane. The observers count the kangaroos they see that fall in a strip that is marked by rods or streamers attached to the wing. In Figure 9.2.18 you can see the red weights hanging from the cords on the wings. These cords trail behind the plane like streamers as it flies.

This strip will be 200 m wide on the ground, as long as the plane maintains its height (Figure 9.2.19). The observers count in sets of 97 seconds. This time must be exact, so that the area that has been counted is then equal to 1 square kilometre.

From the numbers counted in multiple strips, the density of kangaroos can be estimated for a much larger area. Here, density means the number in a particular area. An example would be 20 kangaroos/km^2.

REVIEW

1 Why would ecologists want to count the number of kangaroos in an area?

2 Assess whether knowing the weather patterns in a region would be necessary to predict kangaroo numbers over next few years.

3 Explain why the plane must fly at exactly the correct speed and height and the count must be done for exactly 97 seconds.

4 Calculate the population size if the density of kangaroos was 60 per square kilometre and the total area is 200 square kilometres.

MODULE 9.2

Review questions

Remembering

1 Define the terms:
  a sustainability
  b biodiversity
  c decomposer.

2 What term best describes each of the following?
  a a sequence of organisms feeding on each other
  b the organisms make their own food, and provide all the food for the rest of the ecosystem
  c how well an area supports life.

3 What is the word equation and balanced formula equation for photosynthesis?

4 List four reasons why humans should protect ecosystems.

Understanding

5 Predict what will happen when droplets of sugar water are placed near a trail of ants in the science4fun on page 360.

6 Why is photosynthesis so important to ecosystems?

7 Explain why decomposers are vital for a sustainable natural ecosystem.

8 a How does energy enter an ecosystem?
  b What happens to energy after it enters plants?

9 Why is biodiversity so important in ecosystems?

Applying

10 Use Figure 9.2.12 on page 366 to name:
  a three animals that compete for bees as a food source
  b two animals that compete for leafminers as a food source.

11 Identify the level of consumer you are when you eat:
  a beef
  b an apple.

12 Use a pyramid of energy to describe energy flow for any food chain with five links in it in Figure 9.2.12 on page 366.

13 Use Figure 9.2.6 on page 364 to calculate how much energy would be available to kookaburras that ate snakes.

Analysing

14 Consider the food web in Figure 9.2.12 on page 366. Analyse how the populations of the following species may be affected as seasonal changes cause eucalypts to stop flowering and slow down their growth.
  a bees
  b honeyeater
  c peregrine falcon
  d carnivorous wasps.

15 Contrast the flow of matter and energy through a natural ecosystem.

16 a If you drew a productivity pyramid for a desert, then what differences do you think it may show when compared with a productivity pyramid from a coral reef?
  b How different would productivity pyramids be for a woodland five years before a drought and the year after a drought?

17 If the amount of wattle in the food web in Figure 9.2.20 on page 370 increased, discuss what you think would happen to each of the following populations.
  a leafhoppers
  b wasps
  c honeyeaters
  d beetles.

MODULE
9.2 Review questions

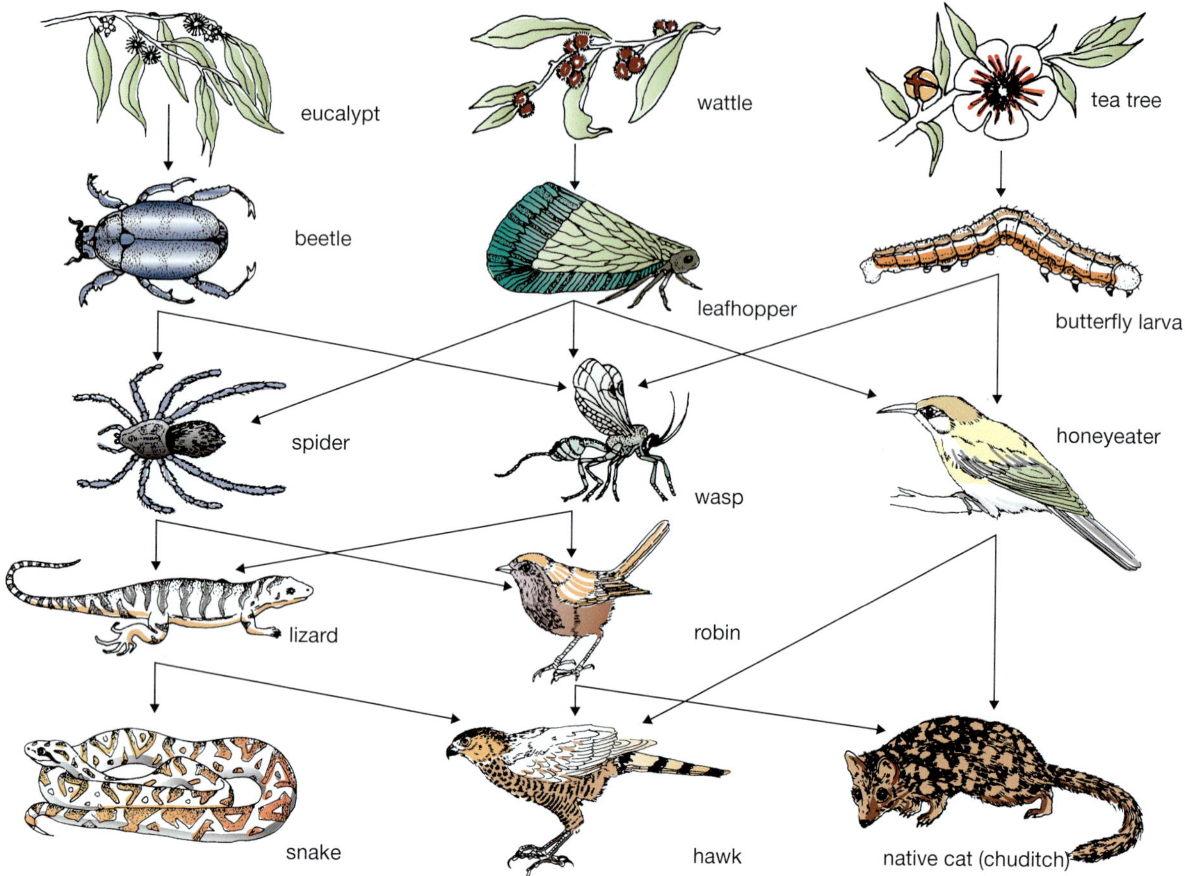

FIGURE 9.2.20

Evaluating

18 Consider the following suggestions from a pamphlet designed to help farmers protect and encourage native animals.

- Fence off from the stock any native vegetation remaining on the farm on land that cannot be used.
- Plant many different trees and shrubs native to the area.
- Leave old trees standing, especially those with hollows in them.
- Form corridors of vegetation between bushland areas on the farm and neighbouring farms.

Discuss how each of these suggestions may help protect and encourage native animals in farming areas.

19 **a** What do you think your chances are of finding a photosynthetic producer in a cave?

b What kind of producer might you find in caves?

Creating

20 Use your knowledge of energy flow in ecosystems to construct a scientific argument that it is more efficient to feed the world's population on plants than on animals.

21 Some people believe that humans should not try to protect and repair natural ecosystems, that humans do not need the natural world and do not 'owe anything' to the planet. Use the information you now have to write a letter or email to a newspaper arguing the opposite point of view.

MODULE
9.2 Practical investigations

1 • Making starch

Processing & Analysing | Communicating

Purpose

To test whether all plants make starch in the light.

Hypothesis

Do you think all plants will produce starch as an energy store if they are grown in the light?

Before you go any further with this investigation, write a hypothesis in your workbook.

Timing 60 minutes

Materials

- plants (e.g. geranium and wheat) that have been growing in the dark for two days
- plant (e.g. geranium and wheat) that have been growing in the light for two days
- iodine solution in dropper bottle
- 40 mL methylated spirits
- 5 mL starch solution
- labels for test-tubes
- 2 Petri dishes
- tripod
- Bunsen burner
- hotplate
- gauze mat/wire mat
- test-tube tongs
- bench mat
- tweezers
- 5 small (10 cm) test-tubes
- 100 mL beaker
- oven mitts

SAFETY

Bromothymol blue is toxic if swallowed. Wear safety glasses and avoid contact with your skin.

Methylated spirits is flammable so do not take the cotton wool plug out of the test-tubes near a Bunsen burner flame.

Procedure

1. Boil about 50 mL of water in a 100 mL beaker. Take one geranium leaf and one wheat leaf from the plants grown in the dark. Remove the Bunsen burner and place the leaves in the water for 2 minutes.
2. Using tweezers, take the leaves out and put them into separate test-tubes with 5 cm of methylated spirits. Plug the test-tubes with some cotton wool. Label both tubes 'Dark' at the top. Tip out the water from the beaker. (Careful: it is hot.)
3. Using leaves from the light this time, repeat steps 1 and 2. Label the test-tubes 'Light'.
4. Put the four test-tubes into about 70 mL of water in the beaker. If you have a hotplate, use that to heat the beaker for 5 minutes on moderate heat, trying not to boil the water in the beaker or dry out the test-tubes. The aim is to remove the green colour from the leaves.

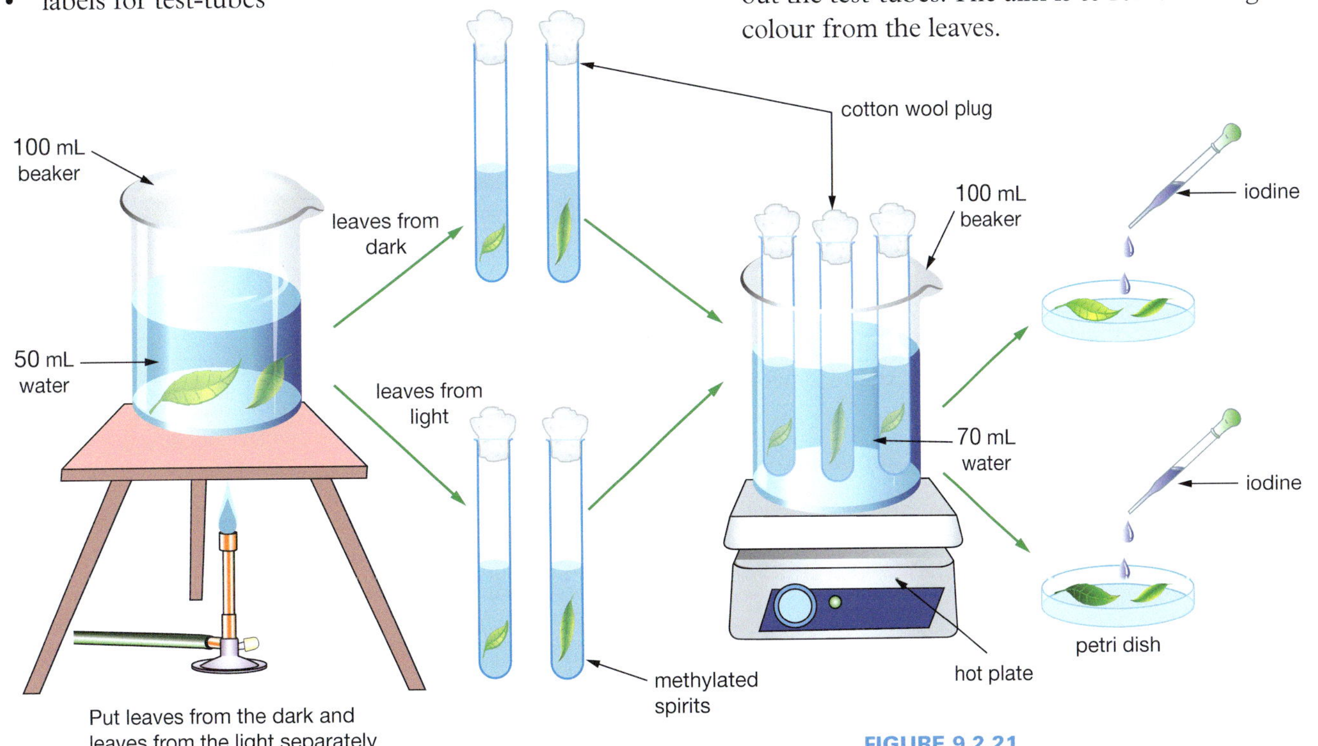

FIGURE 9.2.21

MODULE **9.2** # Practical investigations

If you do not have a hotplate, your teacher may allow you to use a Bunsen burner to heat the beaker. Be very careful that the fumes from the test-tubes and the cotton wool plug are not near the Bunsen flame.

5 Using oven mitts or test-tube tongs, take the test-tubes out of the beaker of water. (Careful: they are hot.) Leave the test-tubes in a rack for 5 minutes to let them cool down. Tip the two leaves from the dark into a petri dish. Tip the two leaves from the light into a different Petri dish. Add 10 drops of iodine solution onto the leaves in each Petri dish. Record your observations.

6 Put 2 cm of starch solution into your remaining clean test-tube. Add a few drops of iodine solution to the starch in the test-tube. This becomes your control. Record your observations.

Results

Record your observations in a table.

Review

1 How did you know if starch was present in a leaf?

2 Why was one leaf from the dark tested as well as one leaf from the light?

3 a Construct a conclusion for your investigation.

b Assess whether your hypothesis was supported or not.

4 Explain where the starch in the leaf came from.

5 What does this practical have to do with matter and energy in ecosystems?

2 • Photosynthesis

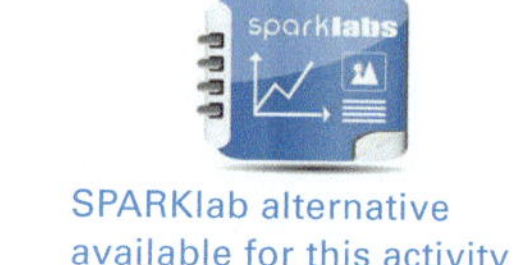

SPARKlab alternative available for this activity

Purpose

To determine whether carbon dioxide and light are necessary for photosynthesis.

Hypothesis

Do you think plants require carbon dioxide and light for photosynthesis?

Before you go any further with this investigation, write a hypothesis in your workbook.

SAFETY

Bromothymol blue is toxic if swallowed.

Wear safety glasses and avoid contact with skin.

Timing

30 minutes on day 1 then 15 minutes on day 2

Materials

- 150 mL of dilute bromothymol blue solution
- 2 small plants or shoots (for example, clover, soursob, geranium)
- 3 × 250 mL conical flasks and stoppers to fit
- aluminium foil
- straw
- lamp

Procedure

1 Your teacher will blow through the bromothymol blue solution until it turns yellow. Blowing adds carbon dioxide to the water.

2 Label the flasks 1–3. Add 50 mL of the yellow bromothymol blue solution to each conical flask.

3 Add a plant to flasks 1 and 3. Stopper all flasks.

4 Wrap flask 3 in aluminium foil so that light cannot enter. Your set-up should look like Figure 9.2.22.

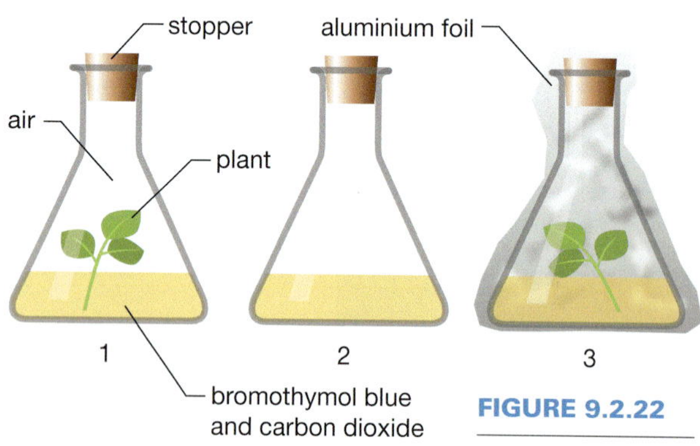

FIGURE 9.2.22

5 Place all three flasks in the light until the next day. If you can, put them in sunlight during the day and leave them under a lamp overnight.

6 Observe the flasks the next day.

Results

Record your results in a table.

MODULE 9.2 Practical investigations

Review

1 a What colour is bromothymol blue when mixed with carbon dioxide?

b What colour is bromothymol blue when there is no carbon dioxide present?

2 Identify the flasks that had no carbon dioxide left at the end of the experiment.

3 a Construct a conclusion for your investigation

b Assess whether your hypothesis was supported or not.

4 a Explain why you needed to compare flasks 1 and 2 to reach your conclusion.

b Explain why you needed to compare flasks 1 and 3 to reach your conclusion.

5 Using this equipment, design a way of testing the hypothesis that respiration produces carbon dioxide and does not require light.

6 Explain how this investigation relates to ecology.

3 • Studying a leaf litter environment

Purpose

To investigate the relationship between the community and the abiotic conditions in a leaf litter environment.

Timing 60 minutes

Materials

- 2 strips of blue cobalt chloride paper in a plastic bag, or a moisture meter
- 3 plastic bags
- 2 margarine containers
- thermometer
- binocular microscope
- Petri dish
- fine forceps
- mounted needle
- diagram sheet of soil organisms

SAFETY

Do not go near long grass or possible hiding places for snakes.

Do not pick up animals with your bare hands, as some may bite or sting. Use containers to pick them up.

Procedure

1 Take the bags, margarine container, thermometer, cobalt chloride paper and forceps to an area around the school that has a natural leaf litter cover.

2 Check the air temperature about 1 m above the soil and record it in your notes.

3 Carefully push the end of the thermometer into the leaf litter but not into the soil. Be very careful not to break the thermometer. Record the temperature in your notes. Replace the thermometer in its case.

4 Remove a piece of cobalt chloride paper with forceps and hold it in the air for 1 minute. Record the colour in your notes. (Alternatively, use the moisture meter.)

5 Place the other piece of cobalt chloride paper deep in the litter layer for 1 minute and then record its colour in your notes.

6 Quietly observe the surface of the leaf litter for a minute or so. Look for any small invertebrate animals moving near the surface. If you can catch them, put them into one of the bags. Do not try to collect vertebrates like lizards or frogs. Try not to kill any of them.

7 Scrape off some of the fresh litter from the surface and place it in the bag with the animals already collected. Try not to dig down into the lower layers of decaying leaves. Now scrape up the darker decaying litter layer into another bag. Finally, scrape up about half a cup of soil and place it into the third bag.

8 Return to the classroom and use the microscope and books to identify any animals in the three layers. Some leaf litter animals are shown in Figure 9.2.23 on page 374.

MODULE

9.2 Practical investigations

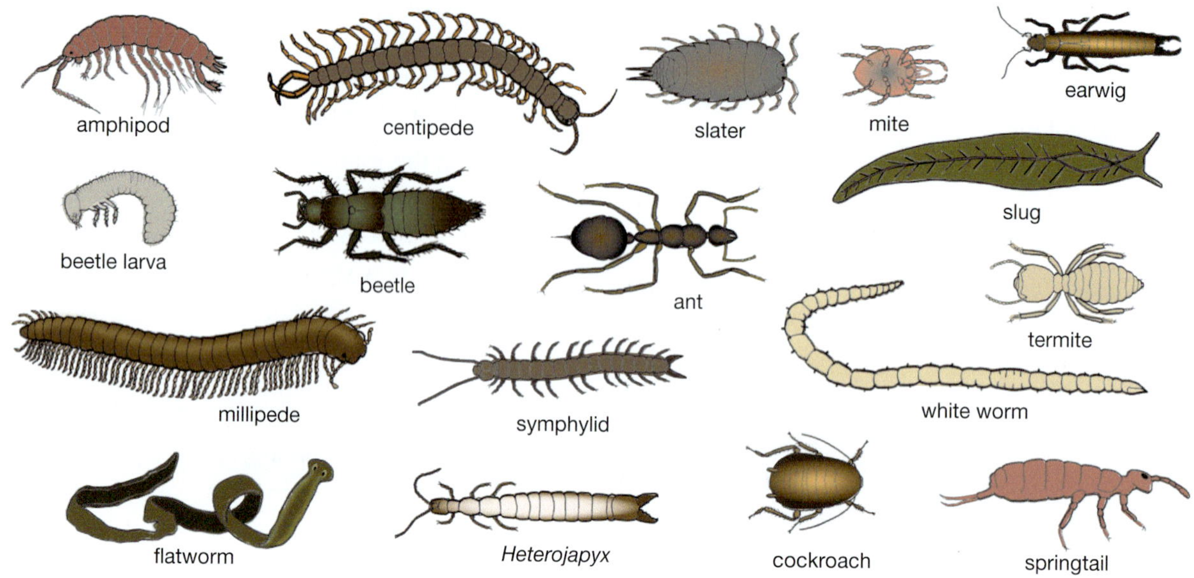

FIGURE 9.2.23

Review

1 Describe the main abiotic conditions of temperature, light and water in the leaf litter layer.

2 Fresh leaves fall each year, but the litter layer tends to stay the same thickness. Explain this observation.

3 The breakdown of the leaf litter begins with fungi, bacteria and certain animals. You probably found some animals from each of the following two groups. Choose one animal in each group and explain why it lives in the litter.

- Group A: springtails, millipedes, symphylids, bristle tails, earwigs, slaters, white worms, amphipods, mites
- Group B: spiders, centipedes (Figure 9.2.24), flatworms, pseudoscorpions, fly larvae, beetle larvae

4 Is biodiversity in the leaf litter important in the survival of this ecosystem?

5 How do you think this activity is relevant to our lives?

FIGURE 9.2.24 Centipede

MODULE
9.3 Natural and human impacts

Natural events such as fire, flood and drought can damage ecosystems, but there can be benefits for ecosystems too. Humans also affect some ecosystems in positive and negative ways. However, humans are a major threat to many ecosystems on Earth.

science 4 fun

Plants and fire

YES

Would some plants survive a fire better than others?

Collect this ...

- hand lens or magnifying glass
- digital camera or mobile phone with a camera
- large plastic bag

Do this ...

1. Find an area at home, in a local park or in part of the school grounds where there are many different plants, ranging from trees to low shrubs.
2. Observe the features of at least five different plants that you think could be affected differently by a fire. Look at features such as leaf position and structure, bark thickness and structure, branch height, seed cases, dead leaves and branches, and any other features that look relevant.
3. Take a photo of each plant and try to find out its common or scientific name. If you can't name it, at least record the type of plant. If possible, collect some parts (such as the bark) of the plant that you think could affect its fire resistance.

Record this ...

1. Describe the features of each plant, in a table.
2. Explain why you think the features would be helpful or not helpful to the plant in surviving fire.

Change due to natural events

Natural events affect ecosystems. For example, seasonal changes affect the organisms of a community because they are influenced by abiotic factors such as changes in rainfall, temperature and the amount of available light. Other natural events, such as bushfires, drought and flooding, also have huge impact on ecosystems.

Bushfires

Bushfires are a common event in Australia and they burn though natural vegetation, plantations and towns. Some are lit accidentally by humans while others are lit on purpose by arsonists. Other bushfires are caused by lightning strikes. Evidence suggests that natural fires have been affecting ecosystems in Australia for over 40 million years, while humans have probably only lived here for around 40 000 years.

Fire has a major impact on ecosystems in Australia because it promotes the germination of many plant species. After a fire, much of the bush regenerates through germination. Many plants are adapted to survive fires and even benefit from fire. Plants such as banksias and hakeas need fire to allow their seed cases to open and release the seeds (Figure 9.3.1). Grass trees and some orchids flower after a fire.

FIGURE 9.3.1 A bushfire has caused the seed cases of this banksia to open up and drop their seeds. Other plant species will do the same.

Eucalypts have oils in their leaves that catch fire easily (Figure 9.3.2). Some also have 'stringy' bark that hangs down to the ground, as in Figure 9.3.3. There is a lot of plant litter that falls to the ground, such as dead leaves and bark. These may seem like strange features for a plant to have in an area where bushfires occur. However, helping a fire to spread probably provides an advantage to eucalypts. After a fire, eucalypts can quickly regenerate, whereas other plants may not. This gives eucalypts a competitive advantage in fire-prone areas.

FIGURE 9.3.2 Eucalyptus leaves catch fire easily because they have glands that make oils. All the light-coloured dots on this leaf are oil glands.

FIGURE 9.3.3 The bark on this Blue Mountains Ash is stringy and allows fire to rise up the trunk and into the tree.

Eucalypts are adapted to survive fires in several ways. Thick bark insulates the living cells beneath it against the heat of the fire and so keeps the growing part of the trunk and branches alive. After a fire, new growth sprouts from the trunk and branches. All the leaves may have died, so new ones must grow to allow the plant to begin photosynthesising again. This growth is known as **epicormic growth** (Figure 9.3.4). The new growth of shoots allows the plants to quickly produce food and gain an advantage over other plant species that were killed and rely on seeds to regenerate.

FIGURE 9.3.4 After a bushfire, burnt eucalypts sprout new leaves from their trunk—they undergo epicormic growth.

Some eucalypt species called **mallees** have **lignotubers**. Ligontubers are swollen stems under the ground. The branches above the ground may have died in the fire, but those below ground are insulated from the heat and quickly sprout after a fire. A mallee is shown in Figure 9.3.5.

FIGURE 9.3.5 The multiple trunks of this Mallee eucalypt have regenerated from an underground stem called a lignotuber.

The effects of fire on animals vary. Slow-moving animals that cannot escape fires by burrowing or migrating may be killed. The dead are replaced by surrounding populations expanding back into the burnt-out area after the fire. Fire promotes the growth of fresh shoots that are highly nutritious for herbivores, which therefore benefit from the fire.

Drought

Australia suffers regular droughts, often lasting five years or more. A drought is a period of no rain or low rainfall. Drought can change ecosystems by increasing the death rate of plants and animals. For example, river red gums require regular water and so will die as the rivers they grow along dry up. Death of these and other plants removes a resource for animals that use the plants for food, shelter and nesting sites. The animals then either migrate out of the area or die (Figure 9.3.6). In this way, entire ecosystems can be destroyed. Loss of plant cover also leads to soil erosion by wind. When the rains do return, water erosion can further damage the land by washing away the soil.

FIGURE 9.3.6 Drought is a time when many animals such as this kangaroo die from lack of food and water.

SciFile

Non-drinker

The spinifex hopping mouse of the deserts of Australia only eats dry seeds and does not drink any water. Instead, it survives on the water it produces by respiration within its own cells. The mouse also has the most efficient kidneys in the world, producing extremely concentrated urine with very little water in it.

Flooding

Heavy rains can result in **floods** where rivers overflow their banks to cover normally dry areas of land. In north-eastern Australia in 2011, there were very heavy rains that led to severe flooding, especially in Queensland. Brisbane and several other cities were flooded.

The flooding also greatly affected natural ecosystems and there were high death rates among native plants and animals. Animals drown from lack of oxygen to breathe and plants drown from lack of oxygen at their roots.

In drier places, the increased rainfall leads to an 'explosion' of life. Animals and plants in such places are adapted to reproduce rapidly to take advantage of the extra food and favourable living conditions. Many small aquatic animals hatch from tiny eggs that were buried in the soil. Some microscopic eggs of small water animals arrive carried by the wind and migrating animals, such as pelicans that fly in to nest in the area. The pelicans arrive because they can feed their young on the abundant animal life that flourishes during the flooding. One place where this occurs is Lake Eyre in South Australia.

Every ten to twenty years, heavy rain in Queensland feeds rivers like the one in Figure 9.3.7 that flow into Lake Eyre. The lake floods and forms a new ecosystem that supports large populations of fish and birds such as the pelicans seen in Figure 9.3.8.

FIGURE 9.3.7 The Warburton River flows into Lake Eyre but rarely carries enough water into the lake to flood it.

FIGURE 9.3.8 Exceptional rains in eastern Australia in 2011 resulted in flooding of Lake Eyre and an explosion of life in the lake.

Change due to human impact

Humans can affect ecosystems in a negative way. Actions such as clearing land and destroying places where species live are negative impacts in a natural ecosystem. So too are disposing of toxic wastes into natural ecosystems. However, humans also try to minimise the negative impacts on ecosystems and make positive contributions to the health of an ecosystem.

Habitat destruction

Habitat destruction is damage done to the environmental factors an organism depends on for survival. Some examples are land clearing, mining, and logging. These activities don't necessarily destroy habitats permanently. Ecosystems can be repaired if there is enough knowledge of the environment and resources available.

The numbers of some animals have declined because of agricultural development. Examples are the rat kangaroo, the potoroo, the Regent honeyeater, woylie and the bilby (Figure 9.3.9). However, some animal populations have increased because of the increased food supply. These animals include kangaroos, locusts and emus.

FIGURE 9.3.9 Bilby populations have declined, partly due to habitat destruction. The bilby is now listed as endangered in Queensland and vulnerable in the rest of Australia.

Introduced species

AB 9.7

Animals and plants brought to Australia from other countries are known as **introduced species**. Introduced animals that have become established in the wild are referred to as **feral animals**. Feral cats and foxes kill native animals and have had a devastating effect on wildlife (Figure 9.3.10). Feral rabbits, cats, camels, donkeys, horses, goats, mice, rats and pigs destroy habitat and compete with native animals for food and shelter. As a result, fewer native animals survive.

FIGURE 9.3.10 Feral cats and foxes kill native wildlife.

Many introduced plants have become weeds and now successfully compete with native plants for the limited resources available. Examples are watsonia, blackberry, veldt grass and lantana shown in Figure 9.3.11. These plants now cover large areas of native bush.

FIGURE 9.3.11 Lantana has become a problem weed in much of northern Australia.

Chemical pesticides

Insecticides are chemicals that kill insects. Some of the first insecticides that were manufactured turned out to be dangerous to other animals as well as the insects. DDT is one example of this. The DDT entered the food chains because it was carried by wind and water onto the land and into lakes and oceans. The chemical was absorbed by plants and animals that lived in the water. For example plant plankton absorbed it with the water that they absorbed.

DDT drifts around in the wind and is carried by waterways to lakes and oceans. The poison passes into cells and accumulated there. Research in the 1970s found that DDT was present in all the organisms in a food chain, like the one shown in Figure 9.3.12. The amount of DDT increases along the food chain. The reason more DDT was found in the clam than in the plankton was that the clam consumed a great deal of plankton. The DDT in the plankton was nearly all stored in the clam's body, and little was lost during consumption. The gull ate many clams and stored most of the DDT in its body. So the gull contained a high concentration of DDT. Poisons that build up in organisms in this way are called cumulative poisons, because they accumulate (build up over time).

The accumulated DDT had serious effects on some types of birds, especially those at the end of the food chain (such as peregrine falcons and eagles). DDT caused these birds to lay eggs with thin shells, which broke when the adult sat down to incubate them.

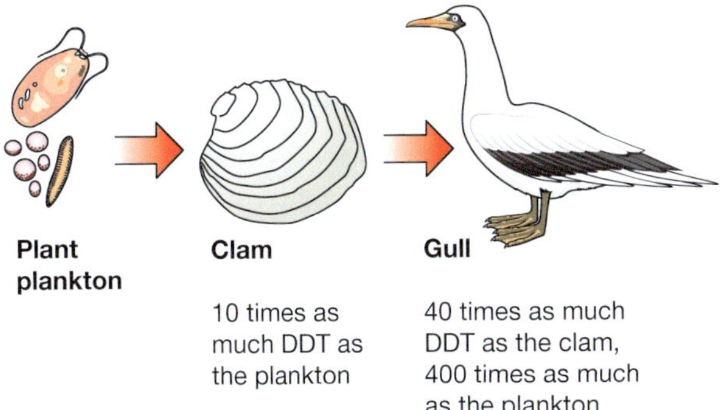

FIGURE 9.3.12 Food chain showing pesticide accumulation

So fewer young were produced, and the population decreased as adults died and were not replaced by young.

DDT also accumulates in humans, especially in breast milk. Research has shown that it is linked to many diseases such as cancer, diabetes, hormonal disruption and reproductive problems. DDT use was banned in Australia in 1987. However, DDT is still used in many poorer countries, particularly those that have a high incidence of the mosquito-borne disease malaria. These countries are more concerned about killing mosquitoes than about the long-term effects of the insecticide DDT. In these countries, DDT levels in breast milk are extremely high.

Not all pesticides accumulate in organisms as DDT does. However, they can still cause problems if they are not used carefully. When the spray drifts away from crops or if the farmer sprays at the wrong time, useful insects can also be killed.

In this way, insecticides can kill predatory wasps that eat insect pests and bees that pollinate flowers (Figure 9.3.13).

FIGURE 9.3.13 Pollination of flowers can be affected when insecticides kill bees.

A more effective alternative to insecticides is biological control. **Biological control** uses natural enemies against the pest species. An example is breeding and releasing wasps to control aphids. Biological control is being used where possible because it does not damage ecosystems as much as chemicals do.

AB 9.8 AB 9.9

Chemical pollution

Chemical pollution is chemicals escaping into the environment that can damage ecosystems. Plastics are becoming a significant problem in oceans and lakes. Many cosmetics now have tiny microbeads of plastic in them to help clean the skin (Figure 9.3.14). Synthetic clothes such as nylon break down in washing machines and become tiny fragments called microfibres. These microbeads and microfibres have recently been shown to be able to enter bodies of lower order consumers such as mussels, clams and shrimps.

FIGURE 9.3.14 This microscopic copepod is eating microbeads.

These plastic fragments absorb heavy metals, such as lead, and chemicals, such as PCBs (polychlorinated biphenyls), which are poisonous to organisms. The plastics and toxic chemicals build up in food chains. The problem is so serious that several countries are banning the sale of products containing microbeads. However, loss of plastic fibres from washing machines is not yet being dealt with. A simple filter on the outlet of washing machines could solve the problem.

Larger lumps of plastic are also dangerous to wildlife. Seabirds that swallow pieces of plastic in the ocean often die. Many other animals such as whales, seals and turtles become entangled in plastic fishing lines, ropes and straps, and drown.

Carbon dioxide absorbs some of the heat radiated from the Earth, warming the atmosphere. Most scientists agree that carbon dioxide is polluting Earth's atmosphere and causing it to get hotter in a process called **global warming**. Increased production of carbon dioxide gas by human activities is thought to be contributing to global warming, but natural factors are also involved.

The problem with raising Earth's temperature is that the polar ice caps may melt, causing sea levels to rise.

Many other chemicals pollute our environment. Oil spills from ships can cause devastation to sea ecosystems (Figure 9.3.15).

FIGURE 9.3.15 This penguin is covered in oil from an oil spill near Capetown South Africa.

Monitoring our environment to protect us against dangerous chemicals is an important job for health authorities and the environmental protection agencies of government.

Prac 1 p. 384 Prac 2 p. 385

Overcropping

Overcropping of animal populations means killing more animals than can be replaced by the normal breeding cycle. This results in a decrease in the population. Many of the world's whale populations have declined dramatically as a result of overcropping. The population of the blue whale in the 1980s was only about 5 per cent of the population earlier in the century. Blue whales are now totally protected, but populations are recovering very slowly.

One of the important roles for ecologists is to find out how many animals can be removed from a population without endangering the survival of the species. Regulations control how many animals may be taken. These regulations vary for different species.

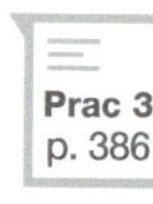

SCIENCE AS A HUMAN ENDEAVOUR — Use and influence of science

Aboriginal people, fire and ecosystems

Australia's Aboriginal people have been here for over 50 000 years. There is debate about how their use of fire to improve hunting affected the vegetation and fauna of the country. Research is continuing into the impact this may have had. Burning of the bush was viewed differently by the early Europeans and the Aboriginal people. The early Europeans seem to have been frightened of fires, whereas in some areas the Aboriginal people burned bushland regularly to improve hunting.

FIGURE 9.3.16 Fire is used to manage the landscape today, much like it was 50 000 years ago.

Today, widespread burning is conducted by government departments to protect houses in fire-prone areas. Burning is also conducted to keep the amount of plant litter low, in order to prevent larger fires that could damage the bush as well as properties. Research indicates that keeping plant litter levels low helps in the management of hot fires.

Research into the effect of fire on native plants such as banksias has revealed that even the earliest plant fossils dating back to 40 million years ago were adapted to fire. Researchers concluded that Aboriginal people could not have affected the characteristics of these plants. However, Aboriginal land management practices may have affected where the plants grew (Figure 9.3.16). Most researchers agree that Aboriginal people in many areas of the country used a hunting method called **firestick farming**. Fires were deliberately lit at certains times of the year, to create conditions suited to plant regrowth. The fresh growth attracted herbivorous animals to the area to feed, and made hunting easier (Figure 9.3.17). The fire also recycled nutrients back into the soil, which helped rejuvenate the land and made it more productive as a food source for humans. In some areas, small fires were lit to produce smoke, which frightened animals out of the undergowth and made them easier to catch.

FIGURE 9.3.17 A painting by Joseph Lycett showing Aboriginal people using fire to hunt in New South Wales in 1817

Some recent research using satellite imaging of land being managed by the Martu people in the Western Desert has demonstrated that using a regular pattern of small fires changes the plant species that grow in grassland areas. The research showed that the vegetation in areas that had regular small burns was different from areas where lightning strikes started the fires. The main effect was that areas subjected to firestick farming had greater diversity of species (Figure 9.3.18 on page 382).

SCIENCE AS A HUMAN ENDEAVOUR

These research results support views proposed by ecologists many decades ago. Early last century, ecologists discovered that fire affects the germination of some species but not others, and that species favoured by fire become more common in areas that are burned.

Some recent evidence contradicts the idea that the use of fire by Aboriginal people had a major impact on the vegetation in forest areas. A research study headed by Dr Scott Mooney from the University of New South Wales looked at 223 sites around Australia and surrounding countries. The study compared charcoal deposits from the past 70 000 years with climate records. The study concluded that the major factor affecting fires was variations in climate between hot and cold periods. Fire activity was high between 70 000 and 28 000 years ago, then decreased between 28 000 and 18 000 years ago. It increased again after 18 000 years ago, and even more after European settlers colonised the country. This evidence seemed to indicate that climate rather than human activity was the main factor in fire history over a long period of time.

Research is continuing, and the debate about burning to control species diversity and to reduce the intensity of fires will probably continue for some time.

REVIEW

1 What differences were there in the views of the early European settlers and the Aboriginal people regarding fire?

2 Government departments these days burn off the bush regularly. What is the purpose of these 'controlled' burns?

3 There is evidence that use of fire by the Aboriginal people of Australia did not change the characteristics of the plants. Describe this evidence.

4 Explain why firestick farming improves the food supply of Aboriginal people.

5 What affects did fires used by the Martu people of the Western Desert have on the spinifex grassland ecosystem?

6 There has been a long held view that Aboriginal people changed many of the forests of Australia into grasslands. What evidence is there that contradicts this view?

FIGURE 9.3.18 An Aboriginal man using firesticks to burn off grassland in Arnhem Land, Northern Territory

MODULE

9.3 Review questions

Remembering

1 Define the terms:
  a habitat destruction
  b overcropping
  c feral animals.

2 What term best describes the following?
  a animals and plants that have been brought from overseas to Australia
  b the increase in the temperature of Earth's atmosphere
  c a chemical that kills insects.

3 What are two characteristics of eucalypts that result in fires spreading?

4 List some ways in which drought and floods affect ecosystems.

5 What are three examples of habitat destruction?

6 Name five introduced species that have caused problems for the native animals and plants in Australia.

7 What is an example of damage to ecosystems caused by the following?
  a chemical pesticides
  b chemical pollution.

Understanding

8 Choose three adaptations that enable eucalypts to survive a fire. How do these adaptations improve the chances of survival?

9 Eucalypts living in places where there are frequent fires often have features that help the fire spread. How can this improve the chances of survival of those eucalypts in the area?

10 Explain how drought could lead to damage to the soil.

11 Why could overcropping result in extinction of a species?

12 Explain how the clearing of land for farming has affected natural communities.

13 Outline how introduced animals (such as rabbits, foxes, cats and wild pigs) affect native animals.

Applying

14 Use your knowledge of the effects of fire to explain how wattle trees may become more common in an area that is burned frequently.

15 Demonstrate that flooding can be an advantage in ecosystems.

16 Use your knowledge of food chains to explain how a cumulative poison such as DDT that is sprayed onto crops could kill predatory birds, such as eagles.

Analysing

17 Salvinia is a plant that floats on the water surface. Each plant is a few centimetres high and grows in clumps about 10 cm wide. It reproduces asexually, spreading rapidly and covering the whole surface of a lake. The weed's roots take oxygen out of the water. Any oxygen produced by photosynthesis in the leaves is released into the air, not the water.
  a Discuss whether this weed would cause problems for other plants in the water.
  b Why could this weed be a problem for consumers that live in the lake?

Evaluating

18 When cropping ('fishing') native species such as western rock lobsters, there are laws that control:
- what size the lobsters must be
- the breeding condition of the lobsters.

How do you think each of these laws may assist in the conservation of the rock lobster?

19 Large areas of the Amazon rainforest are being cut down. What effect do you think this could have on Earth's atmosphere?

20 Australia has laws that count a strip of the seas 12 nautical miles around the Australian continent as our territory. This area is called Australian Territorial Waters. Our navy intercepts any fishing vessels from overseas that come inside this limit. Why do you think they do that?

Creating

21 Many scientists believe that limiting human population growth is necessary to control environmental damage. Construct an argument for or against this statement.

MODULE

9.3 Practical investigations

1 • Wastes

Questioning & Predicting | Planning & Conducting

Yeast is a microscopic fungus found in water and on land. When it has food, it uses oxygen for respiration and to grow.

Purpose

To investigate the effect of wastes on water communities.

Hypothesis

What do you think will happen to the oxygen content of a yeast–water mixture if you give the yeast some milk? Before you go any further with this investigation, write a hypothesis in your workbook.

SAFETY

Methylene blue is toxic if swallowed and irritating to the eyes and skin. Wear safety glasses at all times.

Timing 60 minutes

Materials

- 5 mL of each of 25%, 50%, 75% and 100% milk solutions
- methylene blue indicator solution in a dropper bottle
- dry yeast

- teaspoon
- 20 mL measuring cylinder
- 5 small test-tubes
- test-tube rack
- 100 mL beaker
- marking pen
- timer with a split button
- stirring rod
- graph paper

Procedure

1 Mix half a teaspoon of yeast and 20 mL of warm water in the beaker. Leave this to stand.

2 Mark each test-tube with a line 1 cm from the bottom and a line 3 cm from the bottom. Number the test-tubes from 1 to 5.

3 Pour the 100% milk solution into test-tube 1 up to the 1 cm mark.

4 Repeat the process with the 75%, 50% and 25% milk solutions in test-tubes 2–4 respectively.

5 In test-tube 5 place tap water up to the 1 cm mark.

6 Add 8 drops of methylene blue to each test-tube and mix well. The tubes should appear blue because there is plenty of oxygen in the water. Methylene blue turns colourless when there is no oxygen.

7 Pour the yeast solution into test-tube 1 up to the 3 cm mark. Mix it well and start your timer (using the split function if you have one.) Every 30 seconds add yeast to a different tube (test-tubes 2–5) and mix well.

8 Observe each test-tube to record the time when the blue colour disappears. Remember that each of test-tubes 2–5 actually took 30 seconds less than the time you record for them on the stopwatch.

Results

Record your times in a suitable table. Draw a line graph of the time taken and the concentration (percentage) of the milk.

Review

1 Describe the relationship between milk concentration and time taken to lose the blue colour in the test-tube.

2 a Construct a conclusion for your investigation.
b Assess whether your hypothesis was supported or not.

3 Explain what could happen in a swamp ecosystem if human food wastes got into the water and the microorganisms reacted like the yeast in this investigation.

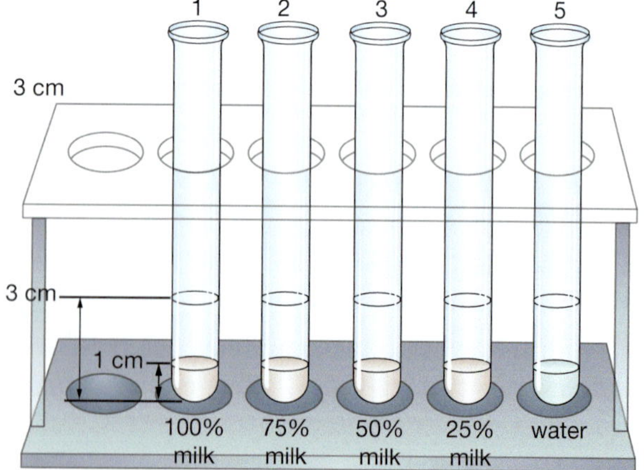

FIGURE 9.3.19

MODULE

9.3 Practical investigations

• STUDENT DESIGN •

2 • Detergents and plants

Planning & Conducting

Purpose

To investigate the effect of detergent on water plants.

Materials

- materials as selected by students

SAFETY

A risk assessment is required for this investigation.

Hypothesis

How do you think detergent entering rivers and lakes will affect aquatic plants living in them? Before you go any further with this investigation, write a hypothesis in your workbook.

Timing

30 minutes – Day 1
30 minutes – Day 2
20 minutes – Day 3

Procedure

1 Researchers suspect that water plants could be damaged by detergents entering our waterways. Design a way of testing whether this is true.

2 Brainstorm in your group and come up with several different ways to investigate the problem. Select the best procedure and write it in your workbook. Draw a diagram of the equipment you need.

3 Before you start any practical work, assess all risks associated with your procedure. Refer to the SDS of all chemicals used. Construct a risk assessment that outlines these risks and any precautions you need to take to minimise them. Show your teacher your procedure and your risk assessment. If they approve, then collect all the required materials and start work.

See Activity Book Toolkit to assist with developing a risk assessment.

Hints

- In your group, decide on aspects of your design, such as:
  - how you will decide that plant health is damaged
  - how you will apply the detegent and how much you will use
  - how many plants to use and how long the experiment will last.
- Use the STEM and SDI template in your eBook to help you plan and carry out your investigation.

Results

Collect all the results in the class.

Review

1 a Construct a conclusion for your investigation.
  b Assess whether your hypothesis was supported or not.

2 Evaluate your procedure. Pick two other prac groups and evaluate their procedures too, identifying their strengths and weaknesses.

MODULE 9.3

Practical investigations

• STUDENT DESIGN •

3 • Testing a germination booster

Questioning & Predicting | Planning & Conducting

Commercial products are available that claim to boost (improve) seed germination for some species. The packs contain granules treated with chemicals from smoke. The seeds that you want to germinate are mixed with the granules and then watered.

Purpose

To test a commercial 'germination booster'.

Hypothesis

Do you think a commercially produced 'germination booster' will be effective?

A risk assessment is required for this investigation.

Before you go any further with this investigation, write a hypothesis in your workbook.

Timing

60 minutes

Materials

- commercial germination booster (such as 'Australian Wildflower Seeds—Seed starter granules')
- materials as selected by students

Procedure

1 Design a way of testing whether the commercially produced 'germination booster' will be effective.

2 Brainstorm in your group and come up with several different ways to investigate the problem. Select the best procedure and write it in your workbook. Draw a diagram of the equipment you need.

3 Before you start any practical work, assess all risks associated with your procedure. Refer to the SDS of all chemicals used. Construct a risk assessment that outlines these risks and any precautions you need to take to minimise them. Show your teacher your procedure and risk assessment. If they approve, then collect all the required materials and start work.

See Activity Book Toolkit to assist with developing a risk assessment.

Hints

- In your group, decide on aspects of your design such as:
  - which species you will test
  - how many seeds you will test
  - how long the experiment will run for
  - how you will decide if the germination booster worked.
- Use the STEM and SDI template in your eBook to help you plan and carry out your investigation.

Results

Record your results in a table.

Review

1 a Construct a conclusion for your investigation.

b Assess whether your hypothesis was supported or not.

2 Evaluate your procedure. Pick two other prac groups and evaluate their procedures too, identifying their strengths and weaknesses.

3 How is this investigation relevant to understanding human impacts in ecosystems?

CHAPTER 9

Chapter review

Remembering

1 What are the components of an ecosystem?

2 List the major ways in which humans are damaging the environment.

3 a Name five feral animals in Australia.
 b What type of damage do they do to ecosystems?

Understanding

4 Why is biodiversity important to ecosystems?

5 Explain why all organisms in an ecosystem are interdependent.

6 Explain how seasonal changes in temperature, water availablity and sunlight can affect ecosystems.

7 Why are termites important animals in Australian ecosystems?

Applying

8 a Identify five biotic factors and five abiotic factors that affect you.
 b Explain how they affect you.

9 Use energy in ecosystems and pyramids of energy to explain why food chains are usually fairly short, with perhaps only four levels of consumer.

Analysing

10 a Contrast a community with an ecosystem.
 b Would you classify rocks, air and water as part of the community?

11 A habitat is the place where an organism lives. Is this just another word for the environment or is there a difference in meaning between these two terms?

12 A dingo and a wedge-tailed eagle both prey on rabbits. The dingo is infested with fleas. Classify the relationships between the dingo, eagle, rabbit and fleas as parasitism, predation or competition.

13 a Choose two biotic factors that would affect you and a plant in your garden. What similarities and differences are there in the effects of these biotic factors on you and the plant?
 b Choose two abiotic factors that would affect you and the same plant. What similarities and differences are there in the effects of these abiotic factors on you and the plant?

14 Figure 9.4.1 shows the changes in oxygen content in a lake. Assume temperature changes had minimal effect on the amount of oxygen dissolved. Analyse and explain the changes in oxygen content over the 24 hours.

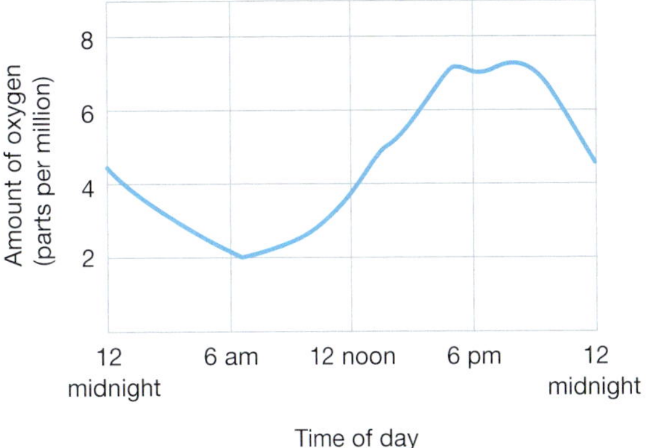

FIGURE 9.4.1

15 Discuss the following statement.

Maintaining a sustainable environment requires scientific understanding of interactions in and of the flow of matter and energy through ecosystems.

Evaluating

16 Imagine a disease came along that wiped out many termites. What effects do you think this may have on ecosystems?

17 Discuss the following statement.

Photosynthesis is the most important process in ecosystems.

18 a Determine whether you can or cannot answer the questions on page 347 at the start of this chapter.
 b Assess how well you understand the material presented in this chapter.

Creating

19 Use the following ten key terms to construct a visual summary of the information presented in this chapter.

biotic factors	adaptation
abiotic factors	environment
interdependence	biodiversity
conservation	human impacts
energy flow	sustainability

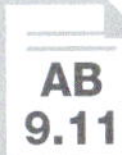

CHAPTER 9 Inquiry skills

Research

1 Processing & Analysing | Communicating

Research the use and humane treatment of animals in scientific research. Find:

- what animals are used in research
- what research is commonly performed
- how researchers try and ensure that the animals are not in pain or suffering
- what alternatives there are to these research methods.

Hold a class debate on why an investigation involving or affecting animals must be humane, justified and ethical.

2 Processing & Analysing | Communicating

Investigate how scientific models can be used to predict the changes in populations of native animals such as rabbits or kangaroos due to environmental changes like fire, flood or drought. Find:

- an example of such a model
- how useful the model is
- the limitations of the model.

Present your findings as a Word document with inserted diagrams or photos.

3 Processing & Analysing | Communicating

Find examples of how Aboriginal and Torres Strait Islanders interact with their environment. Find:

- how they use their understanding of the relationships between organisms, and between organisms and their physical environment
- how their cultural practices and knowledge are used to conserve and manage ecosystems sustainably
- how they thought differently to the early European settlers regarding the use of the environment.

Present your research in digital form.

4 Processing & Analysing | Communicating

Research how scientific and technological advances have been applied to solving the problem of pollution from industry. Find:

- specific examples of five industries each with a different pollution problem from a material such as heavy metals, plastics, poisonous gas, greenhouse gases, dust, smoke, salinity or hot water discharge.
- the ways in which each of these problems was solved by using science and technology.

Present your findings in digital form.

5 Processing & Analysing | Evaluating

People often have different views about human activities in ecosystems. For example, some 'loggers' (forestry workers) in Tasmania were asked by a journalist if they thought cutting down the trees was damaging the forest. The loggers' replied that they thought the 'balance was about right' between cutting down the forest and letting it grow.

Research different views on cutting down forests. In particular, present the views of:

a loggers who are employed to cut down trees
b people labelled 'greenies' by forestry workers
c employers who have businesses in towns that are dependent on logging
d tourism companies that are interested in increasing the tourist trade in natural forests
e tourists who may visit the forests
f biologists trying to protect endangered species
g yourself, based on your research.

Present your research in digital form.

Thinking scientifically 

Questions 1 and 2 are based on the following information.

Interactions between two species of clover (an animal feed) were studied. The two clover species were Yarloop and Tallarook. Seeds of each species were sown in separate trays to see how each species grew on its own. Both species were then grown together in the same seed trays. A few plants were removed every 10 days to measure their growth. The plants were dried out and weighed (dry weight). The average height of the plants when grown on their own was 10 cm for Yarloop and 5 cm for Tallarook. The results are shown in the graphs in Figure 9.4.2. Emergence means when the germinating plants first appeared above ground level.

Inquiry skills

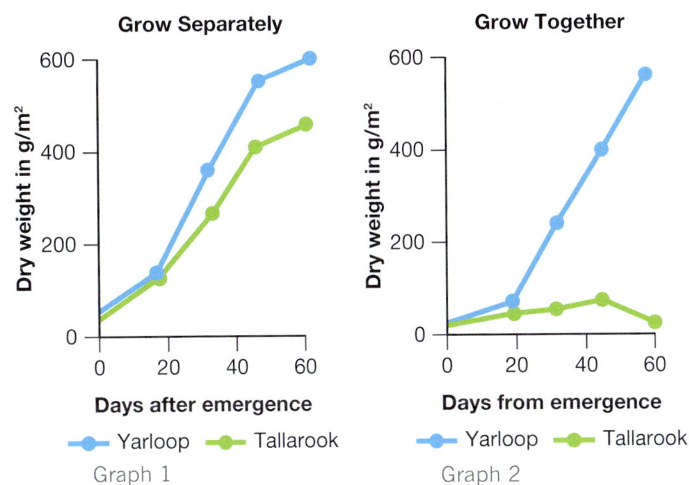

FIGURE 9.4.2

1 Considering graph 1 and the height data, what can you conclude about these two species of clover when they were grown separately (on their own)?
   - **A** Both species grew because the dry weight increased with time.
   - **B** Tallarook grew taller than Yarloop.
   - **C** The soil did not suit Tallarook as much as it did Yarloop.
   - **D** Yarloop naturally grows into a heavier, taller plant than Yarloop.

2 What is the most likely explanation for the results shown in graph 2, when both species were grown together?
   - **A** Tallarook did not grow very well because its dry weight hardly increased with time.
   - **B** Yarloop liked the soil much better than did Tallarook.
   - **C** Yarloop competes better than Tallarook because it captures more sunlight.
   - **D** Yarloop grew faster because it was in the shade.

3 Some students did an experiment to assess a new brand of potting mix and a new brand of fertiliser for growing vegetables. They decided to compare these with the 'old' brand of potting mix and 'old' brand of fertiliser they had been using. They set up their experiment as follows:

Set-up	Conditions in set-up
A	10 tomato seeds, seed tray, old potting mix, old fertiliser
B	10 tomato seeds, seed tray, new potting mix, new fertiliser

Both set-ups were kept in a greenhouse under identical conditions of water, light, temperature etc. The students measured the height of the tomato seedlings after 8 weeks.

What is the most serious fault in the design of this experiment?
   - **A** They should have measured the amount of tomatoes produced, not the height of the seedlings.
   - **B** They did not test the effect of each variable separately, they changed both variables in set-up B.
   - **C** They should have used a lot more seeds than 10 seeds in each set-up.
   - **D** They should have had more replicates of each set-up—at least three of each.

4 Energy and matter flow in a different way through an ecosystem. What is the major difference between them?
   - **A** Matter is lost from an ecosystem as it passes along a food chain because it cannot recycle back to producers.
   - **B** Energy continually cycles through an ecosystem passing between the physical surroundings and the community.
   - **C** Matter does not cycle from the living to non-living surroundings, whereas energy does.
   - **D** Energy is steadily lost from an ecosystem as it passes along food chains because it cannot recycle back to producers.

5 Which of the following is an abiotic environmental factor that affects the population size of a species?
   - **A** predators
   - **B** bushfires
   - **C** competitors
   - **D** introduced species.

6 A bushfire in an area could affect the population of a species that lives there. Which of the following is most likely to be an immediate affect of the bushfire on an animal population in the area?
   - **A** increase in death rate
   - **B** decrease in birth rate
   - **C** rise in immigration
   - **D** decreased in emigration.

CHAPTER 9

Glossary

abiotic factor: a non-living factor of the environment

adaptation: any feature that assists an organism to survive and reproduce in its environment

biodiversity: the range of different species in a community

biological control: the use of a natural enemy of a pest to control the numbers of that pest

biotic factor: a living factor of the environment

chemical pollution: chemicals escaping into the environment that can damage ecosystems

commensalism: relationship where one organism benefits and another organism is unharmed

community: all the living organisms in an ecosystem

competition: relationship between organisms that are trying to use the same limited resource

consumer: an organism in a food chain that feeds on other organisms

competition

decomposers: organisms that break down the bodies of dead organisms and animal wastes and recycle material

ecology: the study of how organisms interact with each other and with their non-living surroundings

ecosystem: a place where organisms and their physical surroundings form an environment that is different from others nearby

ectothermic: animals that obtain body heat from outside their body

endothermic: animals that can generate body heat internally

environment: all the factors in an organism's surroundings that affect its survival

epicormic growth: growth of new shoots from the stems of trees and shrubs after fires

feral animal: an introduced species of animal that has become established in the wild

firestick farming: fires were deliberately lit at certain times of the year, to create conditions suited to plant regrowth

flood: an event where water covers the land and does not soak in for a long time

food chain: a sequence of organisms feeding on each other

food chain

food web: connected food chains showing who eats whom in a community

global warming: the increase in the temperature of Earth's atmosphere

habitat destruction: damage caused to the factors in the environment needed for survival

insecticides: chemicals that kill insects

interdependence: relationship between organisms, where each affects the other's survival

introduced species

introduced species: species not native to Australia

lignotuber: a swollen underground stem of eucalypts that can resist fire

mallee: a type of eucalypt that has lignotubers

mutualism: relationship between two organisms living closely together, where each benefits

mutualism

overcropping: removing more individuals from a population than can be replaced by breeding

parasitism: relationship where one organism lives on or in another organism and feeds off it

photosynthesis: the process by which plants make their food

predation: relationship where one organism kills and eats another organism

producers: organisms that make food for the community

productivity: how well an area supports life

pyramid of energy: a diagram that shows the energy in the food materials at each level in the food chain

sustainability: an ecosystem has the ability to maintain suitable living conditions for the community

AB 9.10

CHAPTER

10

Plate tectonics

Have you ever wondered ...

- why New Zealand and Japan have lots of earthquakes but Australia has few?
- why Australia has no active volcanoes?
- what causes tsunamis?
- whether earthquakes can be predicted?
- what makes mountains form?

After completing this chapter you should be able to:

- identify the major tectonic plates on a world map
- model sea-floor spreading
- explain why earthquakes and volcanic activity are most likely to happen near constructive and destructive plate boundaries
- explain how heat and convection currents are involved in the movement of tectonic plates
- relate the extreme age and stability of the Australian continent to its tectonic plate history
- investigate how the theory of plate tectonics developed, based on evidence from earthquakes, volcanic activity and sea-floor spreading
- investigate and describe technologies involved in the mapping of continental movement
- investigate how living near plate boundaries affects people in Pacific countries such as Japan, Indonesia and New Zealand
- *outline how computer modelling and imaging technologies has improved our ability to predict how tectonic plates move.*

AB 10.1

MODULE

10.1 Moving continents

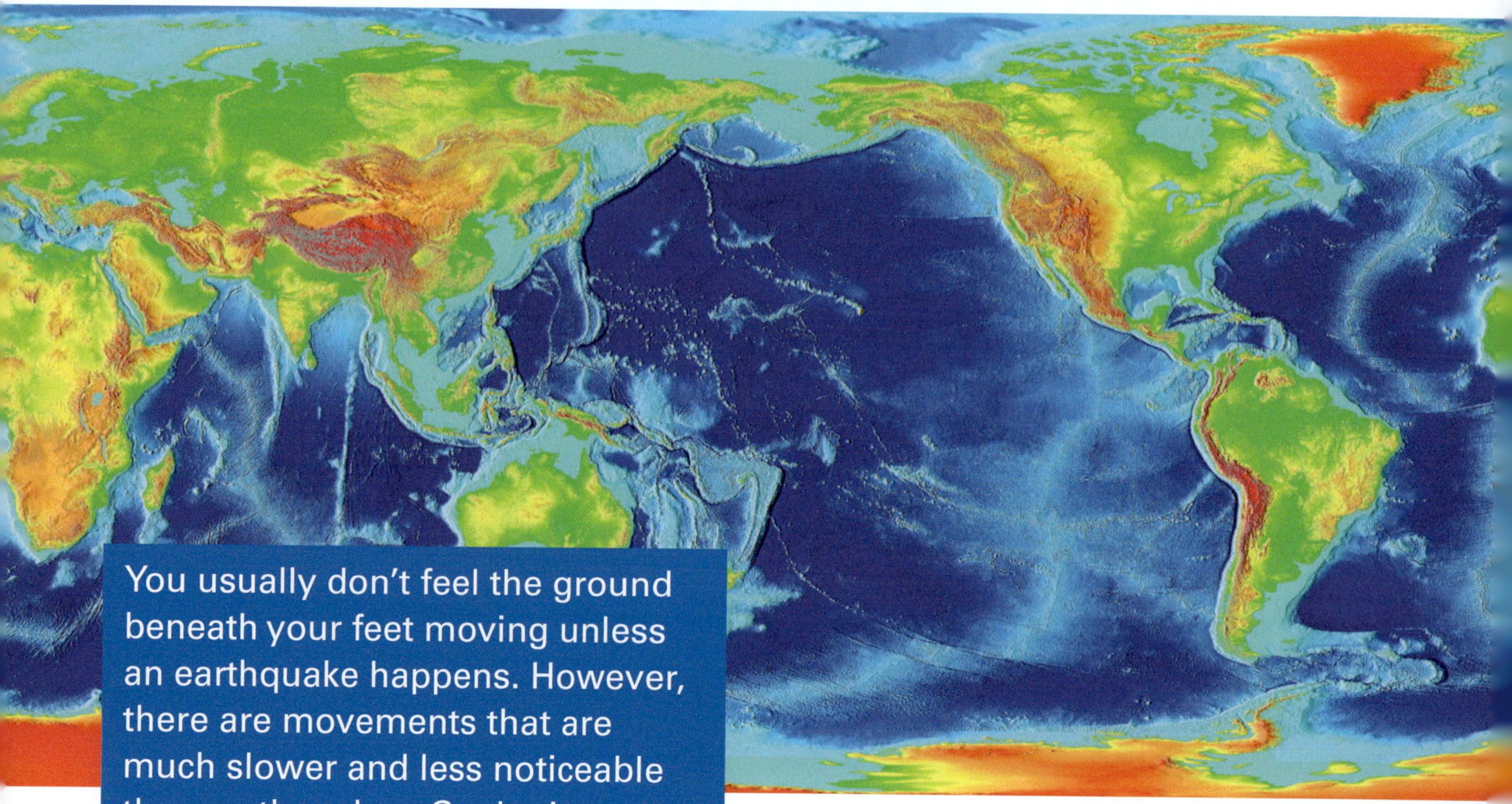

You usually don't feel the ground beneath your feet moving unless an earthquake happens. However, there are movements that are much slower and less noticeable than earthquakes. Geologists have collected a large amount of evidence to show that whole continents are moving.

Continental drift

A revolutionary new theory about Earth was proposed in 1912. In that year, the German meteorologist and geophysicist Alfred Wegener (1880–1930), claimed that the continents were once connected to each other. He called the large landmass Pangaea, and concluded that the continents must somehow have separated and drifted across the oceans. His theory was called **continental drift.**

Wegener based his conclusions on two main observations. The first observation was that the continents seemed to fit together like a jigsaw. You can see this in Figure 10.1.1. The second observation was that fossils of the same prehistoric species were found on continents that were a long way apart. He could see no way that the organisms could cross the oceans to reach all of these places.

Instead, he thought that the continents themselves had shifted. As they shifted, they took the fossils with them. Wegener rearranged the continents, joining them so that the distribution of the fossils matched up across the continents. This can be seen in Figure 10.1.2.

FIGURE 10.1.1 Africa and South America seem to fit together like pieces of a jigsaw.

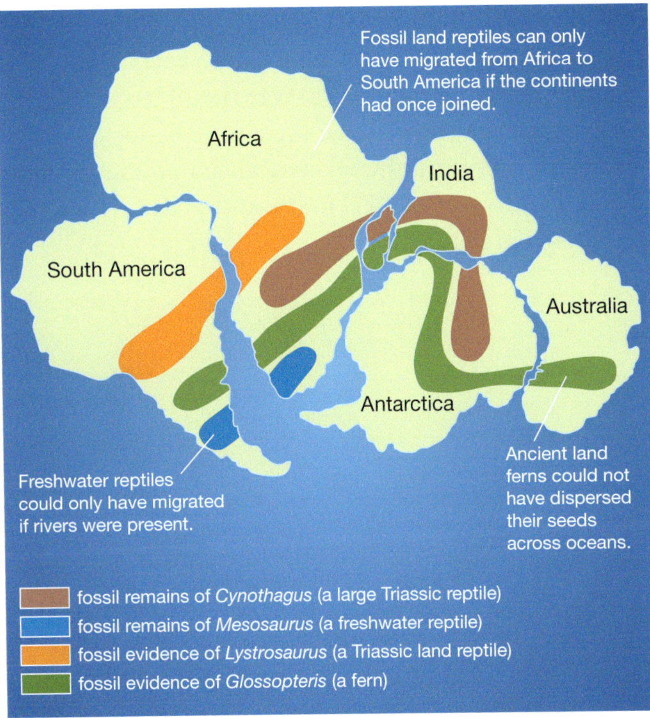

FIGURE 10.1.2 Fossil distribution makes sense if the continents were once connected as shown here.

At the time, few scientists agreed with Wegener's theory. There was no way of measuring whether the continents were moving or not. Also, no mechanism was known that would explain how huge continents could move. Wegener died in 1930, before the discovery of new evidence that would prove him correct.

AB 10.2

Seafloor spreading

In 1872, scientists were surveying the ocean floor for an undersea telegraph cable to be laid between North America and Europe. They measured ocean depth by dropping down very long cables and measuring their length. In the middle of the Atlantic Ocean, they discovered a large mountain ridge that extended a long way south. The extent of the ridge was not investigated then. However, in 1925, German scientists used sonar (echo sounding) to confirm that this ridge ran the entire length of the Atlantic Ocean.

After 1945, further studies using sonar showed that the underwater ridge continued into other oceans and around Earth. In 1953, the ridges were found to have a series of huge cracks along them. The cracks are called **rifts** and so the system was named the Great Global Rift system. Some of it is shown in Figure 10.1.3.

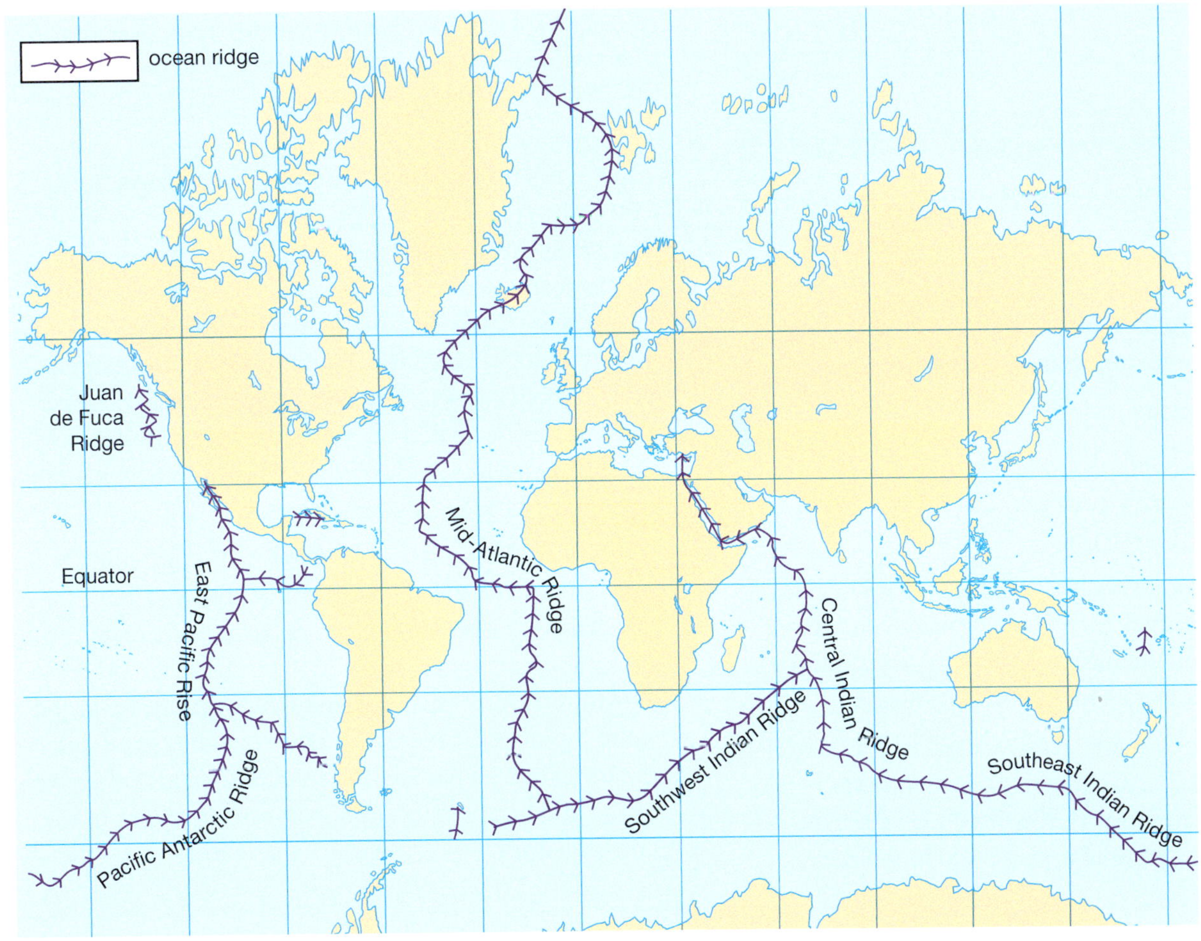

FIGURE 10.1.3 The Great Global Rift system of ocean ridges

In 1962, an American geologist Harry Hess (1906–69), tried to explain what was happening at the ocean ridges. While serving the US navy during World War II (1939–45), Hess had used sonar to map the floor of the Pacific Ocean. There he had discovered many underwater, flat-topped mountains. When the Great Global Rift was found, he thought back to his discoveries of mountains under the sea.

Hess proposed that new rocky crust was being formed at the ocean ridges and spreading outwards. This process was later called **seafloor spreading**. Hess proposed that the crust was sinking down into Earth, forming **ocean trenches**. As the crust sinks, it melts and is destroyed. The process of the crust sinking down like this is called **subduction**. This process is shown in Figure 10.1.4.

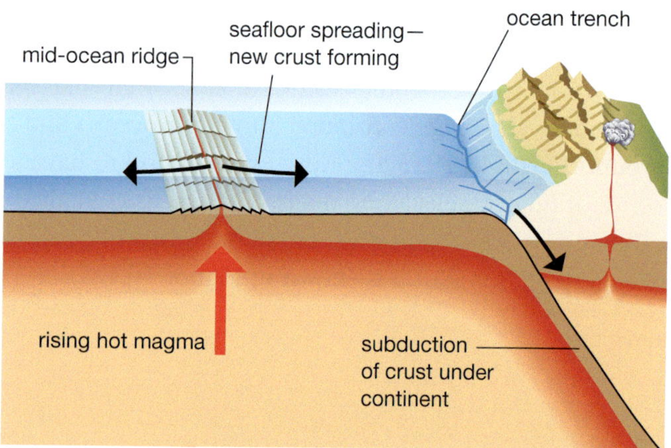

FIGURE 10.1.4 Seafloor spreading forms new crust and subduction eventually destroys it.

The direction of the magnetic field of these magnetite particles can be detected with a device called a magnetometer. During World War II, US navy magnetometers showed that there were bands of alternating strong and weak magnetism on the sea floor. These bands were parallel to the mid-ocean ridges and are shown in Figure 10.1.5.

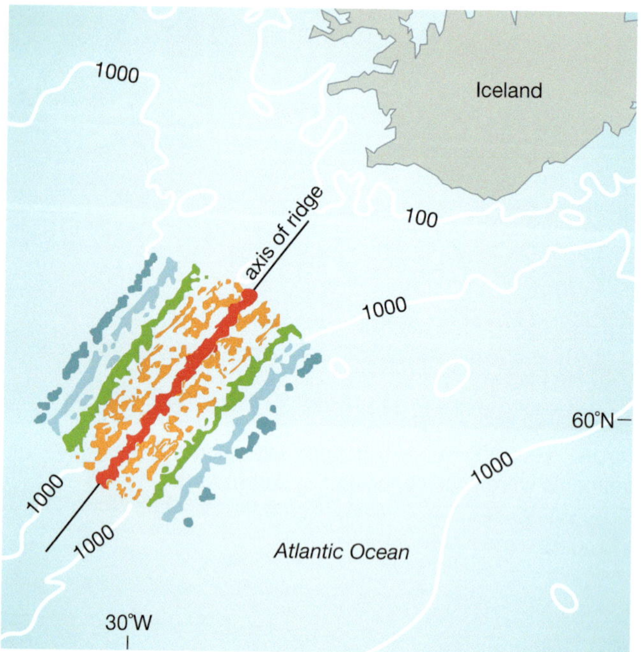

FIGURE 10.1.5 The magnetism in the rocks on each side of the mid-ocean ridges is symmetrical—it consists of alternating bands of strong and weak magnetism.

Support for Hess's theory

The theory of seafloor spreading proposed by Hess is supported by three important pieces of evidence:

- magnetic striping
- age of the sea floor
- sediment thickness.

Magnetic striping

In the 1950s, scientists discovered that many rocks contained the magnetic iron oxide mineral called magnetite. Magnetite has a north-seeking pole and a south-seeking pole, just like a compass needle.

When molten rock solidifies, all the magnetite particles in it line up with Earth's magnetic field to point in the same direction. In this way, the direction of Earth's magnetic field at the time is preserved in the rock.

Geologists found that as they went away from the ridge, the magnetic field preserved in the rocks kept changing. At the start, the rocks had north pointing in the correct direction (towards what is now Earth's north pole). Suddenly the direction of north changed to the opposite direction. North became south, then changed back again. This was very puzzling, because there was no way that the rocks could have spun around. Instead, it indicated that Earth's magnetic field had changed every few million years.

These patterns of strips of rocks with alternating magnetism are called **magnetic striping**. The patterns on either side of the ridge are symmetrical—rocks at a particular distance from the ridge on one side always have their magnetic fields pointing in the same direction as rocks the same distance away on the other side. That is, their magnetic fields both point to the current position of north, or both point to the current south.

More research into magnetic striping led geologists to support Hess's theory about seafloor spreading. They concluded that there were great cracks in the crust and that magma rose up and added to each side of a crack to form new crust on the sea floor. New sea floor was being added equally on each side of the ridge.

This is why the pattern of magnetism was symmetrical—the rocks at equal distances on each side of the ridge were formed at the same time and so had their magnetic fields pointing in the same direction. As Earth's magnetic field changed over many millions of years, so did the magnetic direction preserved in the rocks (Figure 10.1.6).

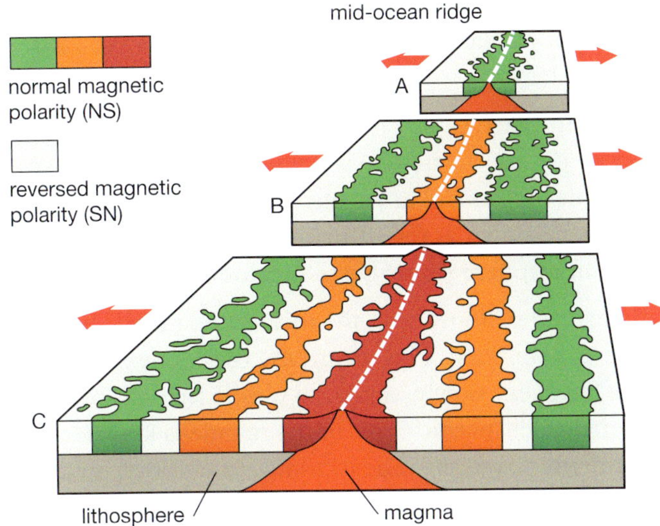

FIGURE 10.1.6 Magnetic striping is due to new sea floor forming on each side of a crack in the crust. The time sequence here goes from A to C.

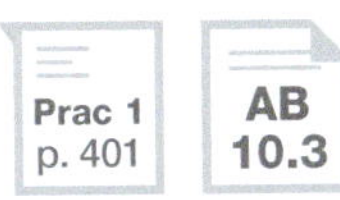

Age of the sea floor

More evidence supporting the theory of seafloor spreading came from the dating of rocks on the sea floor:

- The further the rocks of the sea floor were from the ridges, the older they were. This is exactly what you would expect if new rocks form near the mid-ocean ridge and move outwards away from it.
- The oldest seafloor rocks found were only about 200 million years old. Some of the rocks in the continents were thousands of millions of years old. So the sea floor was very young compared with the continents.

science 4 fun

Making a compass

Can you make a compass?

Collect this ...

- small steel nail or pin
- bar magnet
- 20 cm of cotton
- compass
- sticky tape

Do this ...

1. Place the pin on the desk and stroke it ten times from head to tip with the north-seeking pole of the bar magnet. Lift the magnet up and place it back at the head of the pin after each stroke.
2. Tie or sticky tape the piece of cotton firmly in the middle of the pin so it hangs horizontally, and let it hang down and spin freely. Make sure you are further than 2 metres away from any metal object and the magnet.
3. Note which direction the pin tip points in. Spin it around a few times to see if it always points in the same direction. Compare the direction with a compass if you have one.
4. Repeat step 1 but use the opposite end of the magnet. Note any differences this time.

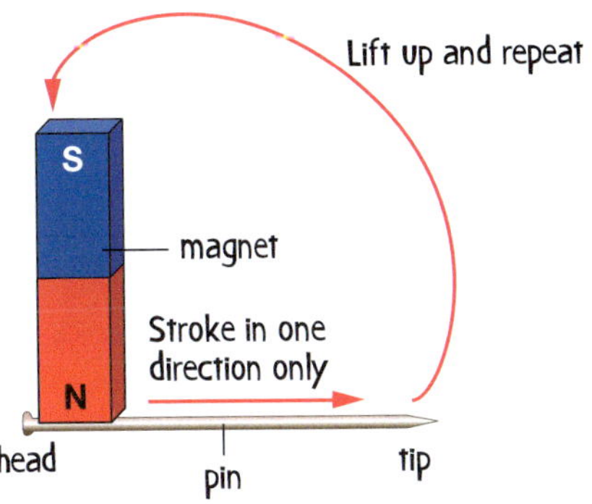

Record this ...

1. Describe what happened.
2. Explain why you think this happened.

Sediment thickness

The sedimentary rock layers on the ocean floor become thicker as you move away from the ridges. This was interpreted by geologists as showing that sediments had been falling for longer on the rocks on the sea floor furthest away from the ridges.

Glomar Challenger

Studying the thickness of sediments on the bottom of the ocean is not easy. Researchers used a special drilling ship called the *Glomar Challenger*. This used extremely long 'drill strings'. The motor was in the ship and the drill bit cut the rock over 5 km below.

Tectonic plates

The picture that scientists now have of Earth's structure supports Wegener's original view of continental drift. Earth's outermost solid layer is known as the **crust**. It is on this layer that we humans and other animals and plants live. It makes up our mountains and plains and makes up the floors of our oceans and seas. The crust varies in thickness from about 5 kilometres under the oceans to about 30 kilometres under the continents.

The crust is rigid, inflexible and is split into many large sections that move about on Earth's surface. Each section is known as a **tectonic plate**.

Tectonic plates can be very large. For example, the whole of Australia and parts of New Zealand are on a single tectonic plate. Most of the crust under the Pacific Ocean is on another plate. There are seven extremely large tectonic plates that are bigger than most continents. There are another ten or so medium-sized ones and about 60 smaller plates. The major plates of Earth are shown in Figure 10.1.7.

Structure of the Earth

Earth consists of several layers, each layer having different physical properties. Its structure is shown in Figure 10.1.8.

Beneath Earth's crust is a layer called the **mantle**. The upper part of the mantle is a rigid solid like the crust. For this reason, the crust and the upper mantle are together known as the **lithosphere** (*litho* means rock). The lithosphere forms Earth's tectonic plates.

Beneath the lithosphere, the rock of the mantle behaves as a 'plastic' solid. This means the rock is flexible and can flow very slowly, similar to plastic that has been heated. This 'plastic' layer of the mantle is known as the **asthenosphere**. The tectonic plates 'float' on the asthenosphere and so can move slowly across it.

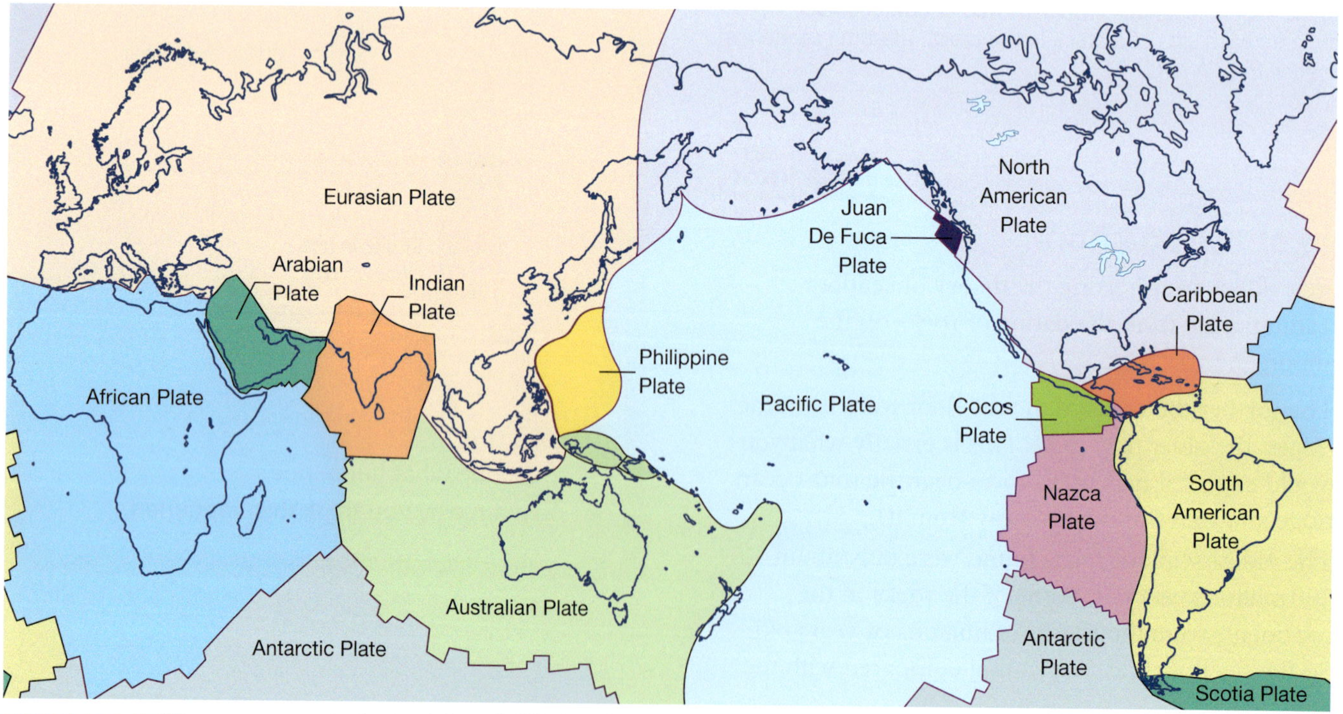

FIGURE 10.1.7 There are many tectonic plates of differing sizes. This map shows 15 of the major plates. Many of these are split into other smaller plates.

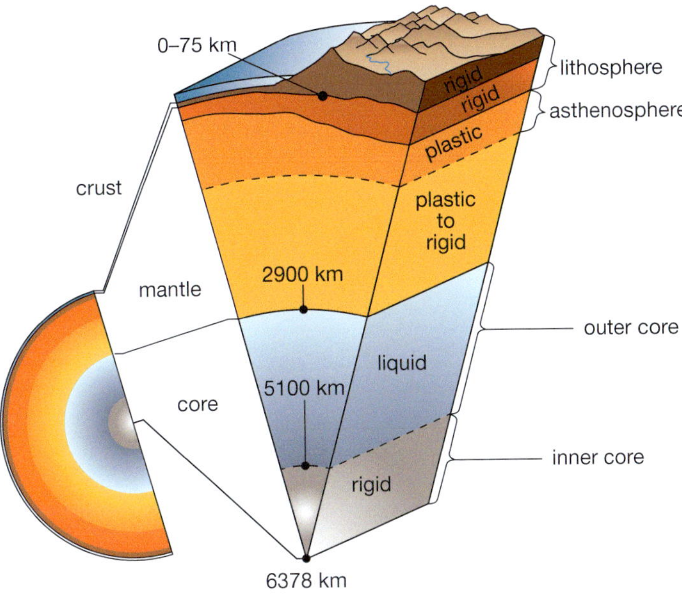

FIGURE 10.1.8 Earth consists of layers. The asthenosphere is like a very thick, slow-moving liquid that allows tectonic plates to move on it.

The mantle is very hot, and becomes hotter as its depth increases. Some of this heat probably dates back to the formation of Earth. Some new heat is created by friction due to moving rocks and also to the decay of radioactive elements. The heat is continually passing from deep in the Earth to the surface.

The **core** of Earth is very dense and is composed mostly of iron and nickel. The inner core is solid, but the outer core is liquid and can flow. Scientists believe this gives Earth its magnetic field.

Why do the plates move?

There are two main theories for how the plates move on the asthenosphere. One is that plates are dragged along as the hot magma in the asthenosphere rises up and then flows under the plates, creating convection currents. As the liquid rock flows, the friction between it and the tectonic plates above may be enough to move them. Figure 10.1.9 shows the convection currents that might carry the plates along.

The other theory is that gravity is involved in moving tectonic plates. This could happen in two very different ways (Figure 10.1.10 on page 398):

- **slab pull**—the ocean crust is denser at the subduction zones near the continents than near the mid-ocean ridges. Gravity pulls the heavy and denser part of the plate downwards at the subduction zone which then pulls the plate into the trench and away from the ocean ridges. This pulls the plates apart at the mid-ocean ridges.
- **ridge push**—new ocean crust forms at the ridges above the rest of the crust and so gravity pulls the new crust downwards. This pushes the older crust below it and squeezes the plates sideways.

Prac 2
p. 402

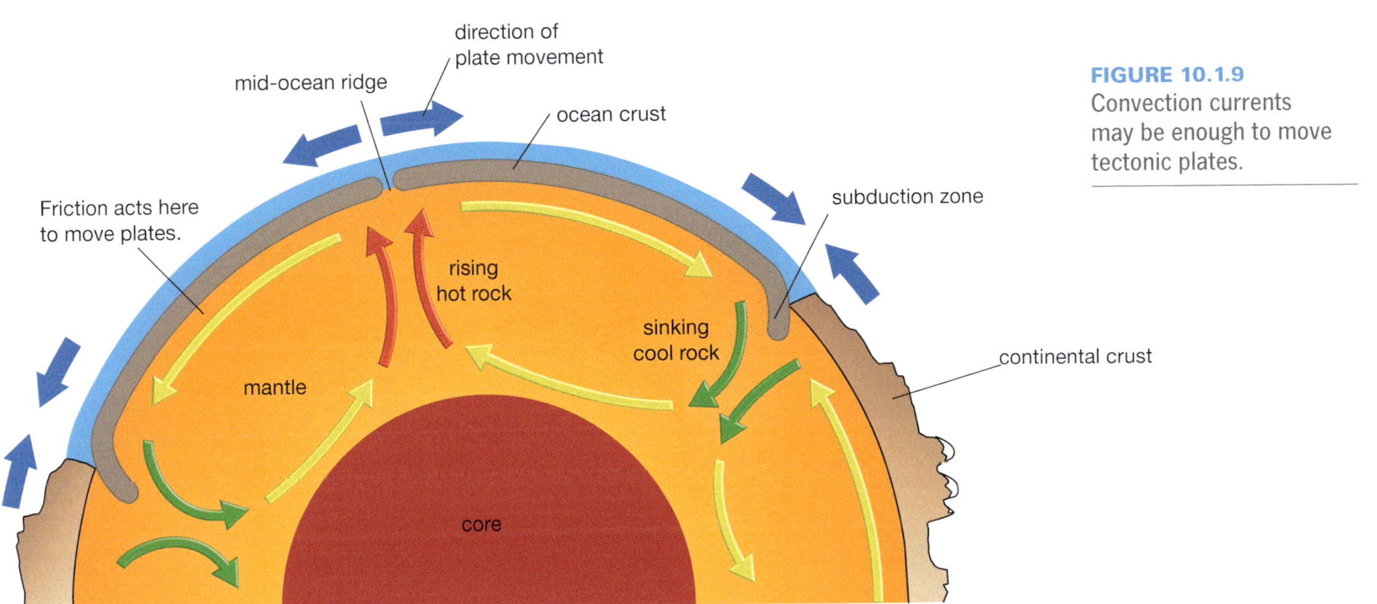

FIGURE 10.1.9 Convection currents may be enough to move tectonic plates.

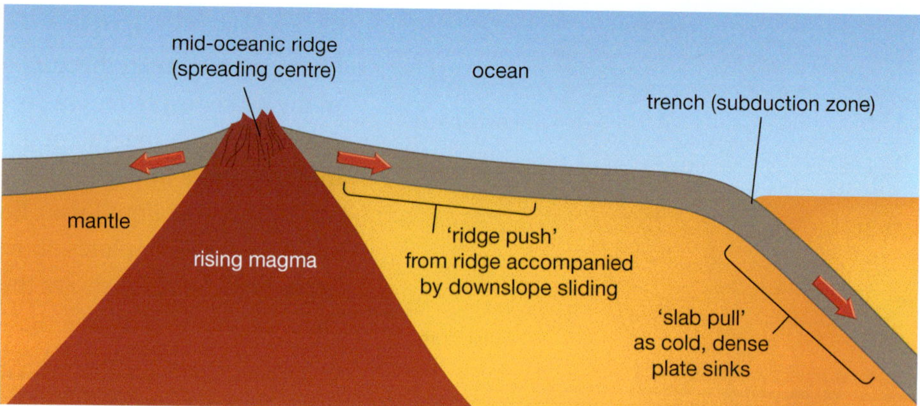

FIGURE 10.1.10 Slab pull and ridge push are two ways in which gravity could move tectonic plates.

Rifting and continental drift

Wegener's theory of continental drift did not explain how the continents moved. Wegener had proposed that gravity and the spinning of Earth could be involved. He thought the continents just scraped across the ocean floors.

Hess's theory of seafloor spreading offered a much more likely mechanism for movement of the continents. He believed that the continents broke up by a process called **rifting**. The crust cracked and subsided, allowing in water from the oceans. One such water-filled rift is the Red Sea, shown in Figure 10.1.11.

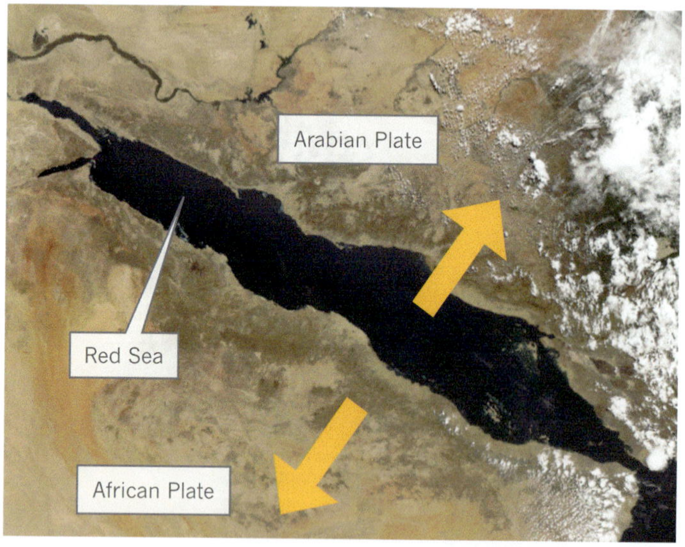

FIGURE 10.1.11 This satellite image shows the Red Sea, a giant water-filled rift valley at the edge of two tectonic plates.

Then the process of seafloor spreading occurred. As new crust formed at the ridges, the continents moved along with the ocean floor as it spread away from the ridges. This is shown in Figure 10.1.12.

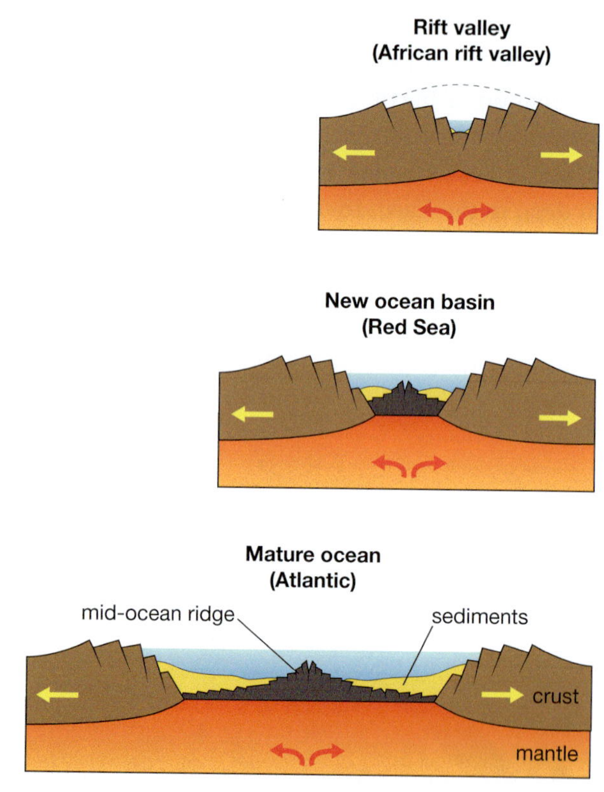

FIGURE 10.1.12 Rifting can cause continents to split apart and new ocean basins to form. This process, called seafloor spreading, helps to explain Wegener's theory of drifting continents.

SCIENCE AS A HUMAN ENDEAVOUR

Use and influence of science

GPlates: mapping Earth through time

GPlates is plate tectonic reconstruction software which is leading the world's research in this field and helping scientists understand more about Earth's geological processes. The EarthByte Group, from the University of Sydney, are the team of geoscientists who led the way in the development of GPlates (Figure 10.1.13).

FIGURE 10.1.13 The EarthByte Group from the University of Sydney

FIGURE 10.1.14 The GPlates software constructs images of Earth's surface topography over time.

The GPlates software helps us visualise the Earth's crust. The software creates images of tectonic plates and other geological features. The user can move these around and manipulate them, to see how the Earth has changed over time. GPlates is accessible through a series of online applications that are easy-to-use and available to anyone.

Scientists from all over the world are using GPlates in their research. The software is capable of fine-scale 3D topographic mapping of Earth's surface, creating animations of the movements of tectonic plates and continental drift over time, realistic images of geological features, composition of the sea floor and mapping of the sea-floor's landscape (Figures 10.1.14 and 10.1.15).

GPlates has recently been used to make important discoveries about the evolution of the Australian continent and has even been used to find potential sites for opal mining. Opal prospectors (miners) were able to map potential opal fields by using GPlates to learn about the conditions required for opal formation and by visualising geological data. GPlates is a powerful tool that can provide insight into the evolution of Earth and enable better modelling for the prediction of future changes in Earth's structure.

FIGURE 10.1.15 GPlates allows users to manipulate and visualise Earth's geological features and changes over time.

REVIEW

1 Why do you think it is important for scientists to research events in Earth's past?

2 How does visualising data (using images rather than numbers) help us to better understand it?

MODULE 10.1 Review questions 

Remembering

1 Define the terms:
   - a continental drift
   - b seafloor spreading
   - c tectonic plate
   - d subduction.

2 What term best describes each of the following?
   - a patterns of magnetism trapped in rocks on each side of plate boundaries
   - b a deep channel in the ocean floor where crust is sinking downwards
   - c the process of continents breaking up, subsiding and allowing in water from the sea
   - d a layer of 'plastic' semi-solid rock in the mantle.

3 What was Harry Hess's theory of seafloor spreading?

4 What observation first led to the theory of seafloor spreading?

Understanding

5 Outline the two main observations that led Wegener to propose the theory of continental drift.

6 How did Wegener work out what Pangaea looked like?

7 Describe three types of evidence that now supports the hypothesis of seafloor spreading.

8 A simple compass can be made by a process like that shown in the science4fun on page 395. Explain how this 'pin' compass could be used to find north.

Applying

9 Use the theory of plate tectonics to explain why Africa and America are older than the sea floor of the Atlantic Ocean.

10 Use Figure 10.1.6 on page 395 to explain the process of magnetic striping.

11 Use Figure 10.1.9 on page 397 to explain how convection and gravity may be involved in the movement of crustal plates.

12 Construct a three-column table to summarise the evidence supporting the theory of plate tectonics. Use the headings *Feature*, *Description* and *Sketch*. A simple sketch of each feature alongside its description will help you remember it. Remember to include Wegener's observations.

13 Consider your answer to question 1c. Rewrite your definition of a tectonic plate but this time use the words lithosphere and asthenosphere.

Analysing

14 Compare the theories of continental drift and plate tectonics.

15 Distinguish between magnetic striping and magnetic field reversal.

16 Compare Figures 10.1.11 and 10.1.12 on page 398, and use them to explain how the Red Sea formed.

Evaluating

17 The large photo on page 392 is a satellite image of Earth's topography using sensors that detect various wavelengths of light. Geographic features are colour-coded by their height. You can see the coding in Figure 10.1.16.

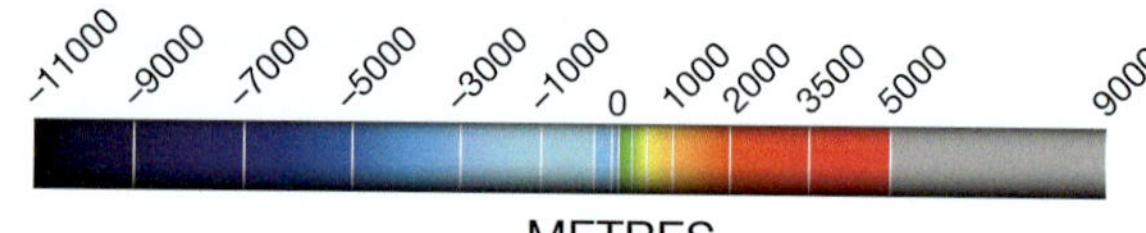

FIGURE 10.1.16

   - a What colour is the Great Global Rift system in the map?
   - b What colour are most of the oceans?
   - c What do the map and Figure 10.1.16 show about the height of the Great Global Rift system and the oceans?
   - d How do you think this fits in with the theories about how the plates move?

18 There is an ocean trench separating Indonesia and Australia. There is an ocean ridge half way between Australia and Antarctica. Geologists suggest that Australia is moving north. What evidence supports this belief?

Creating

19 Consider Figure 10.1.12 on page 398 showing a mature ocean (the Atlantic Ocean). Imagine you had to collect rocks from places in the Atlantic Ocean that could help you test Hess's theory of seafloor spreading. Construct a plan discussing three places where you would look for the rocks and the reasons why you would look in each of those places.

MODULE

10.1 Practical investigations

1 • Magnetic striping

Purpose

To model the magnetic striping patterns found in the rocks from seafloor spreading.

Timing

45 minutes or 30 minutes if coloured paper strips are used

Materials

- 2 × A4 sheets of plain paper
- compass
- bar magnet
- red and yellow pencils OR 2 cm × 21 cm strips of coloured paper – 6 red and 6 yellow
- glue
- 2 clothes pegs
- sticky tape
- 2 desks or tables that can be moved

Procedure

1. Tape the two A4 paper sheets together at one end so you have one long piece of paper. Rule a line across the open end of each sheet 10 cm from the end. Close the paper so the two A4 sheets are face to face with the ruled lines on the inside face of each paper.
2. Push two desks together, leaving a gap of a few millimetres.
3. Push the taped end of the paper down into the gap between the desks until you reach the ruled line. Leave 10 cm of the paper projecting above the desktop.
4. Place a compass on the desktop, next to the top edge of the paper. Place a magnet on the desk about 5 cm away from the compass. Have the north end of the magnet pointing away from the compass.
5. Gently fold the ends of the paper down, one end on each desk, and put a peg on each to weigh each end down (Figure 10.1.17).
6. Under the desk, gently push up on the taped end of the papers until about 2 cm of paper has come out above the desk. Hold the paper still and use the red pencil to colour the 2 cm strip between the opening in the desk and the line on the paper. Alternatively, glue a red strip of paper on either side of the gap in the paper. You should have a 2 cm red strip on each side of the opening in the desk. Your set-up should look like Figure 10.1.17.

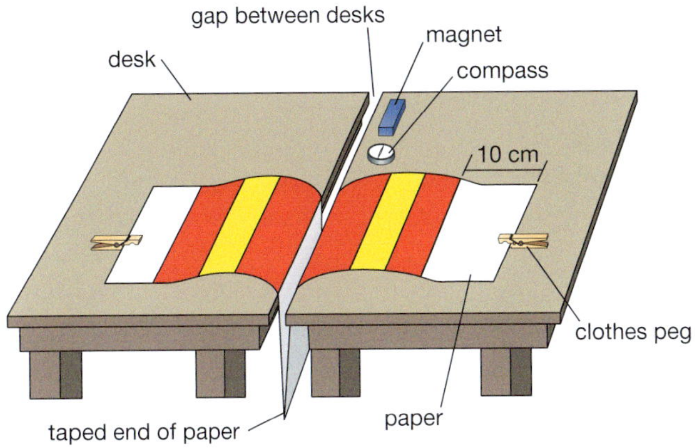

FIGURE 10.1.17

7. Spin the magnet around so that the north end is pointing at the compass.
8. Repeat step 6, but use the yellow pencil to colour in the 2 cm strip. Alternatively, glue a yellow strip of paper to each side. You now should have a red strip and a yellow strip on each side of the opening in the desk.
9. Repeat steps 6–8 until you have three red lines and three yellow lines on each side of the opening in the desk. For step 7, spin the compass around so the north pole is to the top of the desk.
10. On each of your sheets, number the layers 1–6 in the order that they formed. Write on the oldest layer of rock the word 'oldest', and on the youngest layer, the word 'youngest'.

Review

1. In this model, identify what is represented by the:
   - a gap between the desks
   - b two sheets of paper
   - c magnet
   - d red colour on the paper
   - e yellow colour on the paper.
2. Identify the layer on your drawing that represents:
   - a the youngest rocks
   - b the oldest rocks.
3. Explain how this model helps you understand magnetic striping along the ocean ridges.

MODULE

10.1 Practical investigations

2 • Convection  

Purpose

To model the convection currents in the mantle.

Timing 45 minutes

Materials

- 5 potassium permanganate crystals
- small coloured ice cube (water with food colouring)
- 200 mL of water
- 2 × 250 mL beakers
- tweezers
- hotplate or Bunsen burner, tripod, gauze mat and bench mat

SAFETY

A risk assessment assessment is required for this investigation.

Potassium permanganate stains—only handle it with tweezers.

Wear safety glasses.

Procedure

See Activity Book Toolkit to assist with developing a risk assessment.

1 Place 200 mL of water in a 250 mL beaker on the hotplate or gauze mat on the tripod.

2 Using tweezers, gently drop about five potassium permanganate crystals into the centre of the beaker. Do not move the beaker.

3 Gently heat the beaker. (If using a Bunsen burner, use a cool flame directly below the crystals). Figure 10.1.18 shows the set-up.

FIGURE 10.1.18

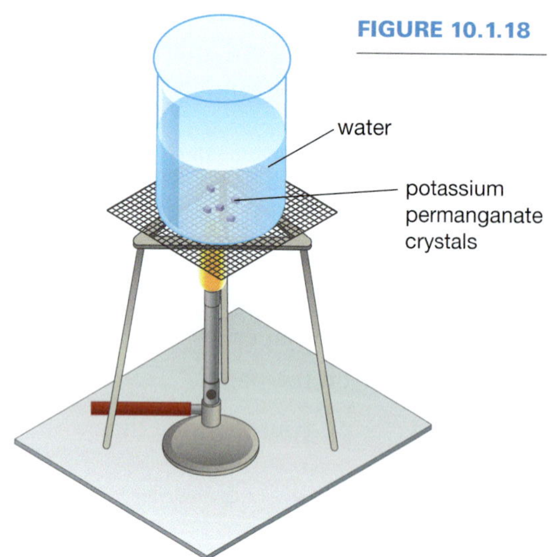

4 Carefully observe what happens to the streams of purple colour that come off the crystals. (It should happen within a minute or two.)

5 Fill the other beaker with 200 mL of tap water.

6 Use tweezers (because food colouring stains) to place a coloured ice cube into the beaker of tap water.

Results

Record your observations for steps 4 and 6 as labelled sketches.

Review

1 Why did the purple colour from the crystals of potassium permanganate rise up to the surface of the water?

2 Explain why the coloured ice cube behaved as it did.

3 a What happened to the purple colour from the potassium permanganate after it reached the surface of the water?

b Use your answer to question 2 to explain why the purple colour did this.

4 Explain how this experiment helps you understand what happens with heat in the mantle.

5 Suggest a way of testing your answer to question 3 by using a bag of ice, purple crystals, a beaker and water.

MODULE 10.2

Plate movements

Tectonic plates move relative to each other in three different ways to produce different landforms. The Himalayas, the world's highest mountains, were formed by the movements of tectonic plates. The theory of plate tectonics helps us understand how our planet changes.

science 4 fun

Future Earth

What might the continents of Earth look like in millions of years from now?

Collect this ...

- a computer or device with access to the internet

Do this ...

1. Find a computer simulation or animation that looks at how plate tectonics will make the future Earth look like.
2. Run the simulation and observe what is happening to particular regions of the Earth such as the continents and the oceans.
3. In particular find out what has happened to Africa, Australia, Antarctica, North America, South America, the Mediterranean Sea and the Indian Ocean.

Record this ...

1. Describe what you found.
2. Explain what you found.

Types of crust

There are two types of crust—continental and oceanic. The **oceanic crust** is found on the ocean floor generally below sea level. The **continental crust** is the crust forming the continents and so it rises above sea level. Oceanic crust is much thinner and much denser than continental crust. Oceanic crust is denser because it contains more of the heavier elements such as iron and magnesium. In contrast, continental crust contains more of the lighter elements such as aluminium and silicon. The densities of the two types of crust are important when tectonic plates are moving near one another.

Types of plate movement

As Figure 10.2.1 on page 404 shows, the tectonic plates move in three different ways at the boundaries between them.

These boundaries are:

- diverging boundaries—where the plates are moving apart from each other
- converging boundaries—where the plates are colliding with each other
- transform boundaries—where the plates are sliding past each other.

The locations of each type of boundary are shown in Figure 10.2.2 on page 404.

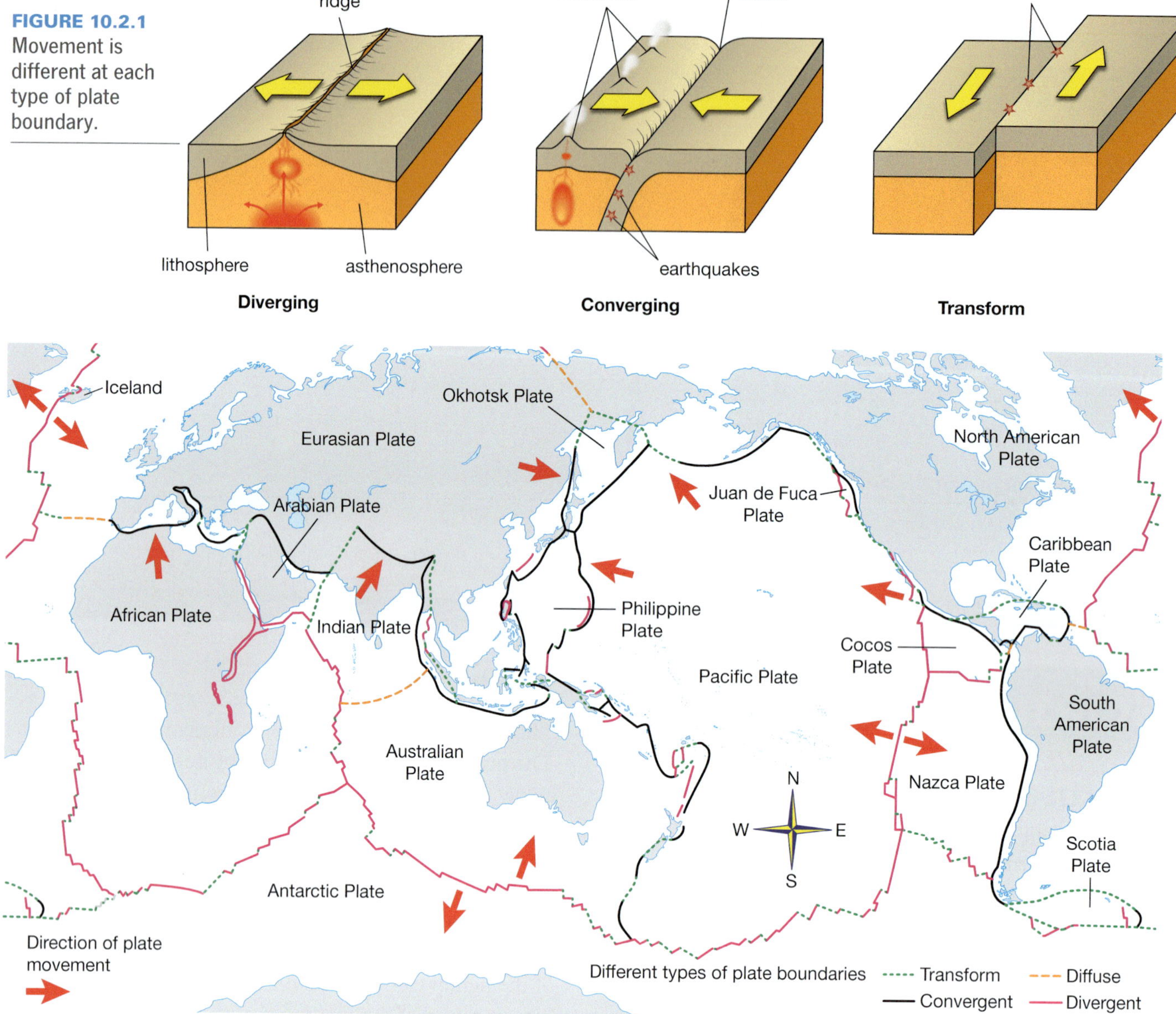

FIGURE 10.2.1 Movement is different at each type of plate boundary.

FIGURE 10.2.2 Locations of the three different types of boundaries. Diffuse boundaries are places where there is no clear boundary.

Diverging boundaries

When things diverge, it means that they separate. For example, cars travelling in a single lane diverge or separate when the road widens into multiple lanes. Tectonic plates can diverge too. **Diverging boundaries** are where tectonic plates are moving apart from each other in opposite directions.

When the plates separate, they leave a rift (deep crack) between them. Magma from the asthenosphere rises up into the rift and solidifies as it cools. This forms new crust, and so diverging boundaries are also known as **constructive boundaries**.

The mid-ocean ridges form a diverging boundary and the ridge itself is evidence that new crust is being constructed.

Diverging boundaries also occur on land. Figure 10.2.3 shows that the Mid-Atlantic Ridge runs right through the island of Iceland. As a result, the island is widening as new crust is formed. Iceland has constant volcanic eruptions as the magma spews up into the rift. The rate of widening has been measured to be about 2–5 centimetres per year. That may not seem like much, but over many millions of years this spread has formed the Atlantic Ocean.

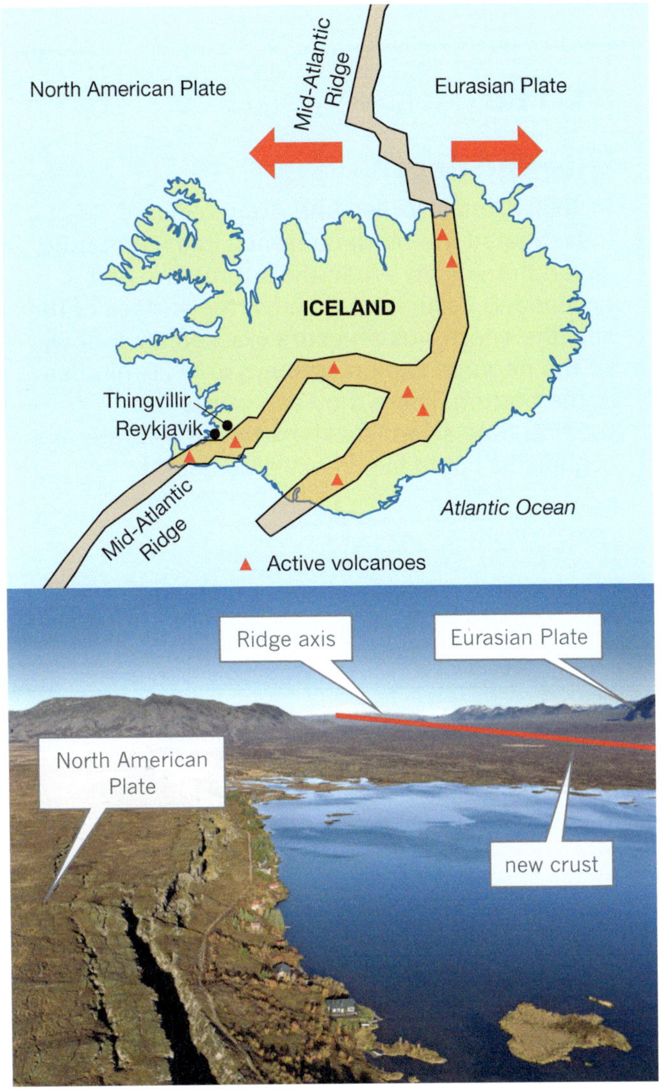

FIGURE 10.2.3 Iceland is located on the Mid-Atlantic Ridge and so the island is steadily growing wider.

SciFile

Deadly volcanoes

Gas and dust from Iceland's volcanoes have often troubled European countries. The most recent was in 2010 when Eyjafjallajökull erupted and shut down all air traffic in northern Europe for a week. An eruption of Laki in 1783 killed many people in Europe with poisonous gases and changed weather patterns for over a year.

FIGURE 10.2.4 Aerial view of Eyjafjallajökull erupting in 2010

Another place where diverging boundaries seem to be forming is in East Africa. As Figure 10.2.5 shows, the Great Rift Valley, runs along the eastern part of the African continent. At present there is no obvious crack or rift in the crust with magma welling up into it, but the land is subsiding and volcanoes occur in several places. Geologists have proposed that this could be where the next ocean will form on Earth.

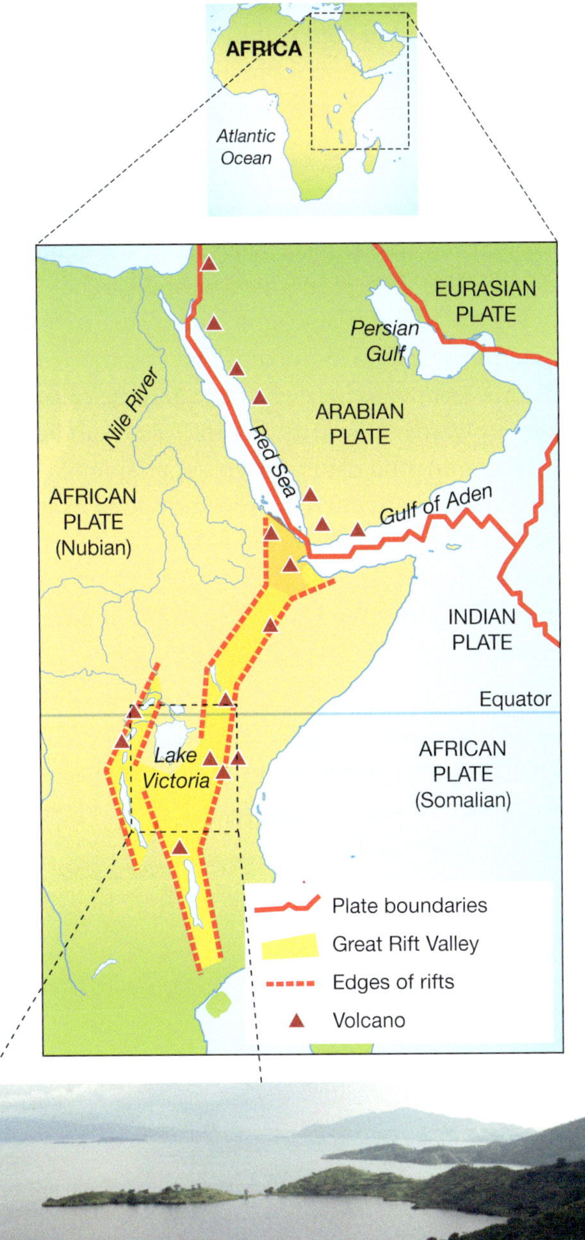

FIGURE 10.2.5 The Great Rift Valley of Africa could be the site of the next ocean to form on Earth.

Converging boundaries

When things converge, it means that they come together. For example, cars entering a busy freeway need to converge and join to form one flow of traffic. **Converging boundaries** occur when two plates are colliding head-on into each other. Rock is destroyed at converging boundaries, and so these boundaries are also known as **destructive boundaries**. These collisions form an assortment of land features such as mountains, chains of islands and underwater trenches. The features that are formed depend on what types of crust collide.

Mountains, volcanoes and trenches

If oceanic crust is colliding with continental crust, then the denser oceanic plate sinks under the lighter continental plate. This is subduction, and can be seen in Figure 10.2.6. The continental plate becomes distorted, forming fold mountains and volcanoes. **Fold mountains** form when plates collide and the crust crumples upwards. Where the oceanic plate subducts under the continental plate, a deep depression in the ocean floor called a **trench** is formed. A good example is where the Nazca Plate collides with South America. The Andes Mountains have been formed along the west coast of South America, and the 8000-metre deep Peru–Chile Trench has formed.

The plate that subducts dives down deep into the mantle, which is extremely hot. The friction of the plates colliding also generates heat. This heat is enough to melt the crust and form magma. So crust is being destroyed as it subducts.

SciFile

Into the depths

In 1960, Donald Walsh and August Picard investigated the Mariana Trench in a submarine called the *Trieste*. When the submarine had descended 9000 metres below the surface of the sea, the enormous pressure cracked a window, so the nervous explorers spent only 20 minutes at the bottom. There they saw some fish—the first time scientists knew that vertebrate organisms could live at extreme depths.

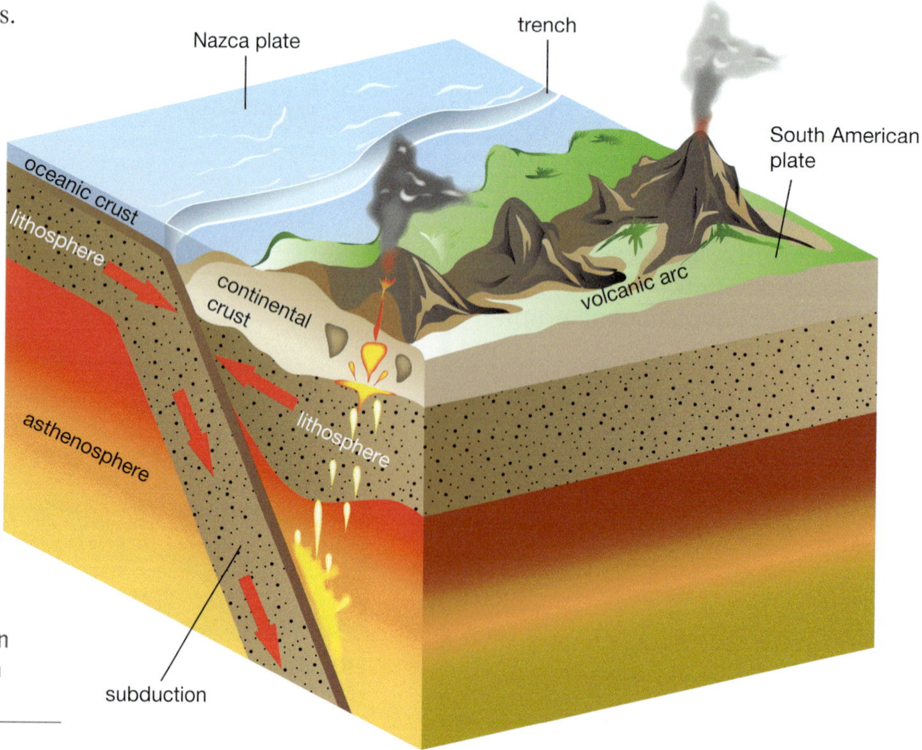

FIGURE 10.2.6 Subduction of an oceanic plate under a continental plate

High mountain systems

When two continental plates collide, both are pushed upwards because both have similar densities. This upwards push forms ranges of very high mountains. The best-known example of this is where the Indian Plate is colliding with the Eurasian Plate (Figure 10.2.7). This has formed the Himalayas, the highest mountain range on Earth. Its highest mountain is Mt Everest at 8848 metres above sea level.

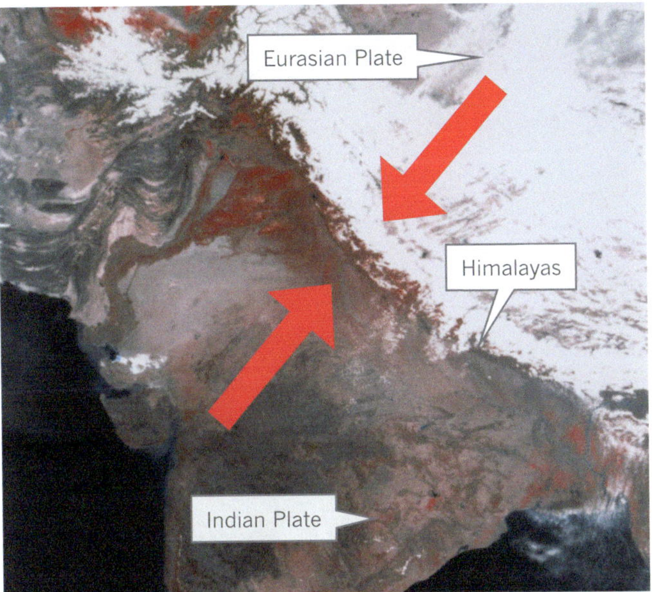

FIGURE 10.2.7 The Himalayas were formed by the collision of two continental plates: the Indian Plate and the Eurasian Plate.

SciFile

Mt Everest is growing

No dinosaur would ever have seen the Himalayas, because dinosaurs were extinct before the Himalayas began forming about 50 million years ago. The Himalayas are still growing at about 1 cm per year. Mount Everest is about 60 cm higher now than when Edmund Hillary and Tenzing Norgay reached the top in 1953.

Island arcs

When two oceanic plates collide, the faster-moving plate always subducts under the other (Figure 10.2.8). This forms a deep trench. The descending plate melts and is destroyed, forming magma. The magma rises to the surface creating a chain of volcanic islands called an **island arc**.

The Mariana Trench is the deepest ocean trench yet discovered. It occurs where the Pacific Plate collides with a small plate called the Mariana Plate (part of the Philippine Plate). The trench is 10 911 metres deep. A string of volcanic islands forms an island arc along the boundary.

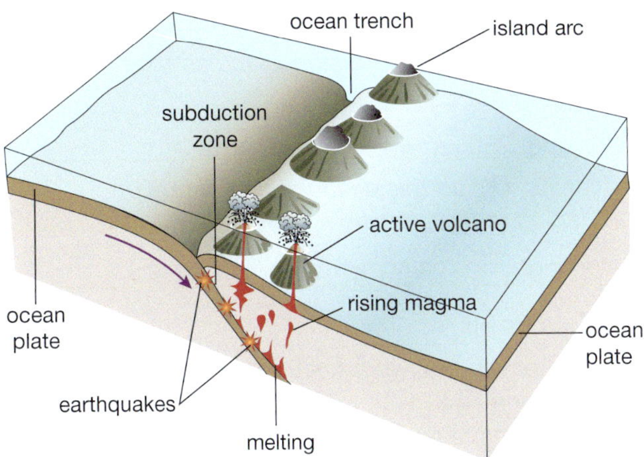

FIGURE 10.2.8 Colliding oceanic plates form ocean trenches and arcs of volcanic islands.

SkillBuilder

Calculating the age of oceans

Geologists can estimate the age of geological features such as the ocean basins, rift valleys or continental islands. To do this, they use the rate of movement of tectonic plates.

For example, geologists can calculate how long it took to form the Atlantic Ocean using two measurements. First, geologists take the width of the Atlantic Ocean between parts of Africa and South America that were once joined. This is about 4000 km. Then geologists consider how quickly Iceland is widening. This is at about 2 cm a year.

First convert the distance across the ocean into centimetres.

$$4000\text{ km} = 400\,000\,000\text{ cm}$$

Now divide this distance by the rate of widening of the plates.

$$\frac{400\,000\,000}{2} = 200\,000\,000\text{ years}$$

$$= 200\text{ million years}$$

So if the rate of movement of the plates has been averaging 2 cm per year, then the Atlantic Ocean is 200 million years old.

Worked example

Calculating the age of oceans

Problem

Iceland is about 250 km wide. Calculate how long the island may have existed if the average rate of movement is 2 cm per year.

Solution

Thinking: Convert the distance across the ocean into centimetres.

Working: 250 km = 25 000 000 cm

Thinking: Divide this distance by the rate of widening of the plates.

Working: $\frac{25\,000\,000}{2}$ = 12.5 million years

Try yourself

1 The ocean between Australia and Antarctica is widening at about 7.5 cm/year. It is about 4000 km wide. Calculate when the ocean may have formed.

2 India and Africa are about 4000 km apart and are moving apart at a rate of about 3.4 cm per year. Calculate how many years ago Africa and India were connected.

3 The distance between Australia and the Indonesian island of Java is about 500 km. The Indo-Australian Plate is moving about 7 cm towards Java per year. Calculate how long it will be before Australia crashes into Java.

Transform boundaries

A **transform boundary** is where two plates are sliding parallel to each other but in opposite directions relative to one another. The plates usually move very slowly past each other but then suddenly slip quickly. When this happens, there is an earthquake. Whether on land or under the ocean, a transform boundary usually has fold mountains along its length. There are also many cracks in the rock, called fault lines. Fault lines usually do not form one continuous crack in the crust along the plate boundary. Instead there are many cracks parallel to each other. Transform boundaries are not constructive or destructive boundaries because they do not make new crust or destroy crust. They can destroy buildings and can buckle land to form mountains, but no rock is destroyed and none is created.

The San Andreas Fault in California (USA) is a transform boundary. You can see it clearly in Figure 10.2.9. This is where the Pacific Plate and the North American Plate move past each other. The cities of Los Angeles and San Francisco are built near the fault line. This fault has moved in the past and has caused massive earthquakes, such as in 1989, and in 1906 when much of San Francisco was destroyed.

FIGURE 10.2.9 The San Andreas fault in California, USA, is a transform fault. Both plates are actually moving north-west but the Pacific plate is moving much faster.

SciFile

Coffee break burns down city

San Francisco was devastated by the earthquake of 1906 but what destroyed even more of the city were fires that raged out of control after the quake. All of downtown and 60% of her suburbs were burnt to the ground. One massive blaze happened after a small fire started by a rescue crew to brew coffee went out of control!

Another transform fault runs right through New Zealand. Meeting at this fault are the Australian Plate (moving north-east) and the Pacific Plate (moving south-west). Movement of this transform boundary created the mountains of the South Island of New Zealand. Transform faults are shown in Figures 10.2.10 and 10.2.11.

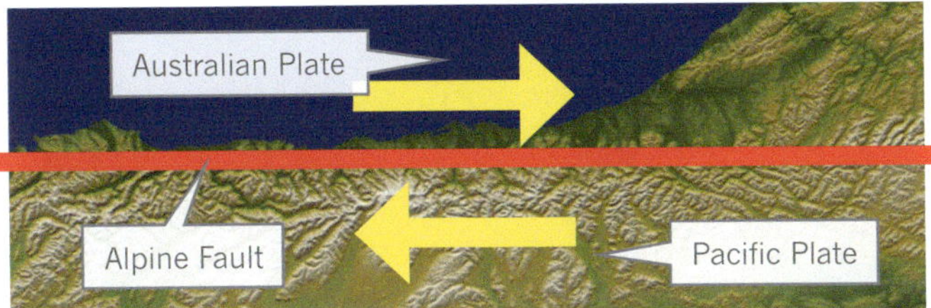

FIGURE 10.2.10 This satellite image shows New Zealand's Alpine Fault.

FIGURE 10.2.11 Milford Sound in New Zealand sits next to the transform fault that runs along the west coast of the South Island.

Prac 1 p. 412 | Prac 2 p. 412 | Prac 3 p. 413 | AB 10.4

STEM 4 fun

Tectonic plate movements

PROBLEM

Can you build a working model to demonstrate tectonic plate movements?

SUPPLIES

- rice
- thin rectangular crackers, like Vita-Weat®
- a range of spreads such as cream cheese, Vegemite®, jam
- food colouring
- container such as an ice-cream tub or plastic drink bottle
- cardboard
- sand

PLAN AND DESIGN Your task is to use some of the materials to represent the plates and select other items to represent the mantle. Test your model out by moving and applying forces to the plates to model different types of plate movements.

Design the solution. What information do you need to solve the problem? Draw a diagram. Make a list of materials you will need and steps you will take.

CREATE Follow your plan. Create your solution to the problem. Take a video.

IMPROVE What works? What doesn't? How do you know it solves the problem? What could work better? Modify your design to make it better. Test it out.

REFLECTION

1 What field of science did you work in? Are there other fields where this activity applies?
2 In what career do these activities connect?
3 What did you do today that worked well? What didn't work well?

Measuring the speed of tectonic plates

Technology now gives scientists the equipment to directly measure how fast tectonic plates are moving. The modern geologist uses several different techniques.

FIGURE 10.2.12 A GPS receiver, solar cells and a satellite link

FIGURE 10.2.13 A satellite laser ranging device

The most widely used technique involves satellites and the global positioning system (GPS). A **GPS ground station** is a receiver and computer that is placed on top of a stand. One is shown in Figure 10.2.12. The legs of the stand are fixed into the rock.

There are many GPS ground stations located on tectonic plates around the Earth. A group of 24 satellites circling the Earth sends radio signals to the GPS ground stations. A GPS ground station can locate where it is accurately if it can detect at least three satellites at the same time. Over time, many measurements are taken of many ground computers. If the location of a ground computer changes, moving further away from other ground computers or closer to them, scientists know that the earth has moved. The computer can determine movement sideways. If four satellites are detected, then up and down movement can also be calculated.

GPS stations can be accurate to within a few millimetres. Measurements from these stations show that rates of plate movement vary greatly between the different plates and even between parts of the same plate. The fastest-moving plate is near Easter Island in the Pacific Ocean. This part of the Pacific Plate is moving at about 15 cm per year. The slowest-moving plate is the Antarctic Ridge, which is moving at only 2.5 cm per year.

Another method that is extremely accurate in measuring plate movement is satellite laser ranging (SLR). This process uses a device that fires a laser beam up to a satellite that is equipped with a reflecting mirror. The laser beam bounces off the mirror and reflects back to the SLR instrument, which measures the time taken for the laser beam to return to it. The computer in the SLR calculates the distance from the SLR to the satellite (Figure 10.2.13). If several SLR instruments are used on the same satellite, they can calculate the amount and direction of movement of the SLR stations.

REVIEW

1 What are two technologies that allow scientists to determine the movement of tectonic plates?
2 GPS ground stations are fixed onto solid rock and do not just sit on it. Explain why.
3 How does the GPS ground station calculate the speed and direction of movement of the tectonic plate?
4 How do scientists determine that a tectonic plate is moving up and down as well as sideways?
5 Compare GPS stations and satellite laser ranging as methods of detecting and calculating plate movement.

MODULE 10.2

Review questions

Remembering

1 Define the terms:
   a diverging boundary
   b destructive boundary
   c island arc
   d ocean trench.

2 What term best describes each of the following?
   a where plates are colliding with each other
   b where plates are sliding parallel to each other but in opposite directions
   c where new crust is created
   d crust that forms the continents.

3 Name the three types of plate boundary.

4 What are the two types of crust that form the tectonic plates?

5 Name one place where the following type of plate boundary can be found.
   a diverging boundary
   b converging boundary
   c transform boundary.

Understanding

6 Describe the process of subduction.

7 When transform boundaries such as the San Andreas fault and the Alpine fault in New Zealand move they destroy buildings with massive earthquakes. Why are transform boundaries not classified as destructive boundaries?

Applying

8 Use plate movement to explain the origin of the Himalayas.

9 Using descriptions of the characteristics of plates, identify the type of boundary where the following processes occur.
   a subduction
   b ocean trenches
   c fold mountains
   d rifting under the sea
   e island arc
   f an ocean trench next to a continental mountain range.

10 Iceland is extremely volcanically active. Iceland is growing at 2–5 cm a year. Explain how these two facts are linked.

11 The Andes is a very high mountain chain along the west coast of South America. The Nazca plate is moving eastwards. Explain how these two facts are linked.

12 Consider your answer to Module 10.1 review question 12 where you constructed a table summarising evidence for plate tectonics. To that table, add any further evidence from Module 10.2.

13 The STEM4fun on page 409 asks you to build a working model to demonstrate tectonic plate movements. Use your knowledge of tectonic plate movements to draw a flow diagram showing how mountains form.

Analysing

14 Draw annotated sketches to show the plate movements between the following pairs of plates.
   a the Arabian Plate and the African Plate near Israel
   b the Australian Plate and the Antarctic Plate
   c the Australian Plate and the Pacific Plate to the north-east of Papua New Guinea.

15 Compare subduction when two ocean plates converge with subduction when an oceanic and continental plate converge.

Evaluating

16 What evidence do you think would help you decide whether two plates had converged in the past to form a continent?

17 At diverging boundaries, seafloor spreading is occurring as new crust is continually created. Earth's crust is not getting any bigger despite this. Propose a reason why.

18 Fossils of sea creatures have been found near the top of Mt Everest (8848 m above sea level). How do you think they could have got there?

19 The oldest known fossils in the ocean floor are about 180 million years old, yet fossils from continents have been found that are 3400 million years old. Propose reasons why.

20 The science4fun on page 403 uses a computer simulation to predict what a future Earth may look like. What three main pieces of information about plate tectonics do you think were probably used to create the simulation?

MODULE

10.2 Practical investigations

1 • Paper plate tectonics  

Purpose

To model plate tectonics.

Timing 45 minutes

Materials

- paper template. To obtain this, type into a web browser: 'Sea-floor spreading and subduction model by John C Lahr'
- cardboard or shoebox
- sticky tape
- scissors
- coloured pencils

Procedure

1 Colour in the paper templates you have downloaded and printed.

2 Construct the model using the instructions provided. Alternatively, make your 'shoebox' from flat sheets of cardboard stuck together.

3 With a partner, operate the model. This represents seafloor spreading where there are also fault lines perpendicular to the rift. The sea floor is separating away from the central rift, but it is doing it in three separate strips. Your partner must gently keep holding the three paper strips below the top. You hold the two side pieces and slowly pull them apart, watching the magnetic strips as they appear.

4 Imagine what would be happening in real strips of ocean crust as their inside edges slide past each other like the paper strips.

Hint

The measurements on your print-out are in inches: 1 inch = 2.54 cm.

Review

1 Describe what the model shows about seafloor spreading.

2 Describe where the three transform faults are on the seafloor spreading and subduction model.

3 a Identify the following features of the seafloor spreading and subduction model.

- i subduction
- ii convection currents
- iii volcanic islands
- iv seafloor spreading
- v magnetic striping

b Justify your answer in each case.

2 • Types of crust  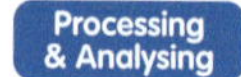

Purpose

To calculate the density of granite and basalt.

Timing 30 minutes

Materials

- granite and basalt (to fit in a displacement can)
- displacement can
- scales or triple-beam balance
- 100 mL measuring cylinder
- beaker
- cotton

Procedure

1 Measure the mass of the basalt and the granite. Record them in a table.

2 Tie the cotton to the granite and the basalt. Set up the equipment as shown in Figure 10.2.14.

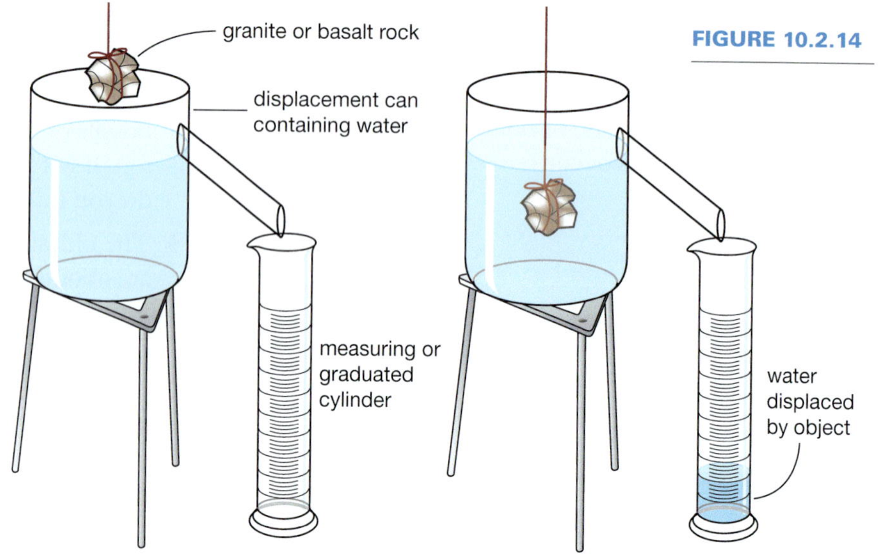

FIGURE 10.2.14

MODULE 10.2
Practical investigations

3 Holding the cotton, lower the basalt to the bottom of the can and hang the string over the edge. Measure the volume of water displaced.

4 Dry the measuring cylinder, remove the basalt and refill the can, then repeat step 3 using granite.

Results

Calculate the density of the basalt and the granite using:

$$\text{density} = \frac{\text{mass}}{\text{volume}}$$

Hint

$1\ cm^3 = 1$ mL water (at 20°C)

Review

1 Compare the density of granite and basalt.

2 The oceanic crust is mainly basalt and the continental crust is mainly granite. Explain why oceanic crust subducts under continental crust at a collision boundary between the two types of crust.

3 • Mobile phones, GPS and tectonic plates

Purpose

To evaluate whether mobile phone GPS apps could be used to track tectonic plate movements.

Timing 20 minutes

Materials

- mobile phone with GPS app

Procedure

1 Find your mobile phone GPS app or install one if necessary. Open the app. The app may ask you to calibrate your phone by moving the phone around. Do this so that the readings will be accurate.

2 Check how the satellite data is displayed. There are probably two different displays. One shows a number of satellites on a circular compass. The other is a table displaying coloured bars indicating signal strength and data such as is the satellite 'On' (probably shown by a tick). You can pick either display for this activity.

3 Find out how to check the number of satellites that are sending data (the satellites that are 'On'). This may appear as a fraction such as ON 8/16. This means only eight satellites of the 16 that are visible to the phone are sending data to your phone. The table probably displays this with coloured bars for 'On' satellites and white bars for 'Off'. The compass display probably does not show which ones are on. Your phone should look similar to that in Figure 10.2.15.

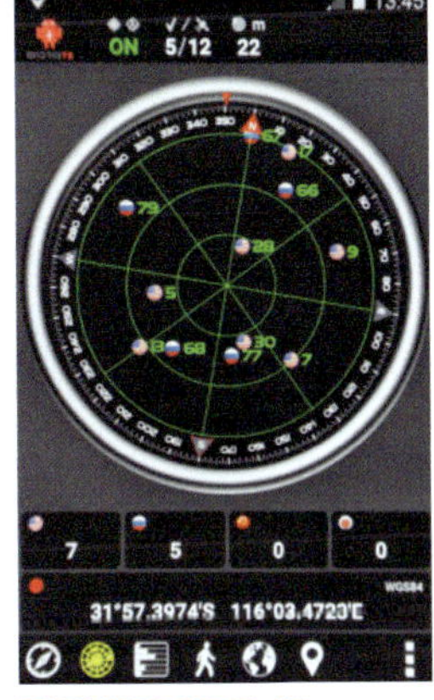

FIGURE 10.2.15

4 Check how to find and read latitude and longitude data from the phone.

5 While staying still in one place, use the GPS app to find your latitude and longitude.

6 Stay in the same place and observe your phone GPS data for at least 5 minutes. During this time, record in a table the latitude, longitude and number of satellites that are sending data (the ON satellites) each minute.

7 Now move your mobile phone a distance of 1 metre and again record latitude, longitude and number of satellites.

Results

Record your results in a table. Give your table a title.

Review

1 What happened to the latitude and longitude readings while you stood in exactly the same place for 5 minutes?

2 How did the number of satellites affect the latitude and longitude data?

3 How did the readings compare when you moved 1 metre from the original spot?

4 Evaluate how useful the GPS is in mobile phones for measuring your position on Earth.

5 Evaluate the possible use of mobile phones to track tectonic plate movement.

6 Considering your answer to question 5, how then is GPS useful for tracking tectonic plates?

MODULE
10.3 Volcanoes and earthquakes

Volcanoes are among the most awesome sights on the planet. The largest explosion in recorded history was from a volcano. Earthquakes can also be deadly. Earthquakes and volcanoes provide evidence of the interior processes of the Earth. They provide information that supports the theory of plate tectonics.

science 4 fun

Google Earth and tectonic plates

Can you use Google Earth to compare the positions of volcanoes, earthquakes and tectonic plate boundaries?

Collect this ...

- a computer or tablet with access to Google Earth

Do this ...

1 Open Google Earth on your device.
2 Find the 'Layers' menu. Click on the + next to the 'Gallery' button. Tick the box against 'earthquakes' and against 'volcanoes'.
3 Find the Earth Gallery. Enter in the search gallery 'tectonic plate boundaries'. Select the file 'Tectonic Plate Boundaries.kml'. Click on the button 'View in Google Earth'.
4 Now find any place you are interested in to start. View the globe and zoom in to your chosen site.

Find many more sites and each time note the position of volcanoes, earthquakes and plate boundaries. If you click on the boundary, it tells you more about the type of boundary and its movement.

Record this ...

1 Describe what you found.
2 Explain what you found.

Volcanic eruptions

A **volcano** is a place where extremely hot material from inside Earth erupts at Earth's surface. This material can include:

- gas such as steam and hydrogen sulfide
- ash made of fine rock particles
- lava
- lumps of solid volcanic rock such as scoria.

Volcanoes form where extremely hot molten rock called **magma** has accumulated below weak spots in Earth's crust. This magma is occasionally pushed upwards with great force into the volcano.

The magma reaches the surface, and is now known as **lava**. It erupts white hot at a temperature of over 1200°C. The lava changes colour as it cools, from white through yellow, orange and red, until it finally hardens and solidifies into rock.

Prac 1
p. 425

SciFile

Victims of Vesuvius

In 79 CE the volcano Mount Vesuvius erupted and covered the Italian city of Pompeii with hot ash. About 16 000 people died. The ash covered bodies and hardened into rock. When each body decayed, it left a cavity in the rock. Scientists poured plaster into the cavity to make a cast of each body. The outside layer was then chipped away, leaving the cast.

Volcanoes and plate boundaries

A comparison of Figure 10.2.2 on page 404 and Figure 10.3.1 shows that most volcanoes are at or near the edges of tectonic plates. This is because the movement of the plates creates weaknesses in the crust and also generates intense heat that can melt rock.

Diverging plate boundaries create weaknesses in the crust because the separating plates thin the crust. This lowers the force on the underlying hot rocks of the asthenosphere and they melt. The magma formed then finds its way up through the weaknesses in the crust.

SciFile

Krakatoa

Krakatoa (Krakatau) is a volcanic island in Indonesia. In 1883, it exploded, blowing two-thirds of the island to pieces. A total of 21 cubic kilometres of land was blown into the air. The sound was clearly heard over 5000 km away and 36 000 people died. The largest ever eruption was 2.1 million years ago in Yellowstone, USA, which released a massive amount of land, around 200 times as much land as Krakatoa.

FIGURE 10.3.1 Most volcanoes form near the edges of the tectonic plates.

Converging plates, create weaknesses in the crust and generate a lot of heat. This is especially evident where subduction occurs. If the colliding boundaries occur under the ocean, then chains of volcanic islands (island arcs) can form at the edges of the tectonic plates (Figure 10.3.2). An example is the Lesser Sunda Islands, Indonesia. They are shown in Figure 10.3.3. Island arcs are common in the ocean because rocks that contain a lot of water melt at a much lower temperature. This assists in creating magma.

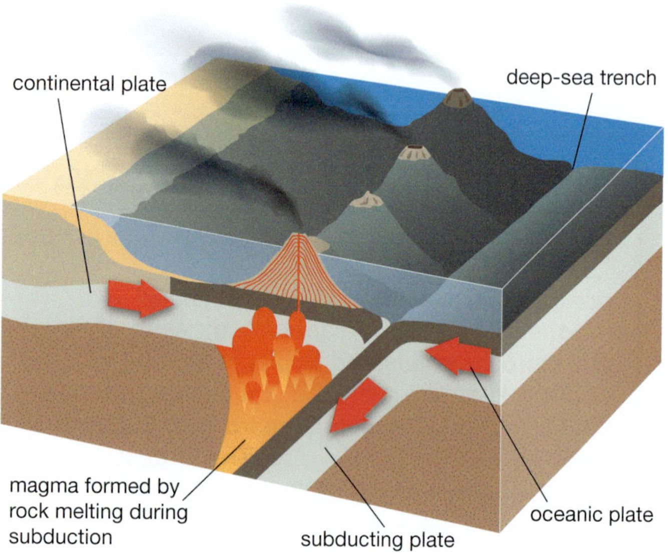

FIGURE 10.3.2 Subduction results in weakness in the crust and generates of a lot of heat. A string of volcanic islands (an island arc) is the result.

Hot spot volcanoes

While most volcanoes form near plate boundaries, some form well away from the edges of the plates. These volcanoes sit over 'hot spots' in Earth's crust. **Hot spots** are isolated weak spots in the crust where a lot of hot magma is being created. They can occur under oceanic or continental plates. Geologists are not sure why these hot spots exist. Research is currently being done to find out where the magma originates in these places.

In the ocean these volcanoes occur in chains of islands. In each island chain, there is always one island with an active volcano, while all the other islands have dormant (inactive) volcanoes. Geologists realised that the formation of these volcanoes could be explained by the theory of plate tectonics. Each island formed as it sat over the hot spot. As the plate moved, the island went with it and so the island no longer sat over the hot spot. A new part of the plate was now above the hot spot and this gradually formed a new volcano. The Hawaiian islands are a good example of an island chain over a hot spot. Their formation is shown in Figure 10.3.4.

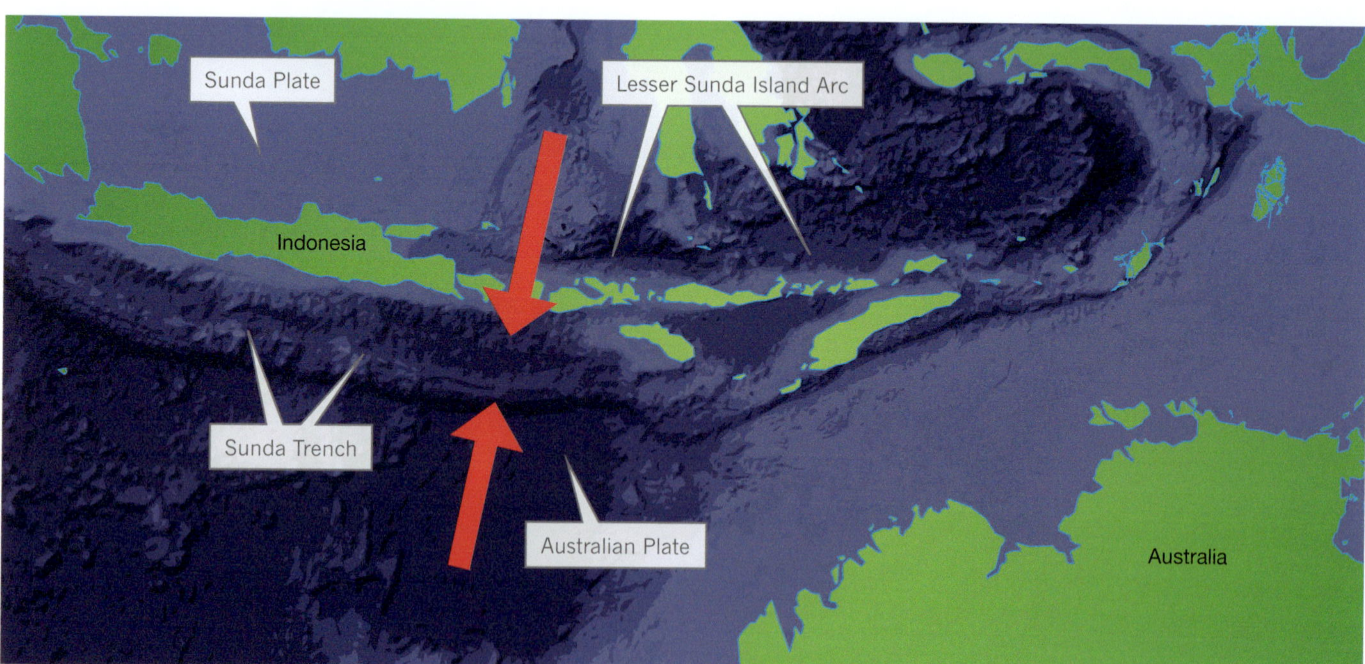

FIGURE 10.3.3 The Australian Plate is moving north and colliding with the Sunda Plate, creating an island arc called the Lesser Sunda Islands (Indonesia).

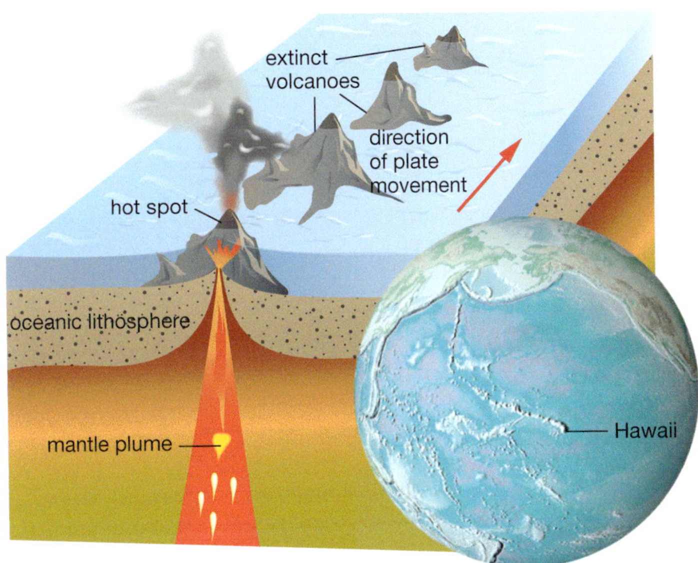

FIGURE 10.3.4 The Hawaiian islands have been formed by movement of the Pacific Plate over a single hot spot in Earth's mantle.

Geologists have proposed two possible origins of the magma formed at hot spots. One is that heat is escaping in narrow plumes from deep in the mantle and stays in a fixed place for many millions of years. Another explanation is that the magma is created in those places because that is where the crust is flexing as the tectonic plates move over the asthenosphere. This flexing is due to the plates twisting as they move, and that the spots where the magma is rising are where the greatest stresses occur in the crust. Whatever the cause, the chain of islands provides evidence that tectonic plates are moving.

Working with Science

SEISMOLOGIST AND MATHEMATICAL GEOPHYSICIST

Rhodri Davies

FIGURE 10.3.5 Dr Rhodri Davies

Seismologists and geophysicists are scientists who study the structure, composition and shifts in Earth's tectonic plates. They also study resulting events, such as earthquakes, tsunamis and volcanoes. Geophysicists use physics to understand these structures and processes, while seismologists specifically study the movement of earthquakes through the Earth through seismic waves. Dr Rhodri Davies combines both these fields as a researcher at the School of Earth Sciences at Australian National University in Canberra (Figure 10.3.5). His work involves modelling volcanic hot spots and the movement of tectonic plates to gain an understanding of Earth's past, and enable more accurate predictions of future volcanic activity.

Dr Davies was recently a lead researcher in a study on the world's longest known chain of continental volcanoes in eastern Australia. The volcanic chain is 2000 kilometres long and stretches from Queensland's central coast to central Victoria. The researchers used a supercomputer and cutting edge technology to model and map variations in the thickness of the Earth's crust and movement of the mantle under eastern Australia. They discovered that the volcanic chain began forming approximately 33 million years ago. At that time, as Australia was drifitng northward over a hot spot. Strong forces caused magma to rise from around 3,000 kilometres below Earth's surface, creating volcanoes. The researchers found that the volcanoes in the south are younger than those in the north, reflecting the northward movement of Australia over the hot spot. They predict that the hot spot is now located between King Island and the Tasmanian mainland, but have not yet found evidence of any volcanic activity there.

To become a seismologist or geophysicist, you will need to complete a Bachelor of Science, majoring in Earth Science. This qualification can lead to careers in mineral exploration, natural resource management, mapping and surveying, consulting, and modelling and forecasting earthquakes, volcanoes and oceanographic activity.

Review

1 Why do you think it is important to research and monitor the movement of the Earth's tectonic plates?

2 Most volcanoes form near the edges of tectonic plates. The volcanic chain that runs from Queensland to Victoria is unusual because it is not near tectonic plate boundaries.

   a Explain the process that formed this unique chain of volcanoes.

   b Explain why there might be stretches of land along this chain without volcanoes.

Earthquakes

An **earthquake** is the rapid movement of the ground, usually back and forth and up and down in a wave motion. It is caused by the rapid release of energy as the tectonic plates move. Tectonic plates are usually held together by the force of friction. Sometimes, however, the forces in the crust become so large that they overcome the force of friction, and the plates suddenly move. This sudden movement sends out waves of energy through the rock and the water. The ground and water then shake as the waves of energy pass through them.

Detecting earthquakes

Earthquakes are measured using an instrument called a **seismometer**. Old-style seismometers consisted of a pen on a moving drum. They used the principle of inertia. The heavy mass attached to the pen has a lot of inertia, meaning it tends to stay still. The rest of the seismometer moves with the vibrations of Earth.

There are many designs for seismometers. Two designs are shown in Figures 10.3.6 and 10.3.7. Modern seismometers are electronic. The trace of a seismometer is called a seismograph.

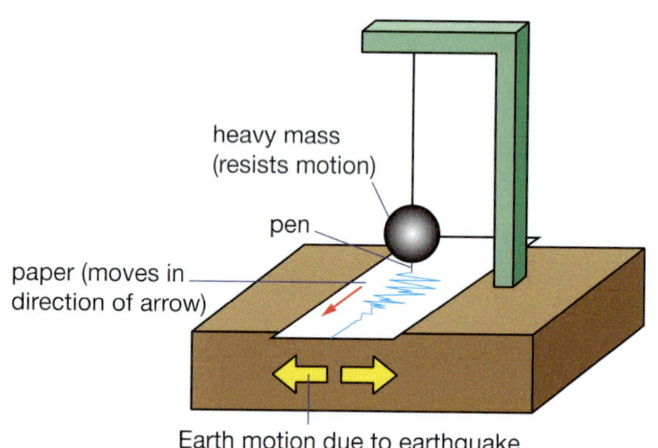

FIGURE 10.3.6 This seismometer uses a swinging pendulum. The fixed supporting arm and the recording paper vibrate back and forth with the ground.

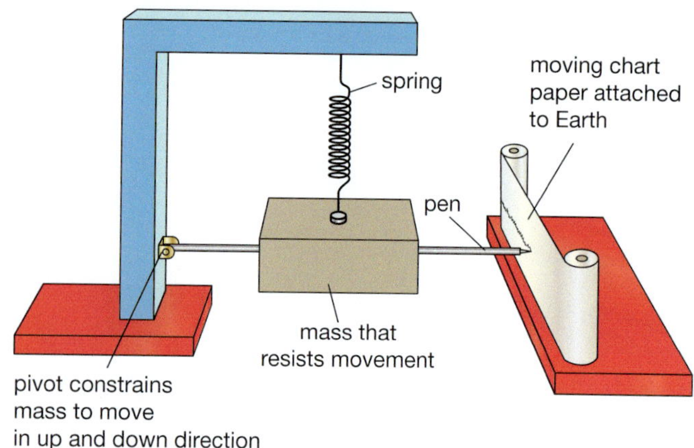

FIGURE 10.3.7 In this vertical-mounted seismometer, the recording paper and the fixed support move up and down with Earth.

Seismic waves

The movement of the ground in an earthquake occurs in a shaking back-and-forth motion called a wave. These waves in Earth caused by earthquakes are called **seismic waves**.

Three main types of seismic waves can be detected on a seismograph (Figure 10.3.8).

- **Primary waves (P-waves)** are longitudinal waves that travel fast through Earth. Longitudinal waves move the ground back and forth.
- **Secondary waves (S-waves)** are transverse waves that travel slightly slower than P-waves through Earth. Transverse waves move the ground up and down.
- **Surface waves** are the slowest waves and cause the most destruction.

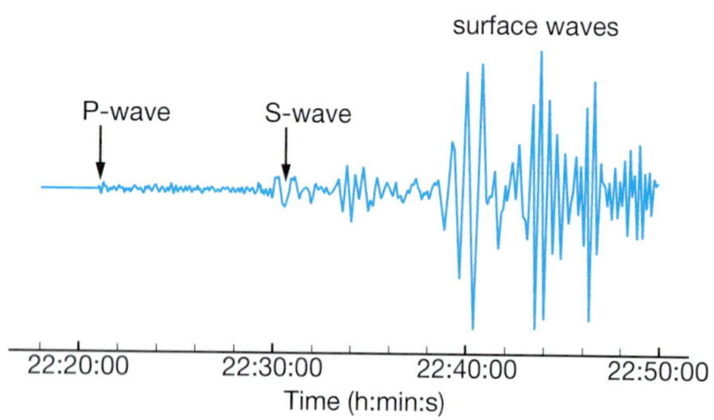

FIGURE 10.3.8 This seismograph shows the pen trace when S-waves, P-waves and surface waves reach the seismometer.

Prac 2
p. 425

S-waves and P-waves travel deep under the ground and then bend upwards to reach the surface of the crust. P-waves shake the ground up and down. S-waves shake the ground sideways, back and forth.

Surface waves travel along the crust near the surface. They travel more slowly than P-waves and S-waves, but they can be much larger. They are particularly destructive if the earthquake is near Earth's surface.

Where earthquakes occur

Nearly all earthquakes start at the edges of the tectonic plates. Compare Figure 10.3.9 with Figure 10.2.2 on page 404 and the locations of earthquakes align with the boundaries of the tectonic plates. The largest number of earthquakes and the strongest earthquakes occur near converging boundaries. This distribution of earthquakes provides further evidence for the theory of plate tectonics.

Epicentre and focus

Earthquakes happen at particular places under the ground where Earth slips, usually along a fault. The place where the quake starts is called its **focus**. This may be many hundreds of kilometres deep in Earth. The point on Earth's surface directly above the focus is called the **epicentre**. Buildings near the epicentre are usually the most heavily damaged.

The severity of an earthquake is calculated in several different ways. One early method, still used today, is measured on the Richter scale.

SciFile

Big quakes

The largest earthquake ever recorded was one of magnitude 9.5 in Chile in 1960.

The world's deadliest recorded earthquake occurred in 1556 in central China. It killed an estimated 830 000 people. In 1976, another deadly earthquake struck in Tangshan, China, where more than 250 000 people were killed.

This scale is an open-ended scale but usually finishes at 10. Each successive number is ten times greater than the previous number. So an earthquake measuring 5.0 is ten times more destructive than one measuring 4.0.

The most commonly used way to measure the strength of an earthquake is the moment magnitude scale (MMS). This scale is based on readings from a seismometer. The number that is calculated from the seismometer gives an idea of how much energy the earthquake had. The scale runs between 1 and about 10, though readings over 10 are possible. The largest ever recorded earthquake was in Chile, South America, in 1960. It had a magnitude of 9.5. This is about 30 times as much energy as an 8.5 reading.

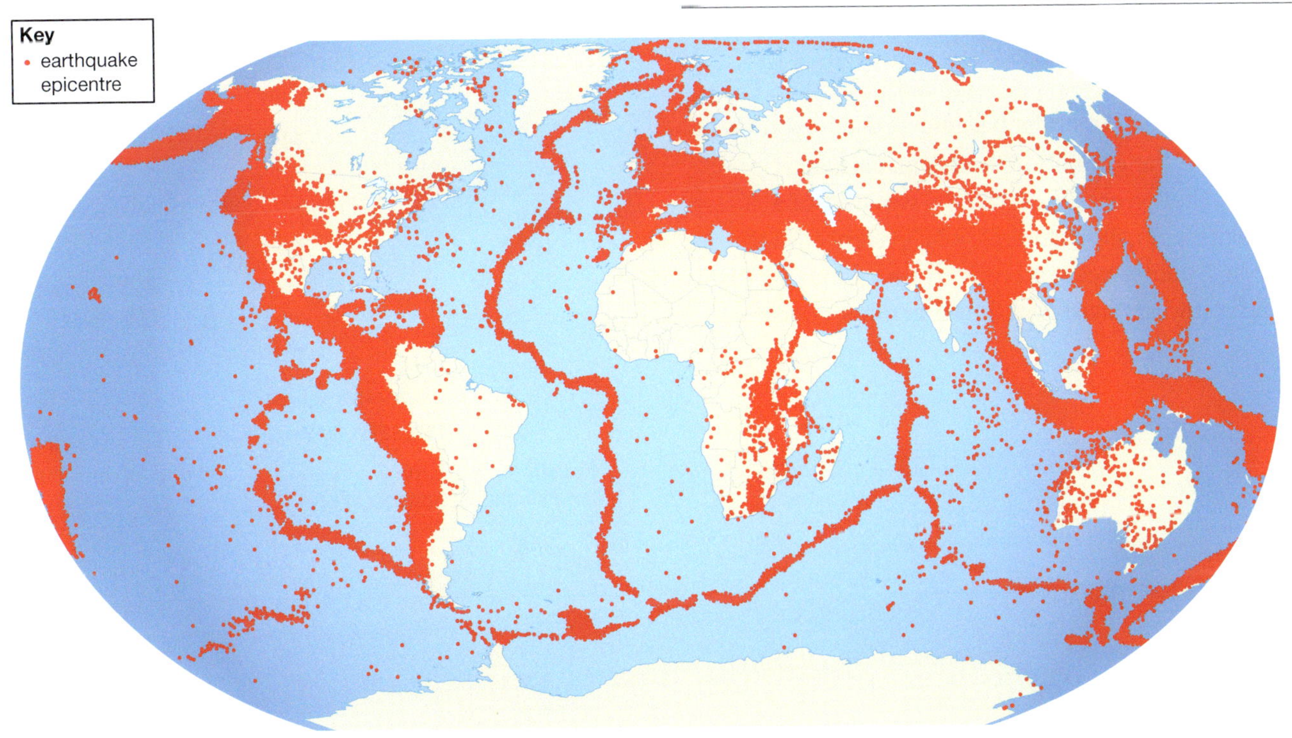

FIGURE 10.3.9 Earthquakes mostly occur near the boundaries of the tectonic plates. However, some types of boundaries have more earthquakes.

Tectonic history of Australia

Pangaea (pan-gee-ah) was an ancient landmass that existed about 225 million years ago. This mega-continent gradually broke into two major landmasses, Laurasia in the north and Gondwana in the south. Australia was part of Gondwana.

Reconstructions of how the land masses appeared in the past are shown in Figure 10.3.10. This diagram also shows that Australia is a very old continent. Its existence dates back to Pangaea, 225 million years ago—the evidence is found in the rocks of Australia's crust. The main events in Australia becoming a separate continent are:

- About 225 million years ago, Australia was connected to the other continents as part of the ancient land mass called Pangaea.
- About 125 million years ago, Australia was still connected to Antarctica. Almost all of the centre of Australia was covered by a sea that separated the country into four landmasses.
- About 85 million years ago, the sea level dropped, exposing more land. Australia slowly moved northwards, but still remained connected to Antarctica at the South Tasman Rise, a small area of land near Tasmania.
- About 65 million years ago, Australia began to separate from Gondwana. This was the Cretaceous period and about the time the dinosaurs were becoming extinct. Australia's landforms and climates changed as it moved northwards on the Indo-Australian tectonic plate.
- About 40 million years ago, the South Tasman Rise separated, and Australia became an island continent.

Long before Pangaea and Gondwana existed, about 1 billion years ago, Australia was part of a supercontinent called Rodinia. The evidence for this is found in 1 billion-year-old rocks from various places in Australia, and similar rocks found in other countries.

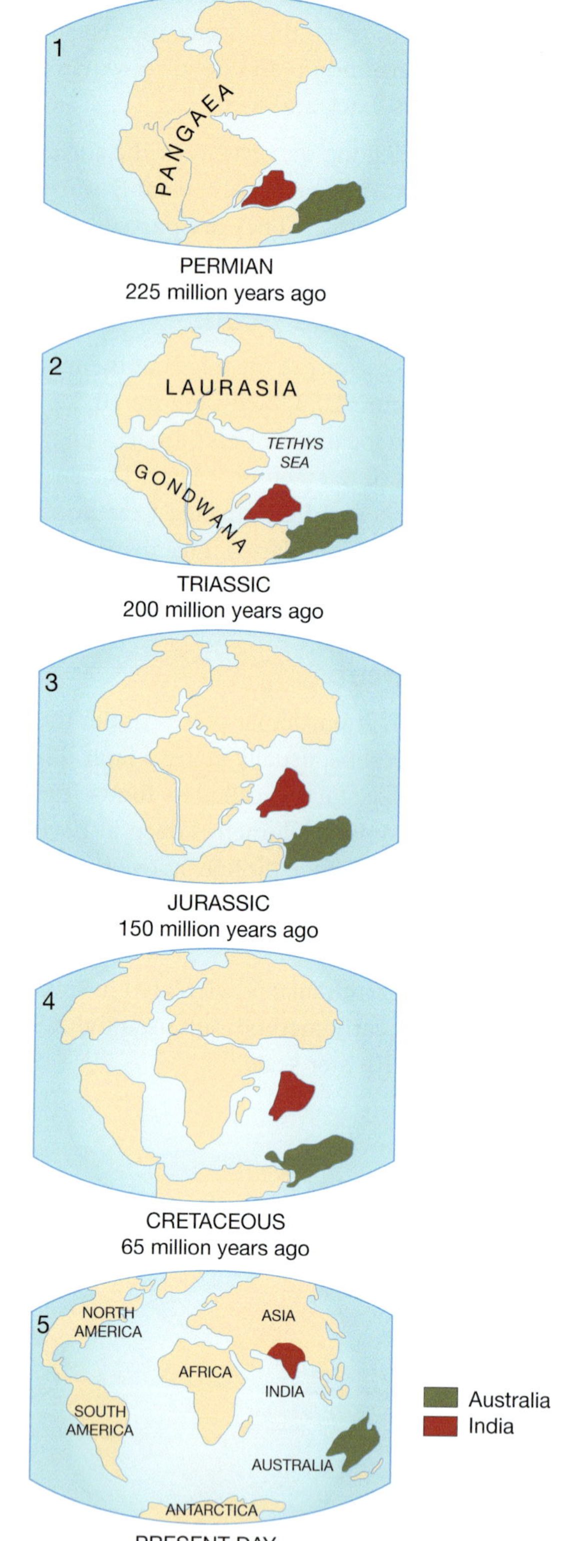

FIGURE 10.3.10 Reconstruction of the break-up of Pangaea, Laurasia and Gondwana

At that time, Australia was in the northern hemisphere connected to Antarctica, parts of North America and China (Figure 10.3.11).

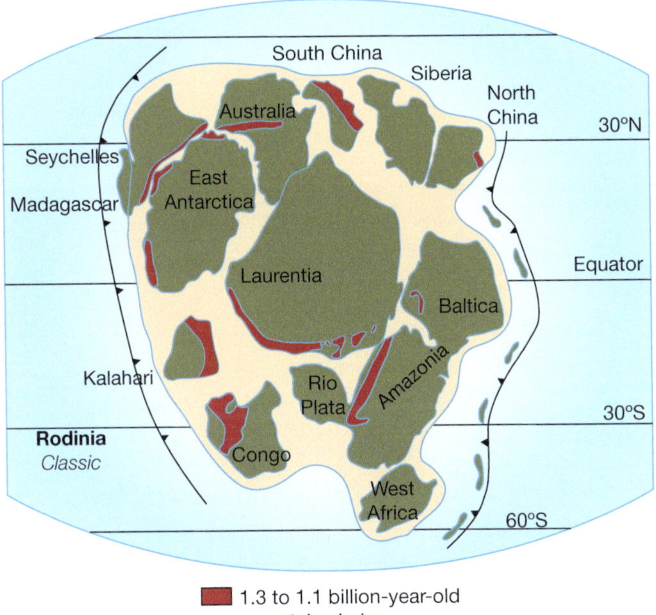

FIGURE 10.3.11 Rodinia was an ancient supercontinent even before Pangea. Rodinia existed 1 billion years ago.

Well before Rodinia, the main parts of Australia that existed were three large areas of igneous rocks, known as cratons. Today, these form a large part of Western Australia and South Australia. The oldest rocks in Australia are found in the Yilgarn craton of Western Australia, dating from 3.7 billion years ago. These cratons have been moving around on tectonic plates since their formation.

Australia has been a very stable continent for many millions of years. There are two main reasons for this. One is that the lithosphere under the continent is about 200 km deep, about double the depth for other continents. This makes the continent resistant to melting and fracture. Figure 10.3.12 shows the thickness of the lithosphere under Australia. The map was prepared by using seismic measurements. Some of these seismic waves were from earthquakes that occurred in Indonesia and the Pacific Ocean. Most of the records are from large machines carried on trucks that vibrate the ground and create P-waves in the rock. Scientists study the P-waves that bounce back off like echoes from rock layers.

The other reason for the stability is that Australia is not near plate boundaries and has therefore had very few earthquakes or volcanoes. About 85 million years ago it was near to a diverging boundary along the east coast, but that boundary now does not exist. Australia has steadily been drifting north towards Indonesia for over 65 million years. There have been few volcanic eruptions.

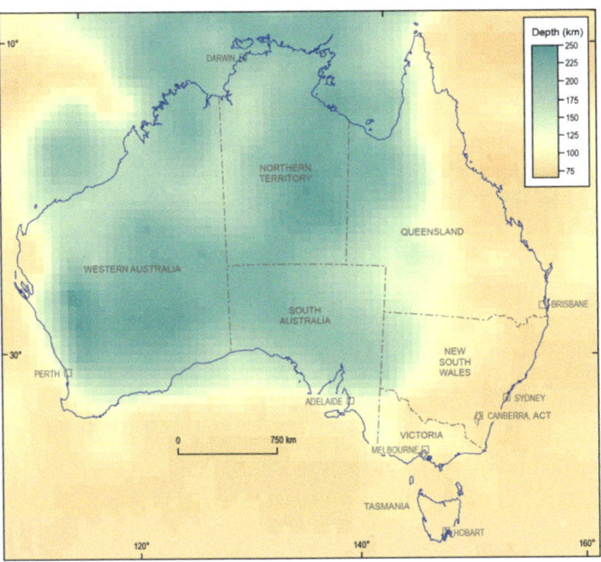

FIGURE 10.3.12 A map showing the thickness of the lithosphere under Australia. It was created by using information from seismic waves.

Most have been because of a now inactive hotspot just off the coast of Victoria. This hot spot created a chain of volcanoes which are now extinct remnants of the original mountains. These range from Cosgrove in Victoria (formed 9 million years ago) to Hillsborough in northern Queensland (formed 33 million years ago). You can see this mountain range in Figure 10.3.13. The last volcano to erupt in Australia was around 5000 years ago at Mt Gambier, South Australia. Australia has few earthquakes and those that we do experience come mainly from stress in the crust due to the bending of the tectonic plate as it pushes into Indonesia and the Pacific Plate.

AB 10.6

FIGURE 10.3.13 The world's longest known chain of continental volcanoes

SCIENCE AS A HUMAN
ENDEAVOUR

Use and influence of science

Science and living near plate boundaries

FIGURE 10.3.14 A tsunami hits Phuket, Thailand on 26 December 2004

Australia lies on one tectonic plate and so it is geologically stable—earthquakes are rare and we have no active volcanoes. However, many millions of people across the world live near plate boundaries and so they need to live with frequent earthquakes and volcanic eruptions. Public safety is a major concern and science is often used to help find solutions.

Earthquakes

An earthquake under the ocean can cause a huge wave called a **tsunami**. On 26 December 2004, one of the largest earthquakes ever recorded (magnitude 9.2) occurred off the west coast of Sumatra in Indonesia (Figure 10.3.14). This earthquake was triggered by sudden movement at a converging boundary between the Australian and Eurasian tectonic plates. A tsunami travelled to the coastlines of 11 countries around the Indian Ocean, and more than 280 000 people died. After this disaster, Australia created an early warning system for tsunamis. This system uses data from 50 seismic stations in Australia and 120 from overseas stations. Computers linked to these stations alert scientists to a potential tsunami. Scientists then input the data into a scientific model to determine how big the tsunami may be, how long it may take to reach Australia and where it could hit the coast. They check with deep ocean detector buoys to see if the tsunami actually forms and if it is acting as expected from the model. The way deep ocean detector buoys work is shown in Figure 10.3.15.

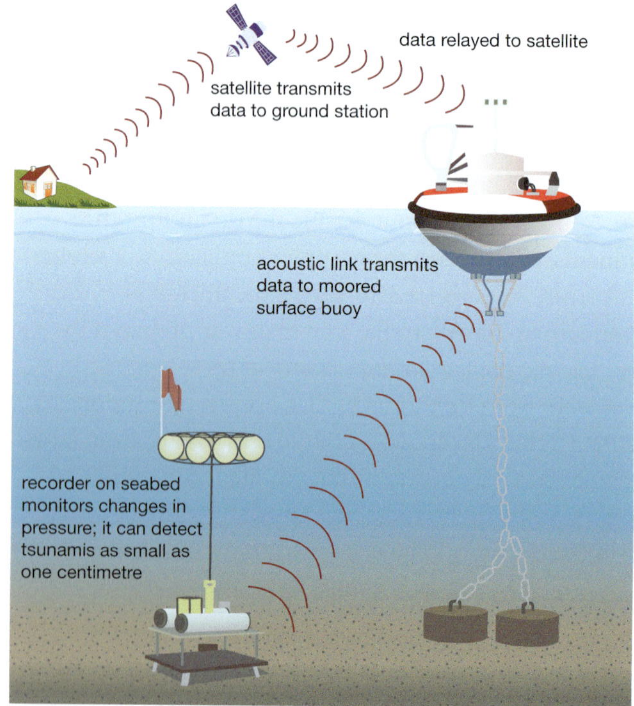

FIGURE 10.3.15 The deep ocean buoy system for early warning of tsunamis

SCIENCE AS A HUMAN ENDEAVOUR

New Zealand, Indonesia, Japan and the west coast of USA all lie on or near major plate boundaries. For this reason, earthquakes are common. Buildings need to be designed and built to withstand earthquakes. Engineers have found that the most effective solution is vibration control. This uses:

- base isolation. Figure 10.3.16 shows how springs, ball bearings and pads can isolate a building from the waves passing through the ground underneath it. This allows the building to remain in position rather than vibrate with the Earth
- dampers. These are structures that move in opposition to the waves and oppose their effect.

FIGURE 10.3.16 Base isolation involves placing structures such as these rubber bearings under a building.

Prac 3 p. 426

Volcanoes

Japan, Indonesia, Mexico and central America all have large numbers of active volcanoes that have formed near major plate boundaries. The volcanoes are monitored in several ways. One is to install seismometers near volcanoes. These create a trace showing the shaking movement of the ground as magma pushes through the rocks. Volcanoes also move by swelling outwards as magma enters. Seismometers can transmit information to computers in volcano monitoring centres. Another way of monitoring volcanoes is to set up GPS ground stations similar to those used when mapping movements of tectonic plates (Figure 10.3.17). The ground stations measure the height of the volcano and indicate if it is swelling.

FIGURE 10.3.17 Setting up a GPS ground station on a volcano

If an eruption appears likely, then warnings can be given to people living nearby and towns evacuated. One recent innovation is a small drone helicopter that carries instruments to measure gas emissions from a volcano. It can be flown up from a safe distance and send back data on gases that can indicate if the volcano is about to erupt.

Volcanoes are also monitored after they erupt. The volcanic ash can rise high into the sky, eventually falling back to the surface, polluting crops and killing farm animals. The ash can also block aircraft engines, leading to engine failure. There are volcanic ash advisory centres in many countries. Australia has one in Darwin and this covers half way across the Indian Ocean, over to New Zealand and north to India. All aircraft in the region are warned of the position of the ash clouds.

REVIEW

1 List four examples of science being used to solve the problems of living near a plate boundary.
2 What is a tsunami and why is it dangerous?
3 List the main features of the tsunami early warning system.
4 Why will GPS monitoring of volcanoes assist in an early warning system?
5 Compare the method of detecting volcanic eruptions using a seismometer and a GPS ground station.

MODULE

10.3 Review questions

Remembering

1 Define the terms:
  a earthquake
  b island chain
  c seismometer
  d volcano.

2 What term best describes each of the following?
  a the point on Earth's surface directly above the focus of an earthquake
  b isolated places away from plate boundaries where a lot of hot magma is collecting
  c the shaking, wavelike movement of the ground in an earthquake
  d molten rock that has erupted onto Earth's surface.

3 List the materials found in volcanic eruptions.

4 Name a tectonic process that generates heat at the boundaries of the plates.

Understanding

5 Explain why volcanoes are often found near the edges of converging tectonic plates.

6 Explain how island chains form from hot spots below the ocean crust.

7 Explain the cause of earthquakes.

8 How could you tell if a line of islands was formed by a hot spot and tectonic plate movement? List the evidence that you would look for.

9 Give two reasons to explain why Australia is such a geologically stable continent.

10 Describe the evidence that supports the view that Australia is an extremely old continent.

11 Explain why Japan has many active volcanoes, whereas Australia has none.

Applying

12 The Australian Plate forms a converging boundary with the Indonesian island of Java at the Sunda Trench. Java is one of the most volcanically active places on Earth. Use your knowledge of plate tectonics to explain the presence of these volcanoes.

13 The Lesser Sunda Islands of Indonesia are an island arc with about seven active volcanoes. Collision boundaries in the ocean often result in volcanic islands. Explain how these two facts are linked.

14 Recent studies of the rocks of Tasmania have confirmed that there are parts of the island that are more like the rocks of early North America (Laurasia) than the rest of Australia. Use Figure 10.3.11 to suggest an explanation.

15 Consider your answers to Module 10.1 review question 12 and Module 10.2 review question 12 where you constructed a table summarising evidence for plate tectonics. To that table, add any further evidence from Module 10.3.

Analysing

16 Compare the formation of an island arc with the formation of an island chain from a hot spot.

17 Consider the following two facts. Collision between oceanic plates produces many volcanoes. Collison between continental plates (such as those that formed the Himalayas) may have few or no volcanoes. How do you explain this difference?

18 There are many earthquakes in the South Island of New Zealand, but the nearest volcanoes are in the North Island, about 700 km away. The North Island has regular earthquakes. How do you explain this difference?

19 Compare S-waves and P-waves.

Evaluating

20 The science4fun on page 414 used Google Earth to compare plate boundaries, and earthquake and volcano locations. A similar result can be obtained by comparing Figures 10.2.2, 10.3.1 and 10.3.9. What conclusions can be drawn from these comparisons?

21 Steam emerging from volcanoes in the Andes mountains of South America contains signs that the water in it came from the Pacific Ocean.
  a How do you think this happened?
  b Use your knowledge of subduction to justify your answer.

MODULE

10.3 Practical investigations

1 • Model volcanoes

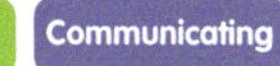

Purpose

To build a model volcano and observe its eruptions.

Timing 45 minutes

Materials

- baking soda
- vinegar
- food dye
- cornflour
- Alka-Seltzer® tablets

- cardboard
- aluminium foil
- newspaper
- plastic tape (for example, duct tape)
- white laboratory tray or newspapers
- glass stirring rod
- scissors
- 100 mL beaker
- reaction vessel (for example, small jar)

SAFETY

A risk assessment is required for this investigation.

Vinegar can damage your eyes, so wear safety glasses at all times. Food dye can stain your skin, so wear gloves.

Procedure

1 Fold the cardboard into a cone shape with a base diameter of about 30 cm, but leave a hole at the top large enough to hold the reaction vessel. Tape the cardboard to form the cone shape.

2 Insert the reaction vessel and tape it in securely. Seal around the edges of the cardboard with plastic tape or aluminium foil.

3 Cover the cardboard with aluminium foil to waterproof it. Tape the foil to seal it.

4 Place the volcano in a laboratory tray or on about 20 sheets of newspaper.

5 Decide how much baking soda and vinegar to use for your first mix. Measure these out. Place the baking soda in the reaction vessel. Add a couple of drops of food dye. Add your measured amount of vinegar.

6 If the volcano does not errupt, explore different amounts of vinegar and baking soda to see if you can improve the effect. Video the experiment on your mobile phone if given permission.

7 An alternative mixture to try is a thin paste of cornflour and water and Alka-Seltzer tablets.

Results

Record your observations.

Review

1 Construct a word equation for the baking soda and vinegar reaction.

2 Compare your model with a real volcano.

• STUDENT DESIGN •

2 • Seismometers

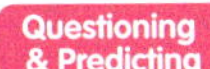

Purpose

To build and test a model seismometer.

Timing 45 minutes

Materials

materials as selected by students

SAFETY

A Risk Assessment is required for this investigation.

Procedure

1 In a team, decide on your seismometer design. Construct a diagram and provide a list of necessary materials to your teacher.

2 Before you start any practical work, assess all risks associated with your procedure. Construct a risk assessment that outlines these risks and any precautions you need to take to minimise them. Show your teacher your procedure and your risk assessment.

If they approve, then collect all the required materials and start work.

See Activity Book Toolkit to assist with developing a risk assessment.

3 Build your seismometer. Test the seismometer by gently bumping the desk without moving the desk.

Hints

- You will have to pull the paper by hand as the pen moves over it.
- Use the STEM and SDI template in your eBook to help you plan and carry out your investigation.

Review

1 Compare your seismographs with those of other groups, and evaluate the effectiveness of the different designs.

2 Propose how you could improve your design.

MODULE

10.3 Practical investigations

• STUDENT DESIGN •

3 • Earthquakes and buildings

Questioning & Predicting | Planning & Conducting

Purpose

To test how three variables may affect the performance of a building in an earthquake area. The three variables are:

1 distribution of weight in a building
2 width of the building or sections of the building
3 base isolation of the building.

Hypothesis

How do you think the three variables will affect the performance of a building in an earthquake area? Before you go any further with the investigation, write a hypothesis for each variable in your workbook.

Timing

60 minutes

Materials

- Plasticine
- hardcover book
- pencils
- wooden blocks, polystyrene blocks, cardboard boxes
- ice-cream container
- ball bearings, marbles, sand, 'hundreds and thousands'
- other materials as selected by students

Procedure

1 In your team, decide how you will make your 'earthquake generator', which is a way of rocking your model buildings back and forth. A hardcover book resting on pencils as shown in Figure 10.3.18 is the simplest solution, but you will need to consider this as an experimental variable and control its effect.
2 Your buildings must be made in three or four sections. Use different sections (floors) about 10 cm high and a base of about 5 cm by 5 cm for each section. You can change this if there is enough material available. You will need to make the sections from different materials such as wood or polystyrene to test variable 1. To test variable 2, some sections will need to be different widths.
3 To test variable 3, base isolation, you can use the ice-cream container and different materials such as ball bearings, sand, marbles and 'hundreds and thousands'. If you want to try anything else, ask your teacher.
4 Build your first model building to test one of the three variables. Place the model building on your earthquake generator and test its performance. Remember that how you operate your earthquake generator is a variable. Record your observations.
5 Test the other two variables.

Use the STEM and SDI template in your eBook to help you plan and carry out your investigation.

Review

1 Explain how you kept the earthquake generator as a controlled variable in your three different tests.
2 a Construct a conclusion for each variable you tested in your investigation.
  b Assess whether each of your hypotheses were supported or not.
3 How important do the three variables seem to be in building design?
4 Evaluate the experimental procedure you used and, if necessary, recommend any improvements.

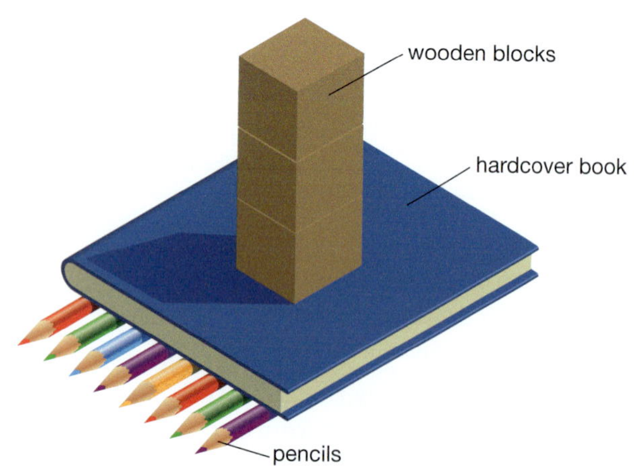

FIGURE 10.3.18

CHAPTER 10

Chapter review

Remembering

1 Who was the first person to propose the theory of continental drift?

2 List ten major tectonic plates.

3 List the evidence that supports the hypothesis of seafloor spreading.

Understanding

4 a Why was Wegener not able to prove his theory?

b Why are scientists now sure that he was correct?

5 Explain the process of seafloor spreading and subduction, and how these relate to convection and gravitational force.

6 Describe the three types of tectonic plate movements and the effects these have on the crust.

7 Why is Australia a very old and stable continent?

8 Describe how technological developments (such as magnetometers, echo sounders and GPS) have increased scientific understanding of global patterns in geological activity.

9 Use the theory of plate tectonics to explain how new landforms such as mountain ranges, volcanic islands and ocean trenches develop.

10 Explain why earthquakes and volcanoes occur more often near convergent plate boundaries.

Applying

11 Use the theory of plate tectonics to explain why geologists predict that the Great Rift Valley in Africa will be the site of an ocean millions of years in the future.

12 Identify the tectonic plates that form the following boundaries.

a a collision boundary between two continental plates creating the highest mountain range in the world

b a constructive boundary forming an island in the north Atlantic Ocean

c a destructive boundary forming many islands in Indonesia.

Analysing

13 a Compare the crust that makes up the Nazca plate and the crust that makes up South America.

b Account for the destructive plate boundary between these two plates.

14 Compare the effects of subduction when oceanic plates meet, with the effects when oceanic and continental plates meet.

15 Mountains can be formed near all three types of tectonic boundaries. In what ways are these processes of mountain building different?

Evaluating

16 What reasons support the conclusion that oceans and some 'mountains' can form from rifting.

17 Rifting of continents creates oceans. Seafloor spreading is the continuing process while the ocean is increasing is size. But oceans can also disappear. What processes do you think would cause this? (Hint: think about collision boundaries such as Australia/Indonesia and Africa/Eurasia.)

18 a Assess whether you can or cannot answer the questions on page 391 at the start of this chapter.

b Use this assessment to evaluate how well you understand the material presented in this chapter.

Creating

19 Use the following ten key terms to construct a visual summary of the information presented in this chapter.

seafloor spreading
magnetic striping
converging boundary
diverging boundary
transform boundary
ocean trench
mountains
island arcs and chains
volcanoes
earthquakes

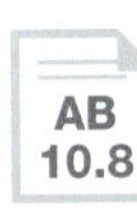

CHAPTER 10 Inquiry skills

Research

1 Planning & Conducting | Communicating

Investigate technologies involved in the mapping of tectonic plate movements. In your research find:

- four different methods by which tectonic plate movement can be detected and measured
- maps showing data for rates of movement of tectonic plates in three different places on the Earth. Examples might be California, New Zealand and Japan
- whether the rate of movement of tectonic plates can be used to predict any problems that may occur for people living nearby.

Present your findings in digital form.

2 Planning & Conducting | Communicating

Research volcanoes in Australia. Find:

- evidence of past volcanoes
- their position on a map
- descriptions of the geology or landforms associated with at least three of the sites
- how the volcanoes are thought to have occurred.

Present your findings as a PowerPoint presentation.

3 Planning & Conducting | Communicating

Research an earthquake such as the 1989 Newcastle earthquake or the 2011 Christchurch earthquake. Find:

- descriptions of the geology of the area around the epicentre
- probable reasons for the earthquake
- whether the disaster had any effect on recommended building designs for the area.

Present your findings as an article for a newspaper or newspaper website.

4 Planning & Conducting | Communicating

Research the Great Rift Valley of Africa, also known as the East African Rift System. Find:

- the general structure of the Rift Valley
- whether the African plate is split through the lithosphere to form separate plates meeting at the Rift Valley
- if scientists are sure what type of plate movement is occurring along the Rift Valley
- theories that have been proposed to explain the formation of the Rift Valley
- evidence that supports particular theories of the origin of the Rift Valley
- whether a new ocean is likely to develop in the Rift Valley.

Present your research in digital form.

5 Planning & Conducting | Communicating

Research the meaning and importance of cratons in the formation of continents. Find:

- the definition of a craton
- the names, ages and locations of the three oldest cratons of Australia
- the age of Australia's oldest cratons compared with other continents
- how cratons are joined to each other to form a continent
- whether there is any link between cratons and important minerals in Australia.

Present your research as answers to the above points.

6 Planning & Conducting | Communicating

Research the following three physical features: Dead Sea, Kermadec Trench, Reunion Island.

- Describe the tectonic plate movement in those areas.
- Name the tectonic plates involved.
- Explain how the physical features formed.
- Discuss which of these physical features may be linked to the extinction of the dinosaurs.

Present your findings as an annotated diagram that includes answers to the above points.

CHAPTER 10

Inquiry skills

Thinking scientifically

1 Consider Figure 10.4.1.

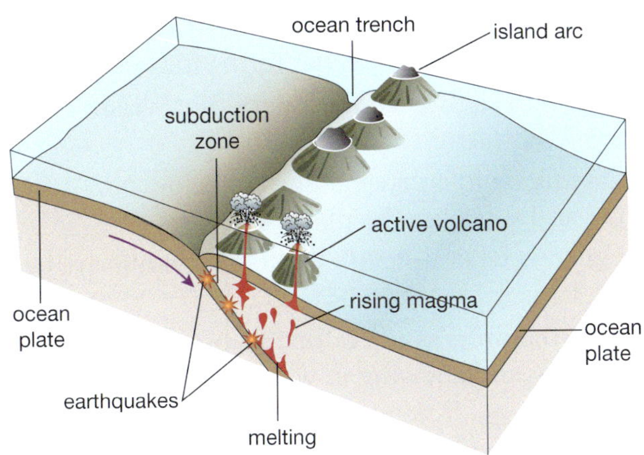

FIGURE 10.4.1

The reason for the island arc is:

A magma cools down fast when it erupts into seawater and so mountains form

B the faster moving oceanic plate always subducts under the slower moving plate

C subduction generates heat and crust melts easier when it contains a lot of water

D mountains form when plates collide and the crust crumples under the force.

Questions 2 and 3 refer to Figure 10.4.2.

2 Figure 10.4.2 is being used to show that:

A Earth is composed of layers

B continents are formed by seafloor spreading

C plate tectonics can change Earth's climates

D tectonic plates move due to convection currents.

3 From Figure 10.4.2, you could deduce that subduction and seafloor spreading together:

A cause the convection currents

B show that the crust is recycled and therefore does not grow larger

C explain the formation of North America and Europe

D explain why volcanoes form from magma originating in the crust.

4 Consider the data on the composition of the crust in Table 10.4.1.

TABLE 10.4.1 Mineral composition of the crust

Mineral group	% of crust	Approximate density of mineral g/cm^3
feldspars	49	2.5–2.7
quartz	21	2.6
pyroxene, olivine and others	15	3–4.3
micas	8	2.7–3
magnetite	3	5.2
other minerals	4	varies greatly

Basalt has a density of about 2.9 g/cm^3, whereas granite is about 2.6 g/cm^3. Which of the following is a likely deduction from this information?

A Continental crust is heavier than ocean crust.

B Basalt probably has more feldspars and quartz than granite has.

C Basalt is largely made of magnetite.

D There is probably more magnetite, pyroxene and olivine in basalt than in granite.

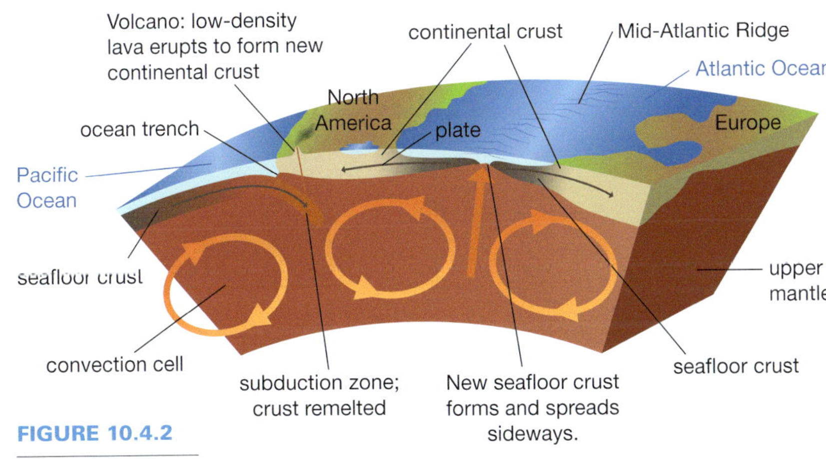

FIGURE 10.4.2

CHAPTER 10

Glossary

asthenosphere: a layer of 'plastic' semi-solid rock in the lower mantle on which Earth's tectonic plates move

constructive boundary: plate boundary where new crust is formed

continental crust: the crust that forms the continents

continental drift: the separating of continents by drifting across Earth's surface

continental drift

converging boundary: where plates are colliding with each other

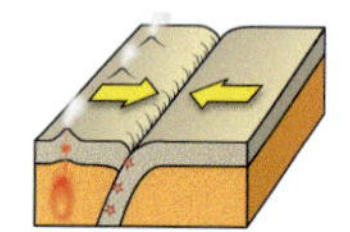

converging boundary

core: the centre of Earth

crust: Earth's outermost solid layer

destructive boundary: plate boundary where lithosphere is destroyed

diverging boundary: where plates are moving apart from each other in opposite directions

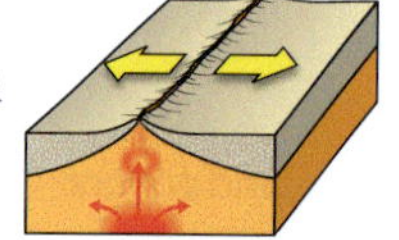

diverging boundary

earthquake: the rapid movement of the ground, usually back and forth and up and down in a wave motion, due to the movement of tectonic plates

epicentre: the point on Earth's surface directly above the focus of an earthquake

focus: the place below ground where an earthquake starts

fold mountain: mountains formed by crust crumpling upwards as plates collide

GPS ground station: a receiver and computer that can detect satellite signals and calculate position on Earth's surface

hot spot: isolated place away from plate boundaries where a lot of hot magma is collecting

hot spot

island arc: a chain of islands formed at the edges of colliding tectonic plates where one plate subducts

lava: molten rock that has erupted onto Earth's surface

lithosphere: name for the crust and the upper mantle together; Earth's tectonic plates

magma: molten rock below Earth's surface

magnetic striping: patterns of magnetism trapped in rocks on each side of plate boundaries

mantle: layer beneath Earth's crust

ocean trench: a deep channel in the ocean floor where crust is sinking downwards

oceanic crust: the crust that forms the ocean floor

primary wave (P-wave): a longitudinal seismic wave that travels fast through Earth

ridge push: older crust is pushed below new ocean crust and squeezes the plates sideways

rift: a zone where Earth's crust and mantle are being pulled apart

rifting: the process of continents breaking up, subsiding and allowing in water from the sea

seafloor spreading: the process of new crust forming at the ocean ridges and spreading outwards

secondary wave (S-wave): transverse seismic wave that travels through Earth

seismic wave: the shaking, wave-like movement of the ground in an earthquake

seismometer: an instrument that detects the seismic waves from an earthquake

slab pull: plates are pulled apart at the mid-ocean ridges

subduction: when one plate sinks below another plate during a collision

surface wave: a seismic wave that travels along the surface of the Earth in the crust

tectonic plate: section of Earth's crust that moves about on Earth's surface

transform boundary: where plates are sliding parallel to each other but in opposite directions

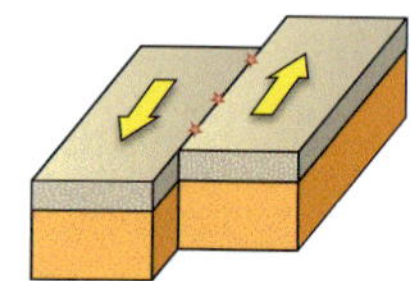

transform boundary

trench: a depression in the ocean floor

tsunami: a huge wave in the ocean caused by an earthquake occurring on the sea floor

tsunami

volcano: a place where extremely hot material from inside Earth erupts at the surface

STEP UP

CHAPTER 11

Psychology

You can access this chapter along with supporting worksheets in your eBook.

This chapter is not core content. It is a useful extension to prepare you with further skills and knowledge for senior studies in Psychology. Before starting this chapter it is recommended that you have completed Chapter 7 Body coordination and Chapter 8 Disease.

Have you ever wondered ...

- how you learn?
- how you can improve your memory?
- why some people are outgoing while others are shy?
- how people can commit terrible atrocities in times of war?

LS

After completing this chapter you should be able to:

- describe some ethical considerations that researchers must follow
- discuss how memories are formed and accessed
- explain different theories of the ways that you learn and describe research that demonstrates these types of learning
- discuss different theories of personality
- discuss how your behaviour is affected by the people around you.

AB 11.1

Index

Page entries marked [e] refer to ebook content

D

E

F

G

H

N

O

P

Q

R